Preservation of Human Oocytes

Reproductive Medicine and Assisted Reproductive Techniques Series

Series editors

David Gardner,
University of Melbourne,
Australia

Jan Gerris,
University Hospital Ghent,
Belgium

Zeev Shoham,
Kaplan Hospital,
Rehovot, Israel

Titles

1. Jan Gerris, Annick Delvigne, François Olivennes
   Ovarian Hyperstimulation Syndrome, ISBN: 9781842143285
2. Alastair G Sutcliffe
   Health and Welfare of ART Children, ISBN 9780415379304
3. Seang Lin Tan, Ri-Chen Chian, William Buckett
   In-vitro Maturation of Human Oocytes, ISBN: 9781842143322
4. Christoph Keck, Clemens Tempfer, Jen-Noel Hugues
   Conservative Infertility Management, ISBN: 9780415384513
5. Carlos Simon, Antonio Pellicer
   Stem Cells in Human Reproduction, ISBN: 9780415397773
6. Kay Elder, Jacques Cohen
   Human Preimplantation Embryo Selection, ISBN: 9780415399739
7. Michael Tucker, Juergen Liebermann
   Vitrification in Assisted Reproduction, ISBN: 9780415408820
8. John D Aplin, Asgerally T Fazleabas, Stanley R Glasser, Linda C Giudice
   The Endometrium, Second Edition, ISBN: 9780415385831
9. Adam H Balen
   Infertility in Practice, Third Edition, ISBN: 9780415450676
10. Nick Macklon, Ian Greer, Eric Steegers
    Textbook of Periconceptional Medicine, ISBN: 9780415458924
11. Carlos Simon, Antonio Pellicer
    Stem Cells in Human Reproduction, Second Edition, ISBN: 9780415471718
12. Andrea Borini, Giovanni Coticchio
    Preservation of Human Oocytes, ISBN 9780415476799

Preservation of Human Oocytes

Edited by

Andrea Borini and Giovanni Coticchio

Tecnobios Procreazione
Centre for Reproductive Health
Bologna, Italy

informa
healthcare

First published in the United Kingdom in 2009 by Informa Healthcare, Telephone House, 69-77 Paul Street, London EC2A 4LQ. Informa Healthcare is a trading division of Informa UK Ltd. Registered Office: 37/41 Mortimer Street, London W1T 3JH. Registered in England and Wales number 1072954.

Tel: +44 (0)20 7017 5000
Fax: +44 (0)20 7017 6699
Website: www.informahealthcare.com

A CIP record for this book is available from the British Library.
Library of Congress Cataloging-in-Publication Data

Data available on application

ISBN-10: 0 415 47679 8
ISBN-13: 978 0 415 47679 9

Distributed in North and South America by
Taylor & Francis
6000 Broken Sound Parkway, NW, (Suite 300)
Boca Raton, FL 33487, USA

Within Continental USA
Tel: 1 (800) 272 7737; Fax: 1 (800) 374 3401
Outside Continental USA
Tel: (561) 994 0555; Fax: (561) 361 6018
Email: orders@crcpress.com

Book orders in the rest of the world
Paul Abrahams
Tel: +44 (0)20 7017 4036
Email: bookorders@informa.com

Composition by Exeter Premedia Servies Private Ltd., Chennai, India
Printed and bound in Great Britain by MPG Books, Bodmin, Cornwall, UK

Contents

Contributors

David F. Albertini PhD Departments of Molecular and Integrative Physiology, Anatomy and Cell Biology, Obstetrics and Gynecology, and Center for Reproductive Sciences, Kansas University Medical Center, Kansas City, Kansas, U.S.A.

Antonia Bazzocchi MD Reproductive Medicine Unit, Infertility and IVF Center, University of Bologna, Bologna, Italy

Maria Antonietta Bonu BSc Tecnobios Procreazione, Centre for Reproductive Health, Bologna, Italy

Andrea Borini MD Tecnobios Procreazione, Centre for Reproductive Health, Bologna, Italy

Yun-Xia Cao MD Center for Reproductive Medicine, Department of Obstetrics and Gynecology, The First Affiliated Hospital of Anhui Medical University, Hefei, P.R. China

Joan Cerdà PhD Laboratory of the Institut de Recerca i Tecnologia Agroalimentàries (IRTA)-Institut de Ciències del Mar, Consejo Superior de Investigaciones Científicas (CSIC), Barcelona, Spain

François Chauvigné PhD Laboratory of the Institut de Recerca i Tecnologia Agroalimentàries (IRTA)-Institut de Ciències del Mar, Consejo Superior de Investigaciones Científicas (CSIC), Barcelona, Spain

Ri-Cheng Chian PhD Department of Obstetrics and Gynecology, McGill University, Montreal, Canada

Ana Cobo PhD Cryobiology Unit, IVI-Valencia, Valencia, Spain

Pierre Comizzoli PhD Department of Reproductive Sciences, Center for Species Survival, Smithsonian's National Zoological Park, Washington D.C., U.S.A.

Jason Conceicao PhD PIVET Medical Centre, Perth, Western Australia, Australia

Giovanni Coticchio PhD Tecnobios Procreazione, Centre for Reproductive Health, Bologna, Italy

Lucia De Santis BSc IVF Unit, Department of Obstetrics and Gynaecology, University Vita-Salute, H. S. Raffaele, Milan, Italy

David H. Edgar PhD Reproductive Services, Royal Women's Hospital/Melbourne IVF and Department of Obstetrics and Gynaecology, University of Melbourne, Melbourne, Victoria, Australia

Ursula Eichenlaub-Ritter PhD Department of Gene Technology and Microbiology, Faculty of Biology, University of Bielefeld, Bielefeld, Germany

Carlo Flamigni MD National Council of Bioethics, Rome, Italy

Barry Fuller PhD Division of Surgery and Interventional Sciences, UCL Medical School, Royal Free Hospital Campus, London, U.K.

David K. Gardner Dphil Department of Zoology, The University of Melbourne, Melbourne, Victoria, Australia

Debra A. Gook PhD Reproductive Services, Royal Women's Hospital/Melbourne IVF and Department of Obstetrics and Gynaecology, University of Melbourne, Melbourne, Victoria, Australia

Maria Köster DVM Department of Gynecological Endocrinology & Reproductive Medicine, University of Bonn, Bonn, Germany

Lilia L. Kuleshova PhD Low Temperature Preservation Unit, National University Medical Institutes, Yong Loo Lin School of Medicine, National University of Singapore, Singapore

Chiara Landolfo MD Reproductive Medicine Unit, Infertility and IVF Center, University of Bologna, Bologna, Italy

Mark G. Larman PhD Department of Zoology, The University of Melbourne, Melbourne, Victoria, Australia

S. Darlene Limback MSc Departments of Molecular and Integrative Physiology, Anatomy and Cell Biology, Obstetrics and Gynecology, and Center for Reproductive Sciences, Kansas University Medical Center, Kansas City, Kansas, U.S.A.

John McGrath PhD Department of Aerospace and Mechanical Engineering, University of Arizona, Tucson, Arizona, U.S.A.

Markus Montag PhD Department of Gynecological Endocrinology & Reproductive Medicine, University of Bonn, Bonn, Germany

Leonardo Notarangelo BSc Reproductive Medicine Unit, Infertility and IVF Center, University of Bologna, Bologna, Italy

Karl G. Nygren MD PhD Queen Sophia Hospital, Stockholm, Sweden

Sharon Paynter PhD Department of Obstetrics and Gynaecology, Cardiff University, Heath Park, Cardiff, U.K.

David E. Pegg PhD Department of Biology, University of York, York, U.K.

Carlos E. Plancha PhD Biology of Reproduction Unit, Institute of Histology and Developmental Biology, Faculty of Medicine of Lisbon, Lisboa, Portugal and Centro Médico de Assistência à Reprodução (CEMEARE), Lisboa, Portugal

Eleonora Porcu MD Reproductive Medicine Unit, Infertility and IVF Center, University of Bologna, Bologna, Italy

Mafalda Rato MSc Centro Médico de Assistência à Reprodução (CEMEARE), Lisboa, Portugal

José Remohí MD PhD Obstetrics and Gynaecology, Faculty of Medicine, University of Valencia, IVI-Valencia, Valencia, Spain

Benjamin Rösing MD Department of Gynecological Endocrinology & Reproductive Medicine, University of Bonn, Bonn, Germany

Nucharin Songsasen DVM, PhD Department of Reproductive Sciences, Center for Species Survival, Smithsonian's National Zoological Park, Front Royal, Virginia, U.S.A.

Audrey Soo BSc General and Adolescent Unit, Institute of Child Health, University College, London, London, U.K.

James J. Stachecki PhD Tyho-Galileo Research Laboratories, Livingston, New Jersy, U.S.A.

Alastair Sutcliffe MD PhD General and Adolescent Unit, Institute of Child Health, University College, London, U.K.

Gabor Vajta PhD PIVET Medical Centre, Perth, Western Australia, Australia and Cairns Fertility Centre, Cairns, Queensland, Australia

Hans van der Ven MD Department of Gynecological Endocrinology & Reproductive Medicine, University of Bonn, Bonn, Germany

Katrin van der Ven MD Department of Gynecological Endocrinology & Reproductive Medicine, University of Bonn, Bonn, Germany

Stefano Venturoli MD Reproductive Medicine Unit, Infertility and IVF Center, University of Bologna, Bologna, Italy

Michael von Wolff MD Department of Gynecological Endocrinology and Reproductive Medicine, Women's Hospital, University of Bern, Bern, Switzerland

Maureen J. Wood MD Department of Obstetrics and Gynaecology, University of Aberdeen, Maternity Hospital, Foresterhill, Aberdeen, U.K.

John Yovich MD PIVET Medical Centre, Perth, Western Australia, Australia and Cairns Fertility Centre, Cairns, Queensland, Australia

Preface

Human in vitro fertilization (IVF) has attained full potential and become widespread following the introduction of reliable ovarian stimulation protocols and methods for the storage and later use of surplus embryos. This has been made possible by the fact that human embryos can be cryopreserved with only limited loss in terms of post-thaw viability and reproductive potential. Embryo cryopreservation is not, however, immune to disadvantages of an ethical and legal nature. The cryopreservation of fully grown oocytes could represent a more widely acceptable option, but until recently it has been applied only sporadically. Efforts carried out in the late 1980s represented a valid proof of principle, but were unable to generate success rates that could justify the adoption of oocyte freezing as a routine procedure. Studies conducted in the last few years based on the use of novel cryopreservation protocols indicate that many of the problems that have originally affected the overall efficiency of oocyte freezing have been solved. This has encouraged the adoption of such a strategy of fertility preservation. As a consequence, over a thousand babies from cryopreserved oocytes have been born so far. Apart from offering a possible alternative to embryo storage, oocyte cryopreservation also has the potential to provide a solution to the effects of changes in social attitudes, such as the increased expectations of women to achieve a full professional career. Those anticipations can involve a delay in the establishment of a family. In a number of cases, though, these women experience reproductive failure, a sort of "biological punishment" for having delayed their chances of conception to later in life. Therefore, storing oocytes can be a tangible possibility for extending the female reproductive life. Chemo- and radiotherapies can also compromise transiently or permanently the reproductively potential. Oocyte preservation can also be critical for young women who are destined to premature ovarian failure as an effect of genetic factors. More generally, gamete storage can be an answer for those women who, for diverse reasons, wish to preserve their reproductive potential for later use.

Despite this, probably the majority of IVF specialists have not yet had the opportunity to fully appreciate the clinical potential and current efficacy of oocyte cryopreservation. Therefore, this text has been conceived with the aim of offering a comprehensive view of the state of the art in oocyte cryopreservation, especially following recent studies that have shown how oocyte cryopreservation is beginning to challenge the supremacy of embryo cryopreservation as the preferred form of fertility preservation in IVF treatment. The book includes chapters on fundamental concepts of low-temperature storage (controlled rate slow cooling and vitrification), aspects of oocyte physiology relevant to the process of cryopreservation, essential biological and clinical evidence, and ethical implications of oocyte cryopreservation. The authors of the chapters are eminent authorities in their respective areas of interest, some of whom have collaborated with the editors over the last several years.

AB
GC

Color plates

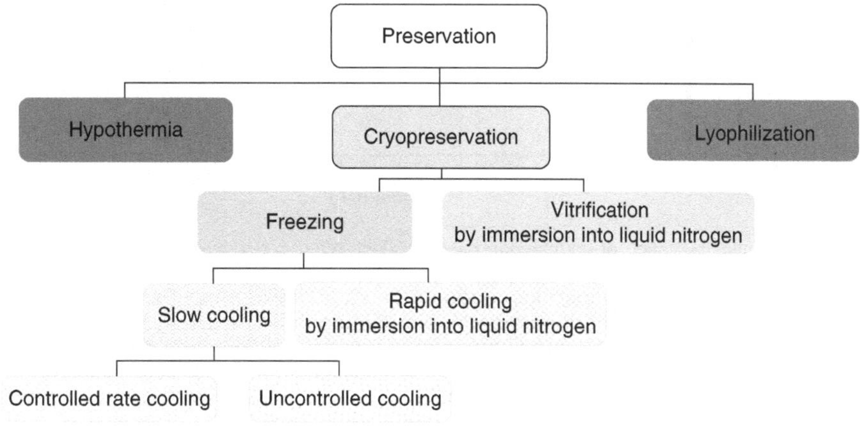

Figure 4.1 Types of preservation. *See Page 37.*

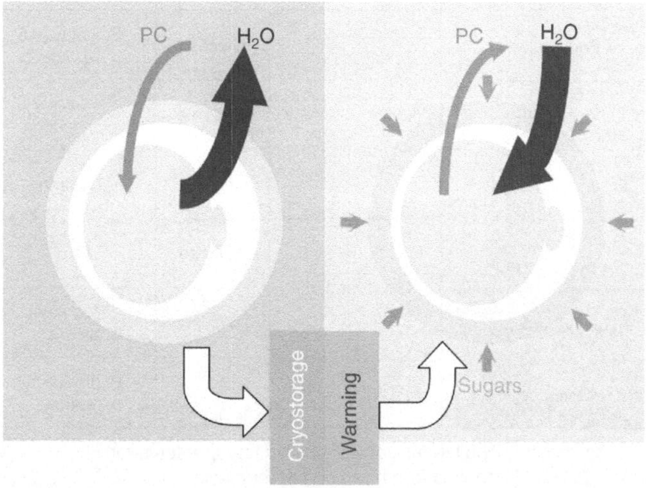

Figure 4.2 Basic representation of oocyte cryopreservation. *See Page 38.*

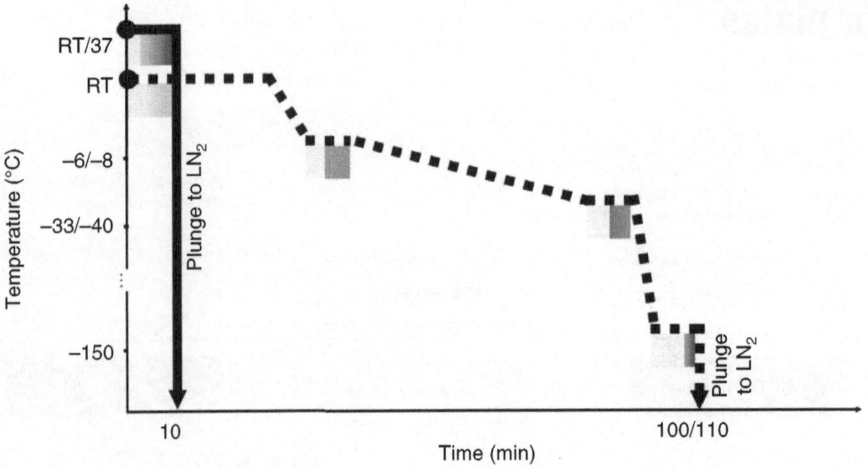

Figure 4.3 Schematic illustration of comparison between vitrification and slow-cooing procedures for human oocytes. Solid line represents vitrification. In a majority of recent protocols, human oocytes are introduced to vitrification solutions and diluted at either room temperature (RT) or 37°C within 10 min. Dashed line red represents conventional slow-cooling protocol (freezing approach). Average duration of slow cooling procedures after loading into freezing machine is about 100–110 min. In both routes, intensity of orange color inside the boxes reflects infiltration of cryoprotectants into oocyte. In the freezing route, white color inside the boxes reflects growth of ice crystals in a straw containing an oocyte. *Abbreviations*: LN_2, liquid nitrogen; RT, room temperature. *See Page 38.*

Penetrating Cryoprotectants	Non-penetrating cryoprotectants		
Increasing Molecular Weight →			
MW <100 Da Low molecular weight agents	180 Da < MW < 594 Da Sugars		MW >1000 Da High molecular weight agents
Formamide	Monosaccharides	Glucose	Polyethylene glycol
Acetamide		Fructose	Polyvinyl pyrrolydone
Ethylene glycol (EG)[a]		Lactose	Dextran
Dimethyl sulfoxide[a]; Propylene glycol[a]	Disaccharides	Sucrose	Ficoll
2,3-Butanediol		Trehalose	Polyvinyl alcohol
Glycerol[b]	Polysaccharides	Raffinose	Hydroxyethyl starch

Increasing molecular weight ↓

[a]commonly applied for oocyte/embryo cryopreservation
[b]commonly applied for sperm cryopreservation

Figure 4.4 Classification of cryoprotectants. *See Page 40.*

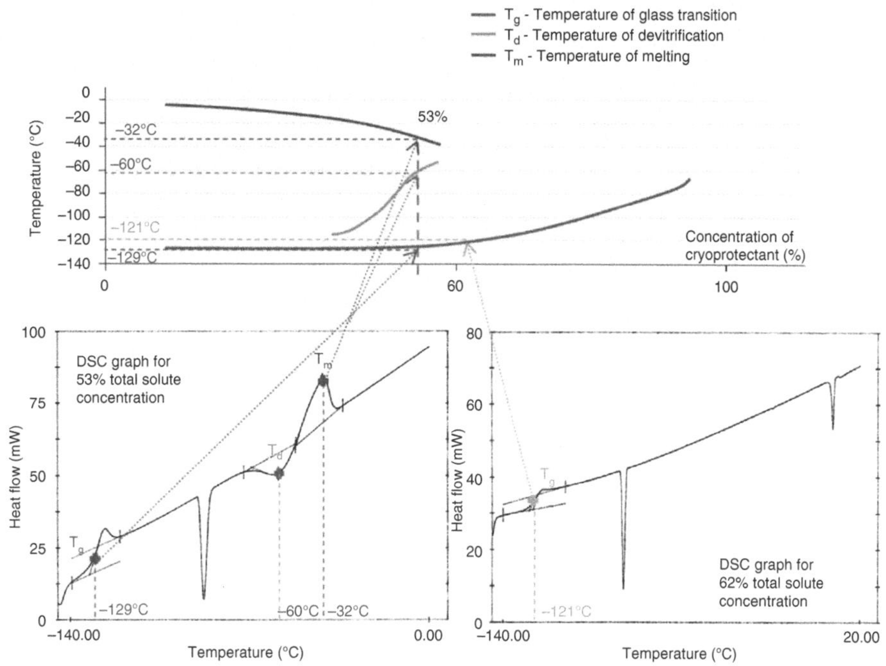

Figure 4.5 Schematic representation of processes that take place in aqueous solutions of cryoprotectants with the temperature and concentration changes, as indicated in the thermograms. *Abbreviations*: DSC, differential scanning calorimetry; T_d, temperature of devitrification; T_g, glass transition temperature; T_m, melting temperature of ice in the solutions. *See Page 41*.

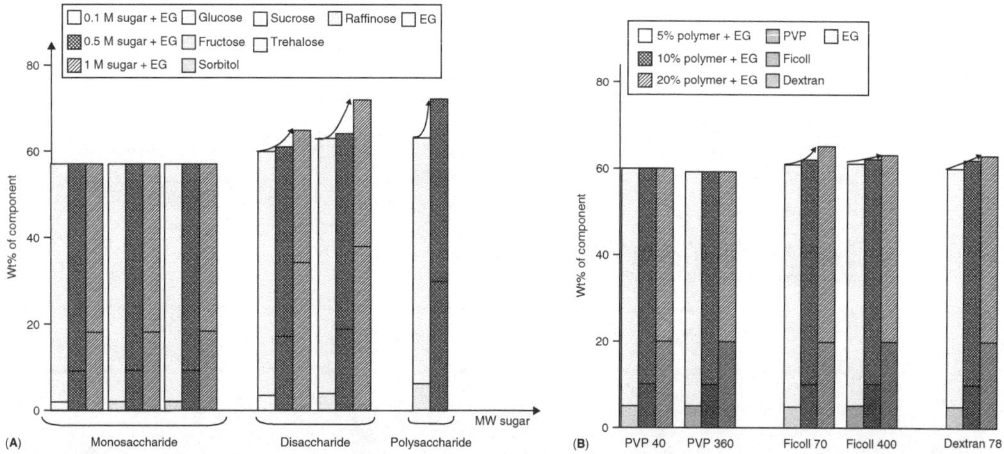

Figure 4.6 Concentrations of vitrification solutions based on DSC measurements: (**A**) ethylene glycol–sugars vitrification solutions; (**B**) ethylene glycol–polymers vitrification solutions. The *upward* sloping arrows show how the type and concentration of sugar and polymer influence the lowest total solute concentration at which vitrification would occur. Molecular weights of polymers (PVP, Ficoll, Dextran) presented in the figure are in 10^3 Da. Data compiled from Refs. 23 and 29. *Abbreviations*: DSC, differential scanning calorimetry; EG, ethylene glycol; MW, molecular weight; PVP, Polyvinylpyrrolidone. *See Page 42*.

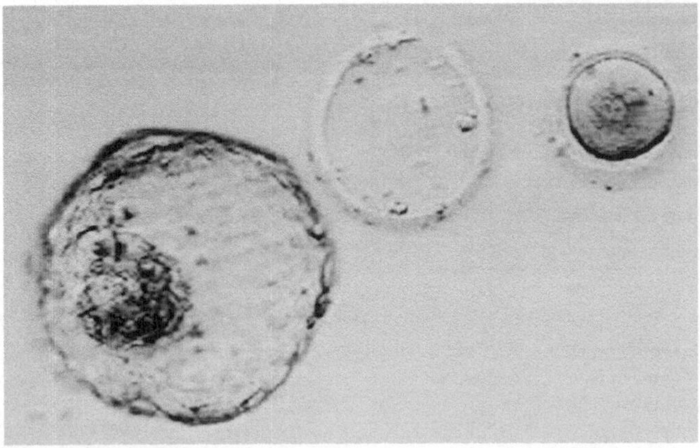

Figure 4.7 Hatched human blastocyst after development from a vitrified oocyte. Original magnification ×200. *See Page 44.*

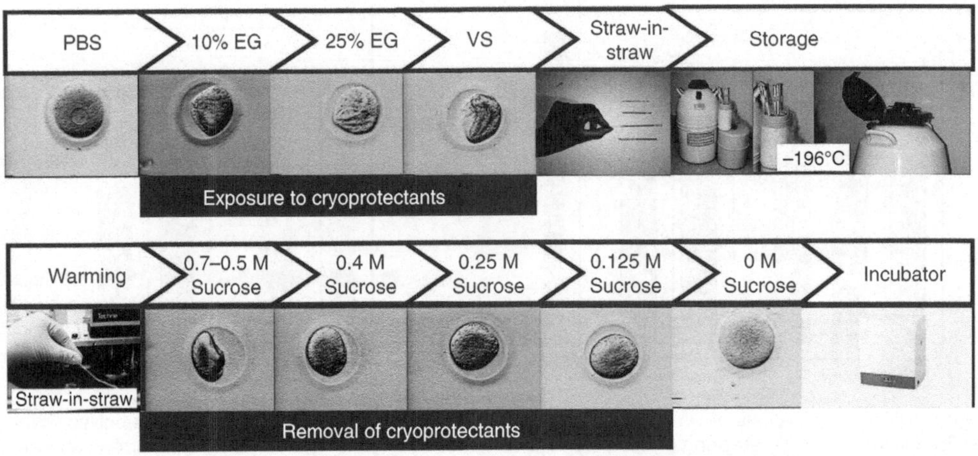

Figure 4.8 Step-wise equilibration and dilution procedures for immature human oocytes involving vitrification solution 40 vol.% EG + 0.6 mol/L sucrose. *Abbreviations*: EG, ethylene glycol; PBS, phosphate buffered saline; VS, vitrification solution. *See Page 47.*

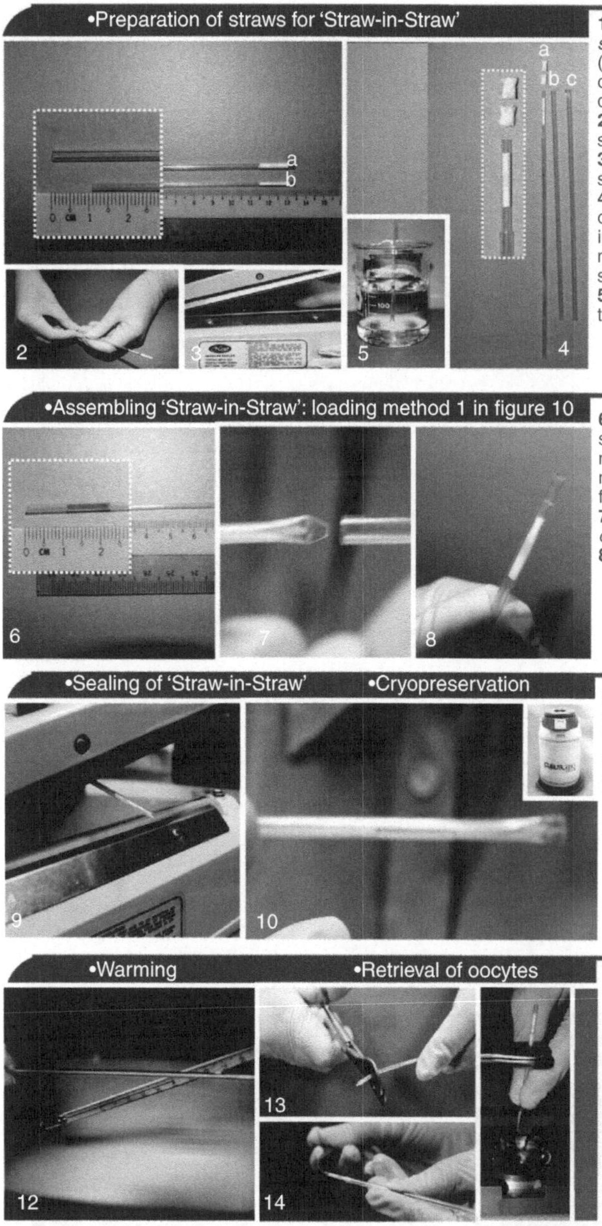

1. **(a)** 500μl french plastic straw (*outer straw*); **(b)** 250μl french plastic straw (*inner straw*) cut to 1cm shorter than outer straw; **insert** detail of straw size difference
2. Removing cotton wool from outer straw.
3. Sealing cotton wool from outer straw.
4. **(a)** Details of cotton wool of the outer straw being pushed out by inserting the inner straw (**insert** – magnified detail); **(b)** unsealed outer straw; **(c)** sealed outer straw
5. Testing outer straw in water to verify there is no leakage

6. *Inner straw* loaded with vitrification solution and oocytes; **insert**: loading method 1 in figure 10. three different methods of oocyte loading are shown in figure 9.
7. Insertion of sealed inner straw in *outer straw.*
8. Unsealed end of 'Straw-in Straw'

9. Sealing of *outer* (500μm) *straw* using heat sealer.
10. The resulting 'Straw-in-Straw' arrangement with oocyte inside the *inner straw.*
11. Direct immersion into LN2. storage in liquid nitrogen (-196°C)

12. Warming by direct immersion to 38°C water bath
13. Cutting of *outer straw.*
14. Removal of *inner straw.*
15. Retrieval of oocytes (pictured: loading case 3 only from figure 10): collection is made by cutting off the top of the inner straw containing wool, the oocyte (in the first dilution solution) then expels onto the dish by gravity.
steps 12–14 are the same for loading cases 1–3 from figure 10.

Figure 4.9 Essential steps of the "straw-in-straw" method: preparation of straws for vitrification (steps 1–5), loading of oocytes (details in Fig. 10), assembling the "straw-in-straw" (steps 6–8), sealing the "straw-in-straw" (steps 9–10) prior to immersing into liquid nitrogen (step 11), post-warming procedures and expelling of oocytes (steps 12–15). *See Page 51.*

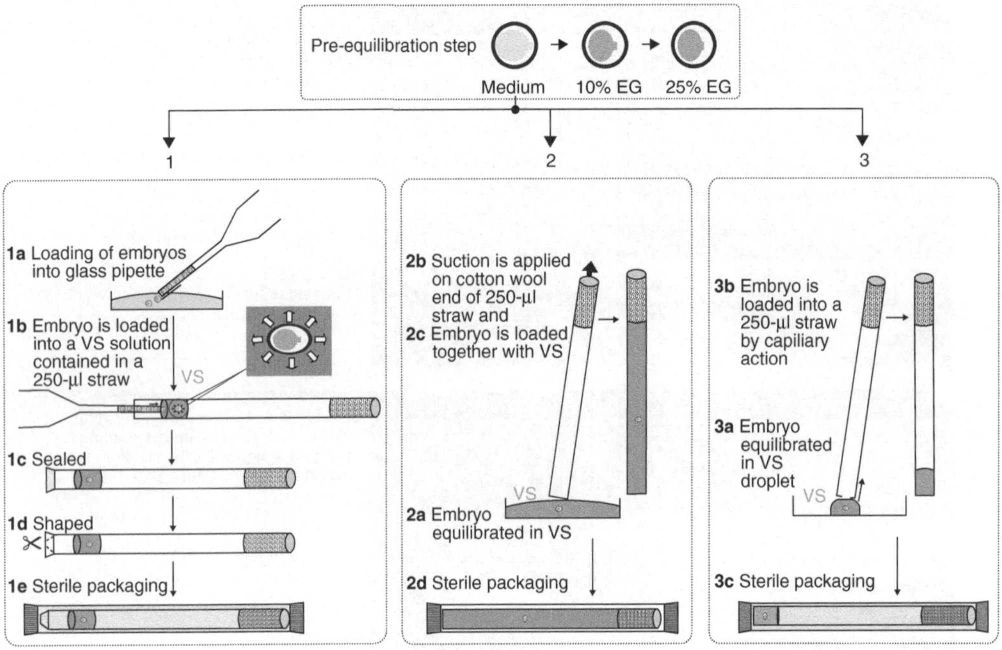

Figure 4.10 Schematic representation of loading embryos/oocytes in the "straw-in-straw" configuration (250- and 500-µl straws). Loading of embryos/ocytes by: (1) the traditional approach, which involves insertion into the pre-prepared column of the final solution using a glass transfer pipette; the 250-µl straw is sealed, shaped, and placed into the 500-µl straw (118,120); (2) suction using an Eppendorf 1000-µl micropipette (91,121–123); (3) capillary action. The insert in step 1b shows a magnified oocyte and illustrates initial shrinkage of oocyte in vitrification solution caused by cell dehydration; the shrinkage of the oocyte in VS is identical for all approaches (steps 1b, 2a, and 3a). *Abbreviations*: EG, ethylene glycol; VS, vitrification solution. *See Page 52.*

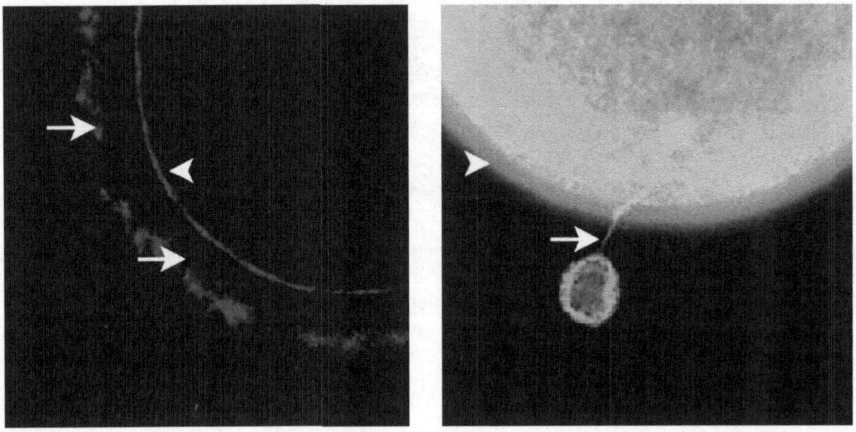

(A) **(B)**

Figure 9.1 (**A**) Fully grown germinal vesicle of hamster oocyte surrounded by cumulus cells exhibiting microfilament-rich trans-zonal processes (arrows). (**B**) Fully grown mouse oocyte with attached cumulus cell projecting a microtubule-rich transzonal process across the *zona pellucida* (arrow). This oocyte was mechanically stripped before labeling. Note a strong microfilament staining at the oocyte cortex (arrow heads in **A** and **B**). *See Page 117.*

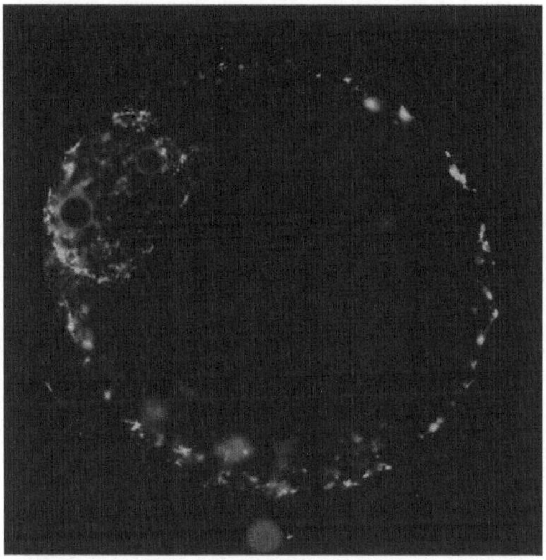

Figure 9.2 Fully grown germinal vesicle of hamster oocyte with microtubule network (green) around the germinal vesicle (blue) and at the cortical region. This oocyte was mechanically stripped before labeling. *See Page 118.*

Figure 9.3 In vitro-matured hamster oocyte depicting a cortical second meiotic spindle with chromosomes equatorially aligned and the first polar body. Note the strong microfilament labeling overlying the spindle at the oocyte cortex. This oocyte was mechanically stripped before labeling. Microtubules (green), microfilaments (red), DNA (blue). *See Page 119.*

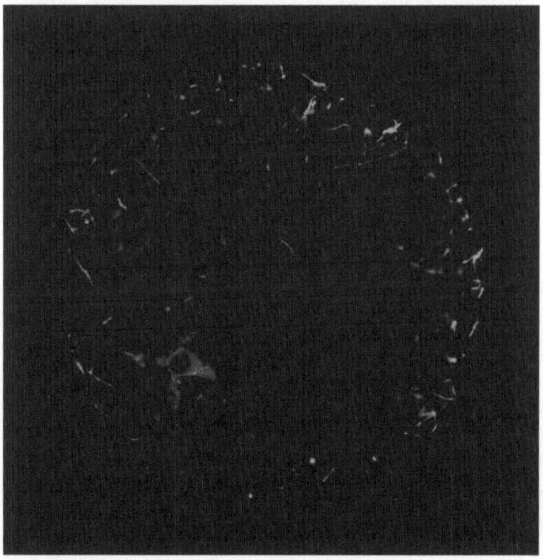

Figure 9.4 Cultured, fully grown hamster germinal vesicle oocyte after slow freezing and fast thawing. In this oocyte the absence of cumulus cells is a consequence of the cryopreservation process, negatively affecting meiotic progression. No mechanical stripping was done. Note the persistence of the germinal vesicle (blue) and the reappearance of a microtubule interphase-like network (green). *See Page 121.*

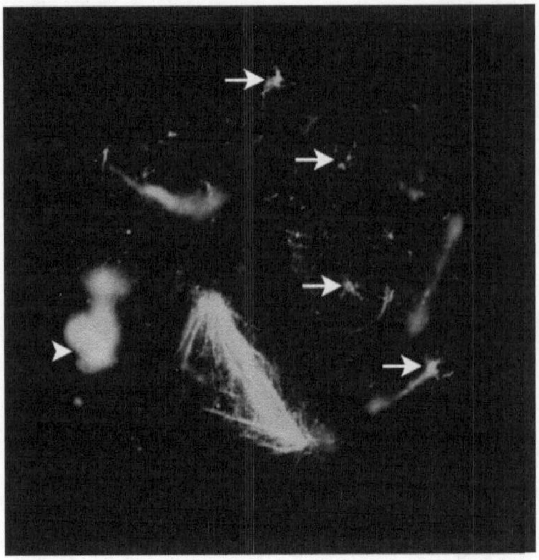

Figure 9.5 In vitro-matured hamster oocyte after slow freezing at the germinal vesicle stage. A distorted second meiotic spindle (green), with non-aligned chromosomes (blue), is clearly observed. Note the first polar body with a microfilament (red) rich region (arrow head), and several foci of premature cytoplasmic microtubule polymerization (arrows). *See Page 123.*

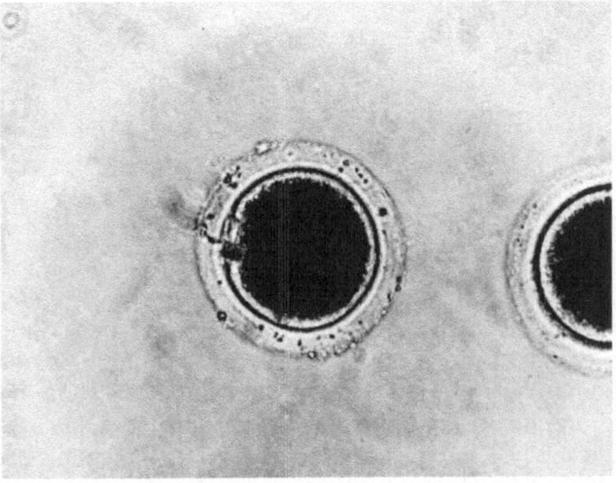

(A)

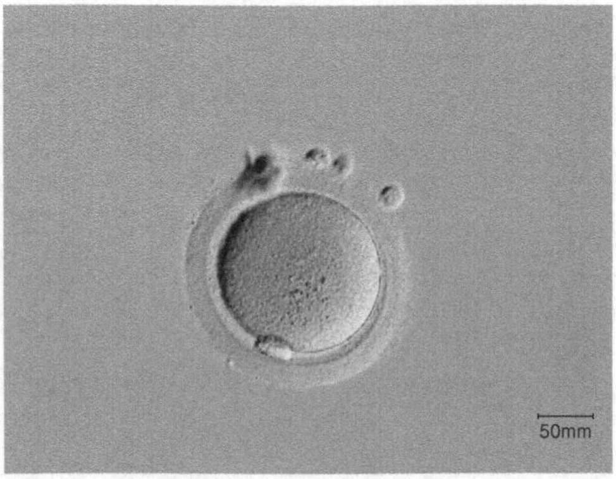

(B)

Figure 11.1 Mature pig (**A**) and human oocytes (**B**). Pig oocytes contain lipid droplets that are believed to increase the sensitivity of these cells to cryodamage. *See Page 143.*

Figure 13.1 Human germinal vesicle (GV)-stage oocyte showing an interphase-like three-dimensional network of microtubules (in green) spanning across the cytoplasm. Within the GV, an uninterrupted ring of heterochromatin is visible around the nucleolus. The image is derived from a collaborative study between Tecnobios Procreazione and the laboratory of Prof. D. Albertini. *See Page 164.*

Figure 13.2 Detail of a human MII oocyte. Actin filaments (in red) are organized in a thin uninterrupted layer beneath the oolemma and do not exhibit a restricted localization coincident with the position of the MII spindle that instead is found in mouse oocytes. The end of the MII spindle oriented toward the center of the oocyte is clearly disorganized, having lost the typical polar convergence of microtubule fibers. The image is derived from a collaborative study between Tecnobios Procreazione and the laboratory of Prof. D. Albertini.
See Page 165.

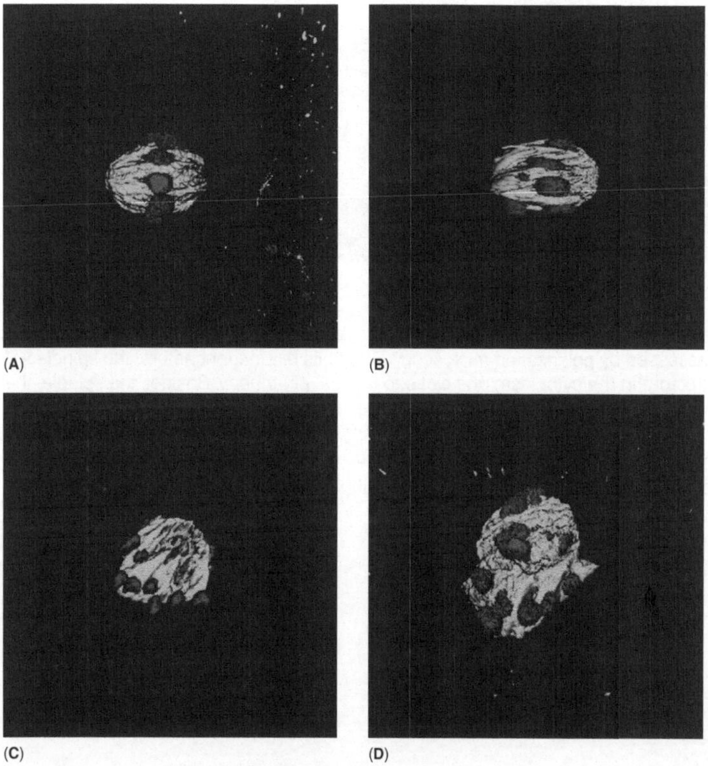

Figure 13.3 Confocal microscopy 3D reconstructions of spindles presenting different microtubule and chromosome configurations: (**A**) bipolar organization, with microtubules converging at both poles and all chromosomes present and evenly aligned at the equatorial plate; (**B**) bipolar spindle, microtubules meeting at both poles, but with chromosomes only partially aligned on the metaphase plate; (**C**, **D**) spindles with microtubules showing signs of disorganization and in part not converging at one or both poles, with chromosomes showing varying degree of misalignment. The images are derived from a collaborative study between Tecnobios Procreazione and the laboratory of Prof. D. Albertini.
See Page 166.

(**A**)

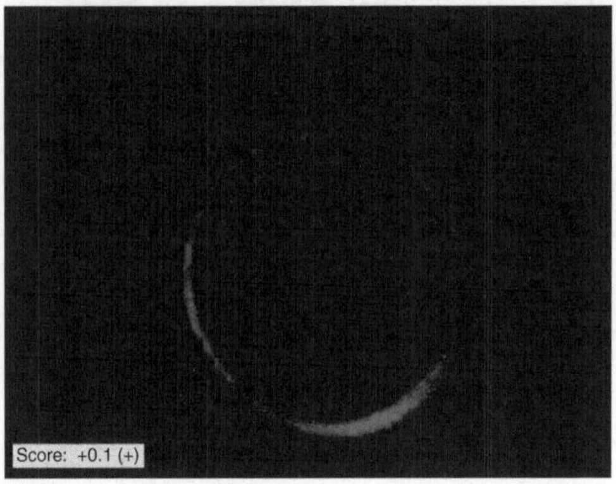

(**B**)

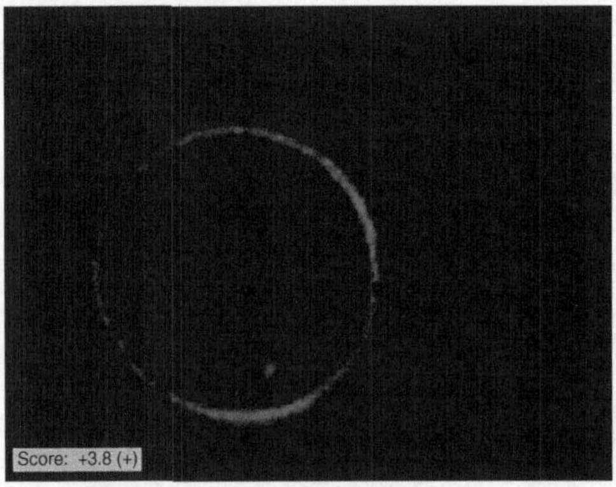

(**C**)

Figure 14.1 Zona and spindle imaging in metaphase II oocytes. The presence and location of a spindle within a metaphase II oocyte can be easily assessed by polarization microscopy. Due to its birefringent nature, the spindle will appear as an intense birefringent structure in the cytoplasm and close to the first polar body. Besides the spindle, the inner ring of the zona pellucida has some birefringent properties and can be visualized. An automatic zona evaluation system (Octax PolarAide™) enables a qualitative analysis of the zona birefringence and thus a subclassification of the quality of the corresponding oocytes, which is displayed as a score. The oocytes shown are characterized by a low score/low quality (**A**), intermediate score/intermediate quality (**B**), and high score/high quality (**C**). For details on zona imaging see Refs. (18) and (19). *See Page 176.*

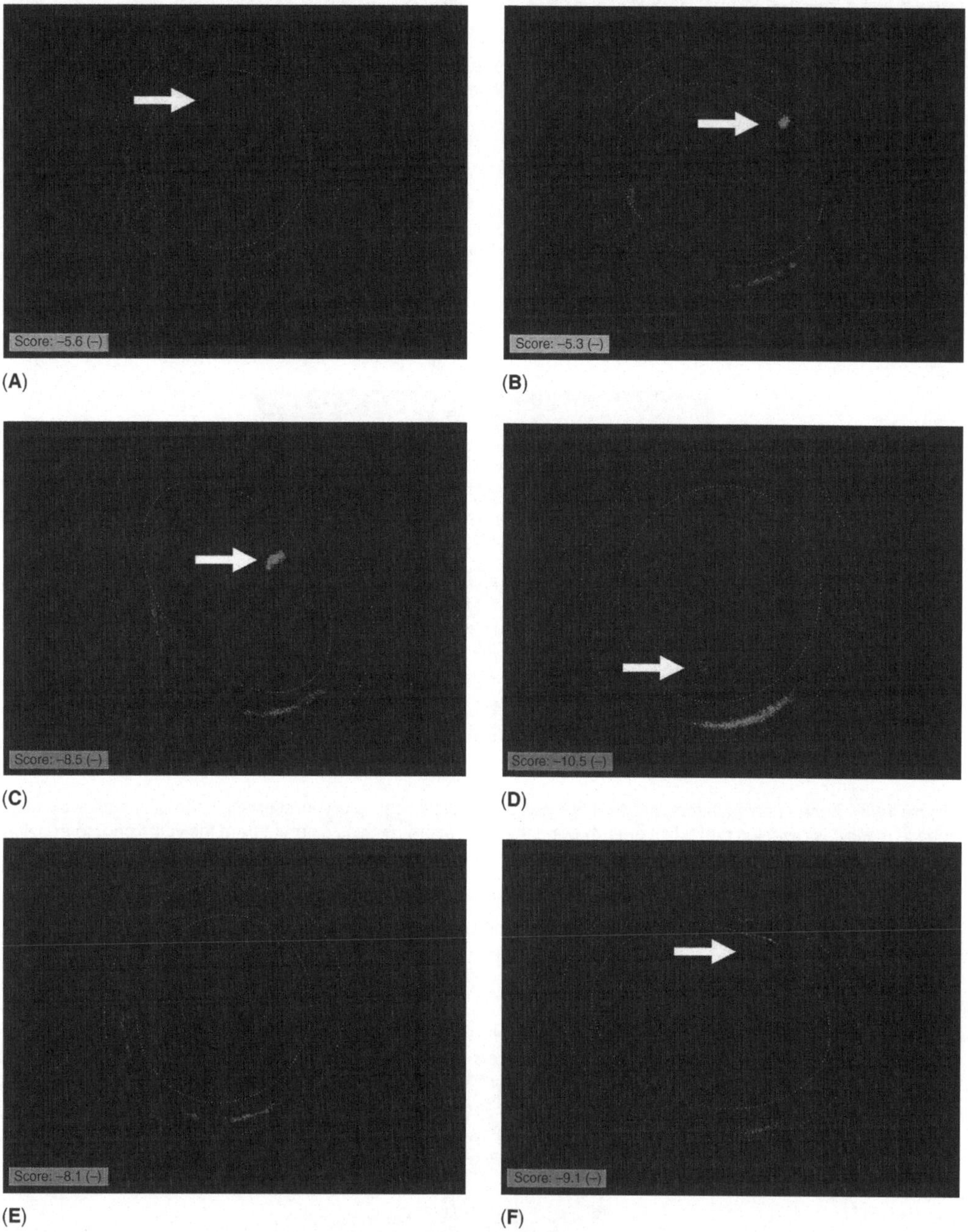

Figure 14.2 Spindle imaging in metaphase II oocytes during slow cooling cryopreservation. The presence and location of a spindle within a metaphase II oocyte was assessed prior to cryopreservation (**A**) and in equilibration solutions 2 (**B**) and 4 (**C**) by polarization microscopy. Following thawing of the same oocyte, the spindle was clearly visible in the first thawing solution (**D**) but disappeared in the following solutions (**E**). After incubation for another 2 hr in culture medium, the spindle did form again (**F**). For slow freezing/rapid thawing of denuded human oocytes we used a commercial kit (OC-Freeze/OC-Thaw; Cook, Limerick, Ireland) and essentially followed the instructions given by the manufacturer. The final freezing solution contains 1.5 M propanediol and 0.2 M sucrose in a HEPES buffered salt solution, while the first thawing solution contains 1.0 M propanediol and 0.3 M sucrose in a HEPES buffered salt solution. A detailed protocol is given in Ref. (46). *See Page 179.*

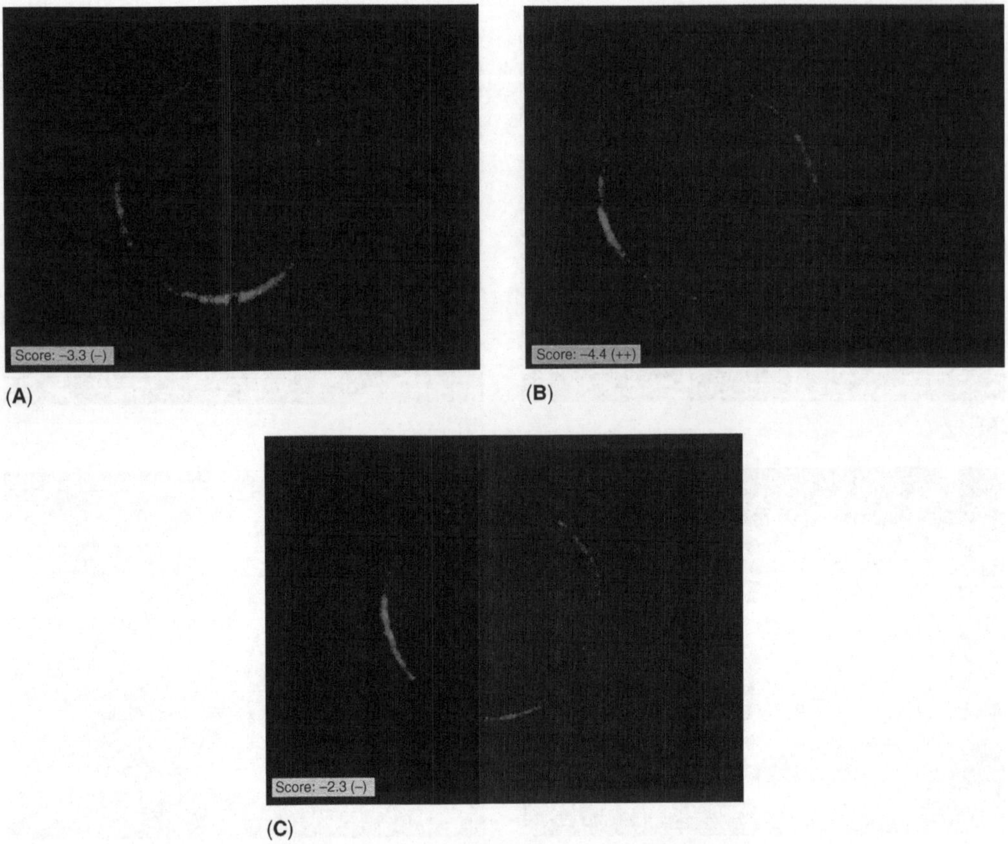

Figure 14.3 Zona imaging in metaphase II oocytes during slow cooling cryopreservation. Zona imaging of a metaphase II oocytes revealed that 3.5 hr after thawing the zona score (**B**) was different from the value prior to freezing (**A**). Following culture for another 18 hr, the zona score was almost similar to that in the beginning (**C**). This indicates that the changes induced by the cryopreservation procedure do have an impact on the architecture of the zona pellucida. The higher value 3.5 hr after thawing may indicate a temporary hardening at least of the inner ring of the zona pellucida. The freeze/thaw procedure was performed as described in the legend to Figure 2. *See Page 180.*

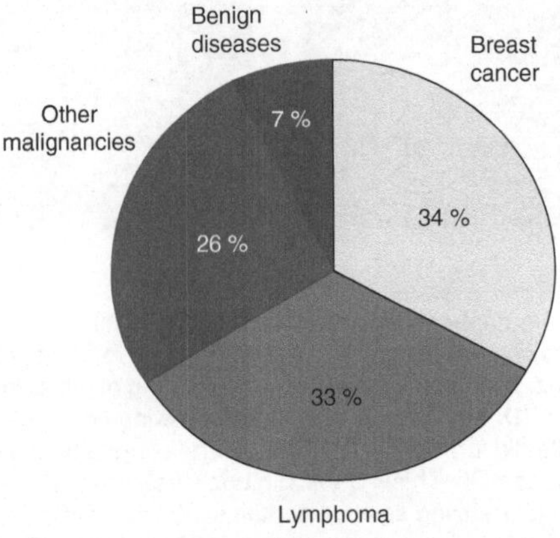

Figure 22.1 Spectrum of patients most frequently counseled for fertility preserving techniques [according to the FertiPROTEKT network (10)]. *See Page 258.*

1 A Historical Overview of Embryo and Oocyte Preservation in the World of Mammalian In Vitro Fertilization and Biotechnology

Nucharin Songsasen and Pierre Comizzoli

Department of Reproductive Sciences, Center for Species Survival, Smithsonian's National Zoological Park, Front Royal, Virginia, U.S.A.

INTRODUCTION

The Benefits of Freezing Embryos and Oocytes for Long-term Storage

In laboratory animals, especially the mouse, advances in molecular genetics has permitted the creation of thousands of mutant mouse strains that are too expensive to maintain as live breeding colonies (1). As a result, an increasing number of genetically valuable mouse strains are preserved and "maintained" by freezing their embryos, oocytes and sperm. Embryo cryopreservation and transfer have become prominent in livestock industry (especially for large and small ruminant breeding) (2) to improve the propagation of genetically valuable individuals. In 2005, more than 370,000 frozen-thawed bovine embryos were transferred worldwide (2). At last, the benefits of cryopreserved oocytes and embryos are widely recognized for wild species conservation because frozen germ cells and embryos could serve as assurances against any unforeseen catastrophes and assist in the management of the populations both in zoos and in nature (3). However, the application of this technology in wildlife conservation is still far from being routinely used (4). So far, live offspring have been produced from frozen-thawed embryos only in less than a dozen of species (felids: ocelots, caracal, African wild cat; primates: macaque, baboon; ungulates: eland, red deer, fallow deer) (4).

Besides semen freezing, oocyte and embryo cryopreservation has become one of the key components in human-assisted reproduction since the 1980s. "Surplus" embryos produced in vitro with oocytes obtained from a single collection can be stored for a later use (5). This considerably reduces the costs and risks of exogenous hormone stimulations associated with multiple oocyte collections. In addition, embryo cryopreservation offers more flexibility in case of ovarian hyperstimulation syndrome related to hormonal treatment before embryo transfer. In such circumstances, which also carry significant health risks (5), the patients may terminate the hormone treatment and embryo transfer can be postponed. In addition, despite its enormous benefits, ethical and legal issues are associated with storage and the ultimate fate of cryopreserved embryos (6). Such concerns may be overcome by the possibility to store oocytes. As a single haploid cell, an oocyte fails to elicit the same emotional, moral, and lawful concerns as an embryo (7). Oocyte cryopreservation also permits fertility preservation for young cancer patients who face early ovarian failure caused by radiation or chemotherapy (5,6). Besides cancer patients, oocyte cryopreservation enables women to voluntarily postpone or expand their reproductive life for social reasons (8).

From Cryotherapy to Modern Cryobiology

The word "cryobiology" originated from the Greek words kryo (cold), bios (life), and logos (study) (9). The history of cryobiology dates to 2500 BC when Egyptians determined how to use low temperatures to stop bleeding, inflammation, and create local anesthesia (cryotherapy). However, the first documented study on the effects of freezing on living animals "New Experiments and Observations Touching Cold" was reported in the 15th century by Sir Robert Boyle (10). This pioneer formulated "Boyle's law" (11), which describes the inverse relationship between the volume of a gas and pressure at a constant temperature—a principle highly relevant to low-temperature biology (11). After the publication of Boyle's monograph, cryobiology did not really progress for several centuries. It could be argued that modern cryobiology began in 1940 when Luyet and Gehenio (12) published a monograph addressing cold injury-induced mechanisms. The authors specifically examined the influence of low temperature on the ability of spermatozoa (from mammals and amphibians) to remain motile, and of eggs (from insects and amphibians) to be fertilized and hatch. Throughout the first half of the 20th century, a growing number of scientists investigated the effect of cold and subzero temperatures on various organisms, including vegetal and animal cells (11). Although several studies on gamete preservation were conducted in early 1900s (11), the discovery of cryoprotective properties of glycerol by Polge et al. (13) bolstered the efforts to preserve cells at subzero temperatures. Basic studies then focused on the cause of cell injury associated with subzero temperature exposure, especially the pioneering work (1957 through 1963) by Peter Mazur (11,14). This investigator formulated mathematical models describing the responses of cells suspended in a cryoprotectant solution and then cooled to subzero temperatures (14). These fundamental studies in the early 1970s have led to the birth of the first live mice after the transfer of frozen embryos (15). As demonstrated in this chapter, such findings were the foundation for the advances in oocyte cryopreservation that began in the 1980s.

HISTORY OF MAMMALIAN EMBRYO CRYOPRESERVATION

Historical Landmarks and Evolution of Embryo Cryopreservation

The first mammalian embryos to be successfully cryopreserved were those from the mouse, in 1972 (15,16). In these milestone studies, mouse embryos were exposed to a medium containing 2.0 M dimethyl sulfoxide (DMSO) or glycerol, seeded (see below), slowly cooled to –40°C to –80°C, and then plunged and stored in liquid nitrogen. Besides the need to identify optimal cooling and warming rates, another essential step for the successful preservation of embryos is the induction of extracellular ice crystals, which is referred to as "seeding" (15). Seeding reduces the unpredictability of nucleation of ice crystals that can occur just a few degrees below the freezing point of the solution or even 10°C to 20°C lower (17). Seeding samples at a temperature below the freezing point of a solution was first shown to reduce survival variability in cryopreserved yeast cells (18), and was later included in embryo cryopreservation protocol (15,16).

One year after the first success in mouse embryo cryopreservation, the first calf produced from cryopreserved embryos, "Frosty," was born (19). This was followed by births of live offspring from frozen-thawed embryos in several other laboratory (e.g., rabbit, nonhuman primates), domestic (e.g., cattle, sheep, goat), and wild animals (e.g., eland, caracal, ocelot) (4,5,20).

The first successful human embryo cryopreservation was reported in 1983 (21). Although pregnancy was obtained after transferring frozen-thawed cleavage stage embryos, it was terminated after 24 weeks due to the development of a septic *Streptomyces agalactiae* chorion amnionitis (21). A year later, the first human baby was born after transferring frozen-thawed embryos (22). Yet, the major step toward the clinical application of cryopreservation of human embryos was made when it was shown that single-cell embryos (pronuclear zygotes) can be successfully cryopreserved using propylene glycol as a cryoprotectant (5). The ability to cryopreserve pronuclear stage embryos alleviated ethical issues because this developmental stage was not defined sensu stricto as an embryo, especially in some European countries (23). Specifically, it was illegal to freeze cleaved stage embryos in Germany, Switzerland, and Italy; hence, only physicians have the choice of freezing either unfertilized oocytes or pronuclear stage zygotes (23).

Cryopreservation of mammalian embryos originally relied on slow cooling between –0.3°C/min and –0.5°C/min to avoid intracellular ice formation, slow warming (25–450°C/min) to prevent abrupt rehydration, and stepwise dilution to remove the cryoprotectants after the embryos were thawed (24). In the early 1980s, Renard et al. (25) as well as Leibo (26) independently developed and introduced a new "one-step" dilution method that used nonpermeating compounds (sucrose and galactose) as an osmotic buffer while removing the permeating cryoprotectants from frozen-thawed embryos. Such an approach permitted cryopreserved cattle embryos to be directly transferred into recipients and become widespread in cattle breeding program (27). Then, in the 1990s, low molecular weight cryoprotectants, including propylene glycol and ethylene glycol, were adapted into protocols because these agents moved rapidly across the cell membranes and hence, decreased cell injury associated with osmotic stress (28–30).

To date, conventional cooling is still routinely used for human embryo cryopreservation in clinical settings (5). One of the most limiting factors has been zona pellucida damages caused by biophysical alterations associated with cryopreservation process that, in turn, are now known to contribute to poor implantation (5). Several alternate approaches, including using polymers, dextran or ficoll or modifying the embryo container type, have been shown to improve clinical outcomes (5). It has also been suggested that cryopreservation compromises the ability of a viable embryo to "escape" from zona pellucida (31). To circumvent this challenge, assisted hatching (e.g., zona drilling by laser) has been shown to be helpful in increasing implantation and pregnancy success using thawed embryos, especially in humans (31,32).

Another major advance to avoid the problems associated with slow cooling and intracellular ice formation was developing "vitrification." This procedure involves brief exposure of cells to a concentrated solution of cryoprotectants before plunging the samples into liquid nitrogen to achieve high cooling rates (–2500°C/min) and solidification without formation of ice (33). This novel approach was first explored successfully in mouse embryos by Rall and Fahy (33). Since then, vitrification has been used to cryopreserve embryos effectively in rat (34), rabbit (35), sheep (36), cattle (37), pig (38), and humans (39). This storage method has been particularly valuable for the embryos of species that have poor tolerance to conventional slow cooling and are hypersensitive to chilling (i.e., pig embryos) and has also found widespread application in difficult-to-store mammalian oocytes (see below). The first human birth from vitrified embryos occurred in the early 1990s (39) and, since then, this technique has been commonly used in clinical practice due to its effectiveness, simplicity and low cost. In fact, a recent meta-analysis across four studies suggested that human embryos stored by vitrification have higher survival than their slow-cooled counterparts (40). Nonetheless, more investigations are required to increase the efficiency of vitrification for producing living, healthy babies.

Challenges in Embryo Cryopreservation (Learned Since the 1970s)

Although now successfully integrated into the management of laboratory mice and breeding of several livestock species, there remain significant challenges in embryo cryopreservation. The highest priorities for future studies should include understanding why embryos from various species, different strains, breeds, or even individuals within a given species (but seemingly of the same gross quality) differ in cryosensitivity. This question extends to sorting out the optimal stage of the embryo to cryopreserve while determining why embryos produced in vitro seem more difficult to cryopreserve than those produced in vivo (27). For example, freezing protocols are well established for the human, mouse, and cow but remain suboptimal in the pig and horse. In addition, early cleaved stage ruminant and pig embryos are more susceptible to damages associated with cryopreservation than blastocysts or hatched blastocysts (41,42). For ruminant embryos, it is well accepted that in vitro-produced embryos are much more difficult to cryopreserve than their in vivo counterparts (27). Furthermore, culture conditions, especially the presence or absence of serum, also affect cryosensitivity of ruminant embryos (43). It has been shown that in vitro-derived bovine embryos cultured in serum-free medium have similar morphology (light color) to their in vivo-produced counterparts and are more tolerant to chilling than those cultured in the presence of serum (43,44). Even in vivo-derived embryos, differences in cryosensitivity among breeds have been observed; embryos from Jersey cows are more susceptible to freezing and thawing than those from Holstein cattle (43). Finally, it has been shown that the genotype of mouse strains and stocks significantly affects cryosurvival of mouse embryos (45).

Chilling injury sensitivity has been particularly relevant to investigators studying the pig, which has resisted many efforts to develop consistently reliable protocols. In this case, a high level of sensitivity to cooling requires such a slow cooling rate to prevent intracellular ice formation that protracted exposure time to a critical temperature range then causes irreparable cell damage (1,43). This damage can be overcome by using a high cooling rate to enable rapid transition through the damaging temperature zone so that there is no time for induced injury (46,47). This challenge was partially resolved in 1997 by using an open pulled straw technique that increased cooling rate approximately eight-fold over conventional 0.25-ml straws, thereby increasing survival of thawed pig embryos (48). However, attempts to increase the cooling by using a superfine pulled straw failed to improve pig blastocysts cryosurvival compared with the open pulled straw technique (49).

Although the specific mechanism explaining species variation to embryo chilling injury has not been elucidated, damage susceptibility appears to be directly related to the amount of intracellular lipid (1,43,44). Early-stage ruminant and pig embryos contain substantial amount of intracellular lipid (characterized by dark cytoplasm) droplets (41,42), and both are highly susceptible to chilling injury. In vitro-derived bovine embryos are much darker, have higher intracellular lipid content, and are more susceptible to chilling and freezing than their in vivo counterparts (44,50). On the contrary, mouse and human embryos contain comparatively less internal lipids (characterized by light cytoplasm) and appear tolerant to cooling (1). But the impact of lipids also appears species-specific as cat embryos that have dark cytoplasm are comparatively resilient to chilling injury and survive slow cooling at an acceptable rate (51). In fact, live kittens have been produced from in vitro-derived early stage embryos cryopreserved by a conventional equilibrium cooling method (51,52).

Nagashima and colleagues (42) were the first to demonstrate that mechanical removal of cytoplasmic lipid by centrifugation and micromanipulation reduced chilling sensitivity in two- to four-cell-stage pig embryos. Soon after, live piglets were produced by the same group of investigators (53) after cryopreserved delipidated pig embryos were transferred to recipients. Since the publication of this milestone research, mechanical delipidation has been used to improve cryosurvival of chilled, sensitive in vitro-produced bovine (54) and porcine embryos (55). However, the disadvantage of this technique is that it requires the removal of lipid via micromanipulating pipettes, which in turn causes damage to the zona pellucida and increases the chance of pathogen transmission. For this reason, the influence of chemical delipidation by a lipolytic agent, foskolin, has been recently studied and found to be beneficial for pig embryos (56). Other noninvasive approaches that have been used to reduce or modify lipid content of embryos include culturing in vitro-derived embryos in *trans*-10 *cis*-12 conjugated linoleic acid (57), linoleic acid–albumin (58), and phenazine ethosulfate (43); these approaches have been shown to improve cryosurvival of pig embryos.

HISTORY OF MAMMALIAN OOCYTE CRYOPRESERVATION

Historical Landmarks and Evolution of Oocyte Cryopreservation

The first exploration into the effect of low temperatures on the survival of a mammalian oocyte (rabbit) was reported by Chang in 1952 (59). In that study, there was a two-fold reduction in the developmental competence of mature oocytes after being held at 0°C for 24 hours. A few years later, similar studies were conducted in the mouse by Sherman and Lin (60,61) who demonstrated that mouse oocytes suspended in a medium containing glycerol and cooled to –10°C and held for one or two hours could be used to produce pregnancies in mated recipients. However, a few oocytes survived, which were then fertilized with the oocytes held at 0°C or –10°C for six hours or longer. After these milestone studies there was little progress until the late 1970s when the first live mice were produced from frozen, and thawed oocytes (62,63). These investigators used the same approach as with mouse embryo cryopreservation (15,16), specifically exposing oocytes to 1.5 M DMSO followed by slow cooling to –80°C before freezing in liquid nitrogen. There have been subsequent efforts to improve survival success. In 1978, Leibo and colleagues (64) performed direct microscopic observation of intracellular ice formation in matured mouse oocytes. They concluded that cell survival after freezing to –150°C in 1 M DMSO depended on the cooling rate and was negatively correlated with intracellular ice formation. Most importantly, survival was poor when oocytes were cooled faster than –1.5°C/min. This observation together with Mazur's mathematical model of the physiochemical process during cooling described earlier, led several other investigators to adopt slow cooling as the best method for cryopreserving oocytes from the human (65,66) and rabbit (67).

Regarding immature intraovarian oocytes, the first offspring (in cattle) from a cryopreserved germinal vesicle (GV) oocyte (that was subsequently matured and fertilized in vitro) was produced in 1992 (68). Likewise, the first human baby born after a similar procedure was reported six years later (69).

While slow cooling had been used somewhat (30–70%) successfully to freeze human oocytes (70,71), poor results were reported for other species including pig, cattle, sheep, and horses (72). As a consequence, causes of damage and poor development of cryopreserved oocytes had become the focus of several studies conducted in the following decade (73).

These studies elucidated that mammalian oocytes hold several unique features that contribute to extreme susceptibility to damage during cryopreservation procedure (see below and the chapters in part 2 of this book). In late 1980s, a vitrification method adopted from that for mouse embryos, described by Rall and Fahy (33), was used to cryopreserve mouse, rat, hamster, rabbit, human, cow, and pig oocytes (73). Although some progress was made in the cryopreservation of mouse oocytes (74), very little improvement was achieved in other species, such as the human (75), cow (76), and hamster (77). Until that point, it was well recognized that mammalian oocytes were sensitive to cooling and the degree of chilling sensitivity was directly correlated with the survival after freezing and thawing. In the early 1990s, Steponkus et al. (46) and Mazur et al. (47) reported that Drosophila embryos, a cell type known to be highly susceptible to chilling injury, could be successfully cryopreserved by the use of an extremely rapid cooling rate. The results of these investigations led Martino et al. (78) to pioneer an ultra-rapid cooling method for bovine oocytes. In that study, oocytes contained in minimal volume of a concentrated cryoprotectant solution were placed on electron microscope grids and plunged directly into liquid nitrogen. Using this approach, the cooling rate was estimated to be more than –100,000°C/min, which was 40 times faster than conventional vitrification (i.e., cooling in plastic straws). Forty percent and fifteen percent of oocytes cryopreserved on the electron microscope grid cleaved after in vitro fertilization (IVF) and developed into blastocysts, respectively, whereas the respective values for oocytes vitrified in straws were only 3% and <1%. This extreme cooling rate appeared to be critical for the success of bovine oocyte cryopreservation, because exposure of bovine oocytes to 0°C for only five seconds significantly reduced developmental competence (42% vs. 20% blastocyst development). Ultra-rapid cooling permits the oocytes to traverse through the critical temperature zone (between 0°C and 15°C) so rapidly that there is no time for chilling injury. This improved survival and function of cryopreserved oocytes has led other investigators to derive alternative cooling devices, including open pulled straws (79), solid-surface vitrification (80), microdrops (81), closed pulled straws (82), cryoloop (83), nylon mesh (84), multi-thermal gradient (85), and cryotop (86), all of which permit cooling rates exceeding –100,000°C/min. Since then, this techniques which is generally referred to as "minimum volume vitrification" (MVV) has been widely adopted to preserve oocytes in several species, including the human (87,88), and it has been recently suggested that MVV is the method of choice for oocyte and embryo cryopreservation (89,90). Quantification of the metabolome and proteome of the oocyte has revealed that slow freezing has a dramatic effect on cell physiology, whereas vitrification appears to have limited detrimental effect (89). This is plausibly achieved by the limited exposure to cryoprotectants. Analyses of meiotic spindle dynamics and embryo development following IVF indicate that vitrification is less traumatic than slow freezing, and therefore, has the greatest potential for successful oocyte cryopreservation (89). Despite these quantitative differences in in vitro assessments, there remain no differences in overall implantation or live birth success in humans after transferring embryos produced by slow cooling versus vitrification (Tables 1 and 2). Novel approaches that are now being evaluated to improve oocyte

Table 1 Clinical Outcomes after Transfer of Embryos Produced from Cryopreserved Human Oocytes (Equilibrium Cooling)

	No. of oocytes			Outcomes			
Investigators (Ref.)	Total	Survived	No. of embryos transferred	% Pregnancy	% Implantation	No. of ongoing pregnancies	No. of births
Porcu et al. (91)	12	4	1	–	–	–	1
Porcu et al. (92)	709	–	–	–	–	–	6
Tucker et al. (93)	16	3	2	–	–	–	1
Fosas et al. (94)	88	79	26	57.1	–	–	5
Tjer et al. (95)	14	10	2	–	–	–	1
Borini et al. (71)	927	678	404	12.3	5.2	11	4
Borini et al. (96)	918	398	176	19.2	12.3	–	–
Bianchi et al. (97)	403	306	180	21.3	13.5	–	4
Parmegiani et al.(98)	6	4	3	–	–	–	1

Table 2 Clinical Outcomes after Transfer of Embryos Produced from Cryopreserved Human Oocytes (Vitrification)

		No. of oocytes		Outcome				
Investigators (Ref.)	Cryopreservation techniques	Total	Survived	No. of embryos transferred	% Pregnancy	% Implantation	No. of ongoing pregnancies	No. of births
Kuleshova et al. (99)	OPS	17	11	5	–	–	–	1
Kuwayama et al. (90)	Cryotop	64	52	32	–	–	3	7
Yoon et al. (88)	Nitrogen slush	364	218	158	–	14.0	–	–
Cobo et al. (100)	Cryotop	30	82	70	65.2	40.8	–	–

Abbreviation: OPS, open pulled straw.

cryopreservation success include intracellular injection of trehalose (101), utilizing choline-substituted medium to reduce the potentially harmful buildup of sodium (102), and treatment with cholesterol-loaded methyl-beta cyclodextrin to stabilize the plasma membrane (103). Basic investigations into the cryosensitivity of mature human oocytes (104) conducted in the 1990s have paid dividends in the birth of several healthy children (87,88). The advent of intracytoplasmic sperm injection (ICSI) to fertilize thawed human oocytes—a process that circumvents zona hardening (preventing sperm penetration) caused by premature release of cortical granules—has significantly contributed to the improved outcomes of human oocyte cryopreservation (92). To date, ~15% of human patients undergoing in vitro maturation and IVF after oocyte cryopreservation become pregnant (105), and there are more than 200 children born worldwide with the use of cryopreserved immature or mature oocytes (106,107). Thus, despite an immense need for further improvement, oocyte cryopreservation has emerged as one of the plausible options for fertility preservation in women in the 21st century.

Challenges in Oocyte Cryopreservation (Learned Since the 1980s)

Over five decades of investigation, it was learned that it is far more difficult to cryopreserve oocytes at the GV or metaphase II (MII) stage than the embryos. The main reasons were first identified and described in the late 1990s. First, the oocyte is actually the largest mammalian cell type because of which it has a relatively lower surface area to volume ratio compared with an embryo (5,73,108). As a consequence, oocytes have low permeability to water and cryoprotectants and are highly susceptible to osmotic damage (73,108). To make it even more complicated, it has been demonstrated that water permeability of mouse oocytes differs among mouse strains (109), indicating the significant role of genotype in dictating oocyte cryosurvival. Second, oocytes are extremely susceptible to chilling injury (110). Cooling MII oocytes to room temperature or below induces irreversible disruption of the meiotic spindle and dispersal of chromosomes (110). Cooling GV or MII bovine oocytes to 4°C or 0°C significantly compromises embryonic development (111,112). Exposure of oocytes to low temperature subtly alters membrane structure and function, caused by thermotropic phase transition and lateral phase separation of membrane lipid components (73). Furthermore, cooling also induces premature exocytosis of the cortical granule, which in turn causes zona hardening (73). Third, the exposure of oocytes to conventional permeating cryoprotectants, such as propanediol, ethylene glycol, and DMSO, disrupts cytoskeletal structures and increases intracellular calcium, which in turn initiates oocyte activation and zona hardening (73,113). Although the issue of zona hardening can be overcome by ICSI, cryopreservation of oocytes still faces several challenges.

SUMMARY AND NEW DIRECTIONS

Since the second half of the 20th century, huge progress has been achieved in cryopreservation in parallel with the development of in vitro culture and manipulation of germ cells. The ability to store gametes and embryos in the frozen stage for indefinite period has made assisted reproduction increasingly practical in human infertility treatments as well as managing and improving genetics of animals. Interestingly, eggs or embryos from nonmammalian species (containing yolk) are too difficult to freeze, and no real progress has been made.

In parallel with significant achievements in mammalian cells, the world of reproductive biotechnologies has also made substantial progress in the in vitro culture and manipulation of gametes. For example, the advent of ICSI has circumvented the issue of zona hardening associated with the exposure of cells to low temperature and cryoprotectants. Furthermore, assisted hatching has been shown to significantly improve clinical pregnancy and implantation rates when frozen-thawed embryos are used. Also, in vitro culture conditions have been improved; for example, the development of sequential culture medium that make it possible to culture human embryos to the later blastocyst stage (114). Therefore, oocyte and embryo cryopreservation are linked with in vitro manipulation techniques and the practical use of one discipline is dependent on the progress of the other. Finally, optimization of current techniques and development of innovative approaches will always benefit from comparative studies conducted in different animal species.

ACKNOWLEDGMENTS

The authors thank Dr. S.P. Leibo for his thoughtful discussion of some historical events presented in this chapter. We are also grateful to Dr. D.E. Wildt for his constructive critique of this manuscript.

REFERENCES

1. Mazur P, Leibo SP, Seidel GE, Jr. Cryopreservation of the germplasm of animals used in biological and medical research: importance, impact, status and future directions. Biol Reprod 2008; 78: 2–12.
2. Thibier M. Transfer of both *in vivo* derived and *in vitro* produced embryos in cattle still on the rise and contrasted trends in other species in 2005. Embryo Transfer Newsletter 2006; 24: 12–17.
3. Wildt DE. The role of reproductive technologies in zoos: past, present and future. Int Zoo Yb 2003; 38: 111–18.
4. Leibo SP, Songsasen N. Cryopreservation of gametes and embryos of non-domestic species. Theriogenology 2002; 57: 303–26.
5. Fuller B, Paynter SJ, Watson P. Cryopreservation of human gametes and embryos. In: Fuller BJ, Lane N, Benson EE, eds. Life in the Frozen State. Boca Raton: CRC Press, 2004: 505–39.
6. Wininger JD, Kort HI. Cryopreservation of immature and mature human oocytes. Semin Reprod Med 2002; 20: 45–9.
7. Bredkjaer HE, Grudzinskas JG. Cryobiology in human assisted reproductive technology. Would Hippocrates approve? Early Pregnancy 2001; 5: 211–13.
8. Woodruff TK. The emergence of a new interdiscipline: oncofertility. Cancer Treat Res 2007; 138: 3–11.
9. Hughesman C. A cold greeting: an introduction to cryobiology. Science Creat Quart 2006. [Available from: www.scq.ubc.ca]
10. Hunter M. Robert Boyle and early Royal Society: a reciprocal exchange of making a baconian science. British J Histo Sci 2007; 40: 1–23.
11. Leibo SP. The early history of gamete cryobiology. In: Fuller BJ, Lane N, Benson EE, eds. Life in the Frozen State. Boca Raton: CRC Press, 2004: 347–70.
12. Luyet BJ, Gehenio PM. Life and Death at Low Temperature. Normandy: Biodynamica, 1940.
13. Polge C, Smith AU, Parkes AS. Revival of spermatozoa after vitrification and dehydration at low temperature. Nature 1949; 164: 666.
14. Mazur P. Kinetics of water loss from cells at subzero temperatures and the likelihood of intracellular freezing. J Gen Physiol 1963; 47: 347–69.
15. Whittingham DG, Leibo SP, Mazur P. Survival of mouse embryos frozen to –196°C and –269°C. Science 1972; 178: 411–14.
16. Wilmut I. The effect of cooling rate, warming rate, cryoprotective agent and stage of development on survival of mouse embryos during freezing and thawing. Life Sci 1972; 11: 1071–9.
17. Mazur P. The role of intracellular freezing in the death of cells cooled at supraoptimal rates. Cryobiology 1977; 14: 251–72.
18. Mazur P, Schmidt JJ. Interaction of cooling velocity, temperature, and warming velocity on the survival of frozen-thawed yeast. Cryobiology 1968; 5: 1–17.

19. Wilmut I, Rowson LEA. Experiments on the low-temperature preservation of cow embryos. Vet Rec 1973; 92: 686–90.
20. Rall WF. Cryopreservation of mammalian embryos, gametes and ovarian tissues. In: Wolf DP, Zelinski-Wooten M, eds. Contemporary Endocrinology: Assisted Fertilization and Nuclear Transfer in Mammals. Twato: Humana Press, 2001; 173–87.
21. Trounson AO, Mohr L. Human pregnancy following cryopreservation, thawing and transfer of an eight-cell embryo. Nature 1983; 305: 707–9.
22. Zeilmaker GH, Alberda AT, van Gent I, et al. Two pregnancies following transfer of intact frozen-thawed embryos. Fertil Steril 1984; 42: 293–6.
23. Orief Y, Nikolettos N, Al-Hassani S. 2005. Cryopreservation of two pronuclear stage zygotes. Rev Gynaecol Prac 2005; 5: 39–44.
24. Leibo SP, Mazur P. Methods for the preservation of mammalian embryos by freezing. In: Daniel JC, Jr, ed. Methods in Mammalian Reproduction. New York: Academic Press, 1978: 179–201.
25. Renard JP, Heyman Y, Leymanie P, et al. Sucrose dilution: a technique for field transfer of bovine embryo frozen in the straw. Theriogenology 1983; 19: 145.
26. Leibo SP. A one-step method for direct non-surgical transfer of frozen-thawed bovine embryos. Theriogenology 1984; 21: 767–90.
27. Massip A, Leibo SP, Blesbois E. In: Fuller BJ, Lane N, Benson EE, eds. Life in the Frozen State. Boca Raton: CRC Press, 2004: 371–92.
28. Suzuki T, Yamamoto M, Ooe M, et al. Effect of sucrose concentration used for one-step dilution upon *in vitro* and *in vivo* survival of bovine embryos refrigerated in glycerol and 1,2 propanedial. Theriogenology 1990; 34: 1051–7.
29. Voekel SA, Hu YX. Direct transfer of frozen-thawed bovine embryos. Theriogenology 1992; 37: 23–37.
30. Songsasen N, Buckrell BC, Plant C, et al. *In vivo* and *in vitro* survival of sheep embryos. Cryobiology 1995; 32: 78–91.
31. Tucker MJ, Cohen J, Massey JB, et al. Partial dissection of the zona pellucida of frozen-thawed human embryos may enhance blastocyst hatching, implantation, and pregnancy rates. Am J Obstet Gynecol 1991; 165: 341–4.
32. Balaban B, Urman B, Yakin K, et al. Laser-assisted hatching increases pregnancy and implantation rates in cryopreserved embryos that were allowed to cleave *in vitro* after thawing: a prospective randomized study. Hum Reprod 2006; 21: 2136–40.
33. Rall W, Fahy G. Ice-free cryopreservation of mouse embryos at –196°C by vitrification. Nature 1985; 313: 573–5.
34. Kono T, Suzuki O, Tsunoda Y. Cryopreservation of rat blastocysts by vitrification. Cryobiology 1988; 25: 170–3.
35. Kasai M, Hamaguchi Y, Zhu SE, et al. High survival of rabbit morulae after vitrification in an ethylene glycol-based solution by a simple method. Biol Reprod 1992; 46: 1042–6.
36. Ali J, Shelton JN. Successful vitrification of day-6 sheep embryos. J Reprod Fertil 1993; 99: 65–70.
37. Tachikawa S, Otoi T, Kondo S, et al. Successful vitrification of bovine blastocysts, derived by *in vitro* maturation and fertilization. Mol Reprod Dev 1993; 34: 266–71.
38. Kuwayama M, Holm P, Jacobsen H, et al. Successful cryopreservation of porcine embryos by vitrification. Vet Rec 1997; 141: 365.
39. Feichtinger W, Hochfellner C, Ferstl U. Clinical experience with ultra-rapid freezing of embryos. Hum Reprod 1991; 6: 735–6.
40. Loutradi KE, Kolibianakis EM, Venetis CA, et al. Cryopreservation of human embryos by vitrification or slow freezing a systematic review and meta-analysis. Fertil Steril 2008; 90: 186–93.
41. Mohr LR, Trounson AO. Structural changes associated with freezing of bovine embryos. Biol Reprod 1981; 25: 1009–25.
42. Nagashima H, Kashiwazaki N, Ashman RJ, et al. Removal of cytoplasmic lipid enhances the tolerance of porcine embryos to chilling. Biol Reprod 1994; 51: 618–22.
43. Seidel GE. Modifying oocytes and embryos to improve their cryopreservation. Theriogenology 2006; 65: 228–35.
44. Leibo SP, Loskutoff NM. Cryobiology of *in vitro*-derived bovine embryos. Theriogenology 1993; 39: 81–94.
45. Schmidt PM, Schiewe MC, Wildt DE. The genotypic response of mouse embryos to multiple freezing variables. Biol Reprod 1987; 1121–8.
46. Steponkus PL, Myers SP, Lynch DV, et al. Cryopreservation of *Drosophila melanogaster* embryos. Nature 1990; 345: 170–2.
47. Mazur P, Cole KW, Hall WH, et al. Cryobiological preservation of Drosophila embryos. Science 1992; 258: 1932–5.

48. Vajta G, Holm P, Greve T, et al. Vitrification of porcine embryos using the Open Pulled Straw (OPS) method. Acta Vet Scand 1997; 38: 349–52.
49. Cuello C, Gil MA, Parrilla I, et al. Vitrification of porcine embryos at various developmental stages using different ultra-rapid cooling procedures. Theriogenology 2004; 62: 353–61.
50. Rizos D, Trudee F, Papadppoulos S, et al. Developmental, qualitative, and ultrastructural differences between ovine and bovine embryos produced *in vivo* or *in vitro*. Mol Reprod Dev 2002; 62: 320–7.
51. Gómez MC, Pope E, Harris R, Mikota S, Dresser BL. Development of *in vitro* matured, *in vitro* fertilized domestic cat embryos following cryopreservation, culture and transfer. Theriogenology 2003; 60: 239–51.
52. Pope CE, McRae MA, Plair BL, et al. Successful *in vitro* and *in vivo* development of *in vitro* fertilized two-to four cell cat embryos following cryopreservation, culture and transfer. Theriogenology 1994; 42: 513–25.
53. Nagashima H, Kashiwazaki N, Ashman RJ, et al. Cryopreservation of porcine embryos. Nature 1995; 374: 416.
54. Tominaga K, Shimizu M, Ooyama S, et al. Effect of lipid polarization by centrifugation at different developmental stages on post-thaw survival of bovine *in vitro* produced 16-cell embryos. Theriogenology 2000; 53: 1669–80.
55. Esaki R, Ueda H, Kurome M, et al. Cryopreservation of porcine embryos derived from *in vitro*-matured oocytes. Biol Reprod 2004; 71: 432–7.
56. Men H, Yuksel A, Riley LK, et al. Improved survival of vitrified porcine embryos after partial delipation through chemically stimulated lipolysis and inhibition of apoptosis. Theriogenology 2006; 66: 2008–16.
57. Pereira RM, Baptista MC, Vasques MI, et al. Cryosurvival of bovine blastocysts is enhanced by culture with trans-10 cis-12 conjugated linoleic acid (10t,12c CLA). Anim Reprod Sci 2007; 98: 293–301.
58. Tominaga K, Hamada Y, Yabuue T, et al. Effects of linoleic acid-albumin on post-thaw survival of *in vitro*-produced bovine embryos at 16 cell stage. J Vet Med Sci 2000; 62: 465–7.
59. Chang MC. Fertilizability of rabbit ova and the effects of temperature *in vitro* on their subsequent fertilization and activation in vivo. J. Exp Zool 1952; 121: 351–81.
60. Sherman JK, Lin TP. Effect of glycerol and low temperature on survival of unfertilized mouse eggs. Nature 1958; 181: 785–6.
61. Sherman JK, Lin TP. Survival of unfertilized mouse eggs during freezing and thawing. Proc Soc Exp Biol Med 1958; 98: 902–5.
62. Parkening AA, Tsunoda Y, Chang MC. Effects of various low temperatures, cryoprotective agents and cooling rates on the survival, fertilizability and development of frozen-thawed mouse eggs. J. Exp Zool 1976; 197: 369–74.
63. Whittingham DG. Fertilization *in vitro* and development to term of unfertilized mouse oocytes previously stored at −196°C. J Reprod Fertil 1977; 49: 89–94.
64. Leibo SP, Mcgrath JJ, Cravalho EG. Microscopic observation of intracellular ice formation in unfertilized mouse ova as a function of cooling rate. Cryobiology 1978; 15: 257–71.
65. Chen C. Pregnancy after human oocyte cryopreservation. Lancet 1986; 1: 884–6.
66. Van Uem JF, Siebzehnrubl ER, Schuh B, et al. Birth after cryopreservation of unfertilized oocytes. Lancet 1987; 1: 752–3.
67. Al-Hasani S, Kirsch J, Diedrich K, et al. Successful embryo transfer of cryopreserved and in-vitro fertilized rabbit oocytes. Hum Reprod 1989; 4: 77–9.
68. Fuku E, Kojima T, Shioya Y, et al. *In vitro* fertilization and development of frozen-thawed bovine oocytes. Cryobiology 1992; 29: 485–92.
69. Tucker MJ, Wright G, Morton PC, et al. Birth after cryopreservation of immature oocytes with subsequent *in vitro* maturation. Fertil Steril 1998; 70: 578–9.
70. Fabbri R, Porcu E, Marsella T, et al. Human oocyte cryopreservation: new perspectives regarding oocyte survival. Hum Reprod 2001; 16: 411–16.
71. Borini A, Sciajno R, Bianchi V, et al. Clinical outcome of oocyte cryopreservation after slow cooling with a protocol utilizing a high sucrose concentration. Hum Reprod 2006; 21: 512–17.
72. Ledda S, Bogliolo L, Succu S, et al. Oocyte cryopreservation: oocyte assessment and strategies for improving survival. Reprod Fertil Dev 2007; 19: 13–23.
73. Critser JK, Agca Y, Gunasena KT. The cryobiology of mammalian oocytes. In: Karow A, Critser JK, eds. Reproductive Tissue Banking: Scientific principles, San Diego: Academic Press, 1997: 329–57.
74. Wood MJ, Barros C, Candy CJ, et al. High rates of survival and fertilization of mouse and hamster oocytes after vitrification in dimethyl sulphoxide. Biol Reprod 1993; 49: 489–95.
75. Al-Hasani S, Diedrich K, van der Ven H, et al. Cryopreservation of human oocytes. Human Reprod 1987; 2: 695–70.

76. Otoi T, Tachikawa S, Kondo S, et al. Developmental capacity of bovine oocytes cryopreserved after maturation *in vitro* and of frozen-thawed bovine embryos derived from frozen mature oocytes. Theriogenology 1992; 38: 711–19.
77. Critser JK, Arneson BW, Aaker DV, Ball GD. Cryopreservation of hamster oocytes: effects of vitrification or freezing on human sperm penetration of zona-free hamster oocytes. Fertil Steril 1986; 46: 277–84.
78. Martino A, Songsasen N, Leibo SP. Development into blastocysts of bovine oocytes cryopreserved by ultra-rapid cooling. Biol Reprod 1996; 54: 1059–69.
79. Vajta G, Holm P, Kuwayama M, et al. Open Pulled Straws (OPS) vitrification: a new way to reduce cryoinjuries of bovine ova and embryos. Mol Reprod Dev 1998; 51: 53–8.
80. Dinnyés A, Dai Y, Jiang S, et al. High developmental rates of vitrified bovine oocytes following parthenogenetic activation, *in vitro* fertilization, and somatic cell nuclear transfer. Biol Reprod 2000; 63: 513–18.
81. Papis K, Shimizu M, Izaike Y. Factors affecting the survivability of bovine oocytes vitrified in droplets. Theriogenology 2000; 54: 651–8.
82. Chen SU, Lien YR, Cheng YY, et al. Vitrification of mouse oocytes using closed pulled straws (CPS) achieves a high survival and preserves good patterns of meiotic spindles, compared with conventional straws, open pulled straws (OPS) and grids. Hum Reprod 2001; 16: 2350–6.
83. Lane M, Gardner DK. Vitrification of mouse oocytes using a nylon loop. Mol Reprod Dev 2001; 58: 342–7.
84. Matsumoto H, Jiang JY, Tanaka T, et al. Vitrification of large quantities of immature bovine oocytes using nylon mesh. Cryobiology 2001; 42: 139–44.
85. Arav A, Yavin S, Zeron Y, et al. New trends in gamete's cryopreservation. Mol Cell Endocrinol 2002; 187: 77–81.
86. Kuwayama M, Vajta G, Kato O, et al. Highly efficient vitrification method for cryopreservation of human oocytes. Reprod Biomed Online 2005; 11: 300–8.
87. Lucena E, Bernal BP, Lucena C, et al. Successful ongoing pregnancies after vitrification of oocytes. Fertil Steril 2006; 85: 108–11.
88. Yoon TK, Lee DR, Cha SK, et al. Survival rate of human oocytes and pregnancy outcome after vitrification using slush nitrogen in assisted reproductive technologies. Fertil Steril 2007; 84: 952–6.
89. Gardner DK, Sheehan L, Rienzi M, et al. Analysis of oocyte physiology to improve cryopreservation procedures. Theriogenology 2007; 67: 64–72.
90. Kuwayama M. Highly efficient vitrification for cryopreservation of human oocytes and emryos: the cryotop method. Theriogenology 2007; 67: 73–80.
91. Porcu E, Fabbri R, Seracchioli R, et al. Birth of a healthy female after intracytoplasmic sperm injection of cryopreserved human oocytes. Fertil Steril 1997; 68: 724–6.
92. Porcu E, Fabbri R, Petracchi S, et al. Ongoing pregnancy after intracytoplasmic injection of testicular spermatozoa into cryopreserved human oocytes. Am J Obstet Gynecol 1999; 180: 1044–5.
93. Tucker MJ, Wright G, Morton PC, et al. Birth after cryopreservation of immature oocytes with subsequent *in vitro* fertilization. Fertil Steril 1998; 70: 578–9.
94. Fosas N, Marina F, Torres PJ, et al. The births of five Spanish babies from cryopreserved donated oocytes. Hum Reprod 2003; 18: 1417–21.
95. Tjer GC, Chiu TT, Cheung LP, et al. Birth of a healthy baby after transfer of blastocysts derived from cryopreserved human oocytes fertilized with frozen spermatozoa. Fertil Steril 2005; 83: 1547–9.
96. Borini A, Lagalla C, Bonu MA, et al. Cumulative pregnancy rates resulting from the use of fresh and frozen oocytes: 7 years' experience. Reprod Biomed Online 2006: 12: 481–6.
97. Bianchi V, Coticchio G, Distratis V, et al. Differential sucrose concentration during dehydration (0.2 mol/l) and rehydration (0.3 mol/l) increases the implantation rate of frozen human oocytes. Reprod Biomed Online 2007; 14: 64–71.
98. Parmegiani L, Fabbri R, Cognigni GE, et al. Blastocyst formation, pregnancy, and birth derived from human oocytes cryopreserved for 5 years. Fertil Steril 2008; 90: 2014.e7–10.
99. Kuleshova L, Gianaroli L, Magli C, et al. Birth following vitrification of a small number of human oocytes: case report. Hum Reprod 1999; 14: 3077–9.
100. Cobo A, Kuwayama M, Perez S, et al. Comparison of concomitant outcome achieved with fresh and cryopreserved donor oocytes vitrified by the Cryotop method. Fertil Steril 2008; 89: 1657–64.
101. Eroglu A, Toner M, Toth TL. Beneficial effect of microinjected trehalose on the cryosurvival of human oocytes. Fertil Steril 2002; 777: 152–8.
102. Stachecki JJ, Cohen J, Willadsen SM. Cryopreservation of unfertilized mouse oocytes: the effect of replacing sodium with choline in the freezing medium. Cryobiology 1998; 37: 346–54.
103. Horvath G, Seidel GE, Jr. Vitrification of bovine oocytes after treatment with cholesterol-loaded methyl-beta-cyclodextrin. Theriogenology 2006; 66: 1026–33.

104. Gook DA, Osborn SM, Bourne H, et al. Fertilization of human oocytes following cryopreservation; normal karyotypes and absence of stray chromosomes. Hum Reprod 1994; 9: 684–91.
105. Chung HM, Hong SW, Lim JM, et al. *In vitro* blastocyst formation of human oocytes obtained from unstimulated and stimulated cycles after vitrification at various maturational stages. Fertil Steril 2000; 73: 545–51.
106. Wright VC, Chang J, Jeng G, et al. Assisted reproductive technology surveillance—United States, 2005. MMWR Surveill Summ 2008; 57: 1–23.
107. Manipalviratn S, Decherney A. Clinical application of human oocyte cryopreservation. Rev Recent Clin Trials 2008; 3: 104–10.
108. Leibo SP. Water permeability and its activation energy of fertilized and unfertilized mouse ova. J Membrane Biol 1980; 53: 179–88.
109. Benson CT, Critser JK. Variation of water permeability (Lp) and its activation energy (Ea) among unfertilized golden hamster and ICR murine oocytes. Cryobiology 1994; 31: 215–23.
110. Pickering SJ, Braude PR, Johnson MH, et al. Transient cooling to room temperature can cause irreversible disruption of the meiotic spindle in the human oocyte. Fertil Steril 1990; 54: 102–8.
111. Martino A, Pollard JW, Leibo SP. Effect of chilling bovine oocytes on their developmental competence. Mol Reprod Dev 1996; 45: 503–12.
112. Wu B, Tong J, Leibo SP. Effects of cooling germinal vesicle-stage bovine oocytes on meiotic spindle formation following *in vitro* maturation. Mol Reprod Dev 1999; 54: 388–95.
113. Vincent C, Pickering SJ, Johnson MH, et al. Dimethylsulphoxide affects the organization of microfilaments in the mouse oocyte. Mol Reprod Dev 1990; 26: 227–35.
114. Geary S, Moon YS. The human embryo *in vitro*: recent progress. J Reprod Med 2006; 51: 293–302.

2 Principles of Cryopreservation

David E. Pegg
Department of Biology, University of York, York, U.K.

INTRODUCTION

Several reviews of the biophysical principles of cryobiology have been published recently and the interested reader is referred particularly to Mazur (1) for a detailed discussion or to Pegg (2) for an introductory account. In this chapter the science of cryopreservation will be approached in a more practical and applied way. We know that freezing living cells is normally lethal, a fact that is put to practical use in cryosurgery. But we also know that cooling slows the chemical processes both of life and of decay and this has lead to the idea that "suspended animation" might be achieved by cooling. Successful preservation will then depend on reducing the destructive action of ice but allowing the protective effect of low temperatures, such that any damaging effects are greatly outweighed by the protective effects. This is a complicated matter: many structures and processes are temperature-dependent and cooling has complex effects that combine to create conditions that are far removed from normal physiology. When cells are cooled much below 0°C, the effects are normally dominated by the freezing of water, which typically constitutes at least 80% of tissue mass. It was generally thought that the ice crystals were directly responsible for damage rather than the concentration of solutes in the progressively diminishing liquid phase as cooling proceeded.

The discovery that changed the whole scene and made practical "cryopreservation" a possibility was the accidental finding by Polge, Smith, and Parkes (3), that the inclusion of 10–20% glycerol enabled the spermatozoa of the cock to survive prolonged freezing at –80°C. It was soon shown that many compounds with similar physical properties had a similar cryoprotective effect (4). By empirical experiment it was also found that cooling and warming rates could have an influence on the extent to which cells that had been frozen with glycerol could be recovered, and gradually a "default" cell cryopreservation technique was established: suspension in a 10–20% solution of a cryoprotectant such as glycerol; cooling slowly (1°C/min); storage at –80°C or lower; and recovery by rapid warming (~100–200°C/min).

These early experiments established the practicality, in concept, of effective preservation of living cells by freezing. Some of the elements required for success had been established—the presence of a cryoprotectant, an appropriate cooling rate, a low enough storage temperature, and an appropriate warming rate—but the underlying mechanisms were poorly understood. Other important factors gradually became apparent: the chemical identity of the cryoprotectant and its concentration, the manner of addition of the cryoprotectant, the temperature at which it was added, the rate at which the cryoprotectant was removed after thawing, and the temperature at which it was removed. Each of these factors will now be discussed in sequence and the underlying science will be developed. The aim is to indicate an approach to the optimization of a cryopreservation protocol for any type of cell.

CHOICE OF CRYOPROTECTANT

Glycerol was the first cryoprotectant to be used in practice, and Lovelock (5) soon showed that its effectiveness could be explained by the simple fact that, by increasing the total solute concentration of the aqueous phase, it would reduce the amount of ice that formed at any given temperature during cooling. It was clear that many other neutral solutes were also likely to be cryoprotective, the required properties being high solubility in water, remaining in solution at low temperatures, low toxicity, and the ability to penetrate cell membranes (6). Other cryoprotectants that have been shown, with various cell types, to be effective include ethylene glycol, propylene glycol, dimethyl acetamide, and even methanol, but, most prominently and commonly, dimethyl sulfoxide (DMSO) (7). Not all cryoprotectants are equally effective for all cells, one important factor being the ability of the cryoprotectant to penetrate the cell membrane. If it does not, then the interior of the cell will simply be dehydrated, as the process of osmosis requires eventually reaching equilibrium by the concentration of its own solutes rather than by having part of its water replaced by the cryoprotectant. This will not be biologically effective. On the other hand, if the cryoprotectant does penetrate, then the protocol that is used to add it must take into account the kinetics of the process. The cells will lose water more rapidly than they gain cryoprotectant, so a characteristic "shrink/swell" curve will be produced and it is particularly important to ensure that the minimum cell volume produced is not less that the minimum volume that the cell in question will tolerate: this is typically, but certainly not invariably, something around a 40% reduction in volume (8). This will be discussed in more detail later. The other important factor is the intrinsic toxicity of the cryoprotectant. The balance between these factors varies between cells and cryoprotectants and at this time there is no alternative to direct experiment to establish the preferred option for any particular cell. Practical experience suggests that, for most cells, DMSO should be the first cryoprotectant to examine because it has often provided the best compromise between cryoprotection and toxicity: but if satisfactory results cannot be obtained with DMSO then other options should be considered.

THE ADDITION OF CRYOPROTECTANTS

As indicated above, the exposure of cells to high concentrations of cryoprotectant causes osmotic dehydration and the cells shrink. If the cryoprotectant permeates the cells, it will enter along with water until the cells reach their final equilibrium volume. The rate of change of cell volume, the minimum volume attained, and the final volume are all determined by the permeability characteristics of the cell membrane with respect to water, the particular cryoprotectant, and the selected temperature. It is usual to add the cryoprotectant at around +4°C to minimize cryoprotectant toxicity. However, the loading process will be more rapid at room temperature, reducing the necessary exposure time, and this temperature may therefore be preferable. For accurate optimization of the process the transport parameters should be determined by experiment for the cell in question; the procedure for doing this is discussed below. With these data, the time course of changes in cell volume can be calculated for any desired scheme of cryoprotectant addition. The final equilibrium volume will depend upon the concentration of impermeant solutes in the solution and will be the same as the normal cell volume only if the concentration of impermeant solutes is isotonic in molar (per liter) terms (9). Using the permeability parameters the minimum volume should be calculated and the solution compositions adjusted such that the cells will not shrink below their minimum tolerated volume. This volume should be determined by direct experiment, using increasing concentrations of an impermeant solute such as sucrose or mannitol and determining the effect by a suitable test of cell vitality (ideally a function test of some sort) until the minimum tolerated volume is identified.

It is often found that it is impossible to achieve the desired concentration of cryoprotectant in a single step: In such cases the permeability parameters can be used to design a multi-step addition protocol such that the calculated cell volume will not fall below the known tolerated volume at any stage. See Figures 1, 2, and 3 for an example. Then, using such a modeled scheme, the actual cells can be exposed to increasing concentrations of cryoprotectant, in a stepwise fashion, until the concentration that results in damage is found. The whole process may then be repeated at a different temperature (but usually only at +4°C and +20°C) and a judgment is made about which temperature to use, +4°C to reduce cryoprotectant toxicity or +20°C to take advantage of the more rapid transport of water and cryoprotectant at the higher temperature.

The quantitative description of the mass transfer processes is well developed (10,11). Thus, the flow of water, J_v, through a cell membrane is given by

$$J_v = L_p \cdot \pi$$

where L_p is a constant (hydraulic conductivity) that is characteristic of the membrane. The osmotic pressure π can be calculated from concentration by multiplying it with the product of the universal gas constant R and the absolute temperature T. The constant RT is 23.235 atm cm^3 mol^{-1}.

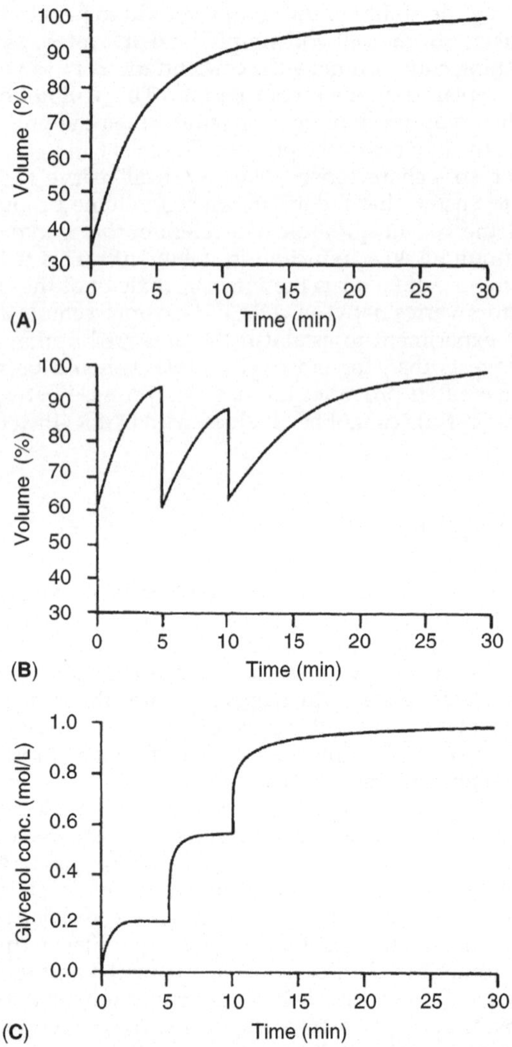

Figure 1 Calculated changes in the volume and the intracellular concentration of glycerol in human thrombocytes during (**A**) the addition of 1 molar glycerol in a single step or (**B**) the addition glycerol in three steps: 0.25, 0.6, and 1.0 molar glycerol. Panel **C** shows the intracellular glycerol concentration throughout procedure **B**. *Source*: From Ref. 13.

If the area of the membrane is A and the internal and external osmolalities are denoted by C_i and C_e, respectively, then we obtain

$$J_v = L_p ART(C_i - C_e)$$

The flux of solute across unit area of a membrane is proportional to the solute permeability ω_s and the difference in concentration of the solute across the membrane. (The more familiar solute permeability P_s (unit cm/sec) is equal to $\omega_s RT$.) The convention for the direction of flux is that outside to inside is positive. The equation describing the solute flux J_s is

$$J_s = \omega_s ART(S_e - S_i)$$

These equations can be solved simultaneously by a numerical method that will run on a standard PC and the values of L_p and ω_s (or P_s) are determined.

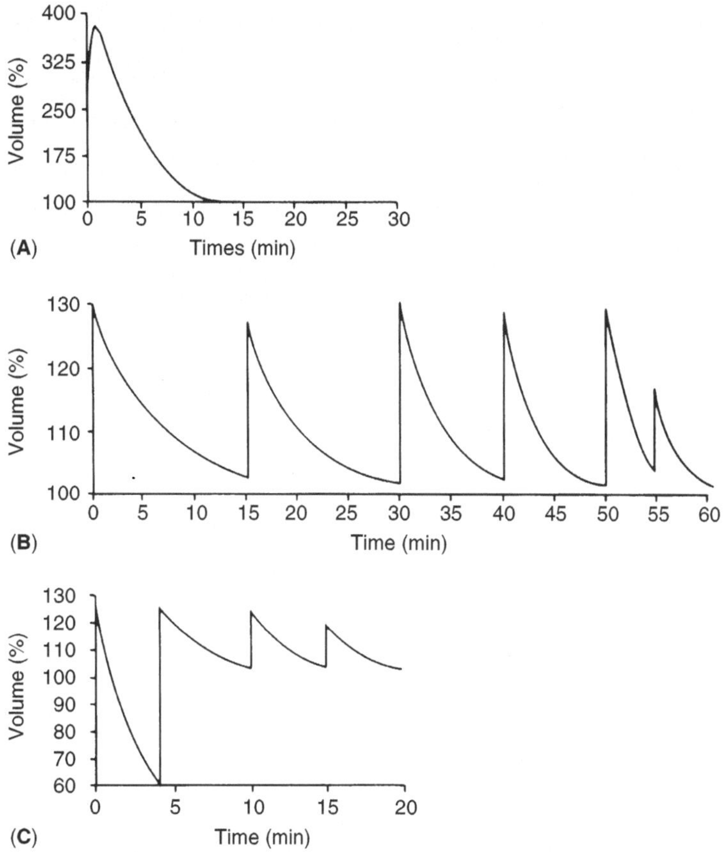

Figure 2 Calculated changes in volume during the removal of 1 molar glycerol from human thrombocytes. (**A**) Dilution to zero glycerol concentration in a single step. This would be lethal. (**B**) Dilution to zero in six steps using isotonic diluents containing 0.7, 0.5, 0.31, 0.17, 0.05, and 0.0 molar glycerol. (**C**) Dilution to zero glycerol concentration using an initial dilution with a hypertonic solution followed by a step with water to restore isotonicity and then finally two isotonic dilution steps. *Source*: From Ref. 13.

A somewhat more complex formalism was elaborated by Kedem and Katchalsky (10) and their equations are often used in cryobiology where they are usually referred to as the K-K equations. Kedem and Katchalsky assumed that the solvent and solute used a common channel through the membrane and they therefore added a solvent/solute interaction term, σ. The K-K formalism is more complex and curve fitting routines can lead to uncertain results due to the fact that the parameter σ is not independent of L_p and ω_s. Kleinhans (11) has argued that the K-K formalism is often invalid because of the presence of separate channels for water and solute. In the author's opinion the simpler so-called "two parameter" formalism developed above is perfectly adequate for the calculations needed to optimize cryopreservation procedures.

Several experimental methods are available for the determination of permeability parameters in cryobiology, but the most commonly used method for isolated cells is to record the time course of cell volume by means of a Coulter counter or by a light-scattering method when the cells are exposed to a known concentration of the compound in question (12,13).

The calculation of changes in cell volume requires one more piece of data in addition to the use of L_p and P_s and that is the so-called "non-osmotic volume" often designated by the letter "b". When cells are immersed in a solution of non-permeating solute they will attain a volume that is an inverse function of the osmolality of the solution. Ideally,

$$V_{rel} = 1/M_{rel}$$

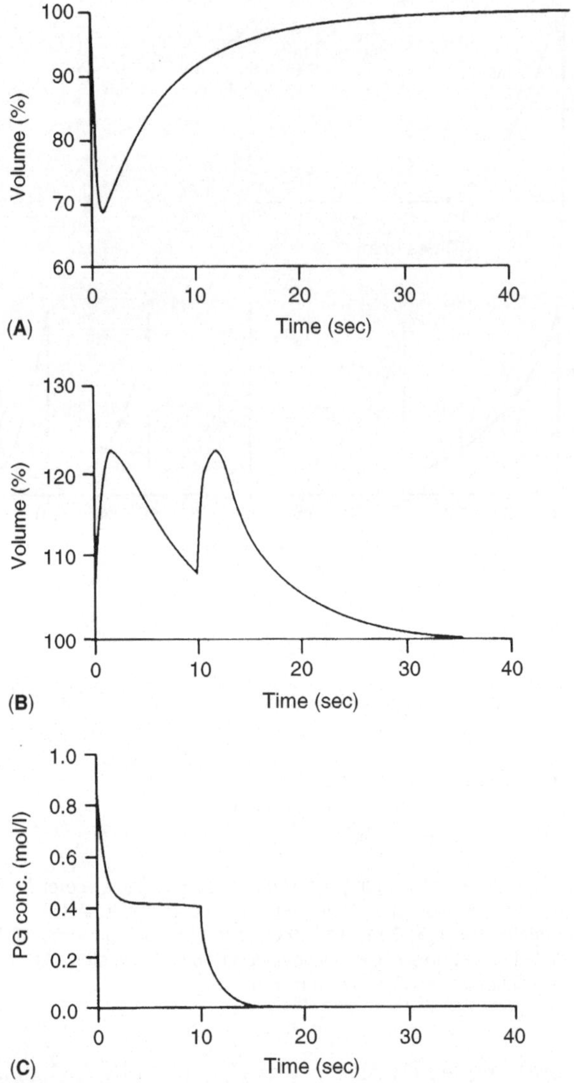

Figure 3 Calculated changes in the volume and the intracellular concentration of propylene glycol (PG) during the introduction and removal of 1 molar PG in human thrombocytes. (**A**) The addition of 1 molar PG in a single step. (**B**) The dilution to zero PG concentration in two steps, 0.4 and 0.0 molar PG. (**C**) The intracellular PG concentration throughout procedure **B**. *Source*: From Ref. 13.

where V_{rel} is the volume of intracellular water relative to the physiological water content and M_{rel} is the external osmolality relative to its physiological value.

A plot of V_{rel} against $1/M_{rel}$, which is known as a Boyle–van't Hoff plot, is a straight line of slope = 1, passing through the origin (as water content is zero at infinite osmolality). In reality there is always an intercept on the y-axis—the so-called non-osmotic water volume or V_{inf}—which probably represents a physically distinct portion of the cell water that is so structured that it does not participate in solution phenomena. For our purposes we need the *cell* volume at infinite osmolality which will comprise V_{inf} plus the volume of the cell solids and is often designated "b". It is the y-intercept on a plot of cell volume against $1/M_{rel}$ (14). In practice the determination b can be combined with the determination of the upper and lower volume limits that the cells will tolerate without damage.

Once the values of L_p, ω_s, and b have been determined, it is possible to predict the changes in intracellular quantities of cryoprotectant and water and hence the time course of cell volume for any scheme of addition or removal of the cryoprotectant, and therefore use these data to design protocols for adding (and removing) cryoprotectants such that cell volume remains within predetermined limits. This was the process used to generate Figures 1, 2, and 3, which show the volume and intracellular cryoprotectant concentrations for selected schemes of addition and removal of two cryoprotectants—glycerol and propylene glycol—for human thrombocytes. Propylene glycol permeated more rapidly and therefore had a lesser osmotic effect. The removal phase is discussed in more detail below.

THE COOLING RATE WHEN ICE IS ALLOWED TO FORM

Shortly after the discovery of the cryoprotective action of glycerol, Lovelock (15) provided strong evidence that the concentration of solutes rather than the ice itself was the cause of freezing injury to cells. Lovelock found no indication in his experiments with human erythrocytes that the rate of cooling had an effect on cryoinjury, but it soon became clear that, first, cooling rate and then warming rate were sometimes important determinants of survival. If freezing damage was indeed caused by solute concentration, then it might be expected to be ameliorated as the cooling rate was increased—on the common sense basis that, whatever the fundamental mechanism, its action would be weakest at the lowest temperatures and therefore survival would increase as cooling rate was increased. Up to a point this was true, as illustrated in Figure 4, where it is shown that for each of the cell types shown, survival improves as cooling rate is increased but then reaches a maximum that is characteristic for each cell type, after which, as cooling rate is increased further, survival decreases. In 1963, Mazur (16) proposed his "two-factor hypothesis" to explain this phenomenon. According to this theory Lovelock's solute damage dominated at the lower cooling rates and was progressively reduced as the cooling rate increased, but then a second damaging mechanism intervened to cause cell survival to decrease with increasing cooling rate. That mechanism was intracellular freezing. Mazur argued that the rate of change of temperature controlled the rate of change of solute concentration outside the cells, which is the driving force for water to be transported out of the cells during cooling. Providing water could leave the cells sufficiently rapidly to maintain thermodynamic equilibrium across the cell membrane, the cytoplasm would not cool below its freezing point and all the ice would be external to the cells. On the other hand, if the cooling rate was too rapid for the permeability of the membrane to allow sufficient water to leave the cell, then the protoplasm would become supercooled, and the greater the extent of supercooling, the more likely it was for the cell to freeze internally. The optimum cooling rate and the magnitude of survival at that rate depended on the intersection of two curves, one describing solute damage and the other

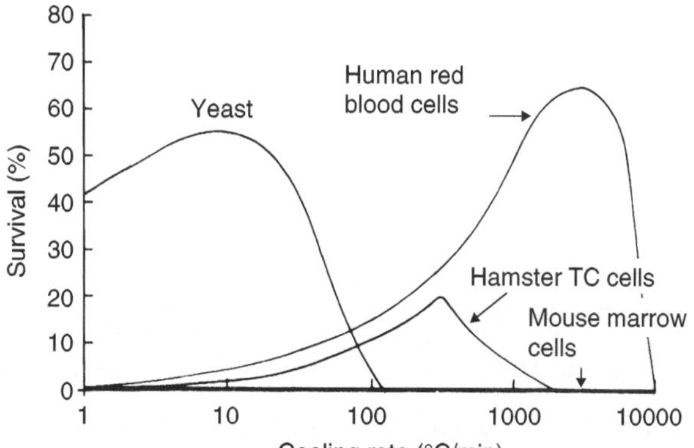

Figure 4 The effect of cooling rate on the survival, following freezing, of four types of cells. *Source*: From Ref. 32.

describing intracellular freezing. Cryoprotectants, by reducing the amount of ice formed at each subzero temperature, reduce the concentration of the other solutes present, notably of salts, and consequently reduce cell damage at low cooling rates. The net effect would depend on where the two curves intersect. With mouse bone marrow stem cells Mazur showed that, in the absence of a cryoprotectant, survival was insignificant (<2%) at all cooling rates: the curves intersected close to the baseline. But as the concentration of glycerol was increased, the solute effect curve moved up and to the left, reducing solute damage and permitting progressively increasing cell survival (Fig. 5). Cryoprotectants have the effect of reducing solution effects and lowering the optimal cooling rate while increasing the maximum survival obtained. They do not, at least to a first approximation, affect the susceptibility of the cell to intracellular freezing.

If the water permeability of the cell membrane is known, and the temperature coefficient of water permeability can be estimated, then it should be possible to predict the effect of cooling rate on the extent of supercooling and therefore the probability of intracellular freezing. Figure 6 shows such data for hepatocytes. The calculated degree of supercooling for different rates of

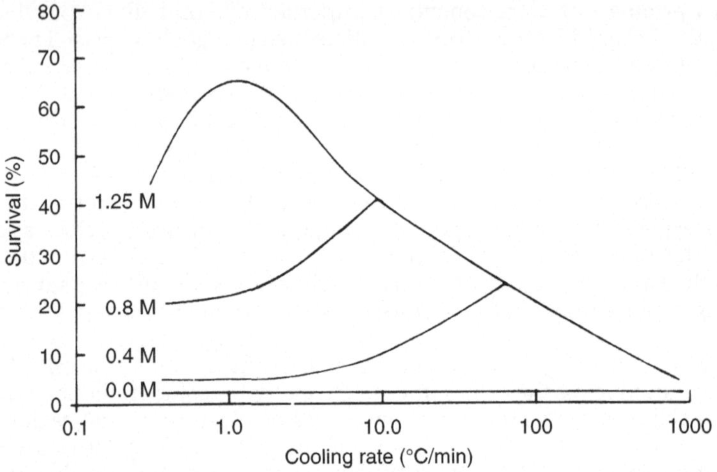

Figure 5 The effect of cooling rate on the cryopreservation of mouse hemopoietic stem cells cooled in the presence of the indicated molar concentrations of glycerol. *Source*: From Ref. 33.

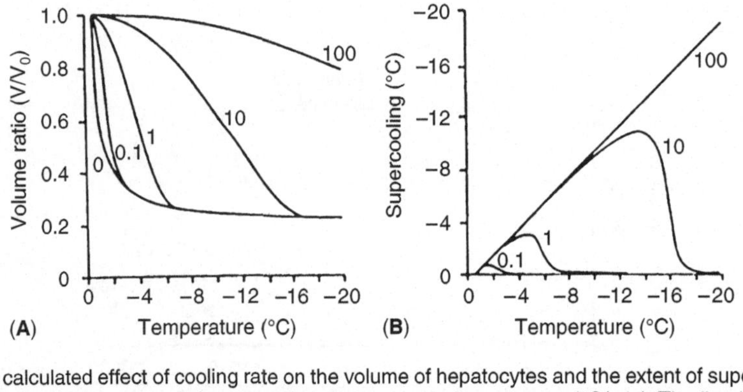

Figure 6 The calculated effect of cooling rate on the volume of hepatocytes and the extent of supercooling of the cell contents. (**A**) Relative volume (V/V_0) of cells cooled at the indicated rates (°C/min). The line labeled 0 is the equilibrium line. (**B**) From the same calculations as in (**A**), the calculated degree of supercooling of the cell contents at the indicated cooling rates. At 10°C/min the cells are supercooled by 10°C and are therefore likely to freeze internally. *Source*: From Ref. 30.

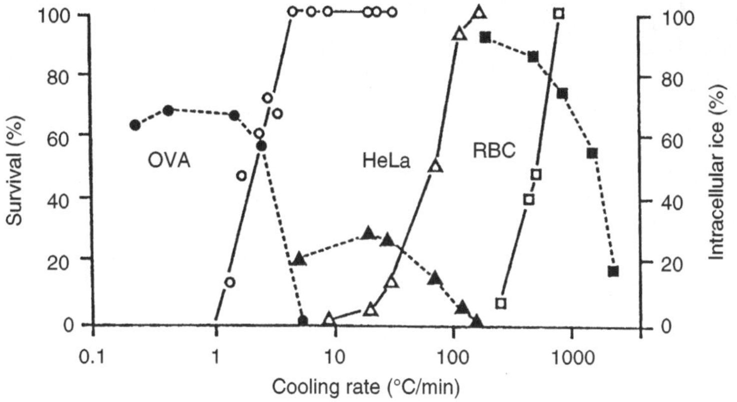

Figure 7 The survival of three types of cells plotted against cooling rate and correlated with the observed occurrence of intracellular freezing. *Source*: From Ref. 17.

cooling of these cells shows that intracellular freezing should be unlikely at 1°C/min but highly probable at 10°C/min. Cooling at 1°C/min should almost eliminate the risk of intracellular freezing and more rapid cooling would be preferred only if solute damage was unacceptably high. Other cells have different water permeabilities, but it has been shown by direct microscopy that the cooling rate that produces intracellular freezing corresponds to the extent of intracellular supercooling that occurs (17) (Fig. 7).

This discussion might suggest that, if all these mechanisms are so well understood and the water and solute permeability of the cells are known, then it should be possible to calculate optimal cooling (and warming) rates with some confidence. Unfortunately, the temperature coefficients of the transport coefficients at very low temperatures and particularly the proper extrapolation rules are not reliably known. Until they are, optimum rates of cooling and warming have to be determined by experiment.

THE COOLING RATE WHEN ICE IS NOT ALLOWED TO FORM

So far we have assumed that ice forms during cooling but does not directly damage the cells if it is exclusively extracellular. However, if cells could be cooled to very low temperatures without any ice being formed at all, then both intracellular ice and the secondary effects of solute concentration would be avoided. Luyet (18) thought that the formation of ice damaged cells directly and he searched for conditions that would produce a vitreous or glassy state with biological systems such that living cells could survive. Aqueous systems are said to be vitrified if the viscosity reaches a sufficiently high value (~10^{13} poises) to make it behave like a solid but without crystallization. In fact, vitrification does occur during conventional cryopreservation where ice is allowed to form. This happens because the concentration of solute in the liquid increases during progressive freezing until the point is reached where the residual solution vitrifies in the presence of ice. The temperature at which this occurs is called T_g' and is illustrated in Figure 8. The problem is that, at temperatures above 0°C, cells will not tolerate exposure to the concentration of cryoprotectant that is required for vitrification without freezing (typically ~80 g% w/w). However, the concentration required to vitrify varies both with cooling rate and cryoprotectant (19) so that it is possible to vitrify solutions that are less concentrated if sufficiently rapid cooling is employed. Hence, it may be possible to achieve the vitreous state in cells using concentrations of cryoprotectant that they will tolerate at temperatures above zero (20,21). Unfortunately, there are two practical problems with this approach: the concentration of cryoprotectant required is very high compared with conventional cryopreservation and toxicity is the limiting factor; also, the resulting vitrified material is unstable, requiring very rapid cooling and even more rapid warming.

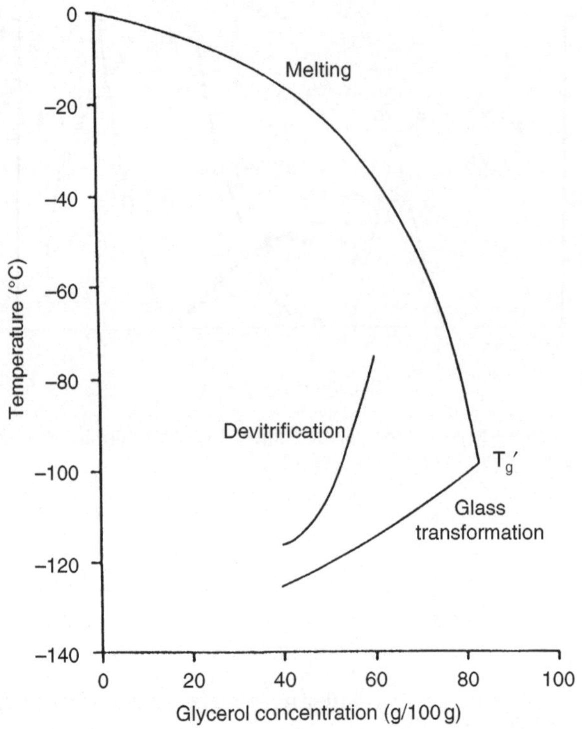

Figure 8 Supplemented phase diagram for glycerol/water. The intersection of the melting curve and the glass transition curve at T_g' indicates the lowest concentration of glycerol that, in theory, will vitrify. In practice, the lower temperatures on the melting curve are unlikely to be reached owing to the high viscosity preventing the crystallization of ice. *Source*: From Ref. 25.

THE STORAGE TEMPERATURE

The temperature that is required for effective long-term cryopreservation is a very important practical question, but is one on which very little work has been done. All the early work on cryopreservation used a storage temperature of −80°C, usually by means of solid carbon dioxide, which has a sublimation temperature of −79°C. This mode of preservation provided long-term, but not indefinite, storage and has now largely been replaced by the use of liquid nitrogen, which has a boiling point of −196°C and provides virtually indefinite storage. Samples that have been vitrified using very rapid cooling with less than the desired concentration of cryoprotectant must be stored at temperatures below the glass transition temperature, which is −123°C for DMSO and −104°C for propylene glycol (22,23).

THE WARMING RATE

In conventional cryopreservation, the warming rate is generally thought to be less important than the cooling rate and generally that is true. However, the conventional rule of thumb (that cooling should be slow and warming rapid) was derived from experiments in which the optimum cooling rate was determined first, using rapid warming. When the optimum warming conditions were then determined using slowly cooled cells it is hardly surprising that rapid warming gave the best results. In fact, the two rates interact and in one study that explored a wide range of cooling and warming rates, two optimal combinations of cooling and warming rates were found—cooling at 1°C/min with warming at 200°C/min and another optimum with cooling at 0.2°C/min and warming at 0.3°C/min (24).

Warming rate is of much greater importance for vitrified samples if suboptimal concentrations of cryoprotectant were used and freezing was prevented by very rapid cooling. This results in an unstable situation and for physical reasons that are discussed elsewhere (25) such vitrified samples are prone to devitrify (i.e., to freeze) during warming. Very rapid warming is essential, which implies the use of very small samples in high-conductivity holders and a relatively high-temperature thawing bath. The effectiveness of this approach was clearly demonstrated by the successful vitrification of *Drosophila melanogaster* embryos (26). These are complex organisms that cannot be preserved by conventional freezing methods. They contain some 50,000 cells and differentiation into organ systems is advanced. Success required careful permeabilization of the waxy vitelline membrane to allow penetration of the cryoprotectant, exposure to 8.5 mol/l ethylene glycol, cooling at 100,000°C/min, storage at ~ –200°C, and warming at 100,000°C/min. The extremely high rate of warming was far more critical than the rate of cooling, which is consistent with the crucial importance of maintaining the vitreous state.

REMOVAL OF CRYOPROTECTANTS

When a permeating cryoprotectant is removed by exposing the cells to a lower concentration of that compound, an osmotic imbalance is produced causing the uptake of water and as a result the cells swell above their initial volume. They then shrink as the cryoprotectant moves out, accompanied by sufficient water to maintain osmotic equilibrium, and they return to their physiological volume only if non-permeating solute has neither been lost nor gained during the process. As cells are generally more sensitive to swelling than to shrinkage, removal of cryoprotectants tends to be more hazardous than their addition. Analogous to the situation when designing an addition protocol, the maximum tolerated volume must be determined by exposing the cells to graded hypotonic solutions of a non-permeating solute. Again, both the rate of change of volume and the final volume must be considered when designing protocols for the recovery of cryopreserved cells, and by using the transport parameters already determined multi-step dilution schedules can be determined. Figures 1, 2, and 3 illustrate some calculated removal procedure for both glycerol and propylene glycol with human thrombocytes. In this case, the removal process for glycerol could be made more practical by including an impermeant solute in the first step solution or alternatively by changing to a more permeable cryoprotectant—propylene glycol.

THE CRYOPRESERVATION OF CELL SUSPENSIONS

The basic cryobiological knowledge reviewed here has made it possible to develop effective methods for the preservation of a very wide range of cells, mostly by conventional cryopreservation in the presence of extracellular ice but also by vitrification methods (see Ref. 27 for a comprehensive survey). For each type of cell, there is a set of conditions that is optimal for preservation, determined by the interaction of the particular properties of the cell in question with the cryobiological factors that have been discussed.

CRYOPRESERVATION OF MULTICELLULAR SYSTEMS AND TISSUES

The situation becomes much more difficult when complex, multicellular systems are considered. Tissues and organs contain a heterogeneous collection of cells, which may have quite different optimum requirements for preservation, yet it is necessary to find a method that will provide adequate survival of all the cells that are important for the function of that tissue. The highest tolerated concentration of cryoprotectant may produce sufficient flattening of the bell-shaped survival curve to enable all the cells present to survive. In organized tissues it may also be important to avoid damage to specific extracellular structures and to retain the normal interconnections between the cells and their basement membranes (28). Extracellular ice in a cell suspension is outside the system that it is desired to preserve and consequently is innocuous, but the situation is quite different for organized tissues: here, the extracellular ice is within the system that is to be preserved and may disrupt the structure of the tissue directly. Taylor and Pegg (29) showed that smooth muscle frozen to –21°C by cooling at 2°C/min in the presence of 2.56 M DMSO was functionally damaged, whereas exposure to the solution composition produced at that temperature was innocuous. Ice had damaged this tissue, and it was shown that the extent of such damage depended upon the specific site of the ice crystals. Damaging effects of extracellular ice have also been demonstrated in the blood capillaries of kidneys and livers (30). The avoidance of freezing or, at least, the limitation of the amount of ice to very small quantities in the least susceptible locations may be particularly important for multicellular systems. Attempts to cryopreserve complex multicellular systems simply by adopting techniques from single-cell systems have generally been unrewarding.

The demonstration that ice forming in tissues produces so much damage has created renewed interest in the possibility of using vitrification with very high concentrations of appropriate cryoprotectants to avoid the formation of ice completely. The problem is that very rapid cooling and certainly very rapid warming simply are not possible for bulky systems, so higher concentrations of cryoprotectant must be used, with consequent toxicity. One approach to this problem is to increase the concentration of cryoprotectant progressively during cooling so that the tissue concentration follows the liquidus curve: ice does not form but the cells do not experience any greater concentration of cryoprotectant than occurs during freezing. This has proved to be entirely practical and very effective for the cryopreservation of articular cartilage, an otherwise recalcitrant tissue (31). A similar method may be effective for other resistant tissues and perhaps even for organs.

CONCLUSIONS

A distinction is made between cooling and freezing. Cooling slows the chemical processes both of life and of decay, but when cells are cooled much below 0°C, the effects are dominated by the freezing of water, which causes the formation of ice crystals and the simultaneous concentration of solutes in the progressively diminishing liquid phase. It was the discovery by Polge, Smith, and Parkes that the inclusion of 10–20% glycerol enabled the spermatozoa of the cock to survive prolonged freezing at –80°C that initiated the era of cryopreservation. After many empirical experiments a "default" cell cryopreservation technique was established: typically, suspension in a 10–20% solution of a cryoprotectant such as glycerol; cooling slowly (1°C/min); storage at –80°C or lower; and recovery by rapid warming (~100–200°C/min). The action of glycerol was explained by the simple fact that it increased the total solute concentration of the aqueous phase and it would therefore reduce the amount of ice that formed at any given temperature. Many other neutral solutes possess the required properties of high solubility in water, remaining in solution at low temperatures, low toxicity, and the ability to penetrate cell membranes. This ability to penetrate the cell membrane is particularly important and the protocols that are used to add or remove it must take into account the kinetics of the process so that the minimum and maximum cell volumes do not transgress the minimum and maximum volumes that the cell will tolerate: this is typically around ± 40%. The effect of cooling rate is explained by Mazur's "two-factor hypothesis," which argues that injury at low cooling rates is due to the concentration of solutes in the remaining liquid and that at high cooling rates is due to intracellular freezing. The optimum cooling rate and the magnitude of survival at that rate depend on the intersection of the curves describing the dependence of these two mechanisms on cooling rate.

We see therefore that ice forming during slow cooling does not damage the cells because it is exclusively extracellular. However, if cells could be cooled to very low temperatures without any ice being formed at all, then both intracellular ice and the secondary effects of solute concentration would be avoided. Aqueous systems are said to be vitrified when the viscosity reaches a sufficiently high value to make them behave as solids but without crystallization. The problem is that cells at ~0°C will not tolerate exposure to the concentration of cryoprotectant that is required for vitrification (typically ~80 g% w/w). The concentration required to vitrify is reduced at high cooling rates, but there are two practical problems with this approach: the concentration of cryoprotectant required is very high, so toxicity is the limiting factor; also, the resulting vitrified material is unstable, requiring rapid cooling and very rapid warming. Such samples must be stored at temperatures below the glass transition temperature, which is –123°C for DMSO and –104°C for propylene glycol.

This basic cryobiological knowledge has made it possible to develop effective methods for the preservation of a wide range of cells, mostly by conventional cryopreservation in the presence of extracellular ice but also by vitrification methods. However, the situation is more difficult for complex, multicellular systems. These usually contain a heterogeneous collection of cells, which may have quite different optimum requirements for preservation and it may be important to avoid damage to specific extracellular structures. Attempts to cryopreserve complex, multicellular systems simply by adopting techniques from single-cell systems have generally been unrewarding. The demonstration that ice forming in tissues produces so much damage has created renewed interest in the possibility of using vitrification with very high concentrations of appropriate cryoprotectants to avoid the formation of ice completely. One approach to this problem is to increase the concentration of cryoprotectant progressively during cooling so that the tissue concentration follows the liquidus curve: ice does not form but the cells do not experience any greater concentration of cryoprotectant than occurs during freezing. This has proved to be entirely practical and very effective for the cryopreservation of articular cartilage, an otherwise recalcitrant tissue. Similar methods may be effective for other resistant tissues and perhaps even for organs.

REFERENCES

1. Mazur P. Principles of cryobiology. In: Fuller BJ, Lane N, Benson EE, eds. Life in the Frozen State. Boca Raton: CRC Press, 2004: 3–65.
2. Pegg DE. Principles of cryopreservation. In: Day JG, Stacey GN, eds. Cryopreservation and freeze drying protocols. Totowa: Humana Press, 2007: 39–57.
3. Polge C, Smith AU, Parkes AS. Revival of spermatozoa after vitrification and dehydration at low temperatures. Nature 1949; 164: 666.
4. Lovelock JE. The protective action by neutral solutes against haemolysis by freezing and thawing. Biochem J 1954; 56: 265–70.
5. Lovelock JE. The mechanism of the protective action of glycerol against haemolysis by freezing and thawing. Biochim Biophys Acta 1953; 11: 28–36.
6. Nash T. The chemical constitution of compounds which protect erythrocytes against freezing damage. J Gen Physiol 1962; 46: 167–75.
7. Lovelock JE, Bishop MWH. Prevention of freezing damage to living cells by dimethyl sulphoxide. Nature (Lond) 1959; 183: 1394–5.
8. Madden PW, Pegg DE. Calculation of corneal endothelial cell volume during the addition and removal of cryoprotective compounds. Cryo-Letters 1992; 13: 43–50.
9. Pegg DE. Red cell volume in glycerol/sodium chloride/water mixtures. Cryobiology 1984; 21: 234–9.
10. Kedem O, Katchalsky A. Thermodynamic analysis of the permeability of biological membranes to non-electrolytes. Biochim Biophys Acta 1958; 27: 229–46.
11. Kleinhans FW. Membrane permeability modelling: Kedem-Katchalsky vs a two-parameter formalism. Cryobiology 1989; 37: 271–89.
12. Wusteman MC, Pegg DE. Differences in the requirements for cryopreservation of porcine aortic smooth muscle and endothelial cells. Tissue Engineering 2001; 7: 507–18.
13. Arnaud FG, Pegg DE. Permeation of glycerol and propane-1,2-diol into human platelets. Cryobiology 1990; 27: 107–18.
14. Pegg DE, Hunt CJ, Fong LP. Osmotic properties of the rabbit corneal endothelium and their relevance to cryopreservation. Cell Biophysics 1987; 10: 169–91.
15. Lovelock JE. The haemolysis of human red blood cells by freezing and thawing. Biochim. Biophys Acta 1953; 10: 414–26.
16. Mazur P. Kinetics of water loss from cells at subzero temperatures and the likelihood of intracellular freezing. J Gen Physiol 1963; 47: 347–69.
17. Leibo SP. Fundamental cryobiology of mouse ova and embryos. In: the freezing of mammalian embryos. Ciba Foundation Symposium 52 (new series) Amsterdam: Elsevier, 1977: 69–92.
18. Luyet BJ, Gehenio PM. Life and death at low temperatures. Normandy, MO: Biodynamica, 1940.
19. Sutton RL. Critical cooling rates to avoid ice crystallization in solutions of cryoprotective agents. J Chem Soc Faraday Trans 1991; 87: 101–5.
20. Fahy GM, MacFarlane DR, Angel CA, Meryman HT. Vitrification as an approach to cryopreservation. Cryobiology 1984; 21: 407–26.
21. Fahy GM, Wowk B, Wu J, et al. Cryopreservation of organs by vitrification; perspectives and advances. Cryobiology 2004; 48: 157–78.

22. Pegg DE, Wusteman MC, Boylan S. Fractures in cryopreserved elastic arteries. Cryobiology 1997; 34: 183–92.
23. Wusteman MC, Pegg DE, Wang L-H, Robinson MP. Vitrification of ECV304 cell suspensions using solutions containing propane-1,2-diol and trehalose. Cryobiology 2003; 46: 135–45.
24. Akhtar T, Pegg DE, Foreman J. The effect of cooling and warming rates on the survival of cryopreserved L-cells. Cryobiology 1979; 16: 424–9.
25. Pegg DE, Diaper MP. Freezing versus vitrification; basic principles. In: Sibinga STh, Das PC, Meryman HT, eds. Cryopreservation and low temperature biology in blood transfusion. Proceedings of the 14th International Symposium on Blood Transfusion. Groningen, 1989. Kluwer Academic Publishers, 1990: 57–69.
26. Mazur P, Cole KW, Hall JW, Schreuders PD, Mahowald AP. Cryobiological preservation of drosophila embryos. Science 1992; 258: 1932–5.
27. Fuller BJ, Lane N, Benson EE. Life in the Frozen State. Boca Raton: CRC Press, 2004.
28. Pegg DE. Ice crystals in tissues and organs. In: Pegg DE, Karow AM, Jr, eds. The biophysics of organ preservation. New York: Plenum Press, 1987.
29. Taylor MJ, Pegg DE. The effect of ice formation on the function of smooth muscle tissue stored at –21°C or –60°C. Cryobiology 1983; 20: 36–40.
30. Rubinsky B, Pegg DE. A mathematical model for the freezing process in biological tissue. Proc R Soc Lond 1988; 234: 343–58.
31. Pegg DE, Wang L, Vaughan D. Cryopreservation of articular cartilage. 3. The liquidus-tracking method. Cryobiology 2006; 52: 360–8.
32. Pegg DE. Cryobiology. In: Proceedings of the Fourth International Cryogenic Engineering Conference, Eindhoven. IPC Science and Technology Press, Guilford, UK, 1972: 47–54.
33. Pegg DE Ice crystals in tissues and organs. In: The Biophysics of Organ Preservation, (Pegg DE. and Karow AM Jr., eds.), Plenum Press, New York, 1987: 117–40.

3 The Rational Basis for Controlled Rate Slow Cooling

Barry Fuller[1] and Sharon Paynter[2]

[1]*Division of Surgery and Interventional Sciences, UCL Medical School, Royal Free Hospital Campus, London, U.K.*

[2]*Department of Obstetrics and Gynaecology, Cardiff University, Heath Park, Cardiff, U.K.*

INTRODUCTION

Storage of unfertilized mammalian oocytes has numerous applications in both human and animal reproductive technologies. Cryopreserved human oocytes can assist in the treatment of human infertility, for example, where loss of fertility may be predicted following ablative anticancer treatments or in patients suffering premature ovarian failure; banking of excess oocytes produced as a result of in vitro fertilization (IVF) treatments; and donation of oocytes to others (storage giving the added advantage of allowing time to screen donors for disease). Oocyte cryopreservation could also serve as a means of delaying child bearing to an age when natural fertility and/or oocyte quality has declined. Applications in animal management include, amongst the domestic species, preservation and cross-nation transport of economically-important strains; the potential to restock depleted herds following outbreaks of diseases such as foot and mouth; preservation of genetically modified strains, thereby reducing the cost of continuous breeding and avoiding problems of genetic drift; and the preservation and management of endangered species. From an ethical and legal standpoint in clinical practice, many people view oocyte cryopreservation as preferable to embryo storage because concerns about parental ownership or consent for use are lessened with unfertilized gametes. In some countries, cryopreservation of human embryos is currently prohibited or strictly limited because of these considerations. In animal management, oocyte cryopreservation gives greater flexibility in breeding programs than does embryo cryopreservation. Cryopreservation of mature oocytes is a relatively young science with a history of only 30 years. The first live births from cryopreserved ovulated mammalian oocytes were reported by Whittingham in 1977 in mice (1) and subsequently by reports in a number of other species including rabbits (2), cows (3), horses (4), and over the past 20 years in humans (5–7). However, in most species success rates for cryopreserved oocytes have remained markedly poorer when compared with the cryopreserved embryos.

As discussed in chapters 6,7,9 and 13 there is a diverse range of morphological and physiological attributes of mammalian oocytes, which have contributed to the difficulty of achieving successful preservation. To recap briefly, the oocyte is a large single cell with low permeability to water, which increases the tendency to retain water in an undercooled state during cooling and leads on to damaging intracellular ice formation (IIF). Membrane permeabilities of oocytes for water and cryoprotectant solutes vary between species, strains, and maturational status of the oocyte, meaning that a universal protocol cannot be easily applied. If cryopreserved oocytes are to be successfully fertilized by natural means, the oocyte must retain the integrity of a number of unique structural features, such as the zona pellucida, the cortical granules, the mitochondrial compartment, and the microtubular spindle. The zona pellucida is the glycoprotein coat surrounding the oocyte; it controls sperm penetration triggered under natural circumstances by the action of a single sperm binding to its receptors, inducing the release of cortical granules which enzymatically crosslink the glycoproteins and prevent further sperm penetration (8). If this physiological machinery is disrupted during cryopreservation, it can act to prevent oocyte–sperm interaction (9) even if other essential cell functions (such as normal membrane functions) have been successfully recovered. In fact, this problem may be logistically countered by application of the intracytoplasmic sperm injection (ICSI) technique in human oocyte cryopreservation (10). The other oocyte-specific concern about cryopreservation has focused on the microtubular spindle, on which the condensed chromosomes are aligned in mature oocytes, and which is responsible for the correct segregation of chromosomes following fertilization and embryo development. Thus, cryopreservation-induced damage could lead to aneuploidy, and there is a well-documented propensity for the microtubular spindle to disassemble on cooling, although there is growing evidence that the spindle is also capable of reversible repair on rewarming (7,11,12). However, it remains unclear whether such reversible "failure and repair" processes have other adverse consequences during subsequent development of the resulting embryo. A potential route to avoid this latter problem has been proposed by choosing to cryopreserve immature oocytes before the nuclear membrane breaks down and the spindle is formed (the so-called "germinal vesicle breakdown stage" in meiosis I), but this raises equally intractable problems of how to access immature oocytes and how to achieve in vitro maturation, even were cryopreservation itself to be achieved successfully. It is fair to say that these difficulties ensure that, at present, cryopreservation of immature oocytes remains very much at the investigative, research-based niche of oocyte storage. Finally, as indicated in chapter 11, species-specific factors play a role in oocyte cryopreservation; for example, porcine oocytes have a high lipid content and are sensitive to chilling (13).

CONTROLLED RATE SLOW COOLING AS AN OPTION FOR OOCYTE CRYOPRESERVATION

The fundamental steps toward oocyte cryopreservation at extreme low temperatures have been described in chapters 2 and 4, and can also be found in other reviews (14,15). For any cell system, these can be viewed in a very simplistic way wherein the ability of water to form ice crystals is restricted, controlled, and compartmentalized to avoid biological injury during the thermal transition from the aqueous state down to temperatures compatible with true long-term biophysical stability. To achieve that stability, all the molecular components of the cell need to exist under extreme dehydration, but at a balanced level of anhydrobiosis where the residual "water of structure" for the biomolecules is still present in the relevant hydration shells to avoid denaturation. In simple aqueous mixtures such as salts in solution, the consequence of this "thermal dehydration" is progression to the solidification of eutectic mixtures at clearly defined end temperatures, dictated by the contents of the particular solvent–solute mixture. In biological systems—containing complex mixtures of macromolecular proteins, carbohydrates, lipids, and nucleic acids, which may be in aqueous solution or may be compartmentalized into ultrastructural entities such as membranes or polymeric aggregates—it is unlikely that such clear eutectic points exist. It is more likely that the mixture will assume a highly viscous amorphous state, acting to kinetically restrict the movement of residual water molecules to ice crystals, and effectively enter a "quasi-glass" state in which molecular motions are inhibited below about –100°C. The term "quasi-glass" is used because there is evidence that such systems may contain minute ice-nucleation centers from which ice crystal of significant size could grow, if conditions are altered to allow this. The concept is supported by biophysical studies using differential scanning calorimetry in the kinds of samples produced during cell cryopreservation (16–18). Also, as discussed in chapters 2 and 4, it is possible to achieve this stable cold dehydrated state by different technical maneuvers which include rapid vitrification techniques or slow, controlled rate cooling.

DEVELOPMENT OF EQUIPMENT FOR CONTROLLED RATE COOLING PROTOCOLS

The ability to control heat transfer depends to a significant degree on the temperature gradient during the cooling process, the thermal characteristics of the sample and housing chamber, and the relative volumes of these compared to that of the coolant. Given that end temperatures around –100°C will be the target, there are limited possibilities for choice of coolant. Passive cooling (by placing samples in a holder, inside a cold chamber such as the cabinet of a –80°C freezer, or in the vapor phase of a dewar of liquid nitrogen) is an inherently nonlinear process because initially temperature gradients will be much higher and thus initial cooling will be much faster. The need for linear cooling has been accepted historically (probably more for reasons of sample monitoring than for adherence to a specific biophysical parameter), and this will be discussed in paragraph Cooling Rates in the High Subzero Range. However, linear slow cooling rates de facto do have the advantage of avoiding potential, injurious rapid rates of cooling in the high subzero range, where there is a high risk of IIF if significant supercooling of the oocyte intracellular environment occurs relative to the surrounding ice matrix (19). Passive slow-cooling chambers are commercially available and may suffice for some routine laboratory cell cryobanking requirements where large numbers of resistant cells are being stored, and a defined degree of attrition during cooling can be accepted, but for sensitive cells such as oocytes, these approaches are not sufficient.

While the degree of complexity (and therefore cost) is increased by choosing to use controlled rate cooling equipment, there are advantages to be gained. Modern machines provide good temperature monitoring and recording of both the cooling chamber and a "dummy" sample tube, so that records of a particular cryopreservation procedure can be collected for quality assurance and sample safety. The equipment is programmed to reproduce set cooling profiles on requirement, irrespective of changes in the local laboratory environment, and is alarmed to alert the staff for any deviation from the protocol. If similar equipment is used, the protocols can be easily repeated in other clinics.

Laboratory equipment to produce slow cooling profiles for oocytes was developed early on by using stirred alcohol baths in evacuated dewars suspended in liquid nitrogen (1). However, a more sensitive system with potential for varying different parts of the cooling profile followed from the work of Hayes et al. (20) which showed that good performance could be achieved down to temperatures below −100°C by balancing flow of cold nitrogen vapor in a closed environment with heating provided by an electrical heater. An example of this early work is shown in Figure 1, and this still remains the underlying principle of most modern nitrogen vapor cooling machines. The constant flow of nitrogen vapor is assured by pressurizing the supply dewar of liquid nitrogen by another small heater and a solenoid valve that permits vapor to flow at a rate dictated by the set cooling profile. As a nonflammable agent, nitrogen vapor is relatively safe to process in this way, as long as the standard concerns about handling cryogenic gasses

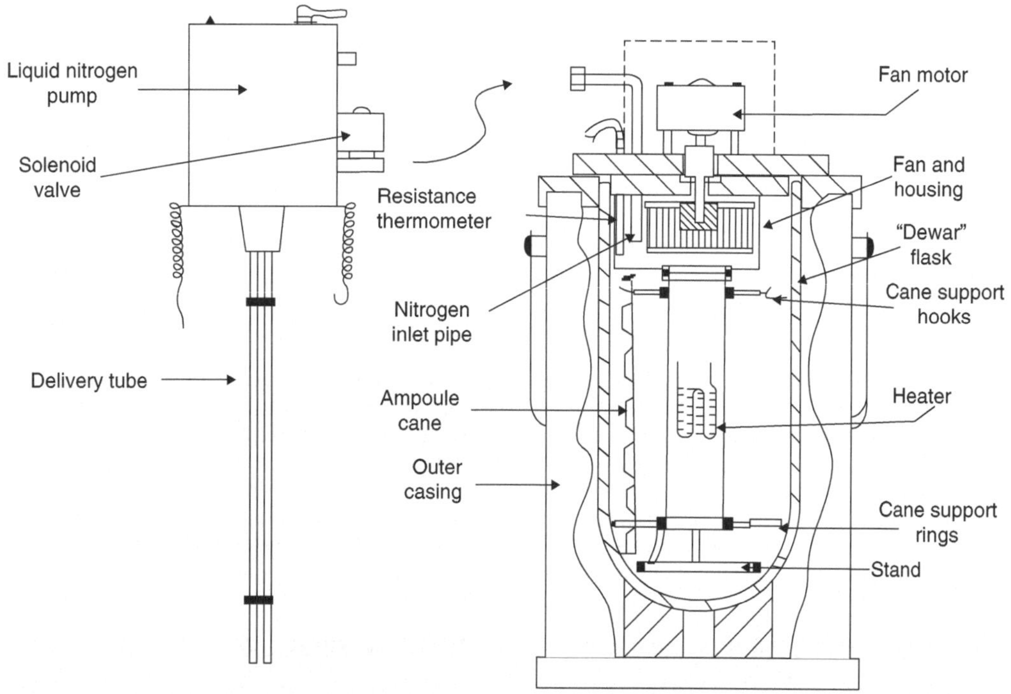

Figure 1 A schematic of an early design for a controlled rate cooling machine, on which most later machines have been modelled. The samples or straws were loaded onto the canes inside the chamber; liquid nitrogen vapor was delivered under pressure into the chamber, mixed by the fan motor; manipulation of the cooling rate was achieved by balancing nitrogen vapor influx with a solenoid valve and warming was provided by an electrical heater, both under a feedback control loop determined by the temperature measurements from the resistance thermometer. *Source*: From Ref. 20.

(21) are complied with. Several manufacturers (e.g., Planer Products, U.K.) produce modern machines based on this principle. It should, however, be remembered that liquid nitrogen is not a sterile commercial product, and cooling equipment should be routinely cleaned along manufacturer guidelines between uses. To our knowledge, there have not been any reports about potential infection of samples during vapor slow cooling, and this is because the samples would be "ice-solidified" and enclosed in vials or ampoules. There is however an issue about using nitrogen vapor (i.e., expelling it as an exhaust from a cooling machine) in environments of regulated air purity such as those encountered in pharmaceutical or cell therapeutic production, where cell cryobanking has to be performed in an adjacent room.

Recently, there has been another approach to slow cooling that employs a linearized motor, which can provide sufficient control to deliver cooling rates appropriate for oocytes (less than 0.5°C/min), based on the free piston Stirling cooler principle (22). The machine (Grant Asymptote EF600) avoids the use of liquid nitrogen as a cryogen. It uses an electrically powered Stirling cycle cryocooler as its cooling source. This small electrically powered freezing unit is distinct from a Peltier or compressor operated freezer in that the Stirling cryocooler has a small oscillating piston that repeatedly expands and compresses helium gas within a cylinder. The piston runs on gas bearings and so has a very long running life. The working fluid, helium gas, is sealed within the cylinder, and is inert and environmentally safe.

The sample plate is machined to accommodate straws (0.25 or 0.5 ml) or cryovials. The temperature of the sample plate is monitored by an internal thermocouple, which is read by a controller, and the voltage to the Stirling cryocooler is adjusted to produce the desired temperature. This simple feedback loop allows accurate temperature control to follow a preselected cooling profile. The cooling rate of the EF600 can be controlled over the temperature range +30°C to −100°C.

Controlled ice nucleation ("seeding") of samples (see section "Control of Other Parameters to Optimize Outcomes During Slow Cooling") can be carried out at a selected temperature plateau in the freezing cycle, without the need to remove sample straws. Two alternative methods of nucleation may be employed, either (a) the conventional method of using forceps and other similar instruments cooled in liquid nitrogen to contact the surface of the supercooled straw or (b) a small nitrous oxide cryosurgical device which uses the local cold spot from the rapid expansion of sterile N_2O to nucleate the samples.

Because this type of equipment does not require direct involvement of a cryogen (only a power source, which can be a portable battery), it is suitable for situations where liquid nitrogen is not available, such as field cryopreservation in veterinary or agricultural practice (22). There may also be situations where the nitrogen vapor expelled from the cooling machine may destabilize clean air environments in sterile rooms, which need to be considered. However, at present, such environments are not routine requirements in clinical IVF laboratories, and as long as the samples are properly packaged in sterile containers (straws or ampoules) before the start of cooling; we could find no reported problems from running the cooling machines in the laboratory suite.

There have been several other developments of laboratory-scale machines for controlled slow cooling, which have not been commercialized, but which find useful application in research groups where a limited range of slow cooling rates is sufficient to meet requirements. For example, Medrano and colleagues (23) employed a stepper motor to slowly lower semen samples in straws in a mobile carrier down into the vapor phase of nitrogen in a dewar of liquid nitrogen. By altering the ramp rate, slow cooling could be achieved, and this could work equally well for oocyte slow cooling. These machines can be run at low cost under laboratory conditions, but in most cases would not fulfill concerns about quality assurance in clinical practice.

CONTROL OF OSMOTIC STRESS BY ANALYSIS OF OOCYTE MEMBRANE PHYSIOLOGY

Exposure of unfertilized oocytes to osmotic stress has been shown to reduce development following fertilization in a variety of species. Developmental stage may be important, and immature bovine oocytes have been shown to be more sensitive to osmotic stress, induced by hypertonic conditions, than mature oocytes (24). Increased proportions of oocytes with disrupted spindle structure have also been observed in porcine oocytes as levels of anisotonic exposure diverged from isotonic (25). Human in vitro matured oocytes have been shown to be capable of a degree of post-ICSI development following exposure to a wide range of osmolalities for five minutes at 37°C (26), but whether further development is impaired is as yet unstudied, as is whether such tolerance applies to longer exposure times at lower temperatures.

In order to control the degree of osmotic stress experienced by a cell on exposure to cryoprotectant or cooling, the permeability of the cell to water and cryoprotectant must be determined. Water permeability (L_p) is determined by measuring cell volume change on exposure to nonisotonic solutions, which evoke osmotic water efflux. Permeability to permeating cryoprotectants (P_{cpa}) is determined by measuring cell volume change (shrink then swell) on exposure to that cryoprotectant. Cell volume can be determined by a number of techniques including several electronic methods. However, such methods require a large cell population. Oocytes are not generally available in sufficiently high numbers for these techniques, this particularly being the case with human oocytes. Fortunately, oocytes are of such a size that direct observation of cell volume change of a single cell using light microscopy is practicable. Their large size not only makes oocytes easy to visualize but also results in water loss at a sufficiently low rate for their cross-sectional area to be determined as a function of time. Most oocytes are spherical in shape making it simple to calculate their volume from their cross-sectional area.

Several methodologies have used light microscopy to determine the volume change of oocytes. One such method is the placement of oocytes into a capillary containing a hypertonic solution and observation through a stereomicroscope, the focus being changed as the oocytes ascend and descend in the tube and photographs being taken at set intervals (27). The movement of oocytes on exposure to anisotonic solutions has been restricted, making observation easier, by placing them in a microscope diffusion chamber (28) (also see chap. 5) which allowed several early studies on oocyte osmotic responses (29). Oocytes were held within a small chamber separated from a bulk flow chamber by a dialysis membrane (30). The oocytes could thereby be held stationary while the extracellular chemical composition was changed in the bulk flow liquid. However, while this technique was useful for the determination of water permeability, difficulties were experienced in modelling the transfer of cryoprotectant solutions through the dialysis membrane. This led to the development of a micropipette perfusion technique (31). Such a technique takes advantage of the fact that oocytes are surrounded by the zona pellucida, a translucent protein matrix which is highly porous and does not serve as a permeability barrier to even large macromolecules or viruses. The oocyte is held in position by suction pressure applied to the zona pellucida through a micropipette, care being taken not to deform the inner oocyte membrane. The oocyte is held in a small volume of initial solution and is perfused with a much greater volume of test solution. A microperfusion chamber has also been developed (32) whereby cells without a zona pellucida can be held within a chamber by a membrane and perfusate aspirated over them.

Input of cell volume change data into mathematical models of solute movement across cell membranes generates values for L_p and P_{cpa}. Such data have been generated for oocytes of a number species. Murine oocytes have been most often used as a model for human oocytes, which, understandably, are rarely available for research. Murine oocytes are approximately half the size of human oocytes but show a similar response under some conditions (Fig. 2). However, the effect of temperature on permeability coefficients can vary between these species (33). Human oocytes have also been shown to display greater variability within permeability characteristics than do oocytes of inbred mouse strains (30). Similar variability has been found on comparison of oocytes of inbred and outbred strains of hamster and mouse, respectively (34). Limited availability of human oocytes for research has led to the study of oocytes which have failed to fertilize following insemination for IVF. Slight differences in permeability have been reported between fresh and failed-to-fertilize human oocytes (35). Permeability of oocytes at various stages of maturation has also been studied. Mature goat oocytes have been reported to have lower water permeability and higher cryoprotectant permeability than immature oocytes, the consequence of which was that mature oocytes underwent lesser changes in cell volume than did immature oocytes in the presence of the cryoprotectant propane-1,2-diol (36). Maturation of bovine oocytes in vitro has been reported to increase water permeability compared with immature and in vivo matured oocytes of the same species (37).

Not only do permeability coefficients, and hence osmotic response, vary not only according to the species and maturational status of the oocyte, they will also vary, for a given oocyte type, depending on the cryoprotectant used. Permeation of the cryoprotectant glycerol, for example, is markedly less than for other commonly used cryoprotectants such as dimethyl sulfoxide and

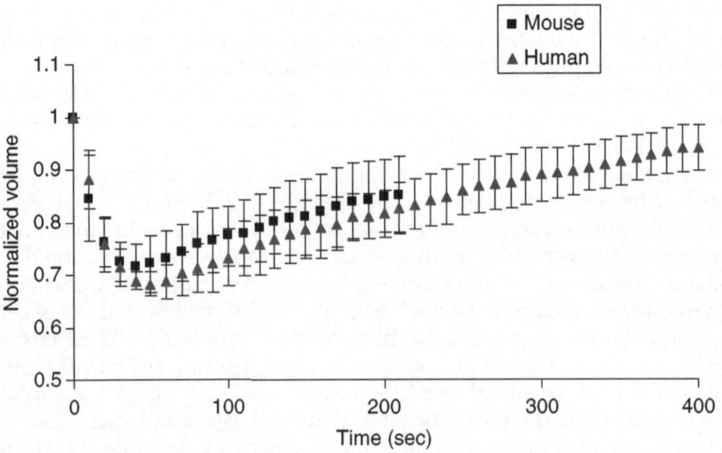

Figure 2 Mean (± SD) volume of mouse (n = 10) or human (n = 9) oocytes during exposure to 1.5 mol/L propane-1,2-diol at room temperature. The volume histories of the oocytes from both species were similar during exposure to cryoprotectant under these conditions.

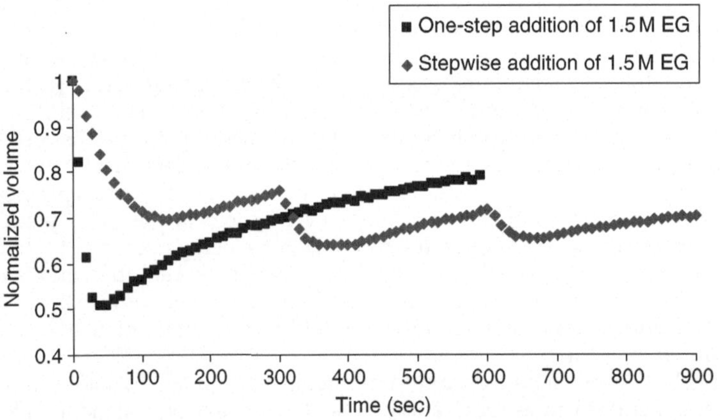

Figure 3 Mean (± SD) volume of human oocytes during exposure to 1.5 mol/L ethylene glycol (EG). Stepwise addition comprised exposure to 0.5 mol/L EG at time 0–300 sec, exposure to 1.0 mol/L EG at time 300–600 sec, and exposure to 1.5 mol/L EG at time 600–900 sec. Oocyte osmotic stress was more severe during one-step addition; in both cases, only partial permeation was achieved after exposure for 10 minutes.

propane-1,2-diol, in murine oocytes (33). Permeation of water and cryoprotectants will also vary with temperature, the lower the temperature, the slower the movement of substances across the cell membrane and hence greater the changes in cell volume prior to equilibration of the cryoprotectant. Thus, the level of osmotic stress experienced by a particular cell type can be modified by changing these factors. Osmotic stress can also be manipulated by changing the concentration of cryoprotectant used. The final cryoprotectant concentration should be sufficient to reduce damage induced by ice formation, but cryoprotectant can be added in a series of steps of increasing concentration in order to reduce cell volume change at each step (38) (Fig. 3).

Protocols developed recently for slow cooling of oocytes employ a mixture of permeating and nonpermeating cryoprotectants, most commonly propane-1,2-diol and sucrose, respectively. Oocytes are typically exposed to stepwise addition of propane-1,2-diol initially and then briefly exposed to propane-1,2-diol plus sucrose prior to cooling. As sucrose does not permeate the oocyte, it causes shrinkage of the oocyte, due to water efflux. Such removal of water prior to cooling is believed to reduce the likelihood of IIF, and the presence of sucrose may also have some protective effect on the cell membrane (39). Increasing the concentration of sucrose from 0.1 to 0.2 mol/L, or even 0.3 mol/L, has been reported to increase survival of human oocytes post-thaw (40), although implantation and live births rates following cryopreservation using such protocols remain variable, and recent results show poor implantation rates using the 0.3 mol/L-sucrose protocol (41). The duration of exposure of oocytes to such sucrose concentrations varies within and between studies with consequent wide variation in the degree of cell shrinkage achieved at the onset of slow cooling (42). This may go someway to explain some of the variability in oocyte function following cryopreservation using these techniques.

CAVEATS ON MEMBRANE PERMEABILITY MEASUREMENTS AS WE CURRENTLY UNDERSTAND THEM

Permeability coefficients for water and various cryoprotectants have been obtained for oocytes of a number of species. However, many studies have reported that not all oocytes remain spherical when exposed to anisotonic conditions. In order to generate permeability coefficients using current measurement and modelling techniques, it is necessary that cells remain spherical in order to apply geometric formulas for the calculation of volume changes based on measurements of areas. Some authors have chosen to discard oocytes that have not remained spherical (43) while others have used chemical treatments (44) or vortexing of cells (26) in order to produce spherical shrinkage. Irregular shrinkage of oocytes is thought to be related to the presence of transzonal projections, which are derived from granulosa cells that surround immature oocytes, and which pass through the zona pellucida to establish contact with the oolemma. Such projections normally breakdown during oocyte maturation but may persist through in vitro maturation. The relevance (if any) for these differences in shrinkage patterns and susceptibility of oocytes to cryopreservation injury are not currently understood. In future, it is possible that more sophisticated imaging and modelling systems will allow calculation of cell volume for nonspherical cells.

The movement of water and cryoprotectant across the cell membrane during cooling, and particularly in the presence of ice, is certainly an area that warrants further study. The use of microscopy to study cell volume change at subzero temperatures is technically difficult, as the presence of ice obscures the images and also causes distortion of the cells. Cooling of the membrane may cause changes to its structure and hence its transport properties. Although a reasonably close correlation between permeability measurements measured at subzero temperatures (0–16°C) and those extrapolated from permeability coefficients measured at suprazero temperatures has been found in the few cases where this has been studied in some species of oocyte (45), there still remains a big question about extrapolating permeability parameters made by measurements at suprazero temperature range for the prediction of osmotic stresses likely to be encountered by the oocytes during cooling in the presence of ice.

So far, this discussion has focused on reduction of osmotic stress created during cryoprotectant addition. However, removal of permeating cryoprotectant will result in cell swelling, particularly when performed in the absence of a nonpermeating solute. As cell swelling is generally less well tolerated than cell shrinkage, the modelling of cryoprotectant removal is important. However, freezing and thawing of oocytes may alter their membrane permeability characteristics. The cell shrinkage that occurs during cooling may also allow irreversible membrane changes to occur. Although no significant differences were found in permeability coefficients of frozen and nonfrozen mouse oocytes (46), greater variability and poorer fits to the mathematical models suggest that the cryopreservation process may induce some physiological/biophysical changes in the osmotic characteristics of oocytes which should be investigated further.

CONTROL OF OTHER PARAMETERS TO OPTIMIZE OUTCOMES DURING SLOW COOLING

Control of Ice Nucleation

Of the remaining biophysical events which can dictate death or survival during oocyte slow cooling, the temperature at which the extracellular medium forms ice crystals during cooling (the so-called nucleation temperature) may have a significant impact. Nucleation in practical cryopreservation can be a random event over a range of temperatures unless it is deliberately induced by "seeding" (see chaps. 1, 2, 5, and 9). As indicated in chapters 2 and 3, and in many previous analyses of the data on oocyte cryobiology, they are large cells with a low surface area to volume ratio and thus a defined kinetic limit at which water is able to leave the intracellular environment during external ice crystal formation (15,47). Oocytes (and indeed most cells) which retain a high intracellular volume of supercooled water during the early stages of a cryopreservation protocol are very susceptible to lethal IIF and Mazur's group (48) has computed the effects of a range of cooling rates (leading to supercooled intracellular water and lethal IIF) for mouse oocytes. Maintaining equilibrium in the osmotic balance between oocyte and external medium can be achieved by cooling at rates below –0.5°C/min in the high subzero range (down to around –30°C). However, this equilibrium needs to be "set" as closely as possible at nucleation point of the mixture. Otherwise, minute ice nucleation centers may be formed inside the initially supercooled oocyte (49). The kinetics of water movement will also depend on the membrane water permeability, which may vary with degree of maturity of the oocyte, and any confounding issues introduced by the presence of cryoprotectants. However, in the study by Trad et al. (19), the degree of maturity had marginal effects on formation of IIF in human oocytes, observed during cryomicroscopy, and cooled with fast rates chosen deliberately to provoke visible IIF. When comparing oocytes from different species (mouse and human), what made a significant difference in IIF nucleation temperature was the size of cell (three-fold volume difference), with human oocytes forming visible IIF at –6.5°C and mouse oocytes forming it at –23°C when cooled under identical conditions. Addition of three different cryoprotectants at 1.5 M concentrations made only a small change (by about –6°C) to the median temperature of IIF formation in human oocytes, while in the mouse these agents depressed median temperature of IIF by about –20°C (19). When human oocytes experienced ice nucleation induced at a range of "seeding" temperatures between –4.5°C and –8°C, then were deliberately cooled at fast rates to facilitate IIF, a strong correlation was established, with the least severe IIF injury in oocytes seeded at –4.5°C. Thus, for human oocytes, seeding as close as possible to the equilibrium freezing point of these mixtures, where the lowest risk for IIF injury was seen, the conditions for subsequent successful equilibrium cooling could be "set." It remains to be further investigated in prospective trials exactly how much this strict control of specific "seeding" temperatures will make routine human oocyte cryopreservation. But until proven otherwise, it would seem prudent to use seeding temperatures as close as practically possible to the equilibrium freezing temperature of the different cryoprotectant mixes used (which can be easily measured and are usually in the range of –2°C to –4°C) to avoid injury.

Cooling Rates in the High Subzero Range

From some of the earliest studies on oocyte slow cooling protocols (1,29), linear cooling rates in the region of –0.3°C/min were used, with cooling progressing down to a perceived safe low temperature (about –70°C) after which the vials could be transferred directly to liquid nitrogen. This value for the rate of slow cooling was selected empirically after demonstrating that viable oocyte recoveries were achievable, however the actual numerical value was dictated more by the physical constraints of the cooling equipment used (evacuated dewar of stirred alcohol suspended in liquid nitrogen) than any other factor. Subsequent analyses of the kinetics of water movement out from mouse embryos (and by implication, oocytes) during slow cooling by Mazur—using his coupled equations which include factors for cell volume, membrane water permeability, temperature, and osmotic potentials—found that a cooling rate in the region of ~0.5°C/min was optimal. Again, by empirical observation of success, and following Mazur's descriptions (50), the end temperature of the slow cooling ramp was moved "upwards" to temperatures between –35°C and –40°C, where most of the "cryoremovable" water should be removed (>90%), and thus risk of IIF during the subsequent plunge cooling was low. (This also had the logistical advantage of reducing the overall time for the protocol). These modifications developed alongside information from the so-called "two-step" cooling protocols developed for tissue culture cells, which also used an end temperature of about –35°C in the first ramp, before transfer to liquid nitrogen (51). However, Mazur did caution that a small fraction of supercooled water would be present in embryos and oocytes even after cooling to –40°C, so the risk of IIF would not be zero. In practice, assuming small sample volumes (as in droplets in plastic straws), sufficiently rapid cooling during the plunge step is likely to be achieved to avoid significant IIF. Equally, even if a small number of ice nuclei persist after the rapid plunge step, rapid warming is more easily achieved in such small sample volumes, which ensures that only a small innocuous or "rescuable" fraction of IIF should develop during warming and thus cell recovery should be high.

This type of linear slow cooling for oocytes has become routine over the past 20 years. However, recently a different approach has been suggested and applied, based on the premise that many of the biophysical properties which alter during ice formation and cooling change in a nonlinear fashion with the applied temperature reduction, including those of mass fraction of ice, viscosity, and osmolality of residual solutions (52). A computation of the rate of change of extracellular solute concentration (in this instance in a glycerol-based mixture) demonstrated that a certain nonlinear cooling profile provided a more constant rate of change, whereas linear cooling resulted in a more variable rate of change (53,54), and applying this approach to sperm cryopreservation resulted in improved recoveries. In practical terms, the computed nonlinear profile results in a more rapid rate of cooling in the higher subzero range. The application of this approach to oocyte cryopreservation is currently under study, but future prospective investigations to compare linear and nonlinear cooling will be required to fully evaluate potential benefit.

Control of Warming After Slow Cooling of Oocytes

It is generally accepted that warming rates should be as fast as practically possible after oocyte cryopreservation, to minimize injury from ice recrystallization as water molecules progressively regain mobility at temperatures above about –40°C. This is achieved by plunging the straws into a water bath between +20°C and +37°C. One practical consideration has been the potential for cracking in the plastic of the straws if they are transferred directly from deep subzero temperatures due to thermal stresses in the plastic, so a brief "hold period" in room air (10 seconds) has been introduced into many protocols (54). There are also issues about decontaminating the outside of the straws before expelling the rewarmed oocytes into dilution media (54). The significance of these small variations in warming protocols have not been prospectively evaluated to any great extent, but in clinical practice, recording and adherence to whichever protocol is selected will again be important for quality assurance.

SUMMARY

The clinical application of oocyte cryopreservation is currently enjoying a renaissance, with many centers now actively promoting the technology as part of a balanced infertility service (55). Reports of successful live births after slow cooling methods continue to grow. The topic about which approach to oocyte cryopreservation (slow cooling or vitrification) yields best results continues to be hotly debated (56), and it will require carefully-constructed multicentre trials before a clear consensus is reached. Particularly for slow cooling protocols, application of fundamental cryobiological principles is leading to a gradual but consistent improvement in outcomes, and promises further advances if the scientific focus can be maintained.

REFERENCES

1. Whittingham DG. Fertilization in vitro and development to term of unfertilized mouse oocytes previously stored at –196°C. J Reprod Fertil 1977; 49: 89–94.
2. Al-Hasani S, Kirsch J, Diedrich K, et al. Successful embryo transfer of cryopreserved and in-vitro fertilized rabbit oocytes. Hum Reprod 1989; 4: 77–9.
3. Fuku E, Kojimat T, Shioya Y, et al. In vitro fertilization and development of frozen-thawed bovine oocytes. Cryobiology 1992; 29: 485–92.
4. MacLellan LJ, Carnevale E, Coutinho da Silva M, et al. Pregnancies from vitrified equine oocytes collected from super-stimulated and non-stimulated mares. Theriogenology 2002; 58: 911–19.
5. Chen C. Pregnancy after human oocyte cryopreservation. Lancet 1986; i 8486: 884–6.
6. Fabbri R, Porcu E, Marsella T, et al. Human oocyte cryopreservation: new perspectives regarding oocyte survival. Hum Reprod 2001; 16: 411–16.

7. Rienzi L, Martinez F, Ubaldi F, et al. Polscope analysis of meiotic spindle changes in living metaphase II human oocytes during the freezing and thawing procedures. Hum Reprod 2004; 19: 655–9.
8. Bleil JD, Wassarman PM. Mammalian sperm-egg interaction: identification of a glycoprotein in mouse egg zonae pellucidae possessing receptor activity for sperm. Cell 1980; 20: 873–82.
9. Carroll J, Depypere H, Matthews CD. Freeze-thaw-induced changes of the zona pellucida explains decreased rates of fertilization in frozen-thawed mouse oocytes. J Reprod Fertil 1990; 90: 547–53.
10. Porcu E, Fabbri R, Serrachioli R, et al. Birth of a healthy female after intracytoplasmic sperm injection of cryopreserved human oocytes. Fertil Steril 1997; 68: 724–6.
11. Stachecki JJ, Cohen J. An overview of oocyte cryopreservation. Reprod Biomed Online 2004; 9: 152–63.
12. Bianchi V, Cottichio G, Fava L, et al. Meiotic spindle imaging in human oocytes frozen with a slow freezing procedure involving high sucrose concentration. Hum Reprod 2005; 20: 1078–83.
13. Isachenko V, Isachenko E, Michelmann HW, et al. Lipolysis and ultrastructural changes of intracellular lipid vesicles after cooling of bovine and porcine GV-oocytes. Anat Histol Embryol 2001; 30: 333–8.
14. Fuller B, Paynter S. Fundamentals of cryobiology in reproductive medicine. Reprod Biomed Online 2004; 9: 680–91.
15. Leibo SP. Cryopreservation of oocytes and embryos: optimization by theoretical versus empirical analysis. Theriogenology 2008; 69: 37–47.
16. Boutron P, Mehl P, Kaufmann A, et al. Glass-forming tendency and stability of the amorphous state in the aqueous solutions of linear polyalcohols with four carbons. I. Binary systems water-polyalcohol. Cryobiology 1986; 23: 453–69.
17. MacFarlane D. Physical aspects of vitrification in aqueous solutions. Cryobiology 1987; 24: 181–95.
18. Baudot A, Alger L, Boutron P. Glass-forming tendency in the system water-dimethyl sulfoxide. Cryobiology 2000; 40: 151–8.
19. Trad F, Toner M, Biggers JD. Effects of cryoprotectants and ice-seeding temperature on intracellular freezing and survival of human oocytes. Hum Reprod 1998; 14: 1569–77.
20. Hayes AR, Pegg DE, Kingston R. A multirate small-volume cooling machine. Cryobiology 1973; 11: 371–7.
21. Fleck R, Fuller B. Cell preservation. In: Stacey G, Davis J, eds. Medicines from Animal Cells. Chichester, UK: John Wiley & Sons, 2007: 417–32.
22. Faszer K, Draper D, Green JE, Morris GJ, Grout BW. Cryopreservation of horse semen under laboratory and field conditions using a Stirling Cycle freezer. CryoLetters 2006; 27: 179–86.
23. Medrano A, Anderson J, Millar JD, Holt WV, Watson PF. A custom-built controlled-rate freezer for small sample cryopreservation studies. CryoLetters 2002; 23: 397–404.
24. Agca Y, Liu J, Rutledge JJ, et al. Effect of osmotic stress on the developmental competence of germinal vesicle and metaphase II stage bovine cumulus oocyte complexes and its relevance to cryopreservation. Mol Reprod Dev 2000; 55: 212–19.
25. Mullen SF, Rosenbaum M, Critser JK. The effect of osmotic stress on the cell volume, metaphase II spindle and developmental potential of in vitro matured porcine oocytes. Cryobiology 2007; 54: 281–9.
26. Van den Abbeel E, Schneider U, Liu J, et al. Osmotic responses and tolerance limits to changes in external osmolalities, and oolemma permeability characteristics, of human in vitro matured MII oocytes. Hum Reprod 2007; 22: 1959–72.
27. Leibo SP. Water permeability and its activation energy of fertilized and unfertilized mouse ova. J Membr Biol 1980; 53: 179–88.
28. McGrath JJ. A microscope diffusion chamber for the determination of the equilibrium and non-equilibrium osmotic response of individual cells. J Microscopy 1985; 139: 249–63.
29. Hunter JE, Bernard A, Fuller B, et al. Fertilization and development of the human oocyte following exposure to cryoprotectants, low temperatures and cryopreservation: a comparison of two techniques. Hum Reprod 1991; 6: 1460–5.
30. Hunter JE, Bernard A, Fuller B, et al. Measurements of the membrane water permeability (Lp) and its temperature dependence (activation energy) in human fresh and failed-to-fertilize oocytes and mouse oocyte. Cryobiology 1992; 29: 240–9.
31. Gao DY, McGrath JJ, Tao J, et al. Membrane transport properties of mammalian oocytes: a micropipette perfusion technique. J Reprod Fertil 1994; 102: 385–92.
32. Gao DY, Benson CT, Liu C, et al. Development of a novel microperfusion chamber for determination of cell membrane transport properties. Biophys J 1996; 71: 443–50.
33. Paynter SJ. A rational approach to oocyte cryopreservation. Reprod Biomed Online 2005; 10: 578–86.
34. Benson CT, Critser JK. Variation of water permeability (Lp) and its activation energy (Ea) among unfertilized golden hamster and ICR murine oocytes. Cryobiology 1994; 31: 215–23.

35. Newton H, Pegg DE, Barrass R, et al. Osmotically inactive volume, hydraulic conductivity, and permeability to dimethyl sulphoxide of human mature oocytes. J Reprod Fertil 1999; 117: 27–33.
36. Le Gal F, Gasqui P, Renard JP. Differential osmotic behavior of mammalian oocytes before and after maturation: a quantitative analysis using goat oocytes as a model. Cryobiology 1994; 31: 154–70.
37. Ruffing N, Steponkus P, Pitt R, et al. Osmometric behavior, hydraulic conductivity, and incidence of intracellular ice formation in bovine oocytes at different developmental stages. Cryobiology 1993; 30: 562–80.
38. De Santis L, Cottichio G, Paynter S, et al. Permeability of human oocytes to ethylene glycol and their survival and spindle configurations after slow cooling cryopreservation. Hum Reprod 2007; 22: 2776–83.
39. Crowe JH, Carpenter JF, Crowe LM. The role of vitrification in anhydrobiosis. Annu Rev Physiol 1998; 60: 73–103.
40. Fabbri R, Porcu E, Marsella T, et al. Human oocyte cryopreservation: new perspectives regarding oocyte survival. Hum Reprod 2001; 16: 411–16.
41. Borini A, Sciajno R, Bianchi V, et al. Clinical outcome of oocyte cryopreservation after slow cooling with a protocol utilizing a high sucrose concentration. Hum Reprod 2006; 21: 512–17.
42. Paynter SJ, Cooper A, Gregory L, et al. Volume changes of mature human oocytes on exposure to cryoprotectant solutions used in slow cooling procedures. Hum Reprod 2005; 20: 1194–9.
43. Paynter SJ, Cooper A, Gregory L, et al. Permeability characteristics of human oocytes in the presence of the cryoprotectant dimethylsulphoxide. Hum Reprod 1999; 14: 2338–42.
44. Younis A, Toner M, Albertini DF, et al. Cryobiology of non-human primate oocytes. Hum Reprod 1996; 11: 156–65.
45. Toner M, Cravalho EG, Armant DR. Water transport and estimated transmembrane potential during freezing of mouse oocytes. J Membr Biol 1990; 115: 261–72.
46. Litkouhi B, Marlow D, McGrath JJ, et al. The influence of cryopreservation on murine oocyte water permeability and osmotically inactive volume. Cryobiology 1997; 34: 23–35.
47. Paynter SJ. Current status of the cryopreservation of human unfertilized oocytes. Hum Reprod Update 2000; 6: 449–56.
48. Mazur P, Pinn IL, Kleinhans FW. The temperature of intracellular ice formation in mouse oocytes vs. the unfrozen fraction at that temperature. Cryobiology 2007; 54: 223–33.
49. Mazur P, Kleinhans FW. Relationship between intracellular ice formation in oocytes of the mouse and Xenopus and the physical state of the external medium—a revisit. Cryobiology 2008; 56: 22–7.
50. Mazur P. Equilibrium, quasi-equilibrium, and nonequilibrium freezing of mammalian embryos. Cell Biophys 1990; 17: 53–92.
51. McGann LE, Farrant J. Survival of tissue culture cells frozen by a two-step procedure to –196°C. I. Holding temperature and time. Cryobiology 1976; 13: 261–8.
52. Morris GJ, Acton E, Avery S. A novel approach to sperm cryopreservation. Hum Reprod 1999; 14: 1013–21.
53. Morris GJ, Goodrich M, Acton E, et al. The high viscosity encountered during freezing in glycerol solutions: effects on cryopreservation. Cryobiology 2006; 52: 323–34.
54. Paynter SJ, Fuller BJ. Cryopreservation of mammalian oocytes. Methods Mol 2007; 368: 313–24.
55. Borini A, Cattoli M, Bulletti C, Coticchio G. Clinical efficiency of oocyte and embryo cryopreservation. Ann N Y Acad Sci 2008; 1127: 49–58.
56. Konc J, Kanyo K, Varga E, Kriston R, Cseh S. Births resulting from oocyte cryopreservation using a slow freezing protocol with propanediol and sucrose. Syst Biol Reprod Med 2008; 54: 205–10.

4 Fundamentals and Current Practice of Vitrification

Lilia L. Kuleshova

Low Temperature Preservation Unit, National University Medical Institutes, Yong Loo Lin School of Medicine, National University of Singapore, Singapore

VITRIFICATION AMONG OTHER APPROACHES TO PRESERVATION

We put forward this chapter with the hope that embryologists and clinicians searching for a cryopreservation approach to be implemented in their clinics will consider the fundamentals and current practices in vitrification. Vitrification, as an emerging approach to cryopreservation, is a subject of enormous interest and complexity, in particular in the context of preservation of embryos and oocytes. This approach has become a hope for improving the success rate of cryopreservation of oocytes, which is of particular importance for infertility treatments of young oncology patients and cases of temporary sperm deficiency. Vitrification produces amorphous solidification of living cells and supporting solutions and completely avoids ice crystal formation during cooling intended for cryostorage. It is equally important that the vitrification process completely avoids ice crystal formation in cryopreserved cells during warming to recover the cells for clinical applications (1).

As there is a delay between the recovery of human oocytes and their re-introduction, there is a necessity to devise and subsequently implement a robust vitrification method in clinical practice to achieve maximum viability of human oocytes. Several major issues still to be outlined are cryoprotectant toxicity versus total solute concentration required for vitrification and elimination of chilling injury (damage caused by reduction in temperature per se). This chapter highlights the importance of the balance between introduction and removal of cryoprotectants during vitrification, including some background to these procedures. Less obvious, and certainly less well recognized by embryologists, is crystallization during warming; this may yet turn out to be of fundamental importance. Therefore, the current state of the art in techniques and carrier systems applied to achieve vitrification of oocytes and embryos in the context of these constraints on safety of in vitro fertilization (IVF) procedures is also reviewed here. While progress has been made on major IVF procedures and human oocytes vitrification, this chapter discusses why understanding the fundamentals of vitrification, rather than of the vitrification procedure alone, is vital for preservation of human oocytes. Moreover, by examining various vitrification protocols, the factors that influence the success of outcomes are also discussed.

The place of vitrification among other approaches in the field of cell and tissue preservation is presented in Figure 1.

Freezing and vitrification are both valuable concepts in cryobiology. The fundamentals of vitrification and current practices of long-term preservation of human oocytes are described in detail in this chapter 3, while freezing has been described in chapter 3. Lyophilization has been developed successfully for microorganisms and was even challenged for mammalian cells, namely erythrocytes. Given the technology available at present, it is not feasible to raise this issue for preservation of oocytes. The purpose of hypothermia is the preservation of cells and tissues for the short term. Hypothermic storage covers the area from subzero temperatures to temperatures less than the physiological temperature. It is mainly applied to prevent ageing of cells when a temporary delay in the usage of cells is required and for transportation purposes. This method may be of interest, for example, for preservation of fish oocytes. On the other hand, it may not be an option for mature mammalian oocytes due to the sensitivity of the meiotic spindle. In contrast to cell suspensions, the concept of vitrification for the preservation of a single object such as an embryo and for oocyte preservation is more definite and more well studied (2–7). It might be thought that a tissue is also a single object, yet vitrification of tissues is comparatively more challenging as it assumes preservation of both cell functions and their

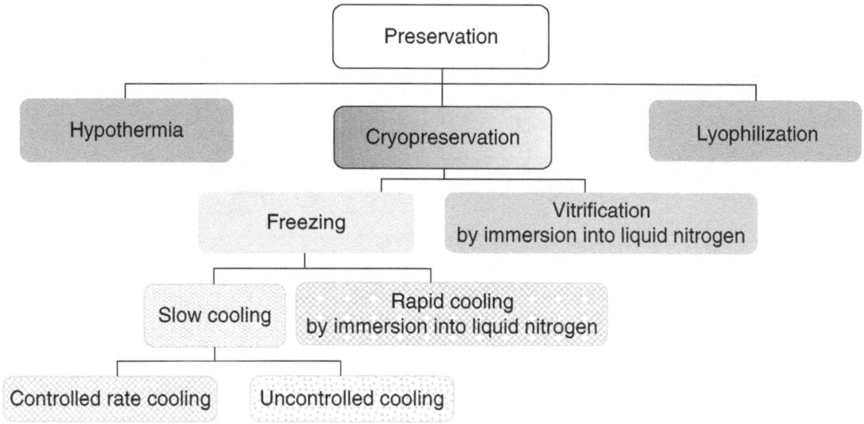

Figure 1 Types of preservation. *See Color Plates on Page xi.*

morphological structure as well as tissue architecture, which may be important for mechanical support and to aid in vital biological functions.

One of the primary issues in embryology is to minimize the duration of exposure of oocytes/embryos to potentially damaging conditions in vitro (1). In relation to such considerations, one of the central issues is whether the fundamentals of vitrification, which have been established in physics, chemistry, and biology, satisfy the primary principle of minimizing damage to oocytes and embryos during the whole cryopreservation cycle.

To prevent damage to oocytes during their exposure to low temperature, cryoprotective agents are used. During the cryopreservation procedure, oocytes should be equilibrated with a solution (cryoprotectants in medium) (Fig. 2). Generally, several equilibration steps in solutions of gradually increasing concentration are recommended, and the number of steps vary depending on the specific protocol (see section "Practical Aspects"). This stepwise procedure results in water leaving the cells and a penetrating cryoprotectant gradually entering the cells by osmosis. After an optimal equilibration period, cells are ready to be introduced to low temperature for cryostorage. Upon warming, the cryoprotectant should be removed gradually to prevent rapid expansion of cells, which would otherwise cause cell damage.

FUNDAMENTALS OF VITRIFICATION

Definition and Historical Overview of Vitrification

The fundamentals of cryobiology and cryomedicine have been developed thoughtfully over decades. For freeze preservation of cells, modern cryobiology recognizes that for each cell type there is an optimum cooling rate with lesser survival above and below the optimum. This is believed to be due to the two-factor theory of cryoinjury: opposing causes of cell injury relating, at a very slow cooling rate, to the so-called solution effects represented by excessively prolonged exposure to high ionic concentration during cooling, and, at a very high cooling rate, to intracellular ice formation whose chance of occurrence increases as the cooling rate increases (8). The optimum cooling rate varies with cell type from >1000°C/min for erythrocytes to <1°C/min for oocytes, depending on cell membrane permeability to water and cryoprotectants, temperature dependence of the permeability coefficients, and the cell surface area (9). These optima are best determined empirically.

It is recognized that cryoprotectant toxicity varies between cell types, but if the toxicity is such that it allows exposure to multi-molar concentrations, an alternative method of cryopreservation is vitrification (Fig. 3). In this process, the suspending solution is loaded with cryoprotectants and the cooling rate is increased to reduce the temperature of the sample below the crystallization temperature before ice nucleation initiates such that glass formation occurs. The advantages of this method include the absence of changes in osmolality and ionic strength dur-

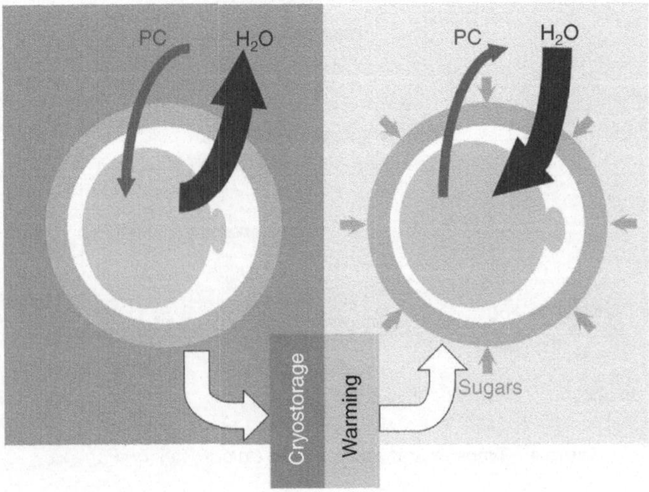

Figure 2 Basic representation of oocyte cryopreservation. *See Color Plates on Page xi.*

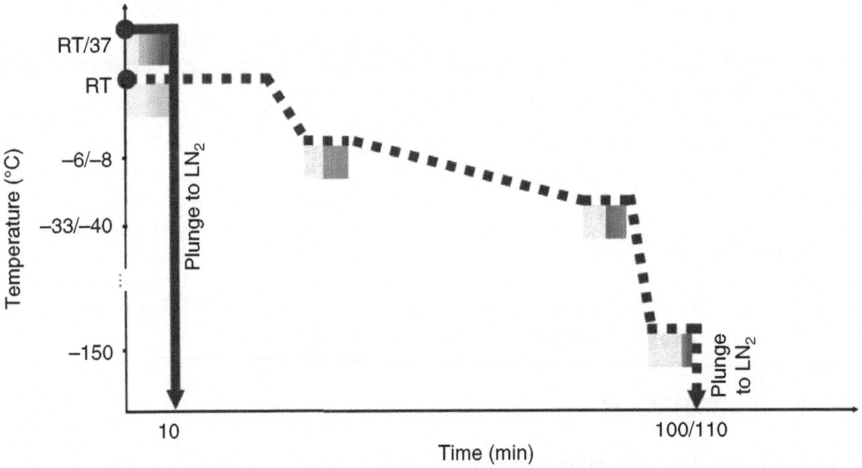

Figure 3 Schematic illustration of comparison between vitrification and slow-cooing procedures for human oocytes. Solid line represents vitrification. In a majority of recent protocols, human oocytes are introduced to vitrification solutions and diluted at either room temperature (RT) or 37°C within 10 min. Dashed line represents conventional slow-cooling protocol (freezing approach). Average duration of slow cooling procedures after loading into freezing machine is about 100–110 min. In both routes, intensity of orange color inside the boxes reflects infiltration of cryoprotectants into oocyte. In the freezing route, white color inside the boxes reflects growth of ice crystals in a straw containing an oocyte. *Abbreviations*: LN_2, liquid nitrogen; RT, room temperature. *See Color Plates on Page xii.*

ing cooling and the ease with which cooling can be achieved. If warming is rapid enough to prevent recrystallization of water, survival is generally high.

Tammann (1898) (10) was the first to report studies on the fundamentals of vitrification from a physical point of view; however, it was Basil Luyet who started the pioneering work on vitrification as a means of cryopreservation (11,12). In 1968, Rapatz and Luyet (13) reported the first successful vitrification of a biological system, namely erythrocytes, with 8.6 mol/L glycerol. Nevertheless, the recent keen interest in vitrification of cells, tissues, and organs was literally spurred by Fahy in the early 1980s when he suggested that ice-free cryopreservation of cells might be achieved through vitrification of the solutions by intra- and extracellular methods (14). Fahy demonstrated that the high concentration of cryoprotectants required to achieve vitrification could be significantly reduced by the use of high hydrostatic pressure. The effectiveness of solutions with concentrations sufficient to achieve vitrification has been investigated and described (15,16). The procedure for successful cryopreservation by vitrification of mammalian embryos and oocytes, including those from humans, has been the subject of intense research over many years. The first successful vitrification performed under practical conditions was reported in 1985 on the study of mouse embryo (2), which until now can be considered as the foundation for vitrification in embryology.

The toxicity of the initial vitrification solution (VS1) was high, but researchers overcame this problem by performing a study in a cold room where the lower temperature would restrict the penetration of the cryoprotectant. In these early studies, some elements of "slow-cooling" procedures, rather than modern vitrification protocols, were used. Embryos were pre-equilibrated to a cryoprotectant solution at room temperature following gradual increase in concentration of cryoprotectants at 4°C. Subsequently, embryos were immersed into liquid nitrogen (2,17). The overall duration of embryo exposure was considerably long (17), and we can consider these studies as the "bridge" between slow cooling and vitrification (1).

After considering all above, the concept of vitrification became misinterpreted. The primary advantage of vitrification is the complete avoidance of ice crystallization during cooling and warming; this is especially true in the latter where the recovery of cells is critical for biological functions. In view of the fact that both vitrification and rapid cooling involve immersing samples into liquid nitrogen, many mistakenly consider these two processes to be synonymous. The difference in the two lies in the former being a process of liquid solidification into stable amorphous state while the latter involves solidification into a crystalline or partially crystalline state (15). Despite the variety of recent assumptions, the stable vitrification state in oocytes can be achieved only by direct immersion into liquid nitrogen using substantially high concentrations of cryoprotectant solution; other criteria need to be fulfilled for the solution to vitrify. By comparing changes in physical properties of ethane-1,2-diol [ethylene glycol (EG)] in a range of cooling rates achievable in standard straws and the majority of specially designed holders that are able to support ultra-rapid cooling, it was found that the difference in concentration of the cryoprotectant necessary for the solution to vitrify is rather small (~2%) (18). Thus, it is a misconception that the concentration of cryoprotectant necessary to maintain stable vitrification can be significantly reduced (by 10% or more) for certain cooling rates, which can be achieved in the recently designed containers/holders for ultra-rapid cryopreservation of human oocytes. As a result of insufficient solute concentration used in ultra-rapid cooling, the growth of small extracellular ice crystals is unavoidable during thawing of cells prior to use.

Vitrification Solutions and Fundamentals of Equilibration/Dilution Procedures

During the early era of vitrification, changes in cell volume during the procedure were noticed and characterized. The moment when the cytoplasm becomes sufficiently concentrated and is capable of being vitrified upon cooling to a low temperature was recognized as the key moment of the procedure (17).

Thus, it became evident that cryoprotectants in vitrification protocols are engaged in more diverse actions than during conventional freezing. The role of cryoprotectants is two-fold: (*i*) they should be able to remove water from oocytes (i.e., sufficiently dehydrate cells); and (*ii*) at the same time replace water with chemicals at the minimal concentration (i.e., be the least toxic to cells) that is still sufficient to form the amorphous state in oocytes, eliminating damage during exposure to low temperature. There is another reason why component selection became crucial with the development of vitrification: with the aid of high concentrations of cryoprotectants it would be easier to achieve vitrification of oocytes. These considerations stimulated a number of studies in which a variety of agents were tested and new formulations were developed for the application in vitrification protocols.

The protective properties of cryoprotectants depend on a number of factors. Adding a cryoprotectant lowers the freezing point of the solution, which decreases with increasing concentration of the cryoprotectant. Although the result is minimal at low solute concentrations, the effect becomes more pronounced after the concentration exceeds 12% to 15%. With increase in concentration of solution, the freezing point decreases progressively followed by its rapid fall in the region of intermediate concentrations. Once the concentration becomes sufficiently high, it leads to the formation of nonstable glass. The basis of vitrification properties lies in a cryoprotectant's ability to form hydrogen bonds with water molecules eliminating ice formation.

It also appears that cryoprotectants reduce negative impacts caused by high levels of salt concentration. This property of cryoprotectants is important for the preservation of oocytes by freezing as the cells are dehydrated and surrounded by concentrated salt. However, this is not an issue during vitrification because oocytes are introduced in a stepwise manner to solutions of increasing concentrations that are made in a medium with balanced salt content.

Cryoprotectants can be classified by their role in cryopreservation as penetrating agents and nonpenetrating agents, which in turn can be further divided by their molecular weight and other properties as described in Figure 4.

During the early studies on vitrification for oocyte and embryo cryopreservation, vitrification solutions that contained a complex mixture of cryoprotective agents have been formulated (19) in an attempt to reduce the solutions' toxicity. In fact, the initial solutions were made up of only penetrating cryoprotectants and therefore were potentially toxic (2,20,21). It became crucial to formulate an efficient vitrification solution that would be nontoxic to embryos/oocytes for a considerably long time. Several groups conducted systematic and extensive investigations involving a number of combinations of cryoprotectants, sugars, and polymers in an effort to identify the most effective and least toxic vitrification solution (6,22–24). EG, as one of the primary permeating cryoprotectants used in vitrification methods, was widely used since the early 1990s, primarily due to its relatively low toxicity compared with other compounds (25,26). At that time, it was demonstrated that mouse oocytes could tolerate 6 mol/L EG for five minutes and 8 mol/L EG for one minute with no adverse effect on their development (27). It was also shown that mouse oocyte developmental potential was reduced by only 30% after exposure to 7 mol/L EG when introduced in a two-step manner (28). Using EG as a base cryoprotectant,

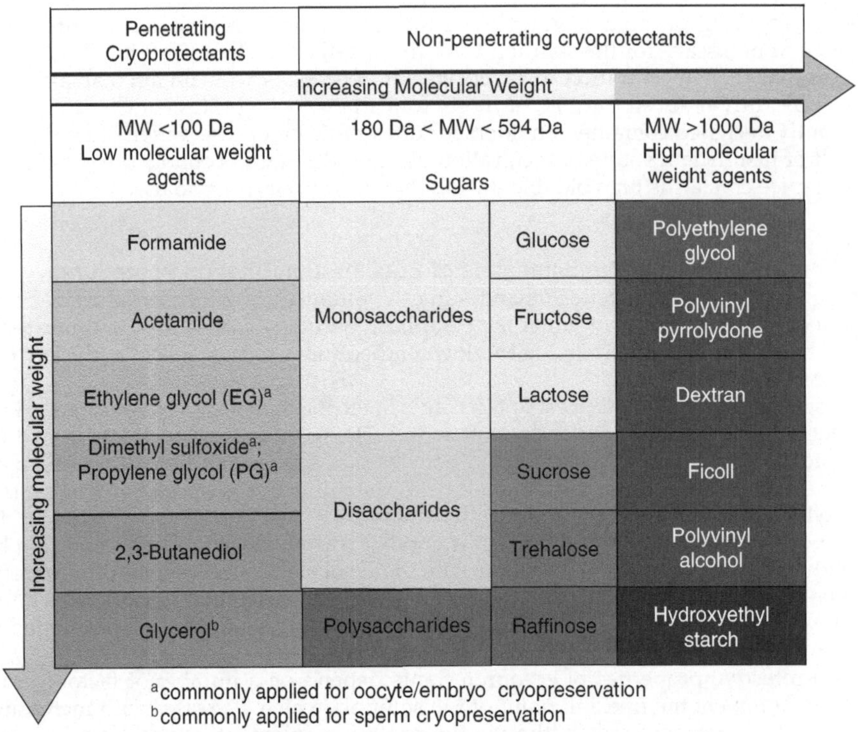

Figure 4 Classification of cryoprotectants. *See Color Plates on Page xii.*

significant improvement in the composition of vitrification solutions has been achieved in the last 15 years (6,23).

There are several methods to examine the physical properties of solutions and systems to verify that they maintain the stable amorphous state during the cooling–warming cycle. The main and most recognized method is differential scanning calorimetry (DSC) or a related technique, namely differential thermal analysis.

The measurements can be done on small volumes (1–50 µl), similar to samples that are used in embryology. The total volume of a sample in a straw and the behavior of a sample in the DSC pan would be similar; however, geometry of the sample and presence of an oocyte/embryo is an additional heterogeneity to the system.

DSC is a very precise method; it measures the heat absorption of a sample during cooling/warming as a function of temperature. In detail, a pair of pans is placed in a thermostatic chamber. The sample pan is filled with an examining solution. The reference pan is filled with an identical volume of known (reference) solution. The two pans are heated with a constant power input to their heaters during a scan. Any temperature difference between the two pans is monitored with a feedback system so as to increase (or decrease) the sample pan's power input. As the masses and volumes of the two pans are matched, the power added or subtracted by the pan feedback system is a direct measure of the difference between the heat capacity of the sample and the reference solutions. Physical (glass transition) and phase (crystallization, melting, etc.) transitions of examined cryoprotectant solutions can be obtained from the DSC thermograms (Fig. 5). By analyzing these thermograms, it can be deduced whether the proposed "candidate" solution can be stated as the vitrification solution (Fig. 6). Solutions with the lowest solute concentration required for vitrification are widely reported in the literature (23,29).

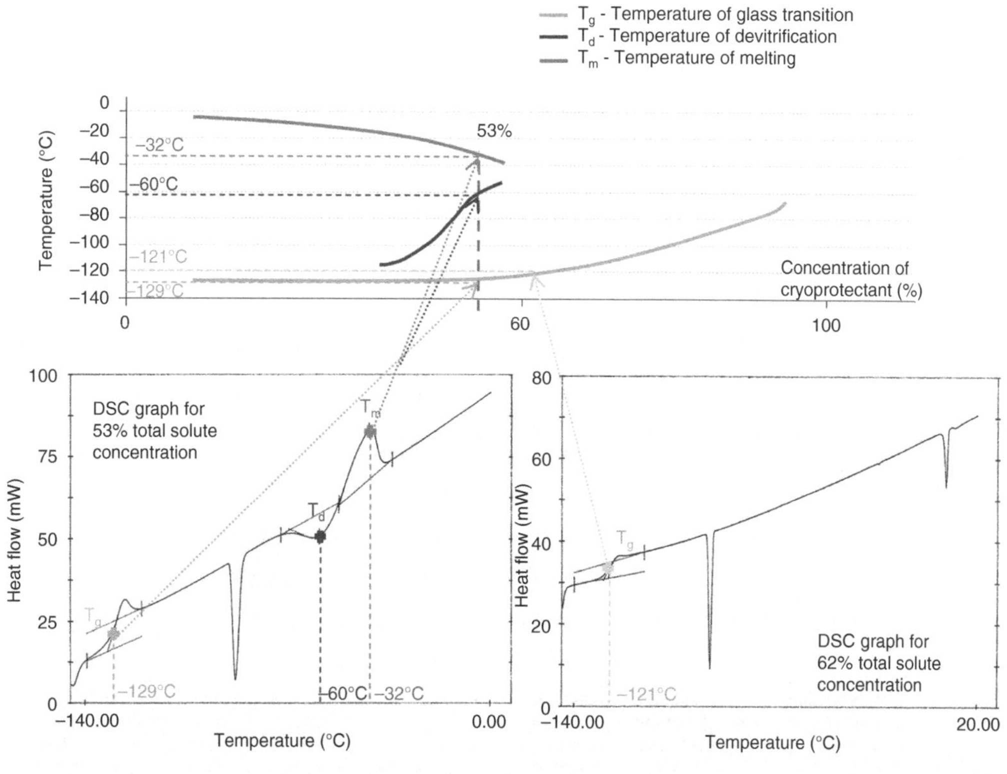

Figure 5 Schematic representation of processes that take place in aqueous solutions of cryoprotectants with the temperature and concentration changes, as indicated in the thermograms. *Abbreviations*: DSC, differential scanning calorimetry; T_d, temperature of devitrification; T_g, glass transition temperature; T_m, melting temperature of ice in the solutions. *See Color Plates on Page xiii*

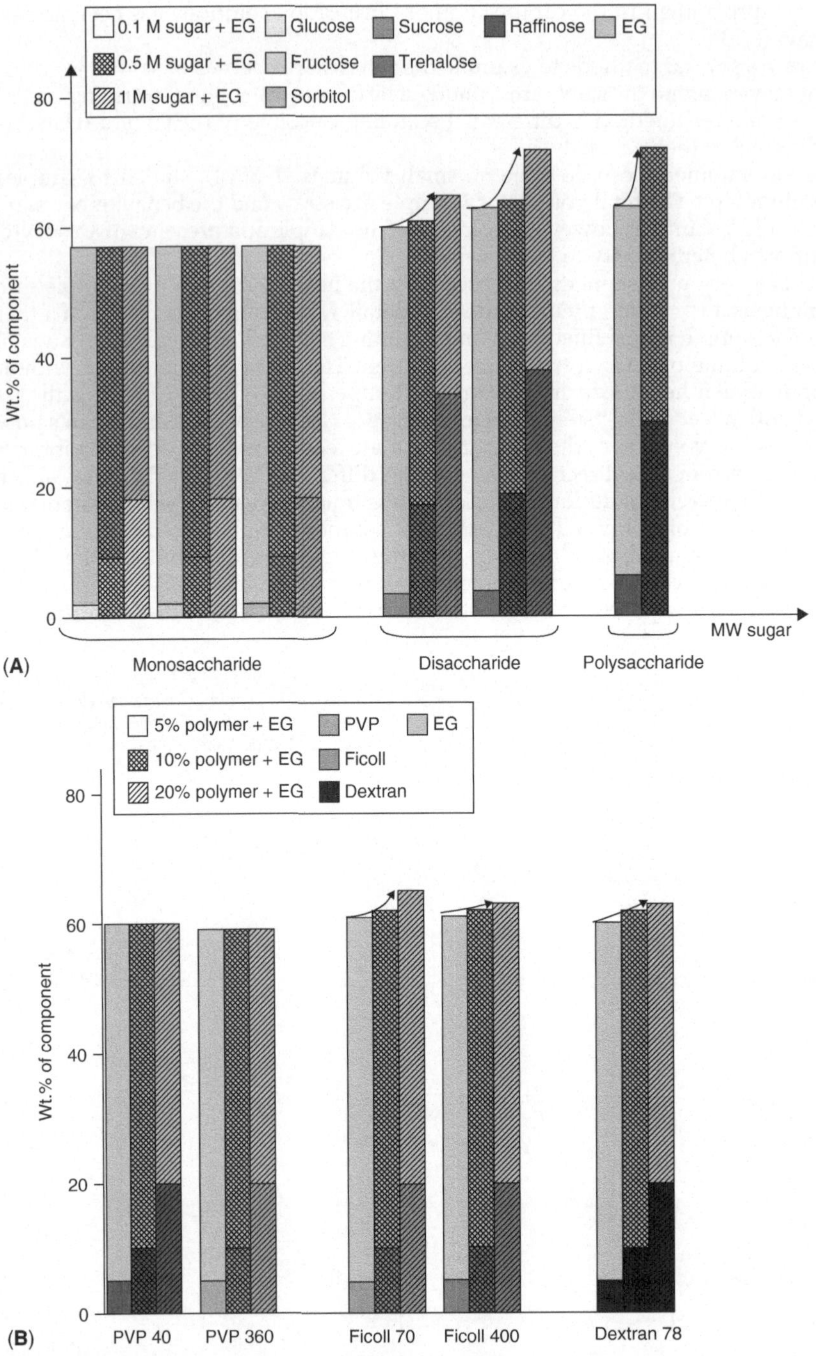

Figure 6 Concentrations of vitrification solutions based on DSC measurements: (**A**) ethylene glycol–sugars vitrification solutions; (**B**) ethylene glycol–polymers vitrification solutions. The *upward* sloping arrows show how the type and concentration of sugar and polymer influence the lowest total solute concentration at which vitrification would occur. Molecular weights of polymers (PVP, Ficoll, Dextran) presented in the figure are in 10^3 Da. Data compiled from Refs. 23 and 29. *Abbreviations*: DSC, differential scanning calorimetry; EG, ethylene glycol; MW, molecular weight; PVP, Polyvinylpyrrolidone. *See Color Plates on Page xiii*

Using the improved vitrification solutions with nonpenetrating additives and technology, embryos can be pre-equilibrated and introduced to the vitrification solutions at room temperatures (4,6,7,24,30–34) or even at physiological temperatures, the latter possibility being particularly important for mammalian oocytes of certain species (e.g., cattle and swine).

Although significant advancement in composition of vitrification solutions have been made, osmotic stress during equilibration and dilution procedures with the resulting cell volume excursions still remains one of the key injury sources during cryopreservation (35,36). It is commonly accepted that addition of cryoprotectant in small steps preferably on a molar basis is less detrimental to cells.

With regard to embryos/oocytes, it was found that they could not be exposed to vitrification solutions in less than two steps without reduction in their viability and that even exposure to 4.0 to 5.5 mol/L of penetrating cryoprotectant in a single step resulted in reduction in their developmental potential (37). Clearly, damage was attributed to osmotic effects given that the exposure was conducted in a one-step manner. To investigate this matter in cryobiology, quantitative permeability values of metaphase II (MII) oocytes with penetrating cryoprotectant were studied (38–40) and solutions to resolve this issue were recommended (40). Although quantitative permeability values to human and mice MII oocytes with propane-1,2-diol (PG) and Me_2SO are higher than those for EG (38,39), it can be easily overcome by exposure of human oocytes to <4 mol/L EG through the additional EG loading steps until transferring the oocytes to the final vitrification solution (40). Dilution of the cryoprotectant from the oocyte after warming is also a challenge. Water enters the oocyte to dilute the cryoprotectant more rapidly than the cryoprotectant's ability to leave the oocyte, primarily due to the difference in molecular weight. This will lead to oocyte swelling, if there is no extracellular osmotic pressure of nonpenetrating cryoprotectant (commonly sucrose) solution is not employed. Sucrose solution prevents swelling by minimizing the difference in osmolarity between the oocyte and its surrounding. Osmotic pressure to the oocyte needs to be reduced gradually for benefits of weakened zona pellucida after cryopreservation. Therefore, a number of dilution solutions of progressively lower concentration of sucrose are able to keep a balance.

PRACTICAL ASPECTS

What Lessons Can Be Learnt from Analysis of Oocyte Vitrification Protocols?

Results of clinical studies on human oocyte vitrification reported for the first 10 years since the first live birth was achieved are presented in Table 1. In total, 1758 oocytes were warmed after vitrification, achieving high survival and fertilization rates of 84% and 82%, respectively (7,41–53). Vitrification is constantly associated with high development capacity judged by percentage of cleavage of embryos and blastocysts derived from fertilized human oocytes post-vitrification (Fig. 7). In the analyzed studies (Table 1), an average cleavage rate of 92% was reported. Embryo transfers were performed in 265 patients and resulted in 100 clinical pregnancies (38% pregnancy rate) providing the yield of 5.7 pregnancies per 100 oocytes thawed. Implantation rates varied from 6.4% (46) in early studies to about 20% (41,50) in recent studies, with an average implantation rate of 13% (41,46,50–52).

Since 1999, when the birth of the first baby girl was reported by Kuleshova and coworkers (7) from an embryo derived from vitrified-warmed oocytes, 31 healthy babies were born after application of this strategy worldwide (7,42,43,46,48,51–55). A total of 1336 oocytes were thawed providing the birth success rate of about 2.3 per 100 oocytes thawed. In fact, the number could be higher because some publications report both births and ongoing pregnancies. For example, births of three healthy babies have been recently reported by Antinori et al. (2007) (51), with 28 ongoing pregnancies at the time of reporting (initially, 39 pregnancies occurred after embryo transfers in 120 patients). In a meta-analysis of vitrification reports published up to 2006 (56), it is mentioned that the rate of live birth in the vitrification cycle was 3.6% but such high results seem to be questionable.

Outcomes of slow-cooling protocols based on extensive literature review (51,57–88) provided the following results: 9369 oocytes were thawed following freezing in PG supplemented with 0.1 to 0.3 M sucrose (average survival rate reported was 66%) and derived embryos were transferred into 1440 patients with 234 implantations and 122 births. Thus, the birth rate of 1.3% for slow-cooling studies is nearly twice lower than the rate for vitrification studies (2.3%, Table 1).

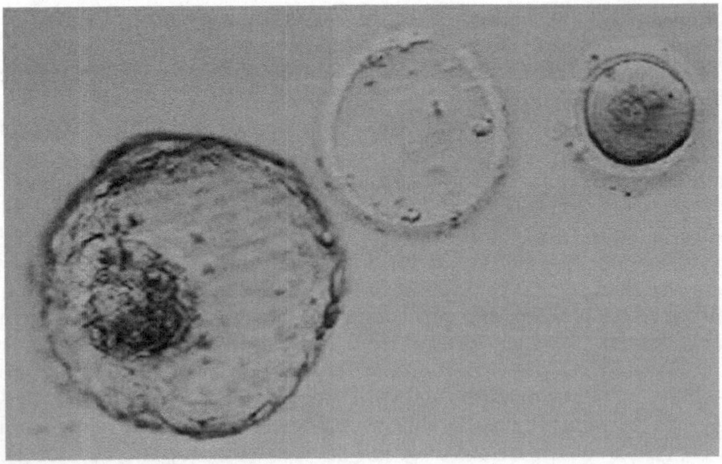

Figure 7 Hatched human blastocyst after development from a vitrified oocyte. Original magnification ×200. *See Color Plates on Page xiv*

Table 1 Summary of Clinical Outcomes of Vitrification of Human Oocytes

	No. of				
Study	Oocytes thawed	Patients	Pregnancies	Births	Ongoing pregnancies
Kuleshova et al. (1999) (7)	17	3	1	1	
Wu et al. (2001) (45)	8	1	1	NR	
Katayama et al. (2003) (43)	46	6	2	1	1
Kim et al. (2003) (44)	51	3	1	NR	1
Yoon et al. (2003) (46)	474	28	6	7	
Chian et al. 2005 (41)	180	15	7	NR	
Kuwayama et al. (2005, 2007) (47,55)	64	29	12	7 + 4 = 11[a]	0[a]
Kyono et al. (2005) (48)	5	1	1	1	
Lucena et al. (2006) (49)	159	23	13	NR	
Selman et al. (2006) (50)	24	6	2	NR	2
Antinori et al. (2007) (51)	330	120	39	3	28
Yoon et al. (2007) (52)	364	28	13	5	7
Chen et al. (2008) (53)	19	1	1	1	
Chian et al. (2008) (42)	17	1	1	1	
Total	1758	265	100	31	39

[a]Initially the births of 7 healthy babies and 3 ongoing pregnancies (47) were reported, eventually 11 healthy babies were born from 10 delivered pregnancies (55).
Abbreviation: NR, number was not reported.

The difference in average survival rates, 84% for vitrification (Table 1) and 66% for slow cooling, also indicates that vitrification is superior to slow cooling. Outcomes of oocyte slow cooling reported in literature for the first 10 years since the first successful cryopreservation of a human oocyte (89), namely from 1986 to 1996, were not included in this analysis.

Recently, the obstetric and perinatal outcomes in 165 pregnancies (200 infants) conceived from oocytes vitrified using a few types of vitrification solutions (EG + Me_2SO/PG + sucrose) and holders in three fertility treatment centers in Canada, Colombia and Mexico were reported (90); it was shown that the mean birth weight and the incidence of congenital anomalies are comparable to those of spontaneous conceptions in fertile women or infertile women undergoing in vitro fertilization treatment. The preliminary findings reported by Chian et al. (90) may provide reassuring evidence that pregnancies and infants conceived following oocyte vitrification are not associated with increased risk of adverse obstetric and perinatal outcomes (a comprehensive report by Dr. Chian is presented in chap. 21).

The wide range of clinical outcomes apparently arises from the different vitrification protocol used in the studies. Each step in the vitrification protocol plays an indispensable role in delivering the optimum efficiency of the protocol. The following sections describe the philosophy of development of each step in deriving some of the vitrification protocols.

Table 2 summarizes 14 known protocols that lead to successful childbirth, and their analysis is given in the following sections.

Selection of Components of Vitrification Solutions

When we began research on vitrification of oocytes, our philosophy was to design a vitrification solution (*i*) consisting of two-thirds of penetrating and one-third of non-penetrating cryoprotectants and (*ii*) with the total solute concentration sufficient to prevent formation of ice during cooling and warming, as measured by DSC. Thus, a vitrification solution 40 vol.% EG + 0.6 mol/L sucrose has been designed and successfully applied for vitrification of human oocytes with subsequent live birth reported for the first time (7).

In later studies on vitrification of human oocytes, proportion of penetrating and nonpenetrating cryoprotectants (sucrose) remains approximately the same, however, in some studies a mixture of penetrating cryoprotectants, for example, 15 vol.% EG + 15 vol.% Me_2SO (48,49,51) or 15 vol.% EG + 15 vol.% PG (41), is used instead of a sole cryoprotectant. Utilization of such a mixture of penetrating cryoprotectants reminds us of early studies in the 1980s, prior to discovering the benefits of nonpenetrating cryoprotectants, when researchers tried to mitigate the negative impact of toxicity of penetrating cryoprotectants by composing vitrification solutions with the aim of achieving a total solute concentration of about 60%. Nowadays, there is no necessity to use a mixture of penetrating cryoprotectants instead of EG, if considerable proportion (e.g., 1/3) of nonpenetrating cryoprotectant is included in composition of the final vitrification solution. Recently, we conducted a study comparing outcomes of vitrification using different vitrification solutions, and found that partial replacement of EG with PG in the original final vitrification solution, 40 vol.% EG + 0.6 mol/L sucrose, did not provide any benefits (91). It also needs to be mentioned that EG is the least toxic cryoprotectant among others as tested on embryos (26). The choice of Me_2SO might be a reflection of early studies, as Me_2SO was employed as a cryoprotectant in conjunction with slow-cooling protocols for human oocytes since 1986 (89). The choice of PG is justified on the grounds that it is an excellent glass-former (92), and it is used in recent slow-cooling protocols for human oocytes (56–87).

In light of awareness of low toxicity of EG, many protocols for vitrification of human oocytes use a sole cryoprotectant EG at concentration of 30 vol.%, supplemented by large amount of sucrose (44–46).

Besides being of low toxicity, EG-based solution has another advantage over Me_2SO-based solution in terms of providing more time for handling and manipulating the procedure due to lower permeability value of EG for human oocytes (40) compared to Me_2SO.

Pre-equilibration

A pre-equilibration step is required for oocytes to adapt to high concentrations of the vitrification solution; it involves selection of the number of solutions and their concentrations. In protocols that use vitrification solutions largely comprising of nonpenetrating cryoprotectant (1 mol/L sucrose), it is logical to use one-step pre-equilibration (10 vol. % EG) because the amount of penetrating cryoprotectant is small in the final solution (30 vol.% EG + 1 mol/L sucrose) (44–46). This was also adopted in protocols where the final solutions consisted of 30 vol.% of penetrating cryoprotectants (41,48).

With increase in concentration of EG in the final vitrification solution to 40 vol.% (7), a two-step pre-equilibration procedure has to be applied, namely 10 vol.% EG and 25 vol.% EG (Fig. 8), which has replaced 20 vol.% EG in our design in short span of time. Another comprehensive study on vitrification of human oocytes in sucrose-EG vitrification solution reports

Table 2 Comparison of Protocols Used for Vitrification of Human Oocytes

	Equilibration procedure			Dilution procedure		
Study	Pre-equilibration	Vitrification solution and duration	Temperature (°C)	First dilution step	Further dilution steps	Temperature (°C)
Kuleshova et al. (1999) (7)	10% EG for 40 sec, 20% EG for 30 sec	40% EG + 0.6 mol/L sucrose for 60 sec	37	0.4 mol/L sucrose for 2–3 min	0.25 mol/L sucrose for 2–3 min, 0.125 mol/L, 0 mol/L sucrose for 3–6 min each step	37
Wu et al. (2001) (45)	NR	5.5 mol/L EG + 1.0 mol/L sucrose for 20 sec	NR	0.5 mol/L sucrose for 10 sec	0.5 mol/L, 0.25 mol/L, 0.125 mol/L, 0.0625 mol/L, 0 mol/L sucrose for 1 min each step	37
Katayama et al. (2003) (43)	NR	15% EG + 15% Me_2SO + 0.5 mol/L sucrose	NR	NR	NR	NR
Kim et al. (2003) (44)	1.5 mol/L EG	5.5 mol/L EG + 1.0 mol/L sucrose	NR	1.0 mol/L sucrose	0.5 mol/L, 0.25 mol/L, 0.125 mol/L, 0 mol/L sucrose	NR
Yoon et al. (2003) (46)	1.5 mol/L EG for 2.5 min	5.5 mol/L EG + 1.0 mol/L sucrose for 20 sec	37	1.0 mol/L sucrose	0.5 mol/L, 0.25 mol/L, 0.125 mol/L, 0 mol/L for 2.5 min each step	37
Chian et al. (2005) (41)	7.5% EG + 7.5% PG for 5 min	15% EG + 15% PG + 0.5 mol/L sucrose for 45–60 sec	RT	1.0 mol/L sucrose for 1 min.	0.5 mol/L,0.25 mol/L, 0 mol/L sucrose for 3 min each step	37
Kuwayama et al. (2005) (47)	NR	Two solutions: 5.0 mol/L EG and 6.8 mol/L EG	NR	NR	NR	NR
Kyono et al. (2005) (48)	7.5% EG + 7.5% Me_2SO for <20 min	15% EG + 15% Me_2SO + 0.5 mol/L sucrose for 60 sec	RT	1.0 mol/L sucrose for 1 min	0.5 mol/L, 0 mol/L sucrose	37
Lucena et al. (2006) (49)	15% EG + Me_2SO for 10 min	15% EG + 15% Me_2SO + 0.5 mol/L sucrose	NR	1.0 mol/L sucrose for 1 min	0.5 mol/L sucrose for 3 min, 0 mol/L sucrose	37
Selman et al. (2006) (50)	7.5% EG + 7.5% Me_2SO for 5 min	15% EG + 15% Me_2SO + 0.6 mol/L sucrose for 5–10 sec	RT	1.0 mol/L sucrose for 2.5 min	0.5 mol/L sucrose for 4 min, 0 mol/L sucrose	NR
Antinori et al. (2007) (51)	7.5% EG + 7.5% Me_2SO for 5–15 min	15% EG, 15% Me_2SO + 0.5 mol/L sucrose for <60 sec	RT	1.0 mol/L sucrose for <60 sec	0.5 mol/L, 0.25 mol/L, 0 mol/L sucrose for 3 min each step	37
Yoon et al. (2007) (52)	1.5 mol/L EG for 2.5 min	5.5 mol/L EG + 1.0 mol/L sucrose for 20 sec	37	1.0 mol/L sucrose for 2.5 min	0.5 mol/L, 0.25 mol/L, 0.125 mol/L, 0 mol/L sucrose for 2.5 min each step	37
Chen et al. (2008) (53)	7.5% EG + 7.5% Me_2SO for 2 min	15% Me_2SO + 15% EG + 10% Ficoll 70 + 0.65 mol/L sucrose	37	0.5 mol/L sucrose for 5 min	0.3 mol/L, 0.1 mol/L, 0 mol/L sucrose for 5 min each step	RT
Chian et al. (2008) (42)	7.5% EG + 7.5% PG for 5 min	15% EG + 15% PG + 0.5 mol/L sucrose for 45–60 sec	RT	1.0 mol/L sucrose for 1 min	0.5 mol/L, 0.25 mol/L, 0 mol/L sucrose for 3 min each step	37

% denotes vol.%.
Abbreviations: EG, ethylene glycol; NR, not reported; PG, propane-1,2-diol; RT, room temperature.

that three-step pre-equilibration is optimal if the concentration of EG exceeds 40 vol.% (40). Generally, analysis of all known reports on vitrification of human oocytes that lead to childbirth (Table 2) imply that the concentration of the solution used in the first step is too high. In the light of recent investigations, it has been found that the concentration of the first solution should not exceed 8%, at least when Me_2SO is involved.

Temperature Factor

In majority of recent protocols, human oocytes are introduced to vitrification solutions and diluted at room temperature (42,48,50,51).

Human oocytes have relatively low level of intracellular lipids than oocytes from domestic animals (93), suggesting lower chilling sensitivity at least during short equilibration/dilution procedures used in vitrification procedures for human oocytes (~10 minutes or less). This strategy is apparently a practical strategy. In the early studies, dilution/pre-equilibration was done at 37°C (7,46,53) mainly due to early reports on sensitivity of the meiotic spindle of human oocytes during prolonged exposure to temperatures below 31°C. Recently, it was found that exposure of human oocytes to 35°C during pre-equilibration/dilution in vitrification protocols produces better outcomes than exposure to either 37°C or room temperature (94).

Dilution

Although oocytes are shrunken at the end of equilibration procedure (Fig. 8), especially in case of short exposure to vitrification solution, yet large amount of penetrating cryoprotectants is still present. This presence makes the oocytes susceptible to osmotic shock during dilution.

The majority of dilution protocols were designed empirically. Most dilution protocols involve high concentrations of the initial dilution solution, namely 1 mol/L [Table 2; (48–50)]. However, in some other reports lower concentrations are used, with the lowest reported concentration of 0.3 mol/L (53). Initially, we also reported low concentration of the first dilution solution; however, later we found benefits in the introduction of an additional step of increased concentration for the first dilution solution. Its concentration may be increased to 0.7–0.5 mol/L sucrose, if accidental overexposure of oocytes to the vitrification solution occurs for a longer time during handling by staff. During overexposure, a considerable amount of penetrating cryoprotectants enters the oocyte. Therefore, water, which rapidly crosses the membrane along with the remaining penetrating cryoprotectant, adds to the volume of liquid inside the oocyte. This may cause swelling of the oocyte, if external pressure of dilution solution is not sufficiently

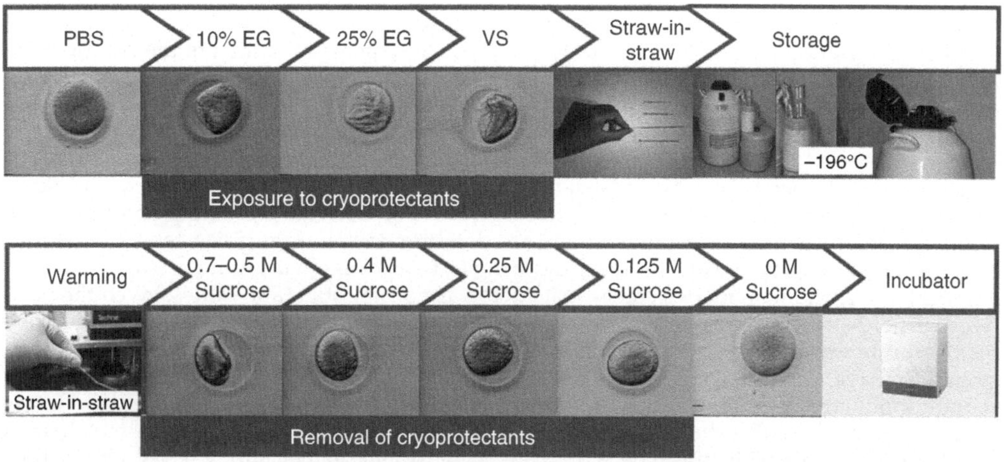

Figure 8 Step-wise equilibration and dilution procedures for immature human oocytes involving vitrification solution 40 vol.% EG + 0.6 mol/L sucrose. *Abbreviations*: EG, ethylene glycol; PBS, phosphate buffered saline; VS, vitrification solution. *See Color Plates on Page xiv*

high. Short exposure of the oocyte to vitrification solution causes its significant shrinkage. Therefore, balance during dilution is easily maintained by hypertonic solution of low concentration. To the best of our knowledge, exposure of oocytes to solution 1 to 0.5 mol/L sucrose does not cause any effect on oocytes (23).Thus, increasing the concentration of the dilution solution does not harm the cells.

We can suggest the application of the following general rule: The higher the proportion of penetrating cryoprotectant in vitrification solutions or the longer the exposure to the final solution, the higher the concentration of the dilution solution that should be used.

Flexible Strategy

Human oocytes are unique cells with regard to cryopreservation. Despite identical morphological appearance, some parameters of human oocytes at the same stage of maturation vary for different infertility patients unlike parameters of reproductive cells of animals belonging to the same species. This may be the critical factor due to which the vitrification of mammalian oocytes, except possibly porcine oocytes, has been successful for a number of years (37,95–98). We also found that the color of the human oocytes during processing indicates the degree of infiltration by the penetrating cryoprotectant: the intense white color indicated overexposure to cryoprotectant and that longer dilution times were required during warming. In light of these observations, we to IVF practitioners a method based on observation, namely instant regulation of the extent of penetration and dehydration of oocytes during exposure to and dilution of cryoprotectants. In detail, the duration of each equilibration and dilution step should be adjusted according to the individual response of oocytes and may be considered flexible within the established range. This approach does not contradict the recommendations of other investigations. It was previously established that the duration of exposure of human oocytes to low concentrations of cryoprotectants and to vitrification solutions, which include a large proportion of polymers and sugars in their composition, could be within quite a broad range without compromising the efficiency of the cryopreservation procedure (6,99). On the other hand, it is rational that many studies focus on determining the accurate length of time for each step of the protocol. It is also our intention to recommend an optimal vitrification procedure for human oocytes of statistically average quality. In our preliminary studies, we usually estimate the appropriate length of time in each solution of the protocols. The time durations are chosen by taking into consideration the toxicity of the solutions to cells and the cells' survivability. Our recent observations support the proposed hypothesis that weaker oocytes require a shorter equilibration time in a solution, whereas stronger oocytes require prolonged exposure to the same solution; it is also beneficial to use the longest equilibration time within the range for successful cryopreservation of oocytes of young patients.

It is evident that dilution solutions, where the concentration of sucrose is decreased in equal quantities, prevent osmotic shock to the weakened zona pellucida of cryopreserved oocytes (7). Application of a flexible strategy additionally helps to eliminate the risk of detrimental expansion or prolonged dehydration of cells. All this leads to development of a procedure that may be considered as balanced. In agreement with the traditional concept, this approach may allow to establish an optimal procedure for vitrification of human oocytes.

We observed that oocyte osmotic response during the equilibration procedure can act as an indicator of the oocyte's viability. Generally, the duration of time and the extent of shrinkage depend on the permeability of the cells to water and to cryoprotectant and the proportion of cell volume that is osmotically inactive, which are determined values at given temperatures for ideal healthy cells. The maximum time and extent of shrinkage within the assessed range of excellent quality human oocytes can be considered as normal, standard values, given the positive signs of better development. We also assume that the oocyte's shape within the zona pellucida during exposure is of importance. We observed that oocytes that shrink with even radial symmetry have a higher chance of survival than oocytes that shrink asymmetrically. Severe radial asymmetry of oocytes during exposure is associated with their poor quality. Hence, observations of anomalies during osmotic response on exposure to cryoprotectants may help to identify weak oocytes with unseen defects, to predict the decline in their survival post-vitrification, and development potential after fertilization. We believe that the present observations may also assist in improving results of a variety of existing vitrification protocols. A comparison of two cryopreservation methods, namely vitrification and slow cooling is presented in Table 3.

Table 3 A Comparison of Two Cryopreservation Methods, Namely Vitrification and Slow Cooling

	Vitrification	Slow cooling
Ability		
To observe oocytes/embryos during procedure	Yes	No
To analyze oocytes/embryos during procedure	Yes	No
To interact with oocytes/embryos	Yes	No
To control penetration of cryoprotectant into oocytes/embryos	Yes	No
To control dehydration rate of oocytes/embryos	Yes	No
To keep oocyte-friendly temperature (e.g., 31–35°C) during the whole equilibration procedure	Yes	No
Other Factors		
Prolonged temperature shock	No	Yes
Duration of time out of incubator	~10 min	~100–110 min
Fracture of zona pellucida	Not possible	Possible
Possibility of oocyte/embryo to be captured by growing ice	Not possible	Possible
Osmotic shock during warming	No	Yes
Length of suboptimal pH	Short term	Long term
Balanced salt content in solutions during cryopreservation	Yes	No

Adapted from Ref. 1.

Vitrification of Immature, In Vitro-Matured, and Biopsied Human Oocytes

Mature human oocytes are challenging to cryopreserve due to several reasons. Oocytes have a complex subcellular structure that can be damaged easily by the brief cooling to room temperature as well as by prolonged exposure to temperatures that are relatively close to the physiological temperature (100,101). This impact on the meiotic spindle and subsequent chromatid nondisjunction is another limitation for application of slow-cooling protocols (101). Most importantly an increase in chromosome abnormalities after fertilization of frozen-thawed human oocytes was also found (100,102). Extended chromosomes held within a nuclear membrane, the absence of a spindle, cortical granules that are not ready to be released, and attached cumulus cell mass characterize oocytes at the germinal vesicle (GV) stage. To prevent spindle disorganization and to decrease the likelihood of aneuploidy, which is associated with cryopreservation of mature oocytes, cryopreservation of immature oocytes at the GV stage appeared advantageous (100,103). In the past, one group has been continuously conducting broad investigations on cryopreservation of immature human oocytes for a number of years (30,104,105). The group reported a high percentage of cleavage of embryos and blastocysts derived from vitrified immature human oocytes, particularly at the GV stage (30,104). In contrast, the same group reported adverse effects of freezing-thawing treatment on the maturation and development rates of immature oocytes (105). It should be noted that the human oocytes used were of good quality because they were retreated from unstimulated patients (105).

A comprehensive study on the vitrification of in vitro-matured human oocytes was reported in 2000 (106). All retrieved oocytes (MII, MI, and GV) were vitrified at the MII stage. The cleavage rate was high up to the eight-cell stage; however, no morula or blastocyst formation was observed. It should be noted that blastocyst formation rate was generally very low at that time due to nonadvanced culture medium (106). In later studies, live birth was achieved from in vitro-matured human oocytes (42). Although the reported survival rate was low (24%), fertilization rate was 75%. All fertilized oocytes produced three embryos, which were in turn transferred into a patient. Transferring the three embryos resulted in a live birth.

Comprehensive research into the mechanisms of oocyte maturation and clinical trials led to the progress of in vitro maturation (IVM) of immature oocytes. It has been demonstrated that IVM of immature oocytes may be promoted by the addition of gonadotropins to the maturation medium, and that the low fertilization rate of in vitro-matured oocytes associated with alteration of the zona pellucida (zona hardening) may be overcome by intracytoplasmic sperm injection. As the progress of IVM of immature oocytes is in demand (107,108), it is necessary to understand the mechanisms that are involved in oocyte vitrification. For instance, cumulus–oocyte interaction of GV oocytes is important for cytoplasmic maturation (109,110). We believe that removal of the cumulus is not necessary for vitrification for the reason described below. Oocytes need to be cooled prior to cryostorage and warmed in order to be used clinically. Vitrification by definition is the transition from amorphous liquid to amorphous solid during cooling and transformation from amorphous solid to amorphous liquid during warming. Consequently, mechanical stress between the cumulus of GV oocytes and zona pellucida does not occur during vitrification. The cumulus–oocyte complex is preserved intact under such conditions; therefore, the oocyte has the ability to be reinitiated and progress toward completion of the first meiotic division from the GV stage to MII.

Oocyte and embryo genetic screening through microbiopsy procedures that analyze their chromosomal status have became a routine in assisted reproductive technologies. As mentioned above, vitrification is known as a physical phenomenon that does not involve phase transitions, and therefore in principle it has to be less challenging for cryopreservation of embryos/oocytes with disturbed integrity of the zona pellucida. A high survival rate of unfertilized MII oocytes after first polar body biopsy has been achieved. Pregnancy has been reported after vitrification of pronuclear stage polar body-biopsied human oocytes (111), which is not surprising because cryopreservation of oocytes after fertilization is not a challenge. However, the success rate after conventional cryopreservation and thawing of biopsied embryos is known to be low (111–113). Conventional freezing involves phase transition, practically, oocytes/embryos moved intensively by ice front inside straw. This is detrimental for oocytes/embryos that lack intactness of the zona pellucida due to artificially created zona openings for the biopsy.

The impact of composition of different EG-based vitrification solutions and equilibration temperatures on post-vitrification viability of oocytes was reported. MII oocytes were vitrified in solutions containing 20 and 30 vol.% EG, which were supplemented with Me_2SO, a 40 vol.% EG-based solution free from Me_2SO, and sucrose and Ficoll (94). The implantation rate increased with increasing concentration of EG in the vitrification solution (94). Furthermore, the highest cleavage rate was found for the 40 vol.% EG-based vitrification solution without Me_2SO when the equilibration temperature was 35°C, which was found to be the optimal temperature. The efficacy of the results in an in vitro study will be confirmed with further implantations. It was reported that vitrification did not impair cleavage rate, mean cell number, embryo quality, and blastocyst development rates of post-vitrified human oocytes (114).

SAFETY OF CRYOPRESERVATION AND CONTAINERS FOR VITRIFICATION

Ensuring sterility is one of the major issues that need to be addressed when designing cryopreservation protocols. Storing reproductive cells in liquid nitrogen has been challenged because of the recovery of viable microbes from the liquid nitrogen tank. These pathogens include fungi (Aspergillus) (115), virus (hepatitis) (116), and bacteria. Viral contamination of cryopreserved embryos has been reported (117). Three types of cryostorage containers are discussed below: "straw-in-straw," arguably the safest method; single straw; and open container, arguably the least safe method in terms of preventing contaminations. Later in the section, we put forward arguments against using serum/proteins in the vitrification procedure for safe cryopreservation.

Closed and Sealed System: The "Straw-in-Straw" Approach

In order to overcome the risk of infection, we have invented a "straw-in-straw" principle. In the next two subsections, we discuss "open" and "closed" types of containers, some of which may provide higher cooling rates but at the expense of a significantly higher risk of contamination. The important distinction that is often misunderstood by the scientific and clinical community is that "closed" and "sealed" containers do not provide the same level of protection; only sealed containers avoid the risk of infection through contact with liquid nitrogen, whereas closed containers do not.

The "straw-in-straw" method uses readily available 250- and 500-µl straws (118), as a simple and cost-effective strategy to eliminate the risk of infection or contamination (Fig. 9). This approach enables samples to be cooled and warmed by direct immersion into liquid nitrogen and a water bath, respectively, with double protection. The idea of this invention is that human specimens are isolated as much as possible from the outside environment and direct contact with liquid nitrogen is eliminated. Therefore, both inner and outer straws (118), or at least the outer straw (119), should be sealed after the smaller inner straw is inserted inside the larger outer straw (Fig. 10).

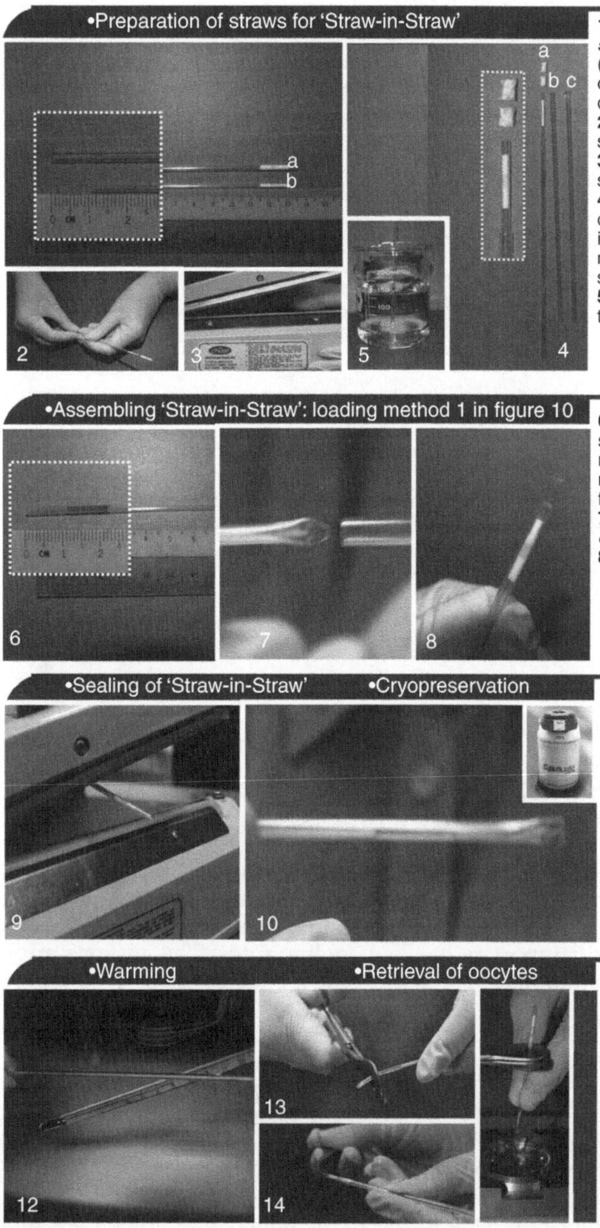

1. **(a)** 500μl french plastic straw (*outer straw*); **(b)** 250μl french plastic straw (*inner straw*) cut to 1cm shorter than outer straw; **insert** detail of straw size difference
2. Removing cotton wool from outer straw.
3. Sealing cotton wool from outer straw.
4. **(a)** Details of cotton wool of the outer straw being pushed out by inserting the inner straw (**insert** – magnified detail); **(b)** unsealed outer straw; **(c)** sealed outer straw
5. Testing outer straw in water to verify there is no leakage

6. *Inner straw* loaded with vitrification solution and oocytes; **insert**: loading method 1 in figure 10. three different methods of oocyte loading are shown in figure 9.
7. Insertion of sealed inner straw in *outer straw.*
8. Unsealed end of 'Straw-in Straw'

9. Sealing of *outer* (500μm) *straw* using heat sealer.
10. The resulting 'Straw-in-Straw' arrangement with oocyte inside the *inner straw.*
11. Direct immersion into LN2. storage in liquid nitrogen (-196°C)

12. Warming by direct immersion to 38°C water bath
13. Cutting of *outer straw.*
14. Removal of *inner straw.*
15. Retrieval of oocytes (pictured: loading case 3 only from figure 10): collection is made by cutting off the top of the inner straw containing wool, the oocyte (in the first dilution solution) then expels onto the dish by gravity.
steps 12–14 are the same for loading cases 1–3 from figure 10.

Figure 9 Essential steps of the "straw-in-straw" method: preparation of straws for vitrification (steps 1–5), loading of oocytes (details in Fig. 10), assembling the "straw-in-straw" (steps 6–8), sealing the "straw-in-straw" (steps 9–10) prior to immersing into liquid nitrogen (step 11), post-warming procedures and expelling of oocytes (steps 12–15). *See Color Plates on Page xv*

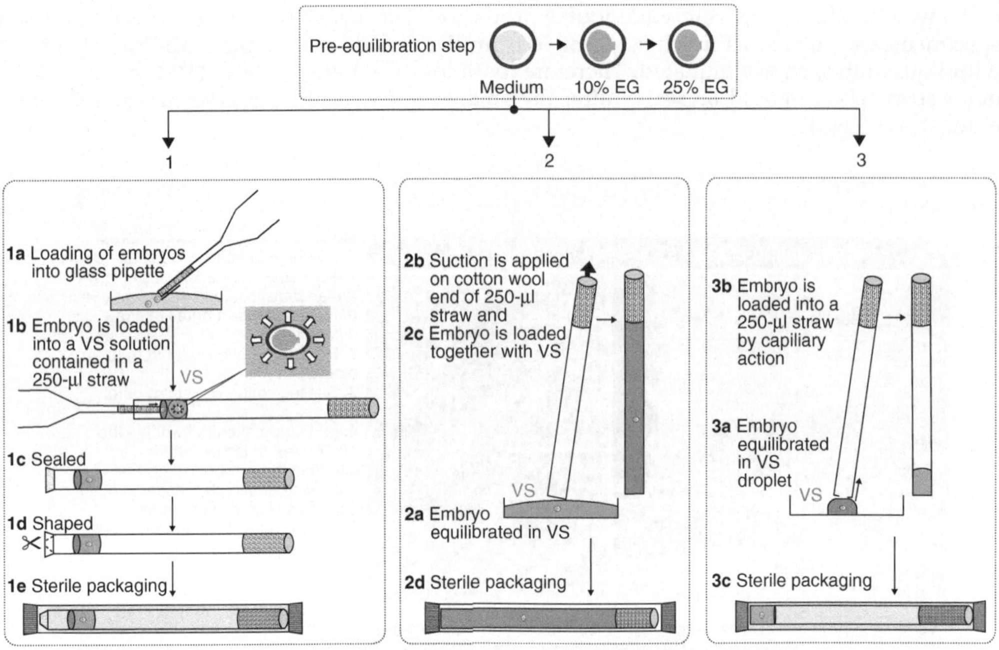

Figure 10 Schematic representation of loading embryos/oocytes in the "straw-in-straw" configuration (250- and 500-μl straws). Loading of embryos/ocytes by: (1) the traditional approach, which involves insertion into the pre-prepared column of the final solution using a glass transfer pipette; the 250-μl straw is sealed, shaped, and placed into the 500-μl straw (118,120); (2) suction using an Eppendorf 1000-μl micropipette (91,121–123); (3) capillary action. The insert in step 1b shows a magnified oocyte and illustrates initial shrinkage of oocyte in vitrification solution caused by cell dehydration; the shrinkage of the oocyte in VS is identical for all approaches (steps 1b, 2a, and 3a). *Abbreviations*: EG, ethylene glycol; VS, vitrification solution. *See Color Plates on Page xvi.*

Initially, the invention of the innovative "straw-in-straw" approach was applied for cryopreservation of embryos (118,120). Later, the robustness of this idea was proven through the success in vitrification of other fragile systems such as cell-matrix systems (121,122), self-assembled cell aggregates cultured as spheroids (91), or sensitive cells such as stem cells (123). Since the first report published in 2000, this idea spread widely and nowadays there are many elaborate containers that have been developed based on this principle. Although these containers may appear to be more sophisticated, they often add only marginally to the efficiency as a variety of low-toxicity vitrification solutions are available that can be used together with containers that can support rather low cooling rates. The author's advice to IVF clinics aiming to minimize contamination risk is to take into consideration the procedure described in Figures 9 and 10.

There are few simple yet effective variants of the basic "straw-in-straw" approach. For example, the "cut standard straw" (CSS) is an elegant invention in which a 250-μl straw is cut at a 45° angle and inserted into a 500-μl straw; the CSS was applied in cryopreservation of human embryos (124). Relatively low cooling rates that are employed under the "straw-in-straw" approach have produced successful outcomes. Cooling rates of 400°C/min [250-μl straw in 500-μl straw (118)], 200°C/min [open pulled straw in 500-μl straws (125)], or 600°C/min [CSS (124)] in double containers were reported. Clearly there is no necessity for the open pulled straw as it decreases cooling rate. Due to the fundamental property of air which is present between two properly sealed containers, namely low heat conductivity, exceedingly high cooling rates are not feasible. Hence, to achieve proper protection against risk of contamination, the cooling rate has to be lowered to a certain extent but undoubtedly it should still remain effective for vitrification of oocytes (118).

Use of a Single Standard Straw

Recently, vitrification of human oocytes in a 250-µl straw without modification has again focused the attention of the scientific community on this simple original design. This design was not discovered accidentally, but resulted from a fundamental and systematic study involving the consideration and analysis of variety of factors (40).

Numerous excellent results have already demonstrated that human oocytes could be cryopreserved in a single 250-µl straw by rapid immersion into liquid nitrogen (58,69). The rationale that a standard 250-µl straw is preferred to the latest ultra-rapid cooling containers is based on higher chances of losing oocytes due to technical malfunctions when using these new containers, which result in oocyte damage (40) or loss of control. After all, extensive training is required to master most of the new techniques.

In contrast to new emerging systems, the IVF clinics have decades of experience in practicing cryopreservation of gametes and embryos using standard straws, for example, freezing of sperm in a 500-µl straw and slow cooling of embryos in a 250-µl straw using a programmable rate freezer. In addition to familiarity with the established technique, sealing of a standard straw is an unsurpassed advantage as the risk of contamination through direct contact with liquid nitrogen is clearly prevented. The availability of numerous solutions for the effective vitrification of embryos/oocytes, which were developed and tested in combination with a 250-µl straw, can also be seen as an advantage.

Cooling in a vapor phase of liquid nitrogen (120°C/min) was proven to be less aggressive for embryo cryopreservation (5). It is particularly effective due to the fact that it reduces mechanical stresses, and therefore cracking of the vitrification solution and damage of zona pellucida is prevented during vitrification. Vitrification of human embryos in liquid nitrogen vapors have resulted in childbirths (6,126). The stable amorphous state was maintained by the use of a vitrification solution with high total solute concentration (74%), while its toxicity was easily overcome by supplementation with a large amount of polymer, namely Ficoll.

Vitrification of oocytes in a 250-µl straw may be considered a semi-safe method, as contamination can still be transferred from the end of the straw during contact with the dilution solution on warming.

Use of "Open" Containers

The quest for high cooling rates, which arises from the intention to avoid the toxicity of the vitrification solutions by decreasing their concentration without compromising the vitrification property, led to popularization of the so-called "open" containers. A common misconception in commercial literature is that these "open" containers are described as "closed" systems, making it appear that the products combine the ideal characteristics, namely sterility and extremely high cooling rates. However, due to fundamental physical property of materials any container that enables a cooling rate of 10,000°C/min and above cannot be as sterile as a sealed container. This is the trade-off that an IVF clinic must be aware of when designing its preservation protocol.

Historically, the idea of developing an "open" container originated from the fact that direct contact of an open container with liquid nitrogen could greatly increase cooling rate (Fig. 11). Though exceedingly high cooling rate may at times be achieved at the expense of risk of contamination over sterility of biological materials, some of these inventions have been applied to IVF practice and yielded fruitful outcomes. One of the first strategies involved the use of electron microscope grid (180,000°C/min). It was initially invented in the study on cattle oocytes (37) and has been subsequently used for vitrification of human oocytes in IVF practice (44–46); these studies reported births of healthy babies (Table 1). However, EM's limitation was cryostorage; advantageously, the cryoloop and cryoleaf then attempted to address this limitation. The former was developed by Lane et al. (1999) (127) whereby exceedingly high cooling rate (720,000°C/min) was reported; recent studies using this system have reported births of healthy babies (51,53). Likewise, a report on live birth using cryoleaf for vitrification of human oocytes has been published (41). In parallel, the hemi-straw, another robust innovation intended for vitrification of oocytes and embryos, was developed. Encouraging reports based on the use of hemi-straw techniques have been published for human oocytes (128) and human blastocysts (129,130). In the past, the author has also conducted studies that employed high-cooling-rate devices and developed designs of metal loop (Fig. 11A); we also carried out a study where stripper tips were utilized as a cooling device (Fig. 11D and E). However, this high-cooling-rate approach was not found to be particularly beneficial, at least in the in vitro trials for vitrification of embryos and human oocytes.

Although adaptation of some "open" containers has demonstrated varying degrees of success for human oocytes, these containers have also been associated with the use of inappropriate vitrification solutions. One of the major limitations of the cryotop (47) and open pulled straw (131) original protocols developed for bovine embryos/oocytes is the insufficient total solute concentration of the final solutions, which may cause devitrification (ice formation) on warming.

Alternatively, it has been demonstrated that a variety of containers, when used as a part of an appropriately designed vitrification protocol, are effective. Therefore, embryologists and clinicians should distinguish the difference between containers and vitrification protocols, which assumes correct pre-equilibration/dilution procedures and vitrification solutions. When examining scientific literature, one must pay attention to the rationale behind the vitrification protocol described in that study and separately to the type of container that has been utilized. In some cases, vitrification protocols and containers that are most appropriate for use can be "mixed and matched" from different studies by an IVF clinic to improve the efficacy of oocyte vitrification.

Armed with the information provided so far in this section, we can now return to exploring the vital difference between "open" and "sealed" containers. While developers of the hemi-straw and cryotop methods honestly alert the readers that their methods do not avoid the risk of contamination, others representing the so-called "sealed" containers have been known to imply that their "closed" container provides the same protection as a "sealed" container; this we now understand to be incorrect. Exceedingly high cooling rates have been reported for certain elaborate containers; however, it seems unlikely that the design of a small container, which comes into contact with liquid nitrogen enclosed in an outer container, could deliver the ultimate contamination-free system. For example, there is a design in which the inner straw is capped by a metal sleeve; the inner straw is then immersed directly into liquid nitrogen and afterwards inserted into an outer straw and sealed. Although the final product is indeed "sealed," the specimen has in fact already been exposed to liquid nitrogen and therefore the pathogens therein; therefore this method cannot be considered as safe as a "sealed" one. In other words, an exceedingly high cooling rate is achievable at the expense of compromising strict sterility.

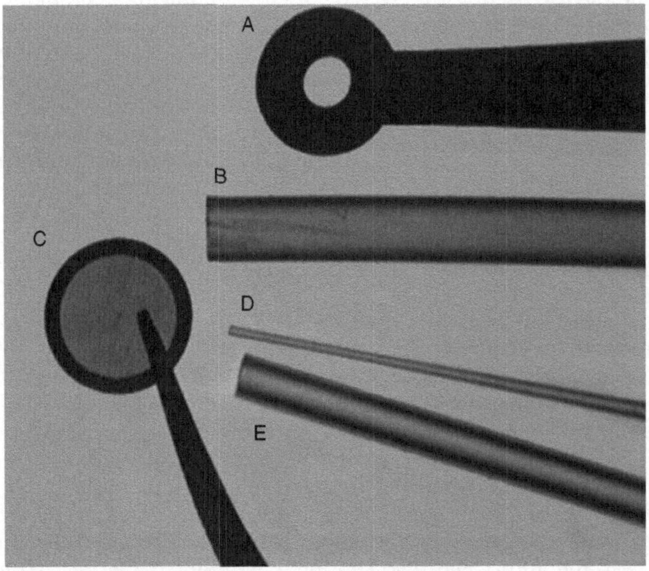

Figure 11 "Open" containers: (**A**) metal loop; (**B**) open pulled straw; (**C**) electron microscopy grid; (**D** and **E**) stripper tips.

Remarks on Containers and Miscellaneous Observations

At the crossroad of choosing between these two options, studies have reported that a very high cooling rate may not be necessary for the cryopreservation of human oocytes as they are more tolerant to chilling than bovine oocytes (132). The interest in achieving ultra-high cooling rates for cryopreservation of human oocytes seems to evolve from the application of this fact to bovine oocytes (37,133), as these mammalian oocytes are chilling sensitive, probably due to their high lipid content (93). As containers intend for human oocytes, their design evolved but became unnecessary complicated.

Storage of specimens in vapors of liquid nitrogen might also prevent cross contamination. Vapor storage presents great advantages in preserving large tissue specimens, but is less appealing for storing human oocytes, which have to be kept in small containers. Pathogenic contamination could be minimized or even eliminated through filtering the liquid nitrogen with a 0.2 μm filter or by UV irradiation of liquid nitrogen. It should be noted that sterilization of straws by irradiation can be considered beneficial in terms of sterility; however, it has negative effects on the straw's properties at low temperatures (134). Contaminating pathogens can be partially removed from cells and tissues by rigorous washing steps. As the methods outlined above have limited effectiveness, the "straw-in-straw" strategy for eliminating contamination appears to be the optimal solution (118).

A common misunderstanding is that specific vitrification procedures have to be employed in conjunction with a specific container; this is not so. Historically, certain studies have shown that in fact selection of an appropriate vitrification procedure is more instrumental to oocyte viability than the type of container. Embryologists and clinicians should be aware that when adopting protocols from specific scientific studies, the use of the recommended vitrification procedure does not mandate them to use the container recommended in the study. A poignant example of the above principle is a study that led to the first live birth from a vitrified oocyte (7). In that study, we used a container that was commonly used at the time; the key to success, however, was the inventive vitrification solution and procedure. The important differentiating factor in our vitrification protocol was that the procedure and solutions were adjusted to optimize shrinkage/expansion of the oocyte (Fig. 8). The total solute concentration (65.4 wt.%) of the final solution [40 vol.% EG + 0.6 mol/L sucrose (7)] was selected for the reason that it achieves a stable vitrification state during possible rewarming in air or during other applications, which is applicable with a wide range of containers. Concentrations, and a number of pre-equilibration solutions were also deduced for the benefit of oocyte physiology rather than in the interest of time and simplicity of the procedure. Our vitrification procedure was almost individually adjusted for each oocyte to determine the optimal duration in each pre-equilibration solution before final immersion into liquid nitrogen, as described in the section "Flexible Strategy." This lets us control the extent of penetration of the cryoprotectant into the cells. We have also deduced dilation steps via the oocyte's osmotic response during the dilution procedure.

Generally, as can be seen from the analysis of all reports on vitrification of human oocytes that lead to childbirth (Table 2), there is no distinct correspondence between success rate and the type of container. Container type is more reflective of chronology and geography; therefore, we believe that all containers are equally effective.

Is the Use of Serum/Proteins Necessary in Vitrification?

Another important issue that needs to be addressed is the employment of products of human origin, mainly serum or proteins, in the cocktail of cryopreservation solutions. The concern with human serum arises from possible involvement of infected donors/sources. Use of large amounts of serum is essential in traditional slow-cooling procedures (58,60,68,69,76), as serum plays an important role in supporting human oocytes during the slow and gradual decrease of temperature in the cryopreservation solution before solidification occurs. As the culture conditions require the exclusion of the use of serum containing medium, using serum as a part of the cryopreservation solution may be disadvantageous in the IVF cycle. During mixing of cryopreservation solution and culture medium, an exothermic reaction occurs. This overheating is particularly challenging for protein/serum-containing medium commonly used to prepare vitrification solutions.

Nevertheless, addition of foreign serum/proteins is still quite common in clinical practice for vitrification of human oocytes (Table 2), just to name a few, human serum (45,53), fetal bovine serum (46), or synthetic serum substitutes (47,50). In contrast to the slow-cooling method, vitrification of oocytes involves direct immersion into liquid nitrogen, which yields cooling rates ranging from hundreds of degrees per minute to thousands of degrees per minute. Hence, we deduced that supplementation with native or synthetic proteins is basically not necessary. We have further tested this theory by developing a novel polymer-based serum/protein-free solution for embryo cryopreservation (24,118) (Table 2). Subsequently, we refined the set of solutions that we developed for vitrification of human oocytes by elimination of serum/protein. We have tested this set of solutions in emerging biomedical areas involving stem cells and tissue engineering. The universality of this set (10 vol.% EG, 25 vol.% EG, 40 vol.% EG, 0.6 mol/L sucrose) was confirmed by application to the vitrification of neuronal stem cells (91,122,123,135) and cell-matrix (121,122), and cell-scaffold systems (91,122,123,135).

Our investigations have proved that this protein/serum-free strategy in combination with "straw-in-straw" vitrification has been successful as high viability and functionality were achieved (91,122,123,135). A combination of both approaches completely eliminates the risk of infections and contaminations.

CONCLUDING REMARKS

We hope that analysis presented in this chapter provides an idea why vitrification as an approach to cryopreservation is expected to be more appropriate for assisted reproductive technology as currently practiced. Therefore, we hope that embryologists will employ the vitrification procedures discussed in this chapter, which are likely to be both embryo and oocyte friendly. Since 1999, when the first baby girl was born from an embryo derived from vitrified-warmed oocytes in our first trial, many healthy babies were born after application of this strategy worldwide. Various procedures for the vitrification of oocytes have been developed, and high viability and pregnancy and birth rates have been achieved. The critical outlook on containers employed for vitrification revealed that they are equally effective. The "straw-in-straw" method using readily available 250- and 500-μl straws is a safe and effective strategy to eliminate direct contact with liquid nitrogen and to avoid risk of infection or contamination. Serum/proteins appear to be unnecessary in vitrification. To conclude, we believe that vitrification is the most promising way to move forward in the cryopreservation of human oocytes.

ACKNOWLEDGMENTS

I gratefully acknowledge the collaboration of my friends and colleagues in the field of cryopreservation who have been involved in my research over the decades. Otani and Odawara Women's Clinics in Japan and SISMER in Italy contributed to the IVF research presented in this chapter.

REFERENCES

1. Kuleshova LL, Lopata A. Vitrification can be more favorable than slow cooling. Fertility and Sterility 2002; 78: 449–54.
2. Rall WF, Fahy GM. Ice-free cryopreservation of mouse embryos at –196°C by vitrification. Nature 1985; 313: 573–5.
3. Kasai M, Komi JH, Takakamo A, et al. A simple method for mouse embryo cryopreservation in a low toxicity vitrification solution, without appreciable loss of viability. J Reprod Fertil 1990; 89: 91–7.
4. Kasai M, Hamaguchi Y, Zhu SE, et al. High survival of rabbit morulae after vitrification in an ethylene glycol-based solution by a simple method. Biol Reprod 1992; 46: 1042–6.
5. Kasai M, Zhu SE, Pedro PB, et al. Fracture damage of embryos and its prevention during vitrification and warming. Cryobiology 1996; 33: 459–64.
6. Mukaida T, Wada S, Takahashi K, et al. Vitrification of human embryos based on the assessment of suitable conditions for 8-cell mouse embryos. Hum Reprod 1998; 13: 2874–9.
7. Kuleshova L, Gianaroli L, Magli C, Ferraretti A, Trounson A. Birth following vitrification of a small number of human oocytes. Human Reproduction 1999; 14: 3077–9.
8. Mazur P. Principles of cryobiology. In: Fuller BJ, Lane N, Benson EE, eds. Life in the Frozen State. Boca Raton: CRC Press, 2004: 3–65.

9. Leibo SP. Preservation of ova and embryos by freezing. In: Brackett BG, Seidel GEJ, Seidel SM, eds. New Technologies in Animal Breeding. New York: Academic Press, 1981; 127–39.
10. Tammann G. Ueber die abhangkeit der Zahl der Kernr, welche sich in verschiedenen underkuhlten flusflussigkeiten. Z Phys Chem 1898; 25: 441–79.
11. Luyet BJ. Differential staining for living and dead cells. Science 1937; 85: 106.
12. Luyet BJ. The case against the cell theory. Science 1940; 91: 252–5.
13. Rapatz G, Luyet B. Electron microscope study of erythrocytes in rapidly cooled suspensions containing various concentrations of glycerol. Biodynamica 1968; 10: 193–210.
14. Fahy GM. Prospect for vitrification of whole organs. Cryobiology 1981; 18: 617.
15. Fahy GM. Vitrification. In: McGrath JJ, Diller KR, eds. In low temperature biotechnology: emerging applications and engineering contributions. New York: ASME, 1988: 113–46.
16. MacFarlane DR, Forsyth M. Recent insights on the role of cryoprotective agents in vitrification. Cryobiology 1990; 27: 345–58.
17. Rall WF. Factors affecting the survival of mouse embryos cryopreserved by vitrification. Cryobiology 1987; 24: 387–402.
18. Baudot A, Odagescu V. Thermal properties of ethylene glycol aqueous solutions. Cryobiology 2004; 48: 283–94.
19. Fahy GM, Levy DI, Ali SE. Some emerging principles underlying the physical-properties, biological actions, and utility of vitrification solutions. Cryobiology 1987; 24: 196–213.
20. Ishimori H, Saeki K, Inai M, et al. Vitrification of bovine embryos in a mixture of ethylene-glycol and dimethyl-sulfoxide. Theriogenology 1993; 40: 427–33.
21. Rall WF, Wood MJ. High in-vitro and in-vivo survival of day-3 mouse embryos vitrified or frozen in a nontoxic solution of glycerol and albumin. Journal of Reproduction and Fertility 1994; 101: 681–8.
22. Ali J, Shelton JN. Design of vitrification solutions for the cryopreservation of embryos. Journal of Reproduction and Fertility 1993; 99: 471–7.
23. Kuleshova LL, MacFarlane DR, Trounson AO, Shaw JM. Sugars exert a major influence on the vitrification properties of ethylene glycol based solutions and have low toxicity to embryos and oocytes. Cryobiology 1999; 38: 119–30.
24. Kuleshova LL, Shaw JM, Trounson AO. Studies on replacing most of the penetrating cryoprotectant by polymers for embryo cryopreservation. Cryobiology 2001; 43: 21–31.
25. Ali J. Highly efficient ultrarapid cryopreservation of established cell lines by vitrification with VS14. Medical Science Research 1996; 24: 837–8.
26. Valdez CA, Abas Mazni O, Takahashi Y, Fujikawa S, Kanagawa H. Successful cryopreservation of mouse blastocysts using a new vitrification solution. J Reprod Fertil 1992; 96: 793–802.
27. Hotamisligil S, Toner M, Powers RD. Changes in membrane integrity, cytoskeletal structure, and developmental potential of murine oocytes after vitrification in ethylene glycol. Biology of Reproduction 1996; 55: 161–8.
28. Bautista JAN, Dela Pena EC, Katagiri S, Takahashi Y, Kanagawa H. In vitro viability of mouse oocytes vitrified in an ethylene glycol-based solution. Japanese Journal of Veterinary Research 1998; 46: 13–8.
29. Shaw JM, Kuleshova LL, MacFarlane DR, Trounson AO. Vitrification properties of solutions of ethylene glycol in saline containing PVP, Ficoll, or dextran. Cryobiology 1997; 35: 219–29.
30. Chung HM, Hong SW, Lim JM, et al. In vitro blastocyst formation of human oocytes obtained from unstimulated and stimulated cycles after vitrification at various maturational stages. Fertil Steril 2000; 73: 545–51.
31. Tada N, Sato M, Amann E, Ogawa S. A simple and rapid method for cryopreservation of mouse 2-cell embryos by vitrification: beneficial effect of sucrose and raffinose on their cryosurvival rate. Theriogenology 1993; 40: 333–44.
32. Saha S, Otoi T, Takagi M, et al. Normal calves obtained after direct transfer of vitrified bovine embryos using ethylene glycol, trehalose, and polyvinylpyrrolidone. Cryobiology 1996; 33: 291–9.
33. Saito N, Imai K, Tomizawa M. Effect of sugars-addition on the survival of vitrified bovine blastocysts produced in vitro. Theriogenology 1994; 41: 1053–60.
34. Chen SU, Lien YR, Cheng YY, et al. Vitrification of mouse oocytes using closed pulled straws (CPS) achieves a high survival and preserves good patterns of meiotic spindles, compared with conventional straws, open pulled straws (OPS) and grids. Hum Reprod 2001; 16: 2350–6.
35. Meryman HT. Absence of unfrozen freezable water in rapidly frozen red cells. Cryobiology 1970; 7: 252–5.
36. Meryman HT. Osmotic stress as a mechanism of freezing injury. Cryobiology 1971; 8: 489–500.
37. Martino A, Songsasen N, Leibo SP. Development into blastocysts of bovine oocytes cryopreserved by ultra-rapid cooling. Biology of Reproduction 1996; 54: 1059–69.

38. Paynter SJ, O'Neil L, Fuller BJ, Shaw RW. Membrane permeability of human oocytes in the presence of the cryoprotectant propane-1,2-diol. Fertil Steril 2001; 75: 532–8.
39. Paynter SJ, Cooper A, Gregory L, Fuller BJ, Shaw RW. Permeability characteristics of human oocytes in the presence of the cryoprotectant dimethylsulphoxide. Human Reproduction 1999; 14: 2338–42.
40. Mullen SF, Li M, Li Y, Chen ZJ, Critser JK. Human oocyte vitrification: the permeability of metaphase II oocytes to water and ethylene glycol and the appliance toward vitrification. Fertil Steril 2008; 89: 1812–25.
41. Chian RC, Son WY, Huang JY, et al. High survival rates and pregnancies of human oocytes following vitrification: preliminary report. Conjoint Annual Meeting of the American-Society-for-Reproductive-Medicine/Canadian-Fertility-and-Andrology-Society; 2005 Oct 15-19; Montreal, Canada.
42. Chian RC, Gilbert L, Huang JY, et al. Live birth after vitrification of in vitro matured human oocytes. Fertil Steril 2008.
43. Katayama KP, Stehlik J, Kuwayama M, Kato O, Stehlik E. High survival rate of vitrified human oocytes results in clinical pregnancy. Fertility and Sterility 2003; 80: 223–4.
44. Kim TJ, Hong SW, Park SE, Cha KY, eds. Pregnancy after vitrification of human oocytes and blastocysts using same cryoprotectant solution, ethylene glycol, and sucrose. 59th Annual Meeting of the American Society for Reproductive Medicine, October 11–15, 2003, San Antonio, Texas.
45. Wu J, Zhang L, Wang X. In vitro maturation, fertilization and embryo development after ultrarapid freezing of immature human oocytes. Reproduction 2001; 121: 389–93.
46. Yoon TK, Kim TJ, Park SE, et al. Live births after vitrification of oocytes in a stimulated in vitro fertilization-embryo transfer program. Fertil Steril 2003; 79: 1323–6.
47. Kuwayama M, Vajta G, Kato O, Leibo SP. Highly efficient vitrification method for cryopreservation of human oocytes. Reprod Biomed Online 2005; 11: 300–8.
48. Kyono K, Fuchinoue K, Yagi A, et al. Successful pregnancy and delivery after transfer of a single blastocyst derived from a vitrified mature human oocyte. Fertil Steril 2005; 84: 1017.e5–6.
49. Lucena E, Bernal DP, Lucena C, et al. Successful ongoing pregnancies after vitrification of oocytes. Fertil Steril 2006; 85: 108–11.
50. Selman H, Angelini A, Barnocchi N, et al. Ongoing pregnancies after vitrification of human oocytes using a combined solution of ethylene glycol and dimethyl sulfoxide. Fertil Steril 2006; 86: 997–1000.
51. Antinori M, Licata E, Dani G, et al. Cryotop vitrification of human oocytes results in high survival rate and healthy deliveries. Reprod Biomed Online 2007; 14: 72–9.
52. Yoon TK, Lee DR, Cha SK, et al. Survival rate of human oocytes and pregnancy outcome after vitrification using slush nitrogen in assisted reproductive technologies. Fertil Steril 2007; 88: 952–6.
53. Chen GA, Cai XY, Lian Y, et al. Normal birth from cryopreserved embryos after intracytoplasmic sperm injection of frozen semen into vitrified human oocytes. Hum Fertil (Camb) 2008; 11: 49–51.
54. Novikov AN, Kuleshova LG, Rozanov LF. Modeling of intracellular crystallization Biofizika 1980; 25: 129–33.
55. Kuwayama M. Highly efficient vitrification for cryopreservation of human oocytes and embryos: the Cryotop method. Theriogenology 2007; 67: 73–80.
56. Oktay K, Cil AP, Bang H. Efficiency of oocyte cryopreservation: a meta-analysis. Fertil Steril 2006; 86: 70–80.
57. Bianchi V, Coticchio G, Distratis V, et al. Differential sucrose concentration during dehydration (0.2 mol/L) and rehydration (0.3 mol/L) increases the implantation rate of frozen human oocytes. Reprod Biomed Online 2007; 14: 64–71.
58. Chen SU, Lien YR, Chen HF, et al. Observational clinical follow-up of oocyte cryopreservation using a slow-freezing method with 1,2-propanediol plus sucrose followed by ICSI. Hum Reprod 2005; 20: 1975–80.
59. De Santis L, Cino I, Rabellotti E, et al. Oocyte cryopreservation: clinical outcome of slow-cooling protocols differing in sucrose concentration. Reprod Biomed Online 2007; 14: 57–63.
60. Fosas N, Marina F, Torres PJ, et al. The births of five Spanish babies from cryopreserved donated oocytes. Hum Reprod 2003; 18: 1417–21.
61. Gook DA, Edgar DH. Human oocyte cryopreservation. Hum Reprod Update 2007; 13: 591–605.
62. Li XH, Chen SU, Zhang X, et al. Cryopreserved oocytes of infertile couples undergoing assisted reproductive technology could be an important source of oocyte donation: a clinical report of successful pregnancies. Hum Reprod 2005; 20: 3390–4.
63. Tjer GC, Chiu TT, Cheung LP, Lok IH, Haines CJ. Birth of a healthy baby after transfer of blastocysts derived from cryopreserved human oocytes fertilized with frozen spermatozoa. Fertil Steril 2005; 83: 1547–9.
64. Tucker M, Wright G, Morton P, et al. Preliminary experience with human oocyte cryopreservation using 1,2-propanediol and sucrose. Hum Reprod 1996; 11: 1513–5.

65. Allan J. Re: case report: Pregnancy from intracytoplasmic injection of a frozen-thawed oocyte. Australian & New Zealand Journal of Obstetrics & Gynaecology 2004; 44: 588.
66. Borini A, Sciajno R, Bianchi V, et al. Clinical outcome of oocyte cryopreservation after slow cooling with a protocol utilizing a high sucrose concentration. Hum Reprod 2006; 21: 512–7.
67. Borini A, Lagalla C, Bonu MA, et al. Cumulative pregnancy rates resulting from the use of fresh and frozen oocytes: 7 years' experience. Reprod Biomed Online 2006; 12: 481–6.
68. Chamayou S, Alecci C, Ragolia C, et al. Comparison of in-vitro outcomes from cryopreserved oocytes and sibling fresh oocytes. Reprod Biomed Online 2006; 12: 730–6.
69. Chia CM, Chan WB, Quah E, Cheng LC. Triploid pregnancy after ICSI of frozen testicular spermatozoa into cryopreserved human oocytes: case report. Hum Reprod 2000; 15: 1962–4.
70. Huttelova R, Becvarova V, Brachtlova T. More successful oocyte freezing. J Assist Reprod Genet 2003; 20: 293.
71. Konc J, Kanyo K, Cseh S. Does oocyte cryopreservation have a future in Hungary? Reprod Biomed Online 2007; 14: 11–3.
72. Kyono K, Fukunaga N, Haigo K, Chiba S, Sato T. Pregnancy and delivery of a healthy female infant after intracytoplasmic sperm injection into cryopreserved human oocytes. Japanese Journal of Fertility and Sterility 2001; 43: 171–7.
73. Levi Setti PE, Albani E, Novara PV, Cesana A, Morreale G. Cryopreservation of supernumerary oocytes in IVF/ICSI cycles. Hum Reprod 2006; 21: 370–5.
74. Miller KA, Elkind-Hirsch K, Levy B, et al. Pregnancy after cryopreservation of donor oocytes and preimplantation genetic diagnosis of embryos in a patient with ovarian failure. Fertil Steril 2004; 82: 211–4.
75. Montag M, van der Van k, Dorn C, et al. Birth after double cryopreservation of human oocytes at metaphase II and pronuclear stages. Fertil Steril 2006; 85:751.e5–e7.
76. Nawroth F, Kissing K. Pregnancy after intracytoplasmatic sperm injection (ICSI) of cryopreserved human oocytes. Acta Obstet Gynecol Scand 1998; 77: 462–3.
77. Notrica J, Kanzepolsky L, Divita A, Neuspiller F, de Fried EP, eds. A healthy female born after ICSI of a cryopreserved oocyte and cryopreserved spermatozoa banked prior to radiotherapy in a patient with a seminoma: A case report. 59th Annual Meeting of the American Society for Reproductive Medicine, October 11–15, 2003, San Antonio, Texas.
78. Polak de Fried E, Notrica J, Rubinstein M, Marazzi A. Oocyte cryopreservation program: Removal vs non removal of cumulus corona complex. Fertil Steril 1998; 70(Suppl 1): S148.
79. Porcu E, Fabbri R, Seracchioli R, et al. Birth of a healthy female after intracytoplasmic sperm injection of cryopreserved human oocytes. Fertil Steril 1997; 68: 724–6.
80. Porcu E, Fabbri R, Ciotti PM, et al. Ongoing pregnancy after intracytoplasmic sperm injection of epididymal spermatozoa into cryopreserved human oocytes. J Assist Reprod Genet 1999; 16: 283–5.
81. Porcu E, Fabbri R, Damiano G, et al. Clinical experience and applications of oocyte cryopreservation. Mol Cell Endocrinol 2000; 169: 33–7.
82. Porcu E, Fabbri R, Ciotti PM, et al. Oocytes or embryo storage? 58th Annual Meeting of the American Society for Reproductive Medicine, October 12–17, 2002, Seattle, Washington.
83. La Sala GB, Nicoli A, Villani MT, et al. Outcome of 518 salvage oocyte-cryopreservation cycles performed as a routine procedure in an in vitro fertilization program. Fertil Steril 2006; 86: 1423–7.
84. Tucker MJ, Wright G, Morton PC, Massey JB. Birth after cryopreservation of immature oocytes with subsequent in vitro maturation. Fertil Steril 1998; 70: 578–9.
85. Winslow KL, Yang D, Blohm PL, et al. Oocyte cryopreservation—a three year follow up of sixteen births. Fertil Steril 2001; 76: S120–1.
86. Wurfel W, Schleyer M, Krusmann G, Hertwig IV, Fiedler K. [Fertilization of cryopreserved and thawed human oocytes (Cryo-Oo) by injection of spermatozoa (ICSI)--medical management of sterility and case report of a twin pregnancy]. Zentralbl Gynakol 1999; 121: 444–8.
87. Yang KS, Goo XL, Meng W, et al. Behavior of hepatocytes inoculated in gelatin-immobilized polyurethane foam. Macromolecular Research 2003; 11: 488–94.
88. Young E, Kenny A, Puigdomenech E, et al. Triplet pregnancy after intracytoplasmic sperm injection of cryopreserved oocytes: case report. Fertil Steril 1998; 70: 360–1.
89. Chen C. Pregnancy after human oocyte cryopreservation. Lancet 1986; 1: 884–6.
90. Chian RC, Huang JY, Tan SL, et al. Obstetric and perinatal outcome in 200 infants conceived from vitrified oocytes. Reprod Biomed Online 2008; 16: 608–10.
91. Magalhães R, Wang X, Gouk S, et al. Vitrification successfully preserves hepatocyte spheroids. Cell Transplantation 2008; 17: 813–28.
92. Boutron P, Kaufmann A. Stability of the amorphous state in the system water 1,2-propanediol. Cryobiology 1979; 16: 557–68.

93. Nagashima H, Kashiwazaki N, Ashman RJ, et al. Removal of cytoplasmic lipid enhances the tolerance of porcine embryos to chilling. Biol Reprod 1994; 51: 618–22.
94. Keskintepe L, Agca Y, Sher G, Keskintepe M, Maassarani G. High survival rate of metaphase II human oocytes after first polar body biopsy and vitrification: determining the effect of previtrification conditions. Fertil Steril 2008.
95. Isachenko EF, Nayudu PL. Vitrification of mouse germinal vesicle oocytes: effect of treatment temperature and egg yolk on chromatin and spindle normality and cumulus integrity. Hum Reprod 1999; 14: 400–8.
96. Miyake T, Kasai M, Zhu SE, Sakurai T, Machida T. Vitrification of mouse oocytes and embryos at various stages of development in an ethylene glycol-based solution by a simple method. Theriogenology 1993; 40: 121–34.
97. Nakagata N. High survival rate of unfertilized mouse oocytes after vitrification. Journal of Reproduction and Fertility 1989; 87: 479–83.
98. Oneil L, Paynter SJ, Fuller BJ, Shaw RW. Vitrification of mature mouse oocytes: improved results following addition of polyethylene glycol to a dimethyl sulfoxide solution. Cryobiology 1997; 34: 295–301.
99. Fabbri R, Porcu E, Marsella T, et al. Human oocyte cryopreservation: new perspectives regarding oocyte survival. Hum Reprod 2001; 16: 411–6.
100. Bernard A, Fuller BJ. Cryopreservation of human oocytes: a review of current problems and perspectives. Hum Reprod Update 1996; 2: 193–207.
101. Pickering SJ, Braude PR, Johnson MH, Cant A, Currie J. Transient cooling to room-temperature can cause irreversible disruption of the meiotic spindle in the human oocyte. Fertility and Sterility 1990; 54: 102–8.
102. Al-Hasani S, Diedrich K, van der Ven H, et al. Cryopreservation of human oocytes. Hum Reprod 1987; 2: 695–700.
103. Van Blerkom J, Davis PW. Cytogenetic, cellular, and developmental consequences of cryopreservation of immature and mature mouse and human oocytes. Microsc Res Tech 1994; 27: 165–93.
104. Hong SW, Chung HM, Lim JM, et al. Improved human oocyte development after vitrification: a comparison of thawing methods. Fertil Steril 1999; 72: 142–6.
105. Son WY, Park SE, Lee KA, et al. Effects of 1,2-propanediol and freezing-thawing on the in vitro developmental capacity of human immature oocytes. Fertil Steril 1996; 66: 995–9.
106. Chen SU, Lien YR, Chao K, et al. Cryopreservation of mature human oocytes by vitrification with ethylene glycol in straws. Fertil Steril 2000; 74: 804–8.
107. Chian RC, Huang JY, Gilbert L, et al. Obstetric outcomes following vitrification of in vitro and in vivo matured oocytes. Fertil Steril 2008.
108. Son WY, Chung JT, Demirtas E, et al. Comparison of in-vitro maturation cycles with and without in-vivo matured oocytes retrieved. Reprod Biomed Online 2008; 17: 59–67.
109. Cha KY, Chian RC. Maturation in vitro of immature human oocytes for clinical use. Hum Reprod Update 1998; 4: 103–20.
110. Goud PT, Goud AP, Qian C, et al. In-vitro maturation of human germinal vesicle stage oocytes: role of cumulus cells and epidermal growth factor in the culture medium. Hum Reprod 1998; 13: 1638–44.
111. Naether OG, Rudolf K, Fischer R, et al. Pregnancy after vitrification of pronuclear stage oocytes biopsied for polar body aneuploidy screening. Reprod Biomed Online 2008; 16: 268–70.
112. Zheng WT, Zhuang GL, Zhou CQ, et al. Comparison of the survival of human biopsied embryos after cryopreservation with four different methods using non-transferable embryos. Hum Reprod 2005; 20: 1615–8.
113. Joris H, Van den Abbeel E, Vos AD, Van Steirteghem A. Reduced survival after human embryo biopsy and subsequent cryopreservation. Hum Reprod 1999; 14: 2833–7.
114. Cobo A, Kuwayama M, Perez S, et al. Comparison of concomitant outcome achieved with fresh and cryopreserved donor oocytes vitrified by the Cryotop method. Fertil Steril 2008; 89: 1657–64.
115. Fountain D, Ralston M, Higgins N, et al. Liquid nitrogen freezers: a potential source of microbial contamination of hematopoietic stem cell components. Transfusion 1997; 37: 585–91.
116. Tedder RS, Zuckerman MA, Goldstone AH, et al. Hepatitis B transmission from contaminated cryopreservation tank. Lancet 1995; 346: 137–40.
117. Bielanski A, Nadin-Davis S, Sapp T, Lutze-Wallace C. Viral contamination of embryos cryopreserved in liquid nitrogen. Cryobiology 2000; 40: 110–6.
118. Kuleshova LL, Shaw JM. A strategy for rapid cooling of mouse embryos within a double straw to eliminate the risk of contamination during storage in liquid nitrogen. Human Reproduction 2000; 15: 2604–9.

119. Kuleshova LL, Otani T, Takeda N, Odawara Y. The evolution and current status in vitrification of oocytes and embryos. Cryobiology 2004; 49: 1.
120. Kuleshova L, Takeda N, Odawara Y. Successful vitrification of human embryos using a modified straw-in-straw method. Cryobiology 2003; 47: 270–1.
121. Kuleshova LL, Wang XW, Wu YN, Zhou Y, Yu H. Vitrification of encapsulated hepatocytes with reduced cooling and warming rates. Cryo Letters 2004; 25: 241–54.
122. Wu YN, Yu HR, Chang S, Magalhaes R, Kuleshova LL. Vitreous cryopreservation of cell-biomaterial constructs involving encapsulated hepatocytes. Tissue Engineering 2007; 13: 649–58.
123. Tan F, Lee K, Gouk S, et al. Optimization of cryopreservation of stem cells cultured as neurospheres: comparison between vitrification, slow-cooling and rapid cooling, "freezing" protocols. CryoLetters 2007; 28: 445–60.
124. Isachenko V, Katkov I, Yakovenko S, et al. Vitrification of human laser treated blastocysts within cut standard straws (CSS): novel aseptic packaging and reduced concentrations of cryoprotectants. Cryobiology 2007; 54: 305–9.
125. Isachenko V, Montag M, Isachenko E, et al. Aseptic technology of vitrification of human pronuclear oocytes using open-pulled straws. Hum Reprod 2005; 20: 492–6.
126. Yokota Y, Sato S, Yokota M, Yokota H, Araki Y. Birth of a healthy baby following vitrification of human blastocysts. Fertil Steril 2001; 75: 1027–9.
127. Lane M, Schoolcraft WB, Gardner DK. Vitrification of mouse and human blastocysts using a novel cryoloop container-less technique. Fertil Steril 1999; 72: 1073–8.
128. Vanderzwalmen P, Bertin G, Debauche C, Standaart V, Schoysman E. "In Vitro" survival of metaphase II oocytes (MII) and blastocysts after vitrification in a hemi-straw (HS) system. Fertility and Sterility 2000; 74(Suppl 1): S215–S6.
129. Vandervorst M, Vanderzwalmen P, Standaart V, et al. Blastocyst transfer after vitrification in a hemi-straw (HS) system. Human Reproduction 2001; 16(Suppl): 153–4.
130. Vanderzwalmen P, Bertin G, Debauche C, et al. Vitrification of human blastocysts with the hemi-straw carrier: application of assisted hatching after thawing. Hum Reprod 2003; 18: 1504–11.
131. Vajta G, Holm P, Greve T, Callesen H. Vitrification of porcine embryos using the Open Pulled Straw (OPS) method. Acta Vet Scand 1997; 38: 349–52.
132. Bernard A, Hunter JE, Fuller BJ, et al. Fertilization and embryonic development of human oocytes after cooling. Hum Reprod 1992; 7: 1447–50.
133. Vajta G, Holm P, Kuwayama M, et al. Open Pulled Straw (OPS) vitrification: a new way to reduce cryoinjuries of bovine ova and embryos. Mol Reprod Dev 1998; 51: 53–8.
134. Shaw JM, Diotallevi L, Trounson A. Ultrarapid embryo freezing: effect of dissolved gas and pH of the freezing solutions and straw irradiation. Hum Reprod 1988; 3: 905–8.
135. Kuleshova LL, Tan CKF, Magalhães R, et al. Effective cryopreservation of neuronal stem or progenitor cells without serum or proteins by vitrification. Cell Transplantation 2009; 18: in press.

5 Predictive Models for the Development of Improved Cryopreservation Protocols for Human Oocytes

John McGrath

Department of Aerospace and Mechanical Engineering, University of Arizona, Tucson, Arizona, U.S.A.

INTRODUCTION

The development of protocols to cryopreserve human oocytes has been dominated by an empirical approach. While the result of this approach has been impressive in the sense that human beings have been born from cryopreserved oocytes, there is much room for improvement in several respects. For material as precious as human oocytes, the fraction of oocytes recovered after cryopreservation must be increased. In addition, the factors leading to non-reproducible results need to be identified with the goal of minimizing or eliminating non-reproducibility. Finally, an improved understanding of the basic principles governing the response of human oocytes to the complex processing involved in cryopreservation protocols is also necessary.

Models based on basic principles have been developed, which can be used to predict the response of cells, including human oocytes, to cryopreservation processing. While these models certainly have limitations, they have enormous potential. The potential benefits include improved recovery and reproducibility, faster processing, and lower cost—all based on a fundamental understanding rather than empiricism.

Computer models exist that can predict the response of cells including human oocytes to the addition and removal of cryoprotectant agents (CPAs) (1–5). The dynamic response as well as the equilibrium state can be predicted. Such models can account for oocyte size, type and concentration of CPA as well as temperature and exposure time. The influence of arbitrary protocols, including single or multiple step CPA addition and removal can be predicted. The presence of impermeable extracellular solutes can be included. The oocyte volume as well as intracellular concentrations of water, CPA, and impermeable solute (modeled here as a salt) can be predicted as a function of time for arbitrary processing conditions of interest.

Similarly, computer models can predict how cells including oocytes will respond to freezing and thawing (6–10). Specifically, it is possible to predict oocyte volume and intracellular water volume as well as the concentration of CPA and impermeable solutes during cooling, storage, and warming. The extent to which intracellular water is supercooled can be predicted. How the seeding temperature and the equilibration time at the seeding temperature influence subsequent response during freezing, storage, and warming can be calculated.

Model predictions of the probability of intracellular ice formation and growth of intracellular crystals for specified freezing protocols are also possible (9). The terms "probability of intracellular ice formation" and "cumulative fraction of intracellular ice formation" in a population are used to mean the same thing in this chapter. Such models make it possible to quantify the amount and size of intracellular crystals. This information can be used to define when vitrification can be achieved as well as provide details concerning devitrification and recrystallization (10). In fact, it is possible to distinguish different types of intracellular ice formation (surface-catalyzed and volume-catalyzed heterogeneous as well as homogeneous nucleation) and the corresponding temperatures using these models (7). There have been reports in the literature that non-linear cooling rates produce improved cell recovery compared to recovery using linear cooling (11). It is possible to predict differences between oocyte responses to linear and non-linear cooling using modeling of the type presented here (12).

The response of human oocytes to the various aspects of cryopreservation processing and how the intrinsic characteristics of the oocytes (e.g., membrane permeability) influence this response is described elsewhere in this book (see chap. 3 by Fuller and Paynter).

In this chapter, a specific cryopreservation protocol developed for human oocyte cryopreservation (13) is used as a vehicle to illustrate what can be done at the present time with respect to predictive modeling and what limitations must be addressed to realize the full potential of such a modeling approach. It should be recognized that a range of the intrinsic properties of human oocytes will lead to a range of oocyte responses in a population. Although this diversity of responses is not modeled in this chapter, it should be clear after reading this chapter that another virtue of the modeling approach presented here is that it is possible to perform sensitivity studies using the computer model to predict how much variation in particular responses of interest would arise due to different oocyte properties or combination of properties.

The specific protocol used to illustrate what can be done using computer modeling is defined in Table 1.

One important point to be made here is that, in principle, the predictive modeling approach described in this chapter links all the cryopreservation processes together. To emphasize this point, this means that the final state at the end of each of the processes defined above represents the initial state of the subsequent process. Most modeling work related to cryopreservation has focused on individual processes or a few coupled processes in the overall protocol rather than linking the entire process. There has been some published work that does link more of the processing steps together (4,8).

Indeed, not all the processing steps defined in Table 1 are modeled here. Part of the reason is that important parameters that are required to model some aspects of the overall process are still not known for human oocytes. Another reason is related to the length limitations defined for this chapter. Nevertheless, it is hoped that the material presented in this chapter will allow the reader to appreciate the potential of this modeling approach and to understand what limitations still exist with regard to realizing the full potential of such modeling.

The addition steps for both cryoprotectant and impermeable solute (sucrose) are included and linked together in the model presented. These two steps are then linked to all five steps involved in the removal of cryoprotectant and impermeable solute. Although the removal of cryoprotectant occurs as the last part of the actual overall cryopreservation process, the predicted response of cryoprotectant removal is coupled directly to the predictions for cryoprotectant addition and presented immediately after the results for the addition of cryoprotectant and impermeable solute. This is done for several reasons. From a practical standpoint, if cryoprotectant cannot be added and removed with little or no damage, the overall cryopreservation process will not be successful. If there is significant damage by simply adding and removing cryoprotectant, the detailed aspects of seeding, cooling, or warming for purposes of modeling or otherwise are of secondary importance until the reasons for damage are defined and minimized or eliminated. Second, presenting all the results at the same time for adding and removing cryoprotectant and sucrose should make direct comparison of some results more

Table 1 Cryopreservation Protocol (13)

Add propane-1,2-diol in a single step at room temperature (22°C assumed)
• 0.0 to 1.5 M, exposure for 10 min
Add sucrose in a single step at room temperature (22°C assumed)
• 0.0 to 0.3 osm, exposure for 5 min
Cool to −7°C at 2°C/min
Seed at −7°C and hold at this temperature for 10 min
Cool at 0.3°C/min from −7°C to −30°C
Cool at 50°C/min from −30°C to −150°C
Hold at −150°C for 10 to 12 min
Transfer to liquid nitrogen
Warm by exposing to air for 30 sec at room temperature and then 40 sec in water at 30°C
Dilute propane-1,2-diol in three steps at room temperature (22°C assumed)
• 1.5 to 1.0 M, exposure for 5 min
• 1.0 to 0.5 M, exposure for 5 min
• 0.5 to 0.0 M, exposure for 10 min
Dilute sucrose in two steps
• 0.3 to 0.0 osm, exposure at room temperature (22°C assumed) for 10 min
• 0.0 osm, exposure at 37°C for 10 min

obvious compared to presenting the results for removal at the end of the entire cryopreservation protocol. The addition and removal of propane-1,2-diol is considered using the membrane permeability parameters published for human oocytes (14). The addition of sucrose is predicted using the membrane permeability parameters published for human oocytes (15). The assumed size of the oocyte (110 μm) and the osmotically inactive volume (19%) are taken from published work (16).

Cooling from 22°C to –7°C is not modeled because it is assumed that extracellular ice does not form during this process until it is seeded. Thus, there would be no driving force for the flow of water or cryoprotectant. It is also assumed that intracellular ice does not form during this cooling process.

The response of human oocytes to extracellular seeding at –7°C followed by holding for 10 minutes at this temperature is modeled. The membrane permeability parameters required for this are defined by extrapolating the published values of human oocyte permeability at higher temperature using the activation energies for water and cryoprotectant permeability (14).

It is possible to predict the probability of intracellular ice formation if two intracellular ice nucleation parameters are known. These parameters have been determined experimentally for several cell types including mouse oocytes in the absence (7) and presence (8,17) of cryoprotectants. Unfortunately, these parameters have not yet been determined for human oocytes. To illustrate what can be achieved with the modeling approach presented here, it is assumed that the nucleation parameters known for mouse oocytes apply to human oocytes as discussed above. It is shown that by making this assumption, most of the general trends known empirically are captured by the model predictions.

To illustrate what can be done for modeling the freezing of human oocytes, it was of interest to include at least the first cooling step in the cryopreservation protocol defined in Table 1. The protocol defined in Table 1 for this first step is cooling at a rate of 0.3°C/min from –7°C to –30°C. However, it was found using the present model that the predicted probability of intracellular ice formation did not match the published experimental results reported by Trad et al. (18), when using the mouse oocyte intracellular ice nucleation parameters in place of the unknown human oocyte parameters. To handle this issue, the model was "tuned" by varying the cooling rate to match the predictions with the experimental results for the probability of intracellular ice formation as a function of seed temperature reported by Trad et al. (18). This exercise indicated that a cooling rate of approximately 1.0°C/min would be required. Thus, the 0.3°C/min cooling rate defined in Table 1 was replaced by a cooling rate of 1.0°C/min to compensate for the unknown human oocyte intracellular ice nucleation parameters.

As will be seen below, this approach produces many interesting and useful results. Using a cooling rate of 1.0°C/min, predictions of intracellular water content as well as salt and cryoprotectant concentrations are made using the permeabilities described above. Modeling the other two cooling processes to the storage temperature and the warming process described in Table 1 are not included, largely due to the limited scope of this chapter.

Model predictions are included that reveal how increasing the concentration of sucrose influences the response of human oocytes to freezing. Results are also presented that predict how the seed temperature as well as the hold time at the seed temperature prior to initiating cooling influence the response of human oocytes to subsequent freezing.

It is well known that the extracellular concentrations of cryoprotectant and sucrose, to which the human oocyte is exposed during the addition and removal of these solution components, play a crucial role in the response of the oocyte to the overall cryopreservation protocol. What may not be well appreciated is how important the details of the small-scale mixing of solutions are during the transfer of oocytes between solutions and how sensitive the oocyte response can be to these details. The lack of control of mixing may play an important role in non-reproducible results. Simple illustrative model predictions are offered to illustrate the point.

RESPONSE OF HUMAN OOCYTES TO THE ADDITION OF PROPANE-1,2-DIOL AND SUCROSE

Although the concentration units of the cryoprotectant in the specific cryopreservation protocol of Fabbri et al. (13) defined in Table 1 are defined in terms of molarity, the computer simulations were performed by converting all molarity units to osmolality by equating molarity and osmolality. For the purposes of the simulations presented here, the difference is not considered important.

As shown in Figure 1, exposure to 1.5 osm propane-1,2-diol causes the human oocyte to shrink to approximately 70% of isotonic volume in approximately 1 minute, followed by swelling to approximately isotonic volume within another 10 minutes. Model results not shown reveal that equilibration would produce an oocyte at 108% isotonic volume after some 30 minutes of exposure to the cryoprotectant.

Following the 10-minute exposure to 1.5 osm propane-1,2-diol, exposure to 0.3 osm sucrose for 5 minutes causes the human oocyte to shrink to approximately 70% of isotonic volume (Fig. 1). Another 5 minutes of exposure would produce the equilibrium volume of 64% of isotonic volume (results not shown).

Other parameters related to the exposure to propane-1,2-diol can be predicted. Intracellular water volume is reduced and then increases, paralleling the cell volume (Fig. 2). Exposure to sucrose squeezes water out of the cell such that about 60% of the isotonic water volume remains in the human oocyte at the end of 5 minutes of exposure to sucrose.

By the end of the 10-minute exposure to propane-1,2-diol, the intracellular propane-1,2-diol concentration nearly reaches the equilibrium value of 1.5 osm (Fig. 3). The subsequent 5-minute exposure to 0.3 osm sucrose produces a rise and fall of the intracellular concentration of propane-1,2-diol, reaching a concentration of approximately 1.6 osm at the end of the 5-minute exposure.

The intracellular salt concentration rises to nearly 0.55 osm and then decreases to near the isotonic value of 0.29 osm during the 10-minute exposure to propane-1,2-diol (Fig. 4). The subsequent 5-minute exposure to 0.3 osm sucrose increases the intracellular concentration of salt significantly, to about 0.5 osm.

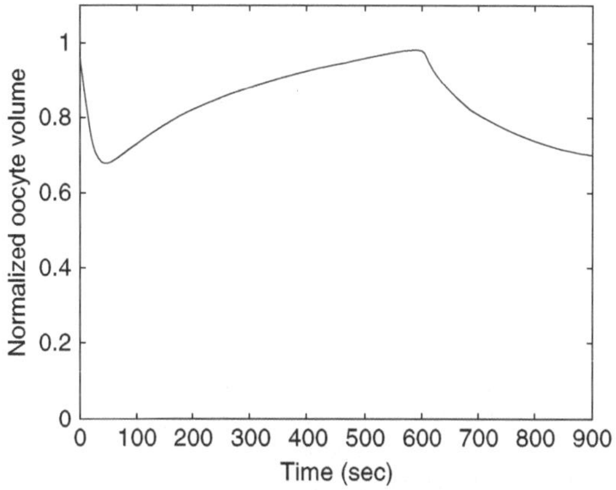

Figure 1 Response of human oocyte normalized cell volume to exposure to 1.5 osm propane-1,2-diol followed by 0.3 osm sucrose.

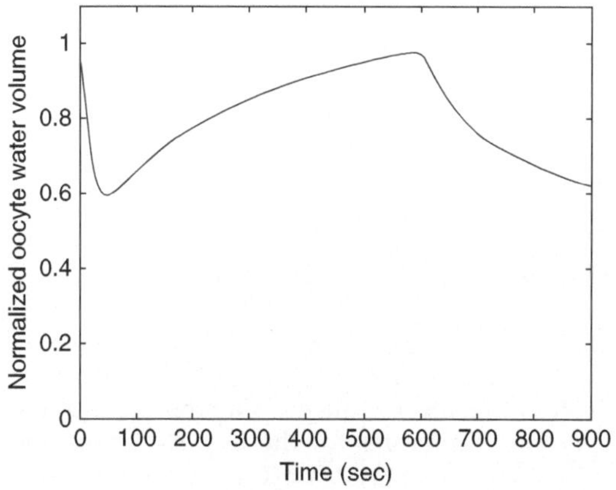

Figure 2 Response of human oocyte normalized water volume to exposure to 1.5 osm propane-1,2-diol followed by 0.3 osm sucrose.

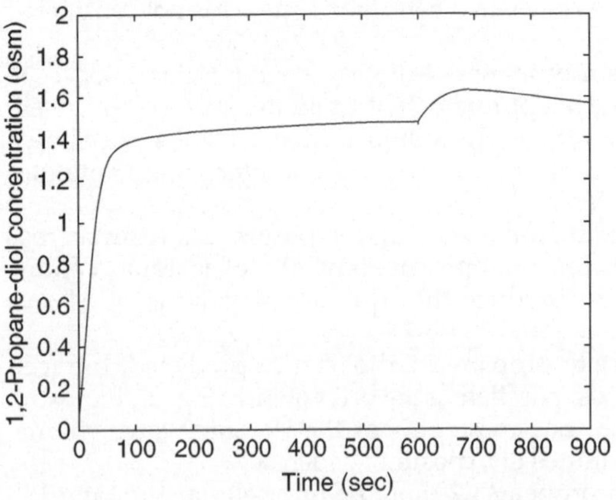

Figure 3 History of intracellular concentration of propane-1,2-diol for human oocyte exposed to 1.5 osm propane-1,2-diol followed by 0.3 osm sucrose.

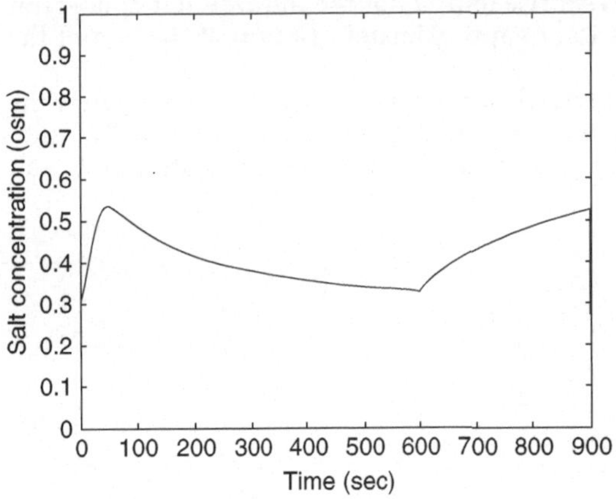

Figure 4 History of intracellular concentration of salt for human oocyte exposed to 1.5 osm propane-1,2-diol followed by 0.3 osm sucrose.

Figure 5 shows two curves. The solid curve describes the rise in intracellular propane-1,2-diol content from 0 moles to approximately 7×10^{-10} moles during the first 10 minutes of exposure to an extracellular concentration of 1.5 osm propane-1,2-diol. It is interesting to note that during the 5-minute exposure to sucrose, propane-1,2-diol is squeezed out of the oocyte (Fig. 5). The second (dashed) curve closely resembles the response of the normalized cell volume (Fig. 1) and represents the absolute cell volume (liters). A comparison of the two curves in Figure 5 is a useful way to appreciate how the intracellular propane-1,2-diol content changes as cell volume changes.

The net result of these two steps of adding propane-1,2-diol and sucrose is to increase the total intracellular osmolality to some 2.1 osm and reduce the intracellular water content by 40%.

The oocyte is reasonable close to, but not completely in, equilibrium at the end of this part of the entire cryopreservation protocol.

RESPONSE OF HUMAN OOCYTES TO THE REMOVAL OF PROPANE-1,2-DIOL AND SUCROSE

For the specific cryopreservation illustrative protocol considered in this chapter (Table 1), the intracellular propane-1,2-diol concentration is reduced in three steps from 1.5 to 1.0 osm to 0.5 osm at room temperature (assumed to be 22°C for modeling) while keeping the sucrose concentration constant at 0.3 osm. The exposure times are 5 minutes for the first two steps and 10 minutes for the third step.

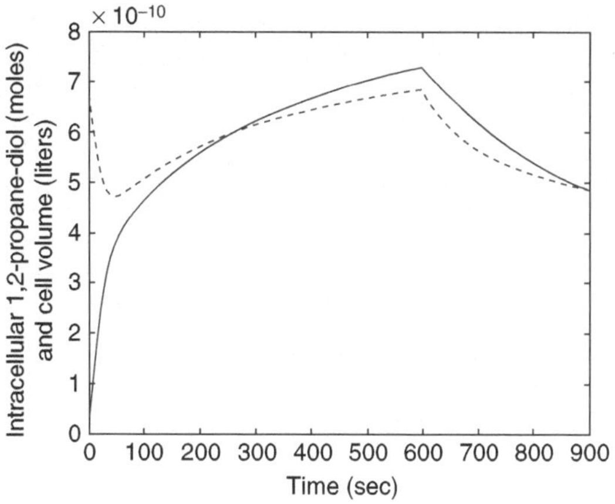

Figure 5 History of intracellular content of propane-1,2-diol (moles; solid curve) and cell volume (liters; dashed curve) for human oocyte exposed to 1.5 osm propane-1,2-diol followed by 0.3 osm sucrose.

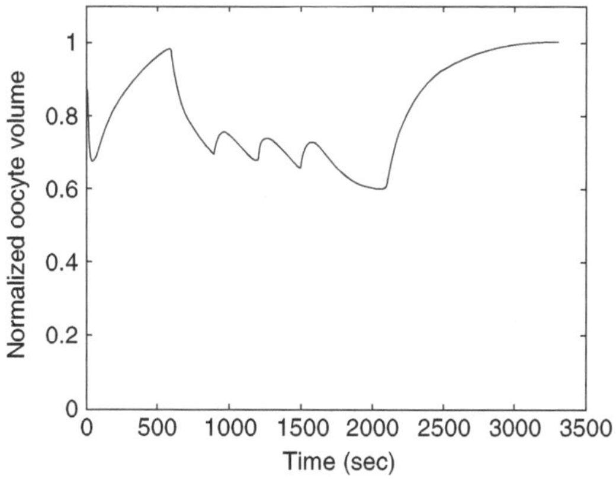

Figure 6 History of normalized human oocyte cell volume in response to the addition and removal of propane-1,2-diol and sucrose as defined in Table 1.

Figure 6 shows the normalized oocyte volume response for the entire process of adding and removing cryoprotectant and sucrose. The first 900 seconds replicates the results shown in Figure 1. Replicating the addition process allows the subsequent response to dilution steps to be compared directly with the addition steps.

During the first two steps of dilution, the oocyte volume increases by approximately 5% for each step and then decreases by approximately 7% for each step (Fig. 6). During the third step the oocyte volume increases by approximately 7% and then decreases by nearly 15%. The oocyte is at some 60% of its normalized volume at this stage. The last two steps involve removing the 0.3 osm sucrose (10 minutes at 22°C followed by 10 minutes at 37°C). At the end of the first sucrose dilution step, the oocyte volume has already returned to nearly isotonic volume.

The response pattern for intracellular water follows that of the cell volume closely (Fig. 7).

Figure 8 describes the intracellular propane-1,2-diol concentration history during the addition and removal of cryoprotectant and sucrose. Once again, the first 900-second period replicates the results presented earlier (Fig. 3). The concentration of propane-1,2-diol has been reduced to 0 osm by the end of the third dilution step. It remains at 0 osm during the subsequent two sucrose dilution steps (20 minutes duration). It would appear that the last 20 minutes of exposure are not necessary to reduce the intracellular cryoprotectant concentration.

The intracellular concentration of salt oscillates during the dilution steps, but never rises above 0.6 osm (twice the isotonic value) during the three steps of propane-1,2-diol dilution (Fig. 9). The concentration of salt is reduced to a value close to the isotonic concentration by the end of the first step where the extracellular sucrose concentration is reduced. It would appear that the final 10-minute step at 37°C is not necessary to reduce the intracellular salt concentration to its isotonic value.

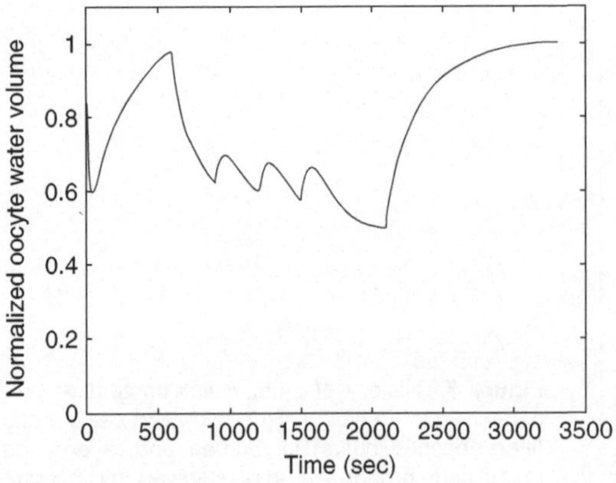

Figure 7 History of normalized human oocyte water volume in response to the addition and removal of propane-1,2-diol and sucrose as defined in Table 1.

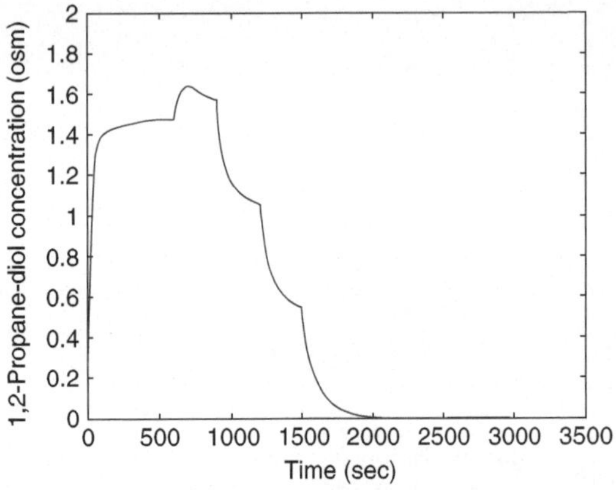

Figure 8 History of intracellular concentration of propane-1,2-diol for human oocyte in response to the addition and removal of propane-1,2-diol and sucrose as defined in Table 1.

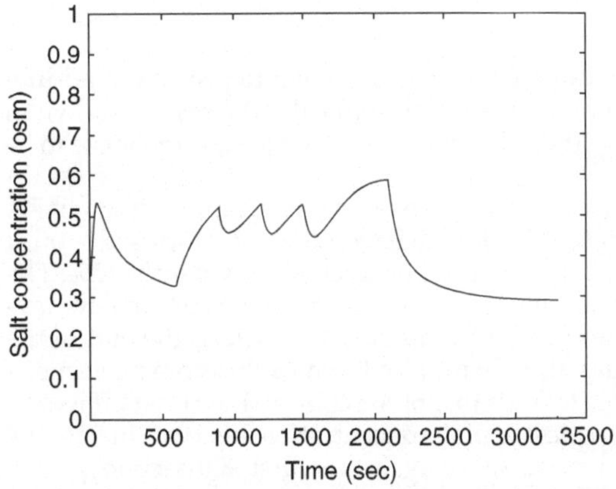

Figure 9 History of intracellular concentration of salt for human oocyte in response to the addition and removal of propane-1,2-diol and sucrose as defined in Table 1.

These predictions suggest that there are some opportunities for optimizing the dilution processing. Indeed, results of predictions not shown here suggest that the entire addition and removal processing could be reduced, without excessive volume excursions, to approximately one-third of the total processing time defined in Table 1. The addition and removal portions would be 10 minutes each, compared to the current total protocol time of 55 minutes defined in Table 1. Even simpler protocols may be possible with a more comprehensive study using the present model. In any case, such model predictions should be coupled with experimental verification.

RESPONSE OF HUMAN OOCYTES TO COOLING RATE DURING FREEZING

The response of human oocytes to freezing requires a number of parameters, including membrane permeability to water and cryoprotectant. The temperature dependence of these permeabilities (activation energy) is also required. These parameters have been determined for human oocytes (14). It is important to recognize that there can be a very significant range of these parameter values in a human oocyte population due to a number of factors as identified by Fuller and Paynter (chap. 3).

In most cases, the membrane permeability to cryoprotectants is significantly smaller than that of water. In addition, the temperature dependence of membrane permeability to cryoprotectants is generally larger than that of water. Consequently, the transport of cryoprotectant during freezing is often ignored in most published models. When the permeability parameters are known, it is possible to predict a number of important responses of the human oocyte during freezing. These include oocyte volume, intracellular water volume as well as intracellular concentrations of cryoprotectant and salt. The amount of supercooling of intracellular water can also be computed. These same oocyte parameters can be computed for hold times at specified isothermal temperatures (e.g., storage) and during warming.

To predict the probability of intracellular ice formation on the basis of physio-chemical principles, it is also necessary to know additional parameters. These include parameters related to the nucleation of intracellular ice as defined by Toner et al. (7). These parameters are typically determined experimentally using cryomicroscopy (19) and they have been determined for hepatocytes (8), mouse oocytes (7), mouse embryos (20), and hematopoietic progenitor cells (21). Unfortunately, these nucleation parameters have not been determined yet for human oocytes in the absence or presence of any cryoprotectants. Furthermore, models for the viscosity of cryoprotectant solutions as a function of temperature and concentration are required to predict the probability of intracellular ice formation using the approaches developed by Toner et al. (7) and Karlsson et al. (8). These models exist for dimethyl sulfoxide (8) and glycerol (9) but not for other cryoprotectants such as ethylene glycol or propane-1,2-diol.

To illustrate what can be done with regard to predicting intracellular ice formation in human oocytes, the two ice nucleation parameters known for mouse oocytes in dimethyl sulfoxide are combined with the known viscosity model for dimethyl sulfoxide. This information for dimethyl sulfoxide is combined with the osmotic properties known for human oocytes. As the membrane permeability properties are known for human oocytes in the presence of propane-1,2-diol and sucrose, these permeability values are used here.

While it is recognized that the resultant predictions are not expected to be completely accurate in a quantitative sense, the approach serves to illustrate the capabilities as well as the current limitations of such modeling. In spite of the current limitations, it is remarkable that the predictions presented in this chapter for human oocytes appear to capture most of the empirically defined major overall trends, while at the same time providing added insight into why the human oocyte responds the way it does. This represents an exciting foundation foreshadowing the more accurate and comprehensive results that can be expected once the necessary viscosity models are defined and intracellular nucleation parameters are determined for human oocytes.

To verify that the computer model itself was properly implemented for predicting the probability of intracellular ice formation, the results of Karlsson et al., (17) which predict the formation of intracellular ice in mouse oocytes cooled rapidly in the presence of 1.5 M dimethyl sulfoxide, were successfully replicated prior to applying the intracellular ice nucleation model to predict human oocyte responses.

As a first step in developing the capability of predicting intracellular ice formation in human oocytes equilibrated in 1.5 osm propane-1,2-diol (no sucrose), the sensitivity to cooling rate was predicted for cooling at various rates to a minimum temperature of –30°C. This temperature corresponds to the minimum temperature of the first of the three cooling steps defined in the specific example cryopreservation protocol defined in this chapter (Table 1). It can be seen in Figure 10 that the current model predicts that the probability of intracellular ice formation will increase from 0% to 100% over the cooling rate range of 0.5°C/min to 2°C/min and that intracellular ice will nucleate at relatively high temperatures at the faster rates.

Specifically, intracellular ice formation is predicted to occur between –10°C and –20°C. While these predictions are not expected to be exactly correct, they do suggest that relatively slow rates of cooling will be required to avoid intracellular ice formation and that intracellular

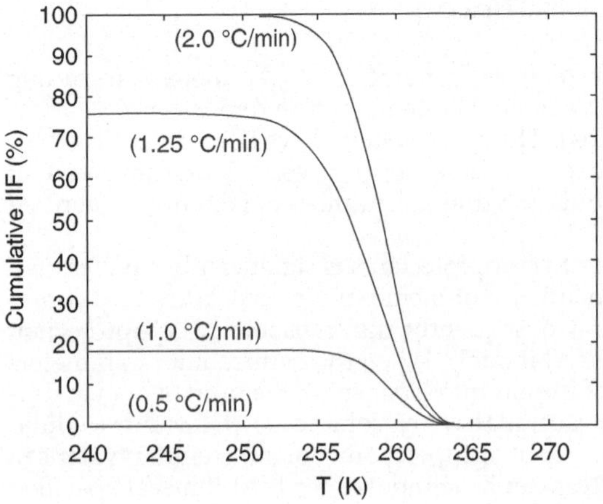

Figure 10 Cumulative percentage of intracellular ice formation (IIF) as a function of temperature for human oocyte cooled at different cooling rates. The human oocyte was equilibrated with 1.5 osm propane-1,2-diol and 0.3 osm salt (no sucrose) at room temperature, then cooled to and seeded at the equilibrium freezing temperature of –3.24°C prior to cooling. The oocyte was allowed to equilibrate fully with the seeded solution prior to subsequent cooling.

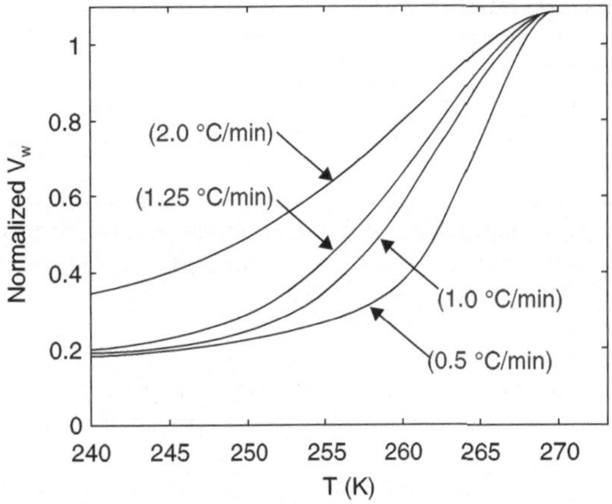

Figure 11 Normalized human oocyte water volume as a function of temperature for human oocyte cooled at different cooling rates. The human oocyte was equilibrated with 1.5 osm propane-1,2-diol and 0.3 osm salt (no sucrose) at room temperature, then cooled to and seeded at the equilibrium freezing temperature of –3.24°C prior to cooling. The oocyte was allowed to equilibrate fully with the seeded solution prior to subsequent cooling.

ice formation will occur at relatively high nucleation temperatures for the faster cooling rates. In spite of the limitations and assumptions made, these results agree well with those of Trad et al. (18). The predictions also illustrate the kind of information that can be obtained once intracellular ice formation nucleation parameters are determined for human oocytes.

Additional parameters can be predicted from such a model. Results are shown that illustrate how faster cooling rates cause oocytes to retain larger fractions of intracellular water (Fig. 11) and produce more supercooling of intracellular water (Fig. 12).

Predictions are not included here for the second and third cooling steps of the specific example cryopreservation protocol defined in Table 1. Obviously, these would be of interest, but given the scope of this chapter and the current limitation regarding the nucleation parameters for human oocytes containing propane-1,2-diol and the lack of a viscosity model for propane-1,2-diol, these predictions have not been included.

INFLUENCE OF SUCROSE CONCENTRATION ON HUMAN OOCYTE FREEZING RESPONSE

It is clear from published work that the addition of the impermeable solute sucrose to the extracellular cryoprotectant solution can have a significant beneficial effect on human oocyte recovery after freezing (15). It is apparent from the modeling results presented above (Fig. 2) that the addition of sucrose reduces the amount of intracellular water significantly before freezing begins. This dehydration is expected to produce a beneficial result by way of reducing the likelihood of intracellular ice and oocyte damage. It is therefore of interest to explore what the

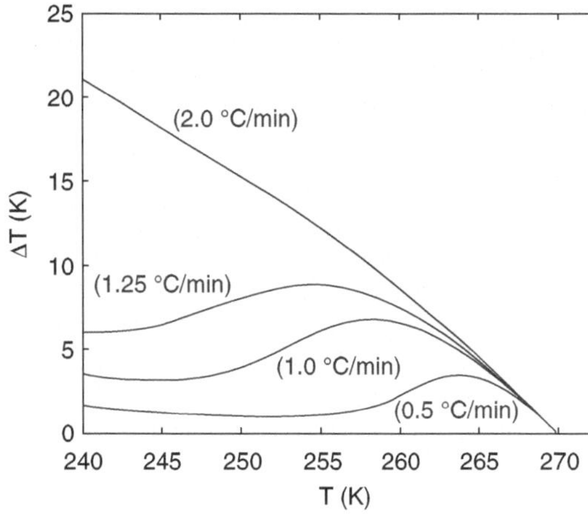

Figure 12 Supercooling of human oocyte intracellular water as a function of temperature for human oocyte cooled at different cooling rates. The human oocyte was equilibrated with 1.5 osm propane-1,2-diol and 0.3 osm salt (no sucrose) at room temperature, then cooled to and seeded at the equilibrium freezing temperature of –3.24°C prior to cooling. The oocyte was allowed to equilibrate fully with the seeded solution prior to subsequent cooling.

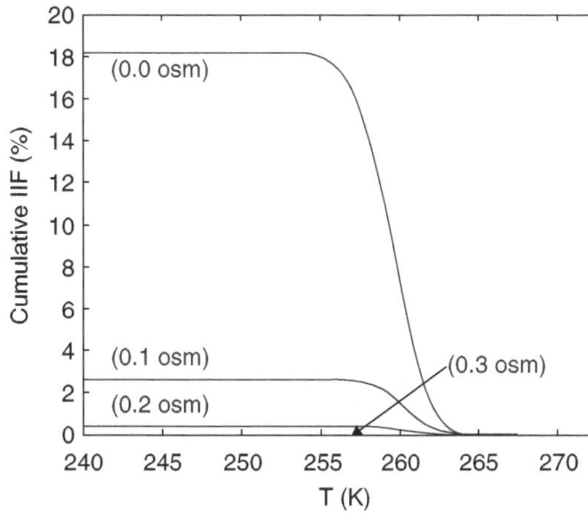

Figure 13 Cumulative percentage of intracellular ice formation (IIF) as a function of temperature for human oocyte cooled at 1 °C/min. The human oocyte was equilibrated with 1.5 osm propane-1,2-diol and 0.3 osm salt with no sucrose in one case. Three other cases with different concentrations of sucrose were also considered. The oocyte was seeded at the respective equilibrium freezing temperature of each solution and allowed to equilibrate fully prior to cooling.

computer predictions reveal concerning how the presence and concentration of sucrose in the cryoprotectant solution influences the oocyte response to freezing.

As can be seen in Figure 13, some 18% of human oocytes equilibrated with 1.5 osm propane-1,2-diol and cooled to –30°C at 1°C/min are predicted to contain intracellular ice at –30°C if no sucrose has been added to the solution.

The effect of adding sucrose to the solution and allowing full equilibration is significant according to the modeling results shown here. In particular, the fraction of oocytes containing intracellular ice at –30°C is predicted to decrease from 18% with 0.0 osm sucrose to 2.5% with 0.1 osm sucrose. Increasing the sucrose concentration further to 0.2 and 0.3 osm reduces the predicted fraction of oocytes with intracellular ice from 2.5% to less than 1%.

Once again, the model predictions provide additional insight with regard to the effect of sucrose addition on intracellular water content (Fig. 14) and intracellular supercooling (Fig. 15).

These figures show that as the sucrose concentration is increased, the amount of intracellular water is reduced at all temperatures, as is the extent of supercooling of intracellular water. It is interesting to note that the magnitude of the intracellular supercooling is a non-monotonic function of temperature and that the maximum value occurs at approximately –15°C (258 K).

Clearly, the probability of intracellular ice formation is sensitive to the reduced water content resulting from sucrose addition. It will be seen below that incomplete mixing of extracellular solutions during oocyte transfer can lead to a range of intracellular water content.

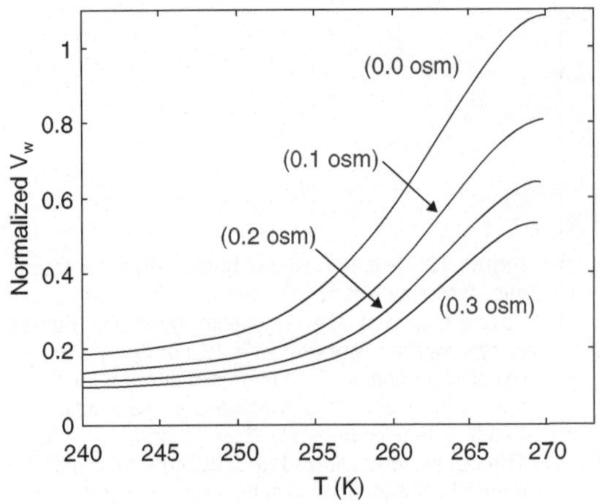

Figure 14 Normalized human oocyte water volume as a function of temperature for human oocyte cooled at 1°C/min. The human oocyte was equilibrated with 1.5 osm propane-1,2-diol and 0.3 osm salt with no sucrose in one case. Three other cases with different concentrations of sucrose were also considered. The oocyte was seeded at the respective equilibrium freezing temperature of each solution and allowed to equilibrate fully prior to cooling.

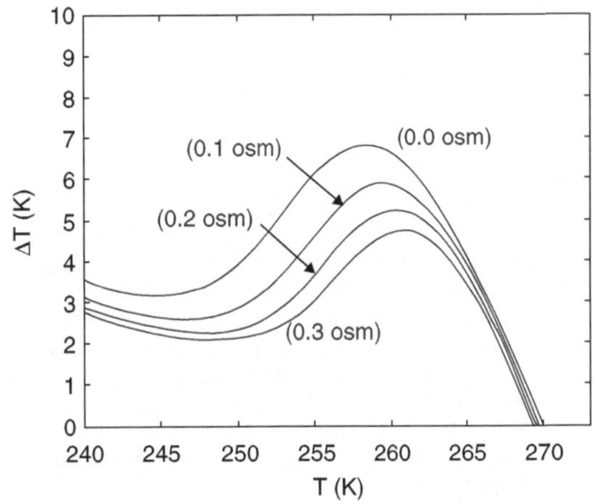

Figure 15 Supercooling of human oocyte intracellular water as a function of temperature for human oocyte cooled at 1°C/min. The human oocyte was equilibrated with 1.5 osm propane-1,2-diol and 0.3 osm salt with no sucrose in one case. Three other cases with different concentrations of sucrose were also considered. The oocyte was seeded at the respective equilibrium freezing temperature of each solution and allowed to equilibrate fully prior to cooling.

This range of water content may not be appreciated by those performing transfer and yet may lead to unexpected and non-reproducible results with regard to intracellular ice formation and oocyte damage.

INFLUENCE OF SEEDING TEMPERATURE AND HOLD TIME ON HUMAN OOCYTE FREEZING RESPONSE

It is important to recognize two extremes with regard to the extracellular seeding of ice. One extreme is that, after seeding of extracellular ice has occurred, sufficient time is allowed for the oocytes to equilibrate fully with the frozen, concentrated extracellular solution. After this equilibration, cooling commences. At the other extreme, seeding of the extracellular ice occurs and cooling begins immediately, with no time for the oocyte to equilibrate with the frozen, concentrated extracellular solution before cooling starts. In the former case, the intracellular water content is reduced and there is no supercooling of the intracellular water when cooling starts after the seeding occurs. In the latter case, the intracellular water content exceeds that corresponding to the equilibrium case and the intracellular water is supercooled even before the cooling begins. The finite supercooling at the beginning of the cooling in the latter case enhances the probability of intracellular ice formation and therefore the likelihood of oocyte damage. Both these extremes are considered below. Before the extremes are considered, results are presented for an intermediate case corresponding to that prescribed in the specific cryopreservation process used here for illustration purposes (Table 1). This intermediate case shows how the human oocyte is predicted to respond to the addition of propane-1,2-diol, followed by the addition of sucrose and then seeding at –7°C

Response to Seeding Process Specified in the Example Cryopreservation Protocol

The response of the human oocyte to the addition of propane-1,2-diol and sucrose has been described earlier in this chapter. When the human oocyte is then cooled and the extracellular solution is seeded with ice at –7°C, it is predicted to shrink significantly in a short period, followed by slow swelling over a much longer time period. Figure 16 shows the human oocyte volume history for the case where it has been exposed to 1.5 osm propane-1,2-diol for 10 minutes followed by exposure to 0.3 osm for 5 minutes, followed by extracellular seeding at –7°C with a hold period of 10 minutes at –7°C.

The first 900-second period of Figure 16 represents the same results as those shown in Figure 1. The significant shrinkage shown in Figure 16 immediately after 900 seconds has elapsed reveals how quickly the oocyte shrinks after seeding occurs. The slow recovery of oocyte volume is indicative of the transport of water and cryoprotectant at the seed temperature, both of which are modeled by using the activation energies of water and cryoprotectant to extrapolate permeabilities to the seed temperature.

Figure 17 shows the human oocyte intracellular water volume history for this same case. Clearly, the water volume response closely resembles that of the oocyte volume. Figure 18 describes the corresponding history of intracellular concentration of propane-1,2-diol for this

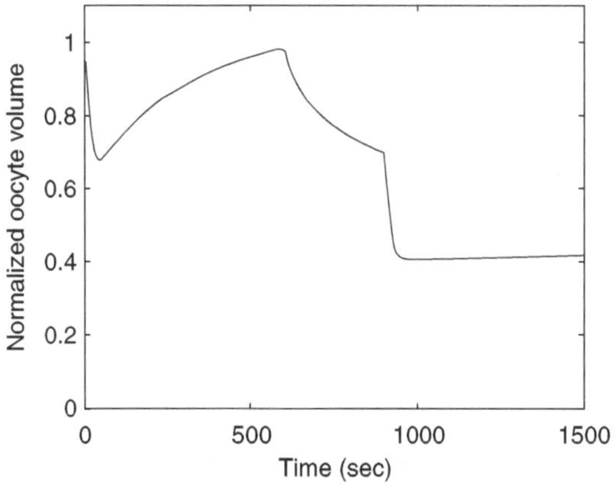

Figure 16 Response of human oocyte normalized cell volume to exposure to 1.5 osm propane-1,2-diol followed by 0.3 osm sucrose and then seeded at –7°C and held for 10 min.

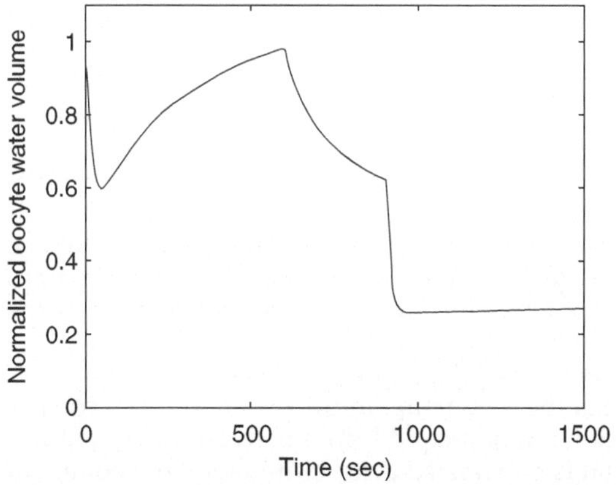

Figure 17 Response of human oocyte normalized water volume to exposure to 1.5 osm propane-1,2-diol followed by 0.3 osm sucrose and then seeded at –7°C and held for 10 minutes.

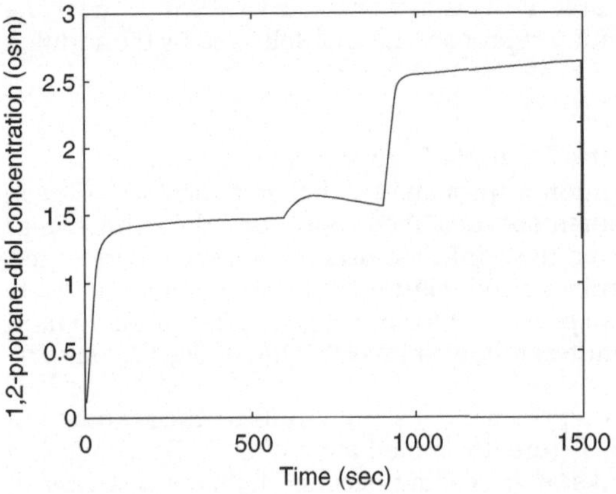

Figure 18 History of intracellular concentration of propane-1,2-diol for human oocyte exposed to 1.5 osm propane-1,2-diol followed by 0.3 osm sucrose and then seeded at –7°C and held for 10 minutes.

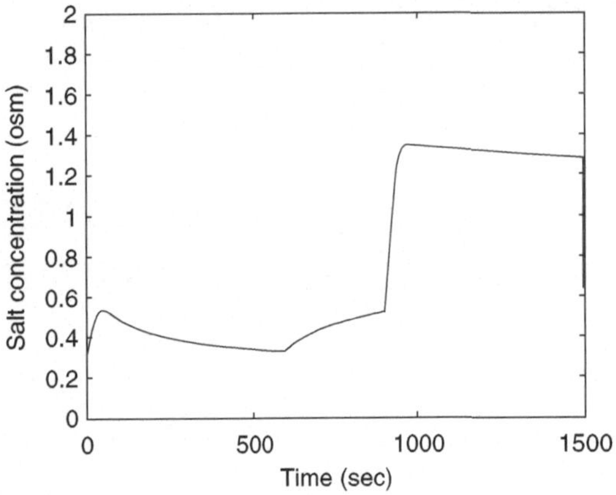

Figure 19 History of intracellular concentration of salt for human oocyte exposed to 1.5 osm propane-1,2-diol followed by 0.3 osm sucrose and then seeded at –7°C and held for 10 minutes.

same case. It is clearly evident that the seeding step increases the intracellular cryoprotectant concentration in a rapid fashion. Figure 19 reveals the history of intracellular salt concentration for this same case. As was true for the cryoprotectant, the seeding step increases the intracellular salt concentration rapidly and significantly.

These results predict that at the end of the 10-minute seeding period, the human oocyte cell volume, normalized to its isotonic value, shrinks to 42%. Similarly, the normalized intracellular water volume has been reduced to 27%. The intracellular propane-1,2-diol concentration has risen to 2.7 osm and the intracellular salt concentration has increased to 1.3 osm, thus producing a total intracellular osmolality of 4.0 osm prior to freezing from this seeded condition.

To predict the dynamics during the isothermal hold period at –7°C, the known values of water and propane-1,2-diol permeability, along with their respective activation energies, were used to define the water and propane-1,2-diol membrane permeabilities at –7°C. The results during the hold period suggest that the human oocyte is not fully equilibrated, but that the hold time of 10 minutes is long enough that the oocyte is reasonably close to equilibrium. It would be interesting to study more completely the full range of seed temperatures and relaxation times to determine their influence on the freezing response of human oocytes to this part of the cryopreservation protocol.

In the next two subsections of this chapter the two limits of the expected responses are considered. In the first case, no relaxation time at the seed temperature occurs prior to cooling below the seed temperature. In this case, the extent of non-equilibrium between the intracellular solution and the frozen extracellular solution is maximized prior to initiation of cooling. In the second case, an infinitely long relaxation time at the seed temperature is considered. In this instance, the oocyte would reach complete equilibrium with the frozen extracellular solution prior to subsequent cooling.

Cooling with No Hold Time: No Equilibration Prior To Cooling

Figure 20 shows the probability of intracellular ice formation as a function of temperature for cooling at a rate of 1°C/min to –40°C for human oocytes fully equilibrated in 1.5 osm propane-1,2-diol (no sucrose) after seeding at four different temperatures (–3.24°C, –4°C, –6°C, and –8°C). The results clearly illustrate that reducing the seed temperature in this case increases the probability of intracellular ice formation significantly.

The cryoprotectant type and concentration, the seeding conditions, and the minimum temperature correspond to published work describing experimental cryomicroscopy intracellular ice formation for human oocytes (18). The cooling rate of 1°C/min is faster than that used in the published work, but the reason for using the faster rate was explained above and is related to the nucleation parameters and viscosity model used for the predictions.

Figure 21 shows the intracellular water volume, normalized to its isotonic value, for the conditions defined for Figure 20.

The depression of the initial temperature prior to cooling as the seeding temperature is reduced is readily apparent. A notable feature of Figure 21 is that the intracellular water content prior to cooling is identical for all seeding temperatures. This same water content is a result of

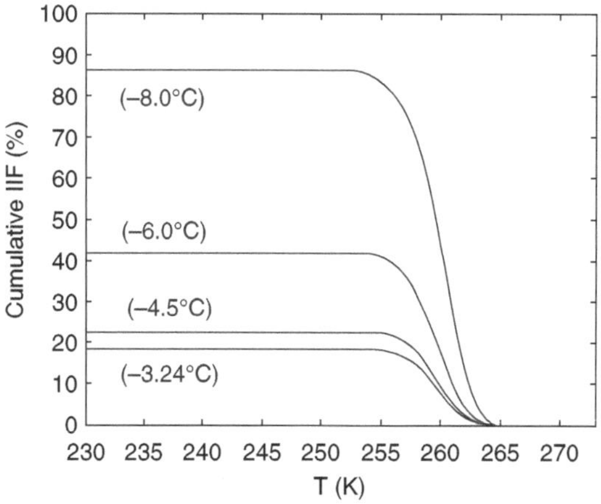

Figure 20 Cumulative percentage of intracellular ice formation as a function of temperature for human oocyte cooled at 1°C/min. The human oocyte was equilibrated with 1.5 osm propane-1,2-diol and 0.3 osm salt (no sucrose) and seeded at four different temperatures prior to cooling immediately after seeding. The equilibrium freezing temperature is –3.24°C for all four cases.

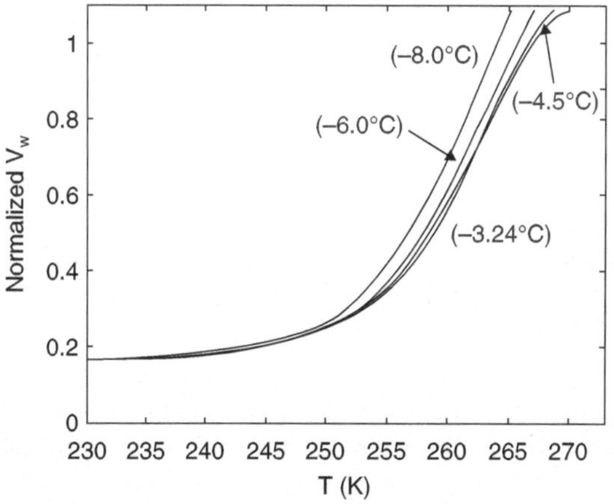

Figure 21 Normalized human oocyte water volume as a function of temperature for human oocyte cooled at 1°C/min. The human oocyte was equilibrated with 1.5 osm propane-1,2-diol and 0.3 osm salt (no sucrose) and seeded at four different temperatures prior to cooling immediately after seeding. The equilibrium freezing temperature is –3.24°C for all four cases.

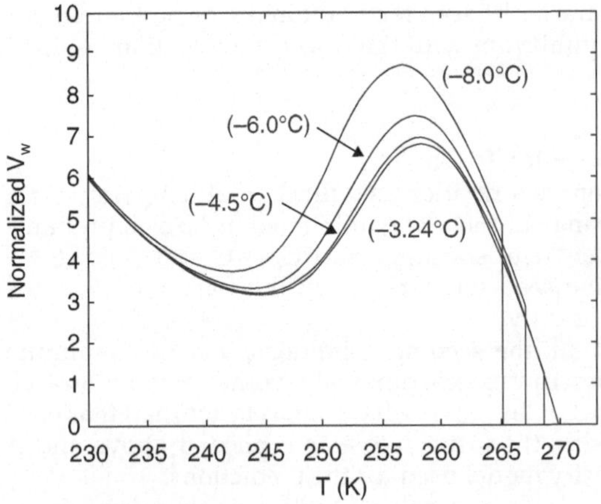

Figure 22 Supercooling of human oocyte intracellular water as a function of temperature for human oocyte cooled at 1°C/min. The human oocyte was equilibrated with 1.5 osm propane-1,2-diol and 0.3 osm salt (no sucrose) and seeded at four different temperatures prior to cooling immediately after seeding. The equilibrium freezing temperature is –3.24°C for all four cases.

the fact that the oocyte has not been allowed any time to equilibrate with the seeded extracellular solution prior to cooling. As will be evident below, this is significantly different from the case when full equilibration is allowed prior to cooling. When no equilibration is allowed before cooling is commenced, an oocyte seeded at a lower temperature will have a larger water content at all temperatures during freezing. Water content approaches the same level at the lowest temperatures modeled here.

Figure 22 shows the extent of supercooling for intracellular water for the conditions corresponding to Figure 20.

At the seed temperature of –3.24°C, there is no vertical jump of supercooling as cooling commences because this temperature matches the equilibrium freezing temperature of the oocyte containing 1.5 osm propane-1,2-diol and 0.300 osm salt. For lower seed temperatures the initiation of cooling produces an initial instantaneous rise in supercooling because the intracellular water is already in a supercooled state as cooling begins. In this instance, as the seed temperature is reduced, the supercooling at the time when cooling commences increases, thereby tending to enhance the likelihood of intracellular ice formation. Once again, it appears that supercooling is a non-monotonic function of temperature for all cases. The maximum amount of supercooling occurs slightly below –15°C (258 K) and the magnitude of the supercooling converges for all seed temperatures at the minimum temperature modeled.

In spite of the difference between the cooling rate used here and that applied by Trad et al. experimentally (18), the predictions clearly illustrate how strongly the probability of intracellular ice formation is correlated to the magnitude of the seeding temperature. This agrees well with the experimental results of Trad et al. It is also clearly evident that when no equilibration time is allowed prior to cooling subsequent to seeding, then a reduction in the seeding temperature increases the likelihood of intracellular ice formation. This is in direct contrast to the influence of seeding temperature on the probability of intracellular ice formation when the oocyte is allowed to equilibrate with the seeded extracellular solution for very long times, as will be seen below.

Cooling After Long Hold Times: Complete Equilibration Prior To Cooling

Figure 23 shows the probability of intracellular ice formation as a function of temperature for cooling at a rate of 1°C/min to –40°C for human oocytes fully equilibrated in 1.5 osm propane-1,2-diol (no sucrose) after seeding extracellular ice at four different temperatures (–3.24°C, –4.5°C, –6°C, and –8°C).

These conditions again correspond to those in the published work describing experimental cryomicroscopy results for human oocytes (18). Here the reduction in the seed temperature from –3.24°C to –8°C decreases the probability of intracellular ice formation from 18% to less than 1%.

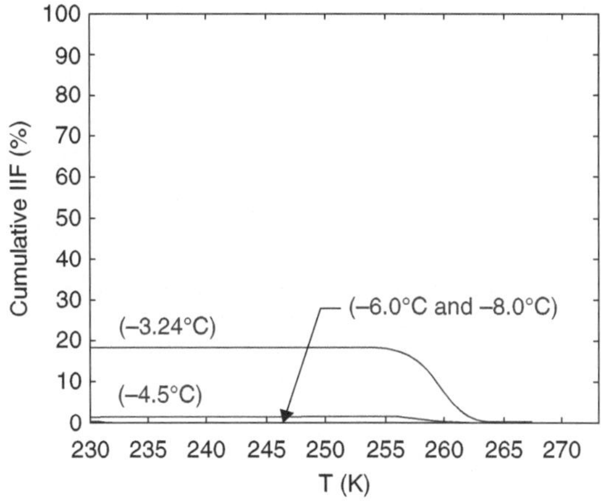

Figure 23 Cumulative percentage of intracellular ice formation (IIF) as a function of temperature for human oocyte cooled at 1°C/min. The human oocyte was equilibrated with 1.5 osm propane-1,2-diol and 0.3 osm salt (no sucrose) and seeded at four different temperatures prior to cooling. The oocyte was allowed to equilibrate fully at the respective seeding temperature prior to cooling for all four cases.

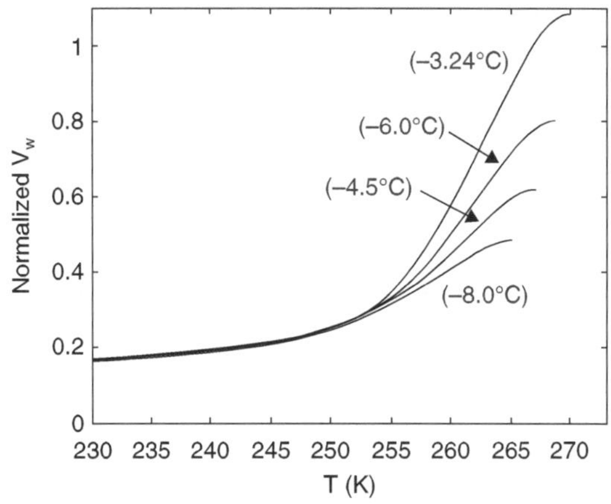

Figure 24 Normalized human oocyte water volume as a function of temperature for human oocyte cooled at 1°C/min. The human oocyte was equilibrated with 1.5 osm propane-1,2-diol and 0.3 osm salt (no sucrose) and seeded at four different temperatures prior to cooling. The oocyte was allowed to equilibrate fully at the respective seeding temperature prior to cooling for all four cases.

Figure 24 shows the intracellular.water volume, normalized to its isotonic value, for the conditions defined for Figure 23.

In this case, when the relaxation time at the seed temperature is very long, water and propane-1,2-diol can move across the membrane to reach equilibrium with the extracellular solution that has been seeded with ice. Consequently, it would be expected that water would leave the oocyte during this extended period where the cell is exposed to a concentrated external solution. Figure 24 reveals that as the seed temperature is reduced, the initial, normalized intracellular water volume decreases, such that it is below 50% at a seed temperature of –8°C. This dehydration prior to subsequent cooling would be expected to reduce the supercooling of intracellular water as well as decrease the likelihood of intracellular ice formation, and thus oocyte damage. These expectations are realized in Figures 25 and 23, respectively.

Note that in Figure 25 in the case of full equilibration prior to cooling, the supercooling does not "spike" immediately after cooling begins as it does when cooling is begun immediately after seeding (Fig. 22). In the case of full equilibration prior to cooling, the intracellular water is equilibrated and not supercooled as cooling is begun. Figure 25 shows once again the non-monotonic nature of intracellular supercooling with peak values in the neighborhood of –15°C (258 K). Interestingly, higher seeding temperatures produce larger supercoolings until the temperature reaches approximately –20°C (253 K). Below this temperature, the supercoolings are larger for the lower seed temperatures.

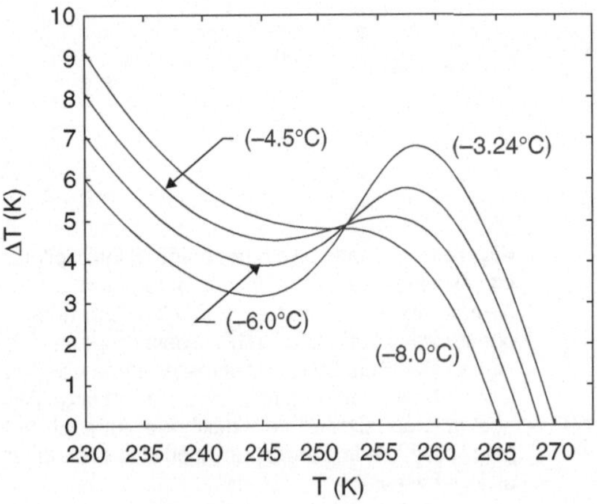

Figure 25 Supercooling of human oocyte intracellular water as a function of temperature for human oocyte cooled at 1°C/min. The human oocyte was equilibrated with 1.5 osm propane-1,2-diol and 0.3 osm salt (no sucrose) and seeded at four different temperatures prior to cooling. The oocyte was allowed to equilibrate fully at the respective seeding temperature prior to cooling for all four cases.

In direct contrast to the results shown in Figure 20, in this case the depression of the seeding temperature has the beneficial effect of reducing the probability of intracellular ice formation. In practice there will be an important interplay between the temperature history produced by seeding because in real systems the straw or vial containing the oocyte may be large enough that the temperature may rise and then fall after seeding and the local concentration around the oocyte may change in a complex fashion.

POTENTIAL PROBLEMS ASSOCIATED WITH OOCYTE TRANSFER BETWEEN MULTIPLE SOLUTIONS OF DIFFERENT COMPOSITIONS

The cryopreservation protocol defined here (Table 1) is representative of many insomuch as it involves multiple steps of adding and removing cryoprotectant and/or impermeable solute (sucrose). This is typically done by pipette transfer of a small volume of solution containing the oocyte at one composition to a significantly larger volume at the next desired composition. Cryopreservation protocols such as the example used here define solutions with specific concentrations of cryoprotectants and impermeable solutes (such as sucrose) that oocytes are exposed to when they are transferred from one solution to another. It is quite natural to assume, for example, that transfer of a human oocyte in an isotonic solution with no cryoprotectant to a solution containing 1.5 osm propane-1,2-diol would expose the oocyte instantaneously to an external concentration of 1.5 osm propane-1,2-diol. Furthermore, the descriptions of transfer techniques provided in the methods sections of published work implicitly suggest that a transfer is a transfer and that this process detail would be quite reproducible from one person to another or from one laboratory to another. Unfortunately, these assumptions may not be true. A simple example of what may happen instead is offered here to alert the reader to the care that will probably be necessary to expect reproducible results. The conclusion is that incomplete mixing at a small scale may create significant and unrecognized consequences regarding the oocyte response to the overall cryopreservation process.

It is well known that non-reproducible recovery from cryopreservation protocols is a shortcoming that requires addressing (15,18,22). The source of this non-reproducibility may be dominated by heterogeneity in the characteristics of the biomaterial (e.g., oocyte size, water permeability, nucleation parameters). On the other hand, factors related to uncontrolled processing may be equally or even more important. The example offered here suggests that the way in which mixing is performed may be an important factor.

The first point to make is that diffusion of cryoprotectants and impermeable solutes (e.g., sucrose) through the solutions that are being mixed represents the mechanism that will determine the characteristic times associated with changing the extracellular concentrations during transfer of an oocyte from one solution to another. The dynamics of the extracellular concentrations then dictate the intracellular state of the oocyte.

By way of example, assume that a human oocyte in an isotonic solution is aspirated into the tip region of a pipette of small diameter. Imagine that a small volume of this solution containing the oocyte is then pipetted into a larger volume of solution containing 1.5 osm propane-1,2-diol. What happens at this point? The answer depends on how the small volume containing the oocyte is mixed with the larger volume. Even if the small volume containing the oocyte is mixed rather vigorously, it is not difficult to imagine that the oocyte may find itself in a small pocket of the initial isotonic solution. If this happens, then for the remainder of the exposure time to the new solution, diffusion of solutes and solvent determines the extracellular concentration history that the oocyte experiences. This, of course, then determines the oocyte response, including volume, the intracellular state of the oocyte, and the response to seeding, freezing, etc.

The following simple relationship defines the characteristic time for solutes to diffuse:

$$\tau \sim \frac{R^2}{D}$$

In words, this expression means that an equilibration time, τ, can be estimated by dividing the square of the length of the domain over which diffusion occurs, R^2, by the diffusion coefficient, D, where the diffusion coefficients of many solutes of interest for cryopreservation are of order 10^{-9} m^2/s.

By way of providing a specific example of this, consider the case of injecting a small volume of isotonic solution containing the oocyte into a much larger volume of solution containing propane-1,2-diol where a small sphere of radius 0.5 mm is produced. This sphere is initially at a uniform concentration of 0 osm propane-1,2-diol (the isotonic solution). It is surrounded by a larger volume of solution at a uniform concentration of 1.5 osm propane-1,2-diol. As soon as these two domains are put into contact, diffusion will begin. The simple expression above reveals that the order of magnitude for the time of equilibration will be 250 seconds. This may seem surprisingly long to some readers, especially in light of the illustrative example protocol considered in this chapter where a human oocyte is exposed to propane-1,2-diol for 600 seconds (Fig. 1). The simple expression above also reveals that the equilibration time is related to the square of the length scale. Thus, if the mixing process is better, such that the characteristic size of the small "pocket" that the oocytes finds itself in is reduced by half (e.g., from 0.5 mm to 0.25 mm), then the characteristic time for equilibration is reduced by a factor of four, from 250 seconds to approximately 60 seconds. Obviously, relatively small changes in the size of any pockets left after injection and macro-scale mixing can be expected to produce significant consequences.

Figure 26 shows a comparison of the extracellular propane-1,2-diol concentration history that an oocyte would experience for three different cases of mixing.

If "perfect" mixing were to occur, the extracellular concentration of propane-1,2-diol would be changed instantaneously from 0 osm to 1.5 osm in a stepwise fashion. This case is compared to the extracellular propane-1,2-diol concentration history that an oocyte would experience if it were at the center of a sphere of radius 0.5 mm or a sphere of radius 1.0 mm. Obviously there is a significant "lag" in the concentration change in the latter two cases compared to the instantaneous case. The larger "pocket" sizes that are produced during mixing produce longer lags in the extracellular concentration history. In addition, note that the correct order of magnitude of the characteristic time to change concentrations is observed in the results produced by computer predictions when those results are compared to the simple estimate made above.

Figure 27 shows the predicted response of the human oocyte water volume (normalized to the isotonic state) for the three cases of interest considered above: instantaneous extracellular concentration change and extracellular concentration changes associated with spherical "pockets" of radius 0.5 and 1.0 mm. Note that the instantaneous case is the same as that shown in Figure 2.

Exposure times of 10 minutes were defined in all cases as this corresponds to the exposure time defined in the example cryopreservation protocol studied here (Table 1). By the end of the 10-minute exposure period there is a difference of some 8% of the normalized water content between the instantaneous case and the "worst" case of a sphere of radius 1.0 mm. The use of sucrose to dehydrate oocytes was shown earlier in this chapter to have a significant impact on the predicted outcome of the probability of intracellular ice formation (Figs. 13 and 14).

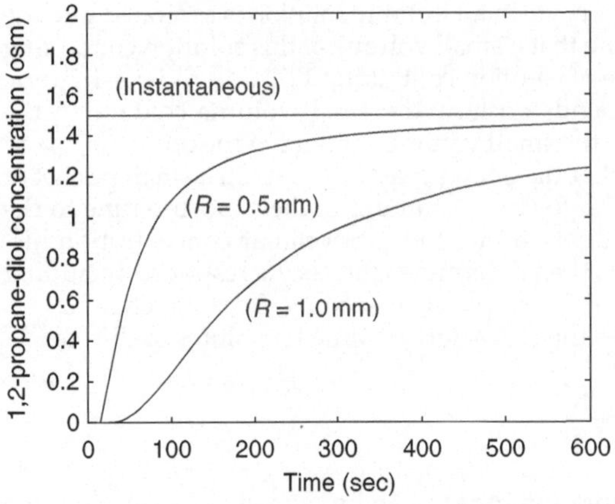

Figure 26 Extracellular concentration history of cryoprotectant for instantaneous mixing and two non-instantaneous mixing cases when an oocyte is transferred from isotonic solution with no cryoprotectant to a propane-1,2-diol concentration of 1.5 osm. The two non-instantaneous cases represent the cryoprotectant concentration histories at the center of a sphere of radius 0.5 mm or 1.0 mm.

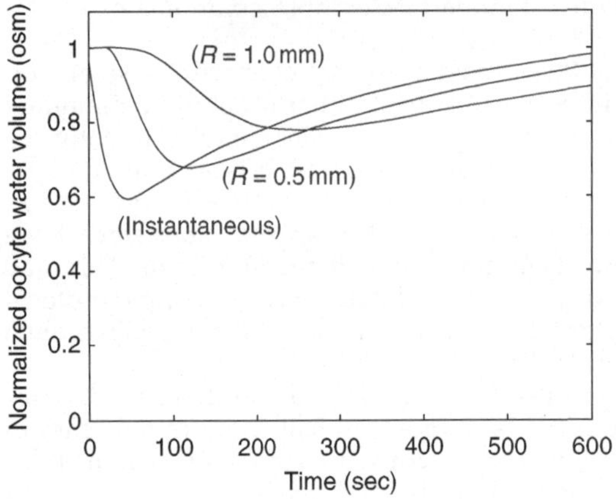

Figure 27 Predicted normalized human oocyte water volume history in response to instantaneous mixing and two non-instantaneous mixing cases for diffusion of propane-1,2-diol to the center of a sphere of radius 0.5 mm or 1.0 mm.

The effect of cellular dehydration produced by exposure to sucrose on the probability of intracellular ice formation is not necessarily directly comparable to that produced by exposure to propane-1,2-diol. However, the point is that seemingly small differences in mixing can produce significant differences in intracellular water content and this is likely to produce rather significant differences in the cryopreservation outcome for human oocytes. Future modeling studies should be undertaken to determine how such variations in mixing for all steps of a specified cryopreservation protocol would influence intracellular water content and the probability of intracellular ice formation, as well as other factors of interest. The results should provide valuable insights concerning this potential problem.

SUMMARY

Computer models have been implemented to predict the response of human oocytes to cryopreservation at controlled cooling rates. A specific, published cryopreservation protocol that involves slow cooling is used as an example to illustrate what can be achieved with this modeling approach as well as what limitations exist.

Unlike most published modeling work of this type, emphasis is placed on explicitly linking the dynamic responses of the oocyte from one part of the protocol to the next. In this manner it is possible to see how previous parts of the protocol influence subsequent steps. Not all of the steps in the published protocol were simulated, partly due to the length constraints of this chapter and partly because not all of the necessary modeling parameters are known yet for human oocytes. Nevertheless, the potential of this type of modeling should be evident.

The predictions made in this chapter appear to capture major trends in the known responses of human oocytes to cryopreservation processing. In addition, the predictions provide insight into mechanisms as well as predict parameters such as intracellular water content and supercooling that would be very difficult, if not impossible, to measure during the dynamics of freezing.

The modeling in this chapter offers some insight into why the details of the seeding process as well as the addition and removal of cryoprotectants and sucrose may have important consequences. For the seeding process, altering the seeding temperature may be beneficial or detrimental, depending on the extent to which equilibration with the extracellular solution is allowed prior to beginning subsequent cooling. It should be recognized that in a relatively large system, the local temperature and concentration may not be equilibrium values, further complicating the situation. An example has also been offered to show why the details of mixing at relatively small scales are important and why what may seem like relatively small differences in mixing are likely to produce what may be significantly different cryopreservation responses for human oocytes. Attention to a level of detail that may have not been appreciated previously by some may be necessary to produce the level of reproducibility desired.

Modeling results not included here suggest that the processes of adding and removing cryoprotectant and sucrose can be shortened dramatically without producing extensive oocyte volume excursions. These predictions should be compared with experimental results to determine whether such modeling based on volume excursion limits alone can be used to further optimize the example cryopreservation protocol used here as well as other protocols.

Determining the ice nucleation parameters and viscosity models for human oocytes is an important priority. Once this is done, the types of studies presented in this chapter can be repeated with nucleation parameters known for human oocytes. After these parameters are known, it will also be possible to predict the amount of ice and the size of ice crystals within oocytes for arbitrary cooling and warming processes. It will therefore be possible to predict when a process is expected to produce vitrification and/or recrystallization.

ACKNOWLEDGMENT

The author would like to thank Mr. Sanket Unhale for developing the original versions of the computer programs that were used in the simulations presented in this chapter.

REFERENCES

1. Kedem O, Katchalsky A. Thermodynamic analysis of the permeability of biological membranes to non-electrolytes. Biochim Biophys Acta 1958; 27: 229–46.
2. Leibo SP. Water permeability and its activation energy of fertilized and unfertilized mouse ova. J Membr Biol 1980; 53: 179–88.
3. Jackowski S, Leibo SP, Mazur P. Glycerol permeabilities of fertilized and unfertilized mouse ova. J Exp Zoo 1980; 212: 329–41.
4. Pfaff RT, Agca Y, Liu J, et al. Cryobiology of rat embryos. I. Determination of zygote membrane permeability coefficients for water and cryoprotectants, their activation energies, and the development of improved cryopreservation methods. Bio Reprod 2000; 63: 1294–302.
5. Kleinhans FW. Membrane permeability modeling: Kedem–Katchalsky vs a two-parameter formalism. Cryobiology 1998; 37: 271–89.
6. Mazur P. Kinetics of water loss from cells at subzero temperatures and the likelihood of intracellular freezing. J Gen Physiol 1963; 47: 347–69.
7. Toner M, Cravalho EG, Karel M. Thermodynamics and kinetics of intracellular ice formation during freezing of biological cells. J Appl Phys 1990; 67: 1582–93.
8. Karlsson JOM, Cravalho EG, Borel Rinkes IHM, et al. Nucleation and growth of ice crystals inside cultured hepatocytes during freezing in the presence of dimethyl sulfoxide. Biophys J 1993; 65: 2524–36.
9. Karlsson JOM, Cravalho EG, Toner M. A model of diffusion-limited ice growth inside biological cells during freezing. J Appl Phys 1994; 75: 4442–55.
10. Karlsson JOM. A theoretical model of intracellular devitrification. Cryobiology 2001; 42: 154–69.
11. Paynter SJ, Ruppert-Lingham CJ, Godfrey J, et al. Cryobiology 2001; 43: 369.

12. McGrath JJ, Unhale SA, Morris GJ. Intracellular ice formation for linear and non-linear cooling of murine oocytes in 1.5 M dimethyl sulfoxide: computational and experimental results. Cryobiology 2008; 57: 335.
13. Fabbri R, Porcu E, Marcella T, et al. Human oocyte cryopreservation: new perspectives regarding oocyte survival. Human Reprod 2001; 16: 411–16.
14. Paynter SJ, O'Neil L, Fuller BJ, et al. Membrane permeability of human oocytes in the presence of the cryoprotectant propane-1,2-diol. Fertil Steril 2001; 75: 532–8.
15. Paynter SJ, Borini A, Bianchi V, et al. Volume changes of mature human oocytes on exposure to cryoprotectant solutions used in slow cooling procedures. Human Reprod 2005; 20: 1194–9.
16. Van den Abbeel E, Schneider U, Liu J, et al. Osmotic responses and tolerance limits to changes in external osmolalities, and oolemma permeability characteristics, of human in vitro matured MII oocytes. Human Reprod 2007; 22: 1959–72.
17. Karlsson JOM, Eroglu A, Toth TL, et al. Fertilization and development of mouse oocytes cryopreserved using a theoretically optimized protocol. Human Reprod 1996; 11: 1296–305.
18. Trad FS, Toner M, Biggers JD. Effects of cryoprotectants and ice-seeding temperature on intracellular freezing and survival of human oocytes. Human Reprod 1998; 14: 1569–77.
19. Toner M, Cravalho EG, Karel M, et al. Cryomicroscopic analysis of intracellular ice formation during freezing of mouse oocytes without cryoadditives. Cryobiology 1991; 28: 55–71.
20. Toner M, Cravalho EG, Stachecki J, et al. Nonequilibrium freezing of one-cell mouse embryos: membrane integrity and developmental potential. Biophys J 1993; 64: 1908–21.
21. Hubel A, Norman J, Darr TB. Cryobiophysical characteristics of genetically modified hematopoietic progenitor cells. Cryobiology 1999; 38: 140–1153.
22. Paynter SJ. Current status of the cryopreservation of human unfertilized oocytes. Human Reprod 2000; 6: 449–56.

6 The Natural Life Cycle of the Mammalian Oocyte

David F. Albertini and S. Darlene Limback

Departments of Molecular and Integrative Physiology, Anatomy and Cell Biology, Obstetrics and Gynecology, and Center for Reproductive Sciences, Kansas University Medical Center, Kansas City, Kansas, U.S.A.

INTRODUCTION

The natural life cycle of the mammalian oocyte exemplifies a continuum in the propagation of the species. It is culminated when a mature female ovulates and supports the term gestation of female offspring that will bear those determinants in the female germ line that bestows on daughters the ability to recapitulate their mothers' ovulatory potential. But on the way to shedding a mature and developmentally competent oocyte, these unique germ line cells pass through a protracted course of differentiation that is punctuated by segments of modification in both mitotic and meiotic states that define the process of oogenesis. Thus, mammalian organisms have evolved complex strategies within the female soma that house, nurture, signal, and release mature oocytes at times that optimize sexual encounters with males and therefore ensure propagation of the species.

Life cycles typically are defined as sequential processes whereby the birth of an organism is followed by development to and through a phase of reproductive competence, and finally as it approaches death, it has either lost its reproductive potential or retained it for varying durations (1). Death before the end of the reproductive life span is the norm in nature for most mammals. Death after the demise of the reproductive competence is the norm for *Homo sapiens*. Mammals, not surprisingly then, vary widely in their duration of life span and in relation to the fraction of their overall life span that is dedicated to acquiring and sustaining their fecundity or reproductive potential (1). The life cycle paradigm can be applied at a cellular and molecular level to the natural history of the female germ line. For example, there is a period of time either during fetal development (primates, ungulates) or postnatally (rodents, lagomorphs) when primordial germ cells initiate oogonial mitotic proliferation in anticipation of their formal entry into meiosis and become arrested in the diplotene stage of meiotic prophase 1 (2). This is defined as the birth of the oocyte. The next stage of the oocyte life cycle is represented by the storage phase during which the germ cells remain in a state of meiotic arrest confined within the primordial follicle. Oocytes proceed through a growth phase during which the conditions of hypertrophy are satisfied by the accumulation of organellar and molecular determinants that support embryogenesis (3). Finally, the oocyte undergoes a definitive period of maturation around the time of ovulation in most mammalian species (4).

These four stages provide a useful framework upon which the legacy and fate of the female germ line can be systematically analyzed. Thus, the purpose of this chapter is to uncover the most recent plausible mechanisms that underlie the fundamental processes operative during the birth, storage, growth, and maturation of mammalian oocytes as we try to understand the consequences of the natural course of evolution of this remarkable cell.

It is hoped that this perspective will guide and inform the major theme of this book as basic and clinical science confronts the challenges that lie ahead in the fields of infertility treatment and fertility preservation. Moreover, our ever-increasing efforts to place unnatural constraints on the progression of oogenesis will hopefully be tempered accordingly for the benefit of humankind.

BIRTH

As mentioned above, the birth of a mammalian oocyte typically occurs when primordial germ cells assume a "female" identity as oogonia. This is but the first step in a process that cannot be disassociated from the gender-specific differentiation of the mammalian gonad and encompasses in most mammals the spectrum of somatic developmental stages from embryo to fetus to neonate (5). Most important is the recent recognition that while genetic sex specifies a testiscular or ovarian pattern of gonadogenesis, birthing of an oocyte is not cell autonomous but rather directed by the somatic environment or niche in which oogonia find themselves (6,7). The most compelling data that bears witness to the importance of the somatic environment for oogenesis comes from work on the mouse (Fig. 1). Incomplete cytokinesis during oogonial proliferation results in the formation of germ cell nests or syncitia (8). A penultimate S phase of the cell cycle readies oogonia for their transition from mitosis into meiosis. In mice, entry into meiotic prophase is initiated at embryonic day 14.5 and appears to be due to both the expression of Stra 8 and the lack of expression of the male determining factor Sry (7). Interestingly, the gender-specific delay in meiosis entry for male spermatogonia is delayed until postnatal day 15 when the central role of Stra 8 is recapitulated for this meiosis inducer. How is meiotic induction controlled? An elegant series of experiments has recently addressed this question (6,7).

It appears that Stra 8 in induced by the production of retinoic acid (RA) produced in somatic cells throughout the embryo. What distinguishes the ovarian environment from its male counterpart in the embryo is its ability to metabolize RA via the Cyp 26b1 gene product that ensures not only germ-cell-specific Stra 8 but also the appearance of meiosis-specific factors such as SCYP3. Thus, the timely expression of somatic factors allows for the RA pathway to induce the onset of meiosis in post-mitotic gonocytes heralding the birth of the mammalian oocyte.

Once the gender-specific assignment of the germ line has occurred in the female, oocyte progression through meiotic prophase can proceed, albeit asynchronously, throughout germ cell nests (8,9). Interactions between somatic cells and germ cells result in the formation of primordial follicles some of which, known as primordial oocytes, will remain arrested through the reproductive life span of the organism. In most mammals, it is well known that the vast majority of germ cells fail to adapt to the safe confines of the primordial follicle and instead are eliminated shortly after birth (9). While the mechanisms of primordial follicle assembly and maintenance of meiotic arrest are only poorly understood in mammals, the factors that establish and maintain a finite reserve of follicles into adulthood are being uncovered (10).

STORAGE AND ACTIVATION

Primordial follicles that survive the postnatal and prepubertal stages of ovarian development establish what is known as the follicle reserve. The follicle reserve represents a finite reservoir of primordial follicles that varies between mammalian species and diminishes with maternal aging. The size of the follicle reserve therefore reflects the particular demands that a given species has for providing an adequate number of ovulations during its reproductive life span. Here too there is great diversity between mammalian species. Both the size of the follicle reserve and the rate of recruitment of primordial follicles into the growing pool define the operational limits of fecundity in mammals. Animal age, diet, number of pregnancies, environmental exposures, and reproductive strategy (monovular vs. polyovular, spontaneous vs. reflex ovulation) all contribute to the rate of consumption of the follicle reserve (11–14). Depletion of the follicle reserve heralds the end of the reproductive life span. Thus, given the progressive decline in primordial follicles, and the added influence that time imposes on the cumulative exposure of oocytes to medications, environmental contaminants, and infectious agents, there is every reason to better understand the factors responsible for maintaining primordial oocytes in a quiescent state from which they become subsequently activated. Thus, it is equally relevant to ask what has been learned about the trigger(s) that result in the activation of the primordial follicle that allows oocytes to enter the growth phase of oogenesis (3).

Current concepts regarding storage of oocytes (i.e., the follicular reserve) have again been buttressed by studies on the mouse ovary (9,11,15). Most interesting are the studies of Pan et al. (16) that have now begun to define the properties of the oocyte genome in relation to its life cycle. For example, when the complexity of oocyte transcription is assessed by microarray studies and confirmed by polymerase chain reaction after reverse transcription of RNA, what appears to be the most active time for gene activation is the time between the end of oogonial proliferation and arrest in the primordial follicle. While it has been quite apparent that both the growth and the maturative phases of oogenesis are, respectively, dependent on transcriptional and post-transcriptional regulation of the oocyte genome, some combination of these are operative

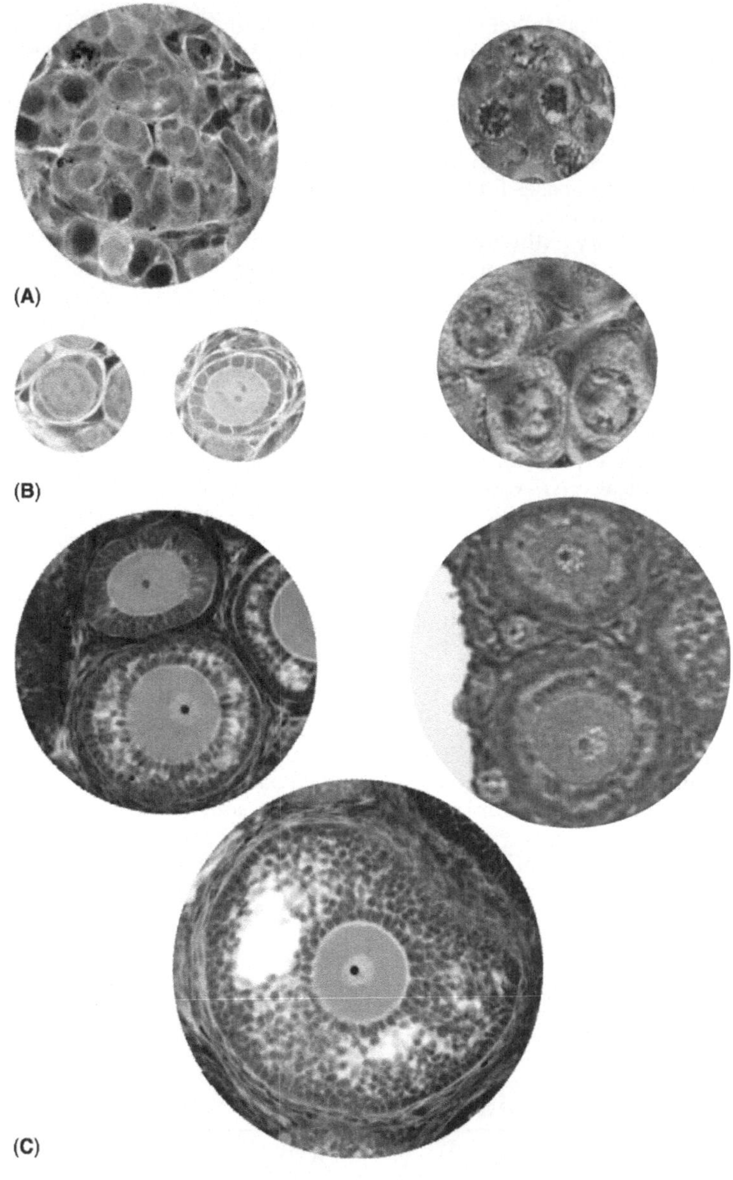

(A)

(B)

(C)

Figure 1 Images demonstrating the sequential stages of oogenesis in the mouse ovary. (**A**) A germ cell cyst (*left*) and examples of synchronous entry into meiosis of germ cells labeled with mouse vasa homologue antibody (*right*). (**B**) Primordial and transitional follicles that represent the major storage form of oocytes in the mammalian ovary. (**C**) Various stages of follicular development through primary, secondary, and early antral types of follicles during which time the oocyte enters into and completes its growth phase. Note the prominent nucleolus that appears as a dark spot within the oocyte nucleus or germinal vesicle.

during the birth and meiotic quiescence intervals. This has been borne out by the identification of many genes, originally thought to be somewhat disparate in their mode of regulation, that if eliminated cause interesting phenotypes in mouse oocytes (Table 1). For example, meiotic failure often results in germ cell loss prior to primordial oocyte formation, as typified in BRAC1 knockout mice (3). Many phenotypes display normal progression through meiotic prophase but oogenesis fails at the time of primordial follicle formation, as exemplified by Figα (17) and Nobox (18). And finally, there are genes that appear to establish and maintain quiescence, as their experimental deletion in both systemic and germ-line-targeted mutations permit breakdown of germ cysts, formation of primordial follicles, but during postnatal life, result in the gradual but accelerated depletion of primordial follicles due to uncontrolled activation (19–27). In essence, these murine mutations phenocopy the human condition of premature ovarian failure and provide provocative insights into the mechanistic bases for ovarian diseases.

Closer inspection of both the types of genes that disrupt quiescence and/or activation and newly discovered regulatory pathways such as PTEN are beginning to suggest that quiescence is an "active" state that in effect deploys as yet unidentified ligand receptor activation events that are responsible for the maintenance of primordial follicles and meiotic arrest during this early stage of oogenesis (28). At least one such pathway is represented by the transcription factor Foxo3. Here, all events leading up to and through the growth phase of oogenesis proceed normally in the ovaries of Foxo3 null females, and yet in the absence of this factor, ovaries become completely devoid of follicles by 15 weeks of age (19,20). The suggestion by these authors and others that a repressive signal is generated in the developing ovary that is used to prevent follicle activation is evocative and bears practical consequences for the practice of fertility preservation. Specifically, it is well known that the placement of the ovarian cortex into culture results in the wholesale activation of primordial follicles, a practice that is widely anticipated for use with frozen cortical slices of ovarian tissue (29). If active release of repression is induced by freezing and/or extirpation, then adequate laboratory conditions need to be established to prevent acute oocyte depletion due to primordial follicle activation following tissue thawing, organ culture, and/or transplantation (3).

GROWTH

Few processes in the life cycle of the oocyte are more important than those that govern meiosis and growth. As the largest cell in the body, and one whose volume expansion approximates a 160-fold increase in mass (bear in mind that an average somatic cell undergoes a two-fold increase in mass), this remarkable example of true hypertrophy underscores the need for mass conservation during preimplantation embryonic development in eutherian mammals (2). The mother's legacy as a provider for her offspring sets the singular charge of the growth phase of oogenesis (4).

Table 1 Genetic Regulation of Early Oogenesis

	Cyst	Primordial				
Gene (location)	Breakdown	Assembly	Precocious activation	Delayed activation	Abnormal activation	Ref.
Sohlh 2 (O)			X		X	22
Sohlh 1 (O)				X	X	23
Lhx8 (O)				X		23
Figα (O)	X	X		X		17
Nobox (O)	X			X	X	18
Foxl2 (pGC)	X			X	X	24,25
Foxo3 (O)			X		X	19,20
PTEN* (O)			X		X	28
GDF9 (O)			X		X	30,31
Retinoblastoma			X			26
AMH			X			27
Smad 3				X	X	33
Kit				X	X	34

*Indicates conditional knockout.

Abbrevations: AMH, anti-müllerian hormone; GDF, growth differentiation factor; O, oocyte; pGC, pregranulosa cell; PTEN, phosphatase and tensin homolog.

Oocyte growth is both highly regulated with respect to its coordination with folliculogenesis and a necessary step in the life cycle of the oocyte before completing meiosis. There have been many recent reviews on the topic of oocyte growth regulation (3,4,11,30). What is being repeatedly reinforced from surveys of this kind is the fact that the growing oocyte derives more of its metabolic support than has been previously appreciated from the neighboring follicular granulosa cells (31,32). Moreover, the notion that feedback regulation between the oocyte and granulosa cells dictates both the rate and the extent of oocyte growth has withstood the test of time, and appears to be mediated by both gap junctional and paracrine forms of cell communication (3,4,15,33). One of the better-characterized examples of paracrine control is that of the oocyte-produced GDF9 member of the TGFβ growth factor family. As shown in Figure 2, mice deficient in this gene exhibit discrete patterns of oocyte growth that reflect a loss of coordination between oogenesis and folliculogenesis. Most striking perhaps is the fact that the ovarian architecture itself, which normally results in the formation of many different somatic tissues, is strikingly simplified when oocyte GDF9 fails to be expressed. Thus, oocyte-specific factors have pronounced effects on the course of ovarian differentiation that can impact the viability of frozen tissues depending on their species- or age-specific properties. A most striking deficit observed in GDF9 null mice is the apparent loss of transzonal projections. As shown in Figure 3, the zona pellucida is laden with extensions from cumulus cells that serve to provide nutrients from the adjoining somatic cells. These structures can vary widely between different mammalian species and may represent different degrees of metabolic dependence for oocytes that undergo varying durations of growth.

Another dramatic and relevant distinction in ovarian architecture that impacts ovarian tissue cryopreservation is the presence of the *tunica albuginea*.Figure 4 illustrates the storage site for primordial follicles in four different species of primates that have adopted different reproductive strategies but share in common with humans the use of the tunica albuginea. This band of dense connective tissue both subtends the ovarian epithelium and maintains primordial follicles in an avsacular environment. Such an arrangement contrasts sharply with the ovary of rodents in which follicle pools persist proximal to both the surface epithelium and the dynamic vasculature. Thus, strategies for cryopreservation must take into account the widely divergent architectural principles that are used in maintaining the primordial follicle reserve.

There are two aspects of oocyte growth that are presently undergoing close scrutiny with respect to the role this stage of the oocyte life cycle plays in the initiation and support of preimplantation embryogenesis. First, the characterization of the maternal inheritance has classically referred to the amplification and storage of organelles; in other words, quantal organellar hyperplasia is achieved without necessarily attaining organellar maturity and functionality. Mitochondria are often the model organelle discussed in this light. A 3 to 4 log order increase in both number and mitochondrial DNA copies is due to the proliferation of mitochondria during oogenesis. This endowment is thought to support the metabolic demands of early embryogenesis at least through the early postimplantation stages of embryogenesis. It now appears that mitochondria have foreshadowed the importance of other organelles whose embryonic functions have been safely ascribed to their production during oocyte growth. Amongst these are nucleoli, centrosomes, and polarity determinants that must function during the allocation of trophectoderm and inner cell mass (3). Thus, the sensitivity of these organelles to cryopreservation at the oocyte stage needs to be more closely evaluated with respect to their embryonic duties.

Second, oocyte growth is associated with both the acquisition of meiotic competence and the imprinting of female-specific genes (4). In the midst of one of the most robust examples of transcriptional activation and maintenance in differentiating cells, the oocyte must revisit its meiotic past and obtain the molecular factors, in the form of proteins or stabilized mRNAs, that will allow it to complete the process of meiosis (nuclear maturation, see below). Furthermore, passage through the germ line is known to result in the demethylation of imprinted loci in the oocyte genome. Thus, as is well documented in the mouse, both re-marking these alleles and assuring that the oocyte bears the methyltranferases that will eventually re-methylate the maternal and paternal genomes are accomplished during oocyte growth (3).

Growth of the mammalian oocyte is then more than an opportunity to "bulk up" on behalf of the embryo. This process utilizes a fundamental signaling mechanism based in the kit ligand and c-kit receptor pathway that is segregated respectively between granulosa cells and the oocyte (21,34,35). It represents a phase in the life cycle whereby intimate communication with the female soma ensures states of mutually agreeable metabolic homeostasis to prevent production

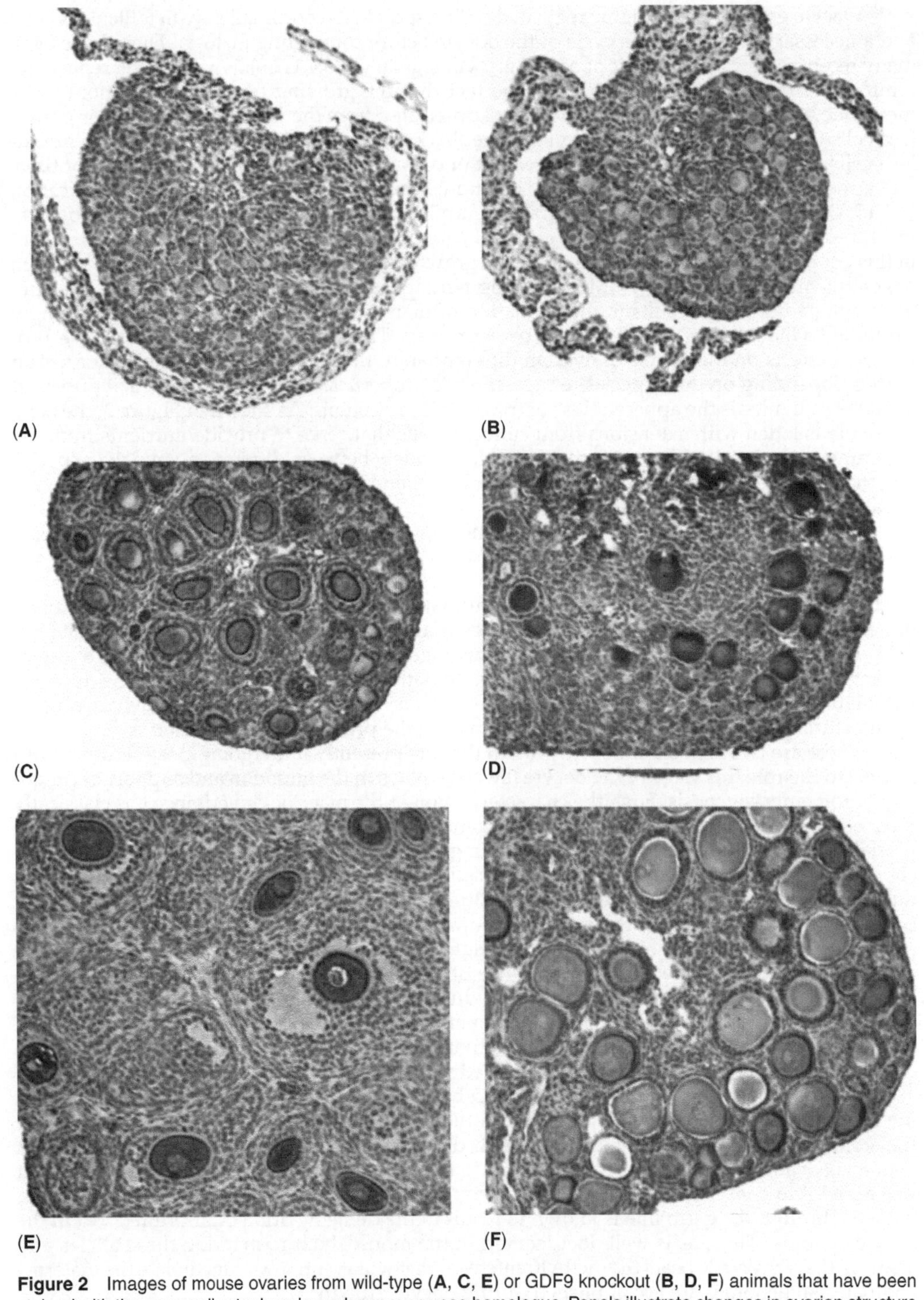

Figure 2 Images of mouse ovaries from wild-type (**A**, **C**, **E**) or GDF9 knockout (**B**, **D**, **F**) animals that have been stained with the germ cell cytoplasmic marker mouse vasa homologue. Panels illustrate changes in ovarian structure from 2-day-old (**A**, **B**), 10-day-old (**C**, **D**), and 21-day-old (**E**, **F**) animals. Note that in controls (**A**, **C**, **E**) there is a progressive dilution of germ cells while somatic compartments are amplified and that in the absence of functional GDF9 (**B**, **D**, **F**) germ cell density remains high and little ovarian volume can be attributed to the somatic compartment. GDF9 depletion causes a remarkable increase in oocyte size due to impaired feedback between oocytes and follicle cells.

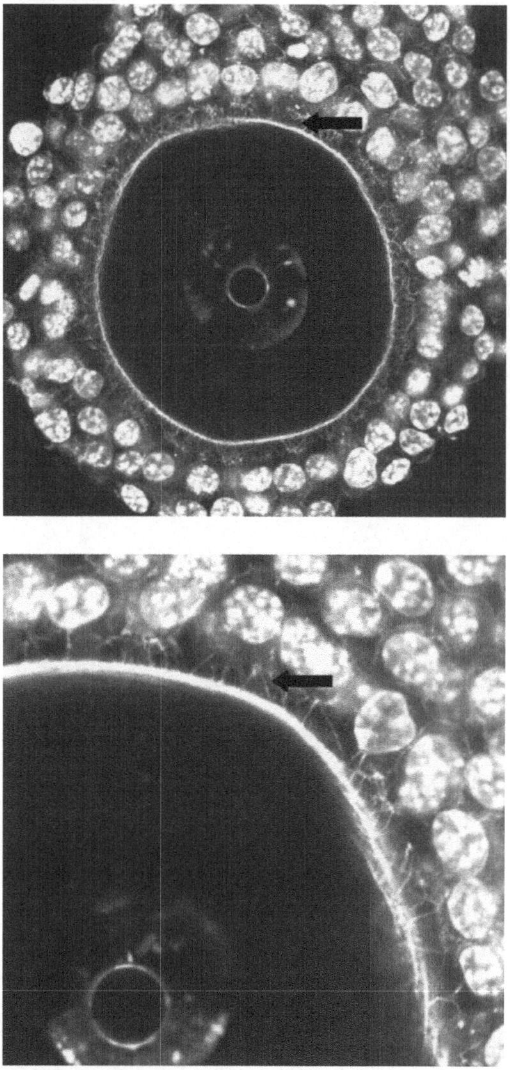

Figure 3 Confocal micrographs of germinal vesicle stage mouse oocyte highlighting the density of transzonal projections at low (*top*) and high (*bottom*) magnification (see arrows). Multiple extensions of cumulus cells traverse the zona pellucida and attach at the oocyte plasma membrane.

of inferior ova. And, it provides for regulation of the completion of the meiotic process. That fundamental alterations in chromatin, such as imprinting and formation of nucleolar precursors, occur during oocyte growth seems to signal the appropriately timed acquisition, in largely epigenetic terms, of qualities that will drive embryonic development (36).

MATURATION

A fully grown mammalian oocyte is not necessarily a developmentally competent one. Rather, it has been appreciated that the oocyte must undergo a maturative phase of development that is traditionally referred to as "meiotic maturation." The simple truth of the matter is that

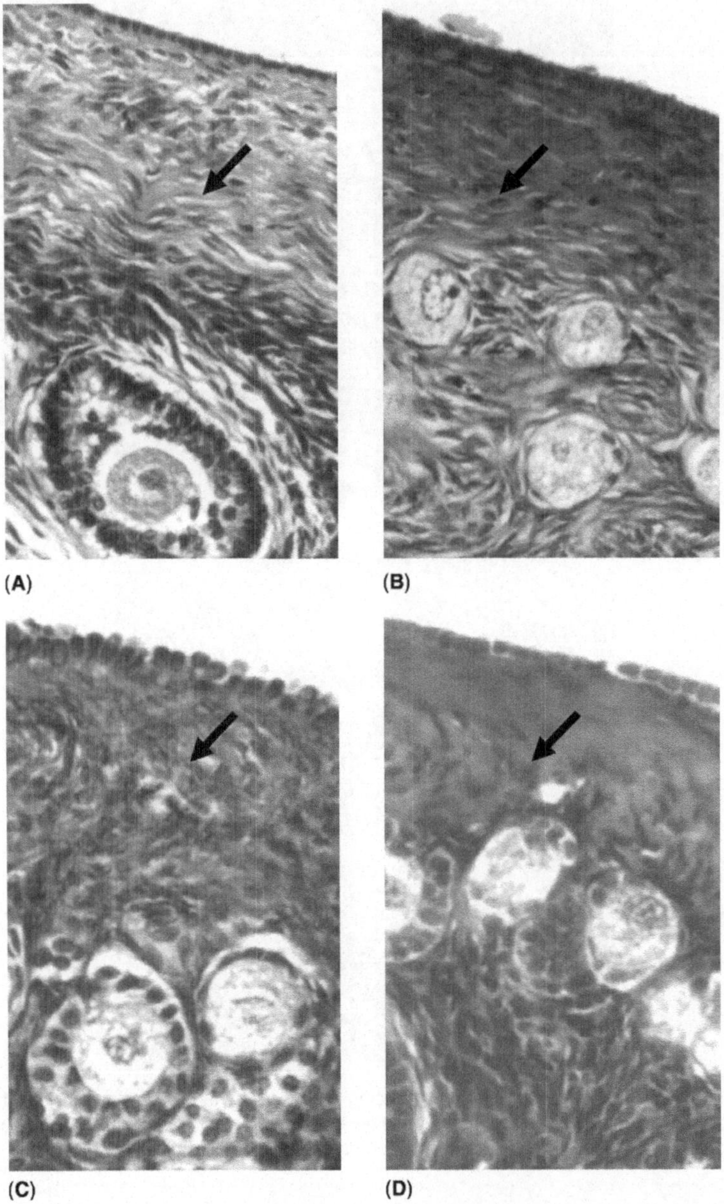

Figure 4 Histological sections of primate ovaries from two old world [*Macaca arctoides* (**A**); *Macaca fascicularis* (**B**)] and two new world [*Cebus albifrons* (**C**); *Saimiri sciureus* (**D**)] primates. Note the dense connective tissue known as the tunica albuginea (arrows) that is oocyte free and subtends the overlying ovarian epithelium. This configuration is typical of human ovary and illustrates the importance of the microenvironment of primordial follicles in ovarian tissue preparations for cryopreservation.

oocyte maturation is a necessity if the oocyte is to effect any degree of development after being fertilized (2). Operationally, oocyte maturation is defined as being either nuclear or cytoplasmic in nature, the former signaling the completion of meiosis and the latter reflecting modifications in cytoplasmic composition and organization (16,37–39). Nuclear maturation encompasses the events associated with the segregation of homologous bivalents during meiosis 1, effecting the haploid condition, and the segregation of sister chromatids during meiosis 2. Cytoplasmic maturation entails a complex series of events that guarantee that the egg will respond correctly to the fertilization stimulus by eliciting departure from the meiotic cell cycle and second polar body extrusion and entry into the embryonic cell cycle. Post-transcriptional and post-translational processes underlie many of these changes in the quality of the oocyte cytoplasm. In addition, for most of the maternal mRNAs that have been in safe storage (37), maturation is the chapter in the oocyte's life cycle when the mRNA by-products of gene expression are either finally translated to make proteins or suffer a timely demise at the hands of the RNA degradation machinery (38). In this way, the embryo inherits what amounts to a clean slate of mRNA species that will be replaced by new templates generated from the zygotic genome (4).

From this very general background, it is worthwhile to examine basic principles in the process of oocyte maturation that are likely to impact oocyte quality determination. Three factors have recently emerged that are likely to bear on human assisted reproductive technologies (ARTs) in general and oocyte cryopreservation in particular.

How does in vitro maturation compare with maturation within the context of the ovarian follicle? It has long been known that the use of controlled ovarian hyperstimulation (COH) adversely effects oocyte maturation in humans. Reduced oocyte quality is thought to be due to the loss of synchrony between nuclear and cytoplasmic maturation during COH and is distinct from the more synchronous course of meiotic progression seen during a natural cycle ovulation. This has now been borne out in many animal models (39). Moreover, it is commonly accepted that in vitro-matured mammalian oocytes are generally of inferior developmental potential when compared to oocytes retrieved after natural cycle ovulations or those stimulated by exogenous gonadotropins (37).

What is the relative contribution of the oocyte and cumulus cells in effecting nuclear and cytoplasmic maturation? Here, too, there is little doubt that cumulus cells play a major role both in controlling the meiotic status of the oocyte and in providing a boost to the metabolism of the oocyte that is needed for egg activation and subsequent development (4,3). How signaling pathways between the oocyte and cumulus regulate the re-initiation and completion of meiosis has been the subject of many studies and excellent reviews on this topic have appeared (40). In contrast, how cumulus cells affect the metabolic performance of the oocyte as it matures has only recently been fully appreciated. The historical roots of metabolic support take their origins in the work of Perreault et al. (41) in showing the importance of cumulus cell-derived glutathione in the control of male pronuclear formation. Since theses studies appeared, compelling evidence has been forthcoming to show that reduced glutathione (GSH) is derived from cumulus cells and passes through gap junctions resulting in ooplasmic accumulations of the order of 5 to 8 mM[3]. It is perhaps not surprising then that besides the role of GSH in pronuclear chromatin modifications, the widely conserved function of GSH as an anti-oxidant emphasizes the importance of metabolic loading for the development and survival of the embryo. It is no wonder then that human ARTs, having recognized the impact of oxidative stress on gamete and embryo performance in vitro, need to engage this aspect of oocyte maturation in future studies.

Finally, at what stage of maturation is the oocyte most likely to retain and express quality determinants needed to establish pregnancy? The question speaks directly to the subject of this book. While conventional oocyte cryopreservation approaches have engaged the use of the mature metaphase arrested oocyte, there are growing concerns regarding the viability and developmental health of thawed oocytes. Given the factors noted above, we propose that more work should be focused on the use of immature oocytes, especially if the problem of metabolic loading can be recapitulated in an in vitro system. Thus, at the end of a long road, oocyte differentiation can be said to have been completed once it has undergone both cytoplasmic and nuclear maturation. What remains in this life cycle is the passage from the ovary and into the purview of embryonic development, a transition that is effected by the process of egg activation.

Egg Activation

In the moment the oocyte receives a signal to launch the conceptus' developmental program, it is said to be activated. Egg activation is both the end of oogenesis and the beginning of embryogenesis and represents a unique point in the continuum that is the female germ line (4). It is most typically defined in terms of acute events that enable formation of the one-cell zygote and the more protracted series of events that lead to the establishment of a viable embryo (41,42). With evidence in hand that subtle modifications in imprinted genes make permissive the embryonic potential of the oocyte, formally known as parthenogenesis, mammals can assume the rightly distinction of having a true continuum between oogenesis and embryogenesis (2). This brief overview of the acute and long-term events resulting from egg activation is meant to serve as reinforcement for the idea that the oocyte is indeed the only truly totipotent cell in the body. It further provides a framework for defining those qualities of the egg that must be scrutinized in order to determine the long-term impact of oocyte cryopreservation on developmental potential.

INTERRUPTING THE OOCYTE LIFE CYCLE: A CLINICAL IMPERATIVE

In conclusion, the oocyte passes through many discrete developmental processes in the course of its birth, growth, maturation, and activation. Each of these steps results from a transition between a quiescent state to the next step in a differentiation pathway. Thus, the female germ line has an intrinsic tendency to pause at key developmental states over time periods that can range from hours to years. How will oocyte cryopreservation impact this property of the mammalian oocyte?

Answers to this question will await further research that should be aimed at identifying the most appropriate stages of oogenesis when cryopreservation can be implemented without seriously compromising initiation and completion of the next step in this process. Logically, then, the later in oogenesis cryopreservation is imposed, the less likely will be the severity of damage to the developmental potential of the female gamete. This limitation should be kept in mind when counseling patients with different requirements for fertility preservation as the menu of cryopreservation options for oocytes continues to grow.

ACKNOWLEDGMENTS

D.F. Albertini dedicates this chapter to inspiring mentors, Professors Everett Anderson and John Biggers, who nurtured his interests in the study of oogenesis. It is a pleasure to acknowledge the continuing support of the NIH, the Hall Family Foundation, the ESHE Fund, and our colleagues in the Center for Reproductive Sciences at Kansas University.

REFERENCES

1. Potts M, Short R. Sex and pregnancy. In: Ever Since Adam and Eve: The Evolution of Human Sexuality. Cambridge: Cambridge University Press, 2000: 104–30.
2. Biggers JD. Oogenesis and ovum maturation. In: Segal SJ, et al., eds. The Regulation of Mammalian Reproduction. Springfield: Charles C. Thomas, 1972: 273–83.
3. Rodrigues P, Limback D, McGinnis L, et al. Oogenesis: prospects and challenges for the future. J Cell Physiol 2008; 216: 355–65.
4. Hutt KH, Albertini DF. An oocentric view of folliculogenesis. Reprod Biomed Online 2007; 14: 758–64.
5. Bainbridge D. The X in Sex: How the X Chromosome Controls Our Lives. Cambridge MA: Harvard University Press, 2003: 125–70.
6. Bowles J, Knight D, Smith C, et al. Retinoid signaling determines germ cell fate in mice. Science 2006; 312: 596–600.
7. Koubova J, Menke DB, Zhou Q, et al. Retinoic acid regulates sex-specific timing of meiotic initiation in mice. Proc Nat Acad Sci 2006; 103: 2474–9.
8. Pepling ME, Spradling AC. Mouse ovarian germ cell cysts undergo programmed breakdown to form primordial follicles. Devel Biol 2001; 234: 339–51.
9. Guigon CJ, Magre S. Contribution of germ cells to the differentiation and maturation of the ovary: insights from models of germ cell depletion. Biol Reprod 2006; 74: 450–8.

10. Begum S, Papaioannou VE, Gosden RG. The oocyte population is not renewed in transplanted or irradiated adult ovaries. Hum Reprod 2008; 23: 2326–30.
11. Skinner MK. Regulation of primordial follicle assembly and development. Human Reprod Update 2005; 11: 461–71.
12. Susiarjo M, Hassold TJ, Freeman E, et al. Bisphenol A exposure in utero disrupts early oogenesis in the mouse. PLoS Genet 2007; 3: e5.
13. McLaren A. Germ and somatic cell lineages in the developing gonad. Molec Cell Endo 2002; 163: 3–9.
14. Motta PM, Makabe S, Naguro T, et al. Oocyte follicle cells association during development of human ovarian follicle. A study by high resolution scanning and transmission electron microscopy. Arch Histol Cytol 1994; 57: 369–94.
15. Ackert CL, Gittens JE, O'Brien MJ, et al. Intercellular communication via connexin43 gap junctions is required for ovarian folliculogenesis in the mouse. Devel Biol 2001; 233: 258–70.
16. Pan H, O'Brien MJ, Wigglesworth K, et al. Transcript profiling during mouse oocyte development and the effect of gonadotropin priming and development in vitro. Devel Biol 2005; 286: 493–506.
17. Soyal SM, Amleh A, Dean J. FIGalpha, a germ cell-specific transcription factor required for ovarian follicle formation. Development 2000; 127: 4645–54.
18. Rajkovic A, Pangas SA, Ballow D, et al. NOBOX deficiency disrupts early folliculogenesis and oocyte-specific gene expression. Science 2004; 305: 1157–9.
19. Castrillon DH, Miao L, Kollipara R, et al. Suppression of ovarian follicle activation in mice by the transcription factor Foxo3a. Science 2003; 301: 215–18.
20. John GB, Shirley LJ, Gallardo TD, et al. Specificity of the requirement for Foxo3 in primordial follicle activation. Reproduction 2007; 133: 855–63.
21. Moniruzzaman M, Sakamaki K, Akazawa Y, et al. Oocyte growth and follicular development in KIT-deficient Fas-knockout mice. Reproduction 2007; 133: 117–25.
22. Choi Y, Yuan D, Rajkovic A. Germ cell-specific transcriptional regulator Sohlh2 is essential for early mouse folliculogenesis and oocyte-specific gene expression. Biol Reprod 2008; 79: 1176–82.
23. Pangas SA, Choi Y, Ballow DJ, et al. Oogenesis requires germ cell specific transcriptional regulators Sohlh1 and Lhx8. Proc Natl Acad Sci USA 2006; 103: 8090–5.
24. Schmidt D, Ovitt CE, Anlag K, et al. The murine winged-helix transcription factor Foxl2 is required for granulosa cell differentiation and ovary maintenance. Development 2004; 131: 933–42.
25. Uda M, Ottolenghi C, Crisponi L, et al. Foxl2 disruption causes mouse ovarian failure by pervasive blockage of follicle development. Hum Mol Genet 2004; 13: 1171–81.
26. Andreu-Vieyra C, Chen R, Matzuk MM. Conditional deletion of the retinoblastoma (Rb) gene in ovarian granulosa cells leads to premature ovarian failure. Mol Endocrinol 2008; 22: 2141–61.
27. Durlinger AL, Kramer P, Karels B. Control of primordial follicle recruitment by anti-müllerian hormone in the mouse ovary. Endocrinology 1999; 140: 5789–96.
28. Reddy P, Liu L, Adhikari D, et al. Oocyte-specific deletion of *Pten* causes premature activation of the primordial follicle pool. Science 2008; 319: 611–13.
29. Braw-Tal R, Yossefi S. Studies in vivo and in vitro on the initiation of follicle growth in the bovine ovary. J Reprod Fert 1997; 109: 165–71.
30. Elvin JA, Clark AT, Wang P, et al. Paracrine actions of growth differentiation factor-9 in the mammalian ovary. Mol Endocrinol 1999; 13: 1035–48.
31. Yan C, Wang P, Demayo J, et al. Synergistic roles of bone morphogenetic protein 15 and growth differentiation factor 9 in ovarian function. Mol Endocrinol 2001; 15: 854–66.
32. Diaz FJ, Sugiura K, Eppig JJ. Regulation of Pcsk6 expression during the preantral to antral follicle transition in mice: opposing roles of FSH and oocytes. Biol Reprod 2008; 78: 176–83.
33. Tomic D, Brodie SG, Deng C, et al. Smad 3 may regulate follicular growth in the mouse ovary. Biol Reprod 2002; 66: 917–23.
34. Kissel H, Timokhina I, Hardy MP, et al. Point mutation in kit receptor tyrosine kinase reveals essential roles for kit signaling in spermatogenesis and oogenesis without affecting other kit responses. EMBO J 2000; 19: 1312–26.
35. Thomas FH, Ismail RS, Jiang J-Y, et al. Kit Ligand 2 promotes murine oocyte growth in vitro. Biol Reprod 2008; 78: 167–75.
36. Zuccotti M, Merico V, Sacchi L, et al. Maternal Oct-4 is a potential key regulator of the developmental competence of mouse oocytes. BMC Dev Biol 2008; 8: 97–111.
37. Kocabas AM, Crosby J, Ross PJ, et al. The transcriptome of human oocytes. Proc Nat Acad Sci 2006; 103: 14027–32.
38. Su Y-Q, Sugiura K, Woo Y, et al. Selective degradation of transcripts during meiotic maturation of mouse oocytes. Devel Biol 2007; 302: 104–17.
39. Johnson LD, Mattson BA, Albertini DF, et al. Quality of oocytes from superovulated rhesus monkeys. Hum Reprod 1991; 6: 623–31.

40. Mehlmann LM, Jones TLZ, Jaffe LA. Meiotic arrest in the mouse follicle maintained by a Gs protein in the oocyte. Science 2002; 297: 1343–5.
41. Perreault SD, Barbee RR, Slott VL. Importance of glutathione in the acquisition and maintenance of sperm nuclear decondensing activity in maturing hamster oocytes. Dev Biol 1988; 125: 181–6.
42. Chatzimeletiou K, Morrison EE, Prapas N, et al. Spindle abnormalities in normally developing and arrested human preimplantation embryos in vitro identified by confocal laser scanning microscopy. Hum Reprod 2005; 20: 672–82.

7 Essential Features of the Mature Oocyte

Ursula Eichenlaub-Ritter

Department of Gene Technology and Microbiology, Faculty of Biology, University of Bielefeld, Bielefeld, Germany

OOCYTE MATURITY RELIES ON ITS HISTORY

As early as in 1988, Paul Wassarman pointed out that embryogenesis begins during oogenesis, emphasizing the essential role of the oocyte and of oocyte quality in the formation of a healthy embryo and offspring (1). David Albertini contradicted this suggestion by stating that "oogenesis begins during embryogenesis," meaning that each individual stage of oogenesis, from primordial germ cell formation to early mitotic and meiotic stages in the embryonic gonad, the long meiotic resting stage followed by follicle recruitment and oocyte growth, and the last stages of maturation until the metaphase II (MII) oocyte becomes ovulated, is important in the genesis of a mature oocyte and may be uniquely susceptible to disturbances that ultimately affect oocyte quality and the health of the embryo and offspring. Studies carried out in the last few years have revealed that there is extensive cross talk between the oocyte and the somatic compartment, particularly when oocytes grow and resume maturation. The oocyte acts as "the captain of the ship" (John Eppig) during folliculogenesis by directing expression in the granulosa cells, resulting in differentiation between the cumulus and mural granulosa cells, antrum formation, and high cooperativity in metabolism between the oocyte and the follicle, thus creating a microenvironment that supports full oocyte maturation (2–4). Only this intricate and integrated network of signaling, via growth factors such as GDF9, their receptors, and cellular signaling pathways, in a concerted manner ensures full growth of the oocyte, as well as its acquisition of competence to resume meiotic maturation and progress to MII and complete nuclear maturation after fertilization. Moreover, it is essential for acquiring the capacity to reach full cytoplasmic maturity, which is characterized by the ability to not only support completion of first and second meiosis but also provide all factors for remodeling of the male chromatin, supporting first mitotic divisions and zygotic gene activation, giving rise to totipotency, and storing all molecules and cell organelles/components for early mitotic divisions until zygotic gene products can eventually replace and provide factors for further development and differentiation. In fact, while "maturity" describes a seemingly static stage of oogenesis, it is derived by a precise series of timed events that orchestrate oocyte growth and maturation, and events such as the maternal imprinting in a sequential, highly ordered manner during oocyte growth (5,6), the conformational changes in chromatin leading to a transcriptionally quiescent stage before resumption of maturation (7–9), and the recruitment and degradation of messages and proteins at specific stages of maturation (10–13). The sequential program of production, activation, degradation, and inactivation of enzymes is closely linked to post-translational modifications of proteins (e.g., by phosphorylation/dephosphorylation) and governs the timed activation of enzymes for cell cycle regulation (14–16) and protein complexes for recruitment, translation, and degradation of maternal messages (10,13,17), and concomitantly for spindle formation and cell cycle progression, the separation of homologous chromosomes and finally cytokinesis, and assembly of MII chromosomes (dyads) with centromerically attached sister chromatids on the meiosis II spindle, making the cell ready to go into anaphase II upon sperm entry (14,16). Maturity also describes the competence of the oocyte to maintain parental imprint marks during early embryogenesis after fertilization, while concomitantly initiating active demethylation of cytosine bases in DNA of heterochromatin and passive demethylation of the maternal heterochromatin during early mitotic divisions (18–20), all in a context of sequential alterations in post-translational modifications of histones and chromosomal proteins modifying chromatin conformation. The environment during maturation or after fertilization can cause disturbances (20). Furthermore, chromosomal stability and DNA integrity are influenced by the repair capacities in mature ooplasm (21), including protection of telomeres from nucleases and repair of lesions in male chromatin (22,23) and possibly elongation of telomeres by factors contained in the ooplasm (24).

Thus, when oocytes are considered "mature," this comprises a number of key features associated with high developmental potential such that the oocyte can be considered:

1. An oocyte that has fully grown in size within a large antral follicle, which has remodeled its chromatin during the growth phase prior to resumption of maturation. All maternal imprints are laid down in a time- and growth-dependent fashion and the chromatin attains a "surrounded nucleolus" configuration, in which it is largely transcriptionally inactive before resumption of maturation (9) and forms a rim around the nucleolus, in preparation to become condensed and progress to first meiosis and condense chromosomes in a chromatin conformation that mediates formation of functional centromeres/kinetochores as attachment sites and hubs of regulatory molecules for establishing a functional spindle apparatus.
2. An oocyte that has the ability to progress to MII of nuclear maturation, has separated the homologous chromosomes in a reductional division at meiosis I, and possesses a highly ordered paracrystalline MII spindle with aligned chromosomes to support normal chromatid separation at anaphase II (and not earlier than anaphase II) (25).
3. An oocyte that not only acquires a large number of relatively inactive mitochondria but is also capable of establishing polarity gradients in its ooplasm, for instance, by formation of microdomains of mitochondrial assemblies with mitochondria of high and low redox potential in the cell cortex or the vicinity of the spindle, respectively (26–28), or by recruiting molecules such as nuclear mitotic antigen (11,29), spindlin (30), nucleoplasmin, protein kinase C, glycogen synthase 3β (31) or transcription factors such as OCT4 (9) to the meiosis II spindle such that they can reach the male chromatin by passing along microtubule tracks from the female to the male chromatin in the shortest, most efficient, and timely manner once the oocyte has been fertilized and the oocyte spindle becomes depolymerized and pronuclei form. Thus, the meiotic spindle is thought to provide a scaffold that mediates spatial and temporal regulation of the signaling pathways orchestrating post-fertilization events.
4. An oocyte that possesses a dense layer of cortical granules and a cytoskeleton that has mediated asymmetric division (32–35). It can be instantaneously remodeled upon sperm entry such that cortical granule exocytosis occurs (36), the membrane undergoes repetitive depolarizations, calcium transients from intracellular stores can activate the oocyte (37) and induce calcium calmodulin-dependent kinase II to release from cytostatic arrest by initiation of proteolysis of meiotic inhibitor EMI2 (early mitosis-like inhibitor 2), and, as a result, activate the anaphase promoting factor APC/C, thus mediating ubiquitinylation of cyclin B regulatory factor of maturation promoting factor (MPF) and inactivating cyclin-dependent protein kinase 1 (Cdk1 kinase) for exit from meiosis II (38,39). This sets the stage for degradation of securin, activation of separase protease, and proteolysis of meiotic cohesin protein, causing loss of cohesion between centromeres of sister chromatids, their detachment, and migration to opposite spindle poles for completion of meiosis II and first polar body formation (40,41).
5. An oocyte that has accumulated proteins, enzymes, and RNAs, in particular messenger RNAs that can be stage specifically recruited for polyadenylation and translation during maturation or early embryogenesis according to consensus sequences in the 3′ untranslated sequence or can be degraded or stored at specific times to orchestrate expression and cell cycle progression and transition from oocyte to embryo (10–13). Such an oocyte should also contain a store of untranslated RNAs and enzymes involved in RNAi-mediated degradation or inactivation of messenger RNAs that modulate gene expression and chromatin conformation in the oocyte and preimplantation embryo and contribute to spindle regulation at oogenesis and control of expression during oogenesis and early embryogenesis (42–44).
6. An oocyte embedded in an expanded cumulus that may contribute to stimulation of oocyte/sperm interactions and an oocyte that possesses a dense and regularly shaped zona pellucida surrounding it, which prevents polyspermic fertilization and also shelters the early embryo from mechanical stresses during passage through the ampullae, creating a microenvironment that may help to enrich molecules in the extracellular space with embryo-derived molecules (45–48).

Thus, a "mature" oocyte that has attained high quality and developmental competence relies on its "history" throughout development rather than presenting just as a specialized cell that possesses a small first polar body, a large cytoplasm, and a zona pellucida, and is arrested at the MII stage of meiosis.

RECOGNITION OF MATURITY FOR CRYOPRESERVATION

An embryologist is faced with the problem of not only selecting the "best," fully mature oocyte for fertilization, but also recognizing the signs of immaturity, overmaturity, and defects associated with aberrant nuclear or cytoplasmic maturation, for instance, by dysmorphisms and aberrations, which may adversely affect development, cause generation of chromosomally aberrant and/or developmentally incompetent embryos, and lead to implantation failure, congenital abnormalities, spontaneous abortion, or disease in the offspring. For cryopreservation it is essential that the most robust oocytes are recognized and selected as they are the most likely to cope with the stress associated with exposure to cryoprotectants, mechanical and osmotic stress, and changes in temperature and media, and with re-forming a spindle and cellular three-dimensional organization in a short period of time in order to stay within the strict time window for optimal fertilization that is related to the fertilization-independent and fertilization-dependent changes in translation, degradation and recruitment of maternal messages and molecules, such as in the spindle apparatus (49–51), or developmental competence (52). The period of thawing and recovery that prepares oocytes for fertilization can be associated to some extent with "postovulatory aging" and therefore has to be kept short. Conversely, it has to be long enough for recruitment of molecules to sites such as the spindle, the cell cortex or sub-plasmalemma, and microdomains within the ooplasm if they have been diffused from their original site during any step of the cryopreservation procedures. Currently, it is largely unknown to what extent preservation might lead to transient degradation of regulatory and structural molecules of the spindle, chromosome condensation, or transient loss of control over the regulation of cellular and mitochondrial redox regulation.

When optimizing cryopreservation, one has to take into account that subpopulations of oocytes may exist that may react differently to cryopreservation. For instance, oocytes from young patients may be protected from loss of cohesion between sister chromatids because they still possess abundant cohesin complexes at their centromeres whereas aged oocytes may have much more fragile centromere cohesion (53–55). As maternal and postovulatory aging can increase the loss of chromosome cohesion and reduce molecules in cell cycle control and cytoskeleton, in addition to maternal factors (53,56–59), aged oocytes can be considered to be especially sensitive to "prolonged" procedures. Healthy young oocytes appear to posses a remarkable plasticity to deal with disturbances such as DNA breaks or disturbances in chromosome congression because they have still functioning checkpoint controls. Aged oocytes may be more susceptible to changes in the environment, temperature, and handling because of permissive cell cycle control and therefore are at increased risk for disturbances ultimately affecting survival and developmental capacity.

As most of the features that are characteristic of a mature oocyte cannot be analyzed directly, especially those related to the expression of maternal factors and chromatin conformation (60), the relevance of detecting non-invasive markers of maturity before and after cryopreservation is therefore of utmost significance for assisted reproduction. Analysis of morphology before and after freezing and of the presence of a spindle can help to identify mature oocytes and the best cryopreservation protocols. Cryopreservation by itself may increase success rates in ART by providing time for completion of chromosome analysis and detection of molecular markers (e.g., in follicular fluid or cumulus) before oocytes are fertilized in a cycle that optimizes uterine receptivity. However, so far only very limited information is available on how maturity, immaturity, or overmaturity and quality before cryopreservation predicts and determines maturity after freezing, and much more research is required in this area.

OOCYTE SIZE AND STATE OF CHROMATIN IN A MATURE OOCYTE

The human oocyte is one of the largest cells in the body and its growth takes place during a prolonged period, which is initially independent of gonadotropic hormones when primordial follicles are recruited, followed by an extensive growth phase in tertiary follicles until oocytes resume maturation downstream of the luteinizing hormone (LH) surge. From studies in human and animal models it is clear that follicle size is related to growth and maturity of oocytes. As growth is associated with sequential alterations in chromatin and maternal imprinting, there has been a concern that hormonal stimulation or in vitro maturation may have adverse effects by preventing completion of imprinting of the maternal DNA for monozygotic gene expression in the embryo and offspring (60–62). Similarly, cryopreservation might adversely affect imprint control. This could either interfere with maintaining maternal imprints during cryopreservation, similar to postovulatory aging (63), or affect the stability and regulation of maternal factors that control preservation of parental imprints in the embryo, when remodeling of paternal and maternal heterochromatin occurs and zygotic gene activation takes place after fertilization (64). Currently, there is no indication that cryopreservation leads to imprint defects in embryos obtained from cryopreserved oocytes or children born from cryopreserved embryos (e.g., from vitrified oocytes) (65,66). Oocyte spindles appear to possess normal shapes (67–69), as discussed in more detail in other chapters of this book. However, alterations in the proteome and ionic homeostasis have been observed in suboptimal cryopreservation protocols (70). Further follow-up studies in human and animal models can hopefully provide more evidence for the safety of procedures on this aspect of oocyte maturity.

OOCYTE NUCLEAR MEIOTIC MATURATION

The presence of a first polar body is an indicator of progression from first to second meiosis, but several studies using non-invasive polarization microscopy revealed that not all oocytes with a polar body have reached MII (71,72). Instead, some may be incompetent to establish a MII spindle whereas others are delayed in meiotic progression and therefore are still in the telophase I or prometaphase II stage of maturation. Obviously, fertilization would pose an increased risk for errors in segregation of chromosomes in the absence of a functional spindle or even failure to extrude one set of chromosomes with the risk of forming a digynic triploid embryo. Freezing oocytes at this stage might affect completion of cytokinesis, but it is also possible that the telophase I/prometaphase II stages might be suitable for cryopreservation as they have not progressed to MII and might do so during recovery from freezing. Further research is needed to show whether this improves or rather interferes with recovery to form a normal MII spindle and mediate chromosome congression.

From animal models it is known that meiotic progression is fine-tuned by opposing events, including, for instance, a partial degradation of cyclins after resumption of maturation until thresholds of activity of Cdk1/MPF have been reached and oocytes progress to metaphase I, followed by rapid degradation when they progress to anaphase I and second meiosis (14,16). Inhibiting protein synthesis in MII-arrested oocytes eventually results in chromosome decondensation due to reduced MPF activity (73). Alterations in the expression and activity of kinases, phosphatases, and components of the anaphase promoting factor can therefore significantly delay or advance meiotic maturation as well as affect spindle re-formation after cryopreservation. In addition, ablation or reduced expression of checkpoint proteins will advance meiosis whereas adverse exposures including depolymerization due to temperature shifts that interfere with chromosome attachment can lead to a transient meiotic arrest prior to progression to anaphase I (14,74). As second meiosis is completed rather rapidly after fertilization, the timing of events such as pronuclear apposition and time to cleavage might present an indirect marker of oocyte maturity, quality, and capacity to reorganize the cytoskeleton and chromatin after cryopreservation. Therefore, it is imperative to first analyze the timing of spindle re-formation in a routinely used cryopreservation protocol, not only to optimize the time for fertilization but also to find the best window for spindle analysis in frozen/thawed oocytes (e.g., by non-invasive spindle analysis), and retain a strict protocol to fertilize and assess pronuclear formation/scores and early cleavage according to these fixed points in order to obtain information on kinetics and improve outcomes by selection of mature, undamaged oocytes and good quality embryos.

Hormonal homeostasis and genetic background can profoundly modulate cell cycle progression in animal models (75). Consequently, it may be assumed that "delayed" or "advanced" human oocytes may come from a suboptimal follicular environment due to an inappropriate stimulation protocol or reflect a patient-specific disturbance leading to total or transient meiotic arrest/delay. Rescuing such oocytes by in vitro maturation for few hours may increase the chances for a conception, particularly when few mature oocytes are available. In cases where immature germinal vesicle-stage oocytes are retrieved in intracytoplasmic sperm injection cycles, chances that they will form a chromosomally balanced embryo appear low in view of studies showing increased aneuploidy when compared with mature oocytes from the same patients (76), and only a small proportion possesses spindles when matured to MII (23). In addition, spindles in such oocytes have a tendency to degenerate in a fairly short time period, suggesting that they reached nuclear maturity but are incompetent to support normal development (25). Cryopreservation of such immature oocytes from stimulated cycles that progressed to MII in vitro should increase the risks for abnormalities due to synergistic effects of oocyte immaturity and stresses by cryopreservation.

In IVF cycles where maturity cannot be easily assessed by analysis of oocyte polar body extrusion, cumulus expansion can help to identify oocytes that progress to meiosis II. However, it is now believed that analysis of apoptosis or expression of certain mRNAs related to stress by hypoxia, suboptimal follicular support, or up- or downregulation of genes in response to oocyte-derived growth factors may be predictive of oocyte maturity (77). Unfortunately, so far there are no fast, inexpensive, and uncomplicated tests with high predictive value that use cumulus or markers in follicular fluid for molecular analyses in the selection of mature oocytes, apart from the common, well-known morphological criteria. Cryopreservation has the potential to improve the selection of mature oocytes by cumulus or follicular markers once appropriate methodologies exist as it provides time for such analysis. Although LH induced signaling branches for induction of cumulus mucification and expansion, inflammatory signaling for ovulation, and resumption of oocyte maturation (78,79), testing for the cumulus proteome or metabolome might also reflect some oocyte-derived, maturation-dependent changes. Accordingly, more information on signaling pathways and molecules obtained from cumulus in cryopreservation cycles may eventually help to design "custom-made" treatment protocols for (*i*) cryopreservation; (*ii*) selection of mature, high quality oocytes before and after freezing; and (*iii*) identification of patients with a specific problem or defined mutation, which may help improve outcomes or overcome the disturbance in maturation to MII or sensitivity to freezing.

OOCYTE CYTOPLASMIC MATURATION: MITOCHONDRIA AND POLARITY

Although it is still unclear when and to what extent spatial organization and polarity influence and determine the fate of the oocyte and early embryo, it is predicted that subzonation and establishment of gradients and microdomains enriched with molecules and organelles in the large ooplasm may facilitate and thus optimize meiotic progression, egg activation, and embryogenesis. For instance, the spindle acts as a sink for molecules that are passed to the embryo and translocate before or along with the female pronucleus to a central location in the vicinity of the chromatin to be either integrated into the pronuclei or at a position where the first mitotic spindle will be formed (11). Therefore, the factors influencing the remodeling of chromatin, activation of transcription, or regulation of transport from the cytoplasm to the nucleus should be present in the right location upon fertilization, thus avoiding long periods to "collect" them from all over the ooplasm. Cryopreservation of oocytes may result in reversible or irreversible redistribution of spindle-associated factors. This does not necessarily result in fertilization failure but is expected to delay development and interfere with the developmental program, and possibly even cause arrest or abnormalities. Thus, removal of the spindle or checkpoint ablation in cloning is implicated in partial depletion of the ooplasm from factors re-setting the chromatin of the somatic cell nucleus and providing totipotency (80). Further studies are needed to analyze the effects of different cryopreservation protocols on the redistribution of spindle-associated factors, evaluate the benefits and disadvantages of the procedures, and assess the impact on oocyte quality.

Unfortunately, it is impossible to visualize the distribution of mitochondria and defined maternal factors on the spindle or in the ooplasm non-invasively, but it is clear that features of dysmorphic oocytes such as accumulation of large vacuoles, a highly granular cytoplasm, or a very large polar body is an indicator of a disturbed distribution of cellular components (81,82). A robust spindle characterized by highly ordered paracrystalline arrays of microtubules can be taken as an indicator of the capacity of the oocyte to establish polarity and recruit molecules to this site, which supports early embryogenesis (25), but it does not predict chromosome congression or the presence of maternal factors required for embryogenesis, as, for instance, evident from the good morphology of oocytes from PCOS patients having a limited developmental potential (83). However, as spindle size and density of fibers are influenced by chromosome-derived gradients (84,85), they should not only relate to maturity but also reflect the state of chromatin-assisted spindle stabilization (e.g., before or after cryopreservation). Spindles are highly dynamic and undergo continuous depolymerization/repolymerization, so even slight changes in temperature or culture conditions may alter their morphology. Given the dynamic nature of spindles and their susceptibility to change over time, the analysis of the spindle (e.g., by non-invasive polarization microscopy) should always be kept under a stringent time schedule to obtain meaningful information.

CONCLUSIONS

In conclusion, mature oocytes should possess full nuclear and cytoplasmic maturation competence. Any disturbance in the synchrony of events affecting the concerted nature of cell cycle regulation, chromatin conformation and chromosome segregation, spatial organization, and distribution and functionality of organelles as well as expression and degradation of maturation and maternal molecules can not only pose hazards for normal development of the embryo after fertilization and health of the offspring, but is also likely to synergistically affect robustness to cryopreservation, survival, and developmental potential. The procedures in cryopreservation should therefore be optimized to minimize aging and oocyte overmaturity and/or alterations in spatial organization of cell organelles and maternal products. Changes in epigenome and maternal products stored within the ooplasm due to exposures to cryoprotectants are critical and may accelerate age-related deterioration of oocytes and overmaturity. However, safe and efficient cryopreservation procedures can also provide options for molecular analyses of non-invasive markers in cumulus and follicular fluid for oocyte selection and detection of aberrations such as cytoplasmic immaturity, errors in chromosome segregation, incompetence to support embryogenesis, or mutations involved in subfertility. The time gained for discourse between embryologists, geneticists, and gynecologists by safe cryopreservation will be valuable to decide on options to improve treatment, counseling, and outcomes in routine assisted reproduction.

REFERENCES

1. Wassarman P, The mammalian ovum. In: Knobil E, Neill J, eds. The Physiology of Reproduction. New York: Raven Press Ltd, 1988: 69–102.
2. Sugiura K, Su YQ, Diaz FJ, et al. Oocyte-derived BMP15 and FGFs cooperate to promote glycolysis in cumulus cells. Development 2007; 134: 2593–603.
3. Su YQ, Sugiura K, Wigglesworth K, et al. Oocyte regulation of metabolic cooperativity between mouse cumulus cells and oocytes: BMP15 and GDF9 control cholesterol biosynthesis in cumulus cells. Development 2008; 135: 111–21.
4. Gilchrist RB, Lane M, Thompson JG. Oocyte-secreted factors: regulators of cumulus cell function and oocyte quality. Hum Reprod Update 2008; 14: 159–77.
5. Lucifero D, Mann MR, Bartolomei MS, Trasler JM. Gene-specific timing and epigenetic memory in oocyte imprinting. Hum Mol Genet 2004; 13: 839–49.
6. Schaefer CB, Ooi SK, Bestor TH, et al. Epigenetic decisions in mammalian germ cells. Science 2007; 316: 398–9.
7. Tan JH, Wang HL, Sun XS, et al. Chromatin configurations in the germinal vesicle of mammalian oocytes. Mol Hum Reprod 2009; 15: 1–9.
8. Kageyama S, Liu H, Kaneko N, et al. Alterations in epigenetic modifications during oocyte growth in mice. Reproduction 2007; 133: 85–94.
9. Zuccotti M, Merico V, Sacchi L, et al. Maternal Oct-4 is a potential key regulator of the developmental competence of mouse oocytes. BMC Dev Biol 2008; 8: 97.
10. Mendez R, Richter JD. Translational control by CPEB: a means to the end. Nat Rev Mol Cell Biol 2001; 2: 521–9.
11. Eichenlaub-Ritter U, Peschke M. Expression in in-vivo and in-vitro growing and maturing oocytes: focus on regulation of expression at the translational level. Hum Reprod Update 2002; 8: 21–41.

12. Guzeloglu-Kayisli O, Pauli S, Demir H, et al. Identification and characterization of human embryonic poly(A) binding protein (EPAB). Mol Hum Reprod 2008; 14: 581–8.
13. Radford HE, Meijer HA, de Moor CH. Translational control by cytoplasmic polyadenylation in Xenopus oocytes. Biochim Biophys Acta 2008; 1779: 217–29.
14. Vogt E, Kirsch-Volders M, Parry J, et al. Spindle formation, chromosome segregation and the spindle checkpoint in mammalian oocytes and susceptibility to meiotic error. Mutat Res 2008; 651: 14–29.
15. Vardy L, Pesin JA, Orr-Weaver TL. Regulation of Cyclin A protein in meiosis and early embryogenesis. Proc Natl Acad Sci U S A 2009; (in press).
16. Holt JE, Jones KT. Control of homologous chromosome division in the mammalian oocyte. Mol Hum Reprod 2009; (in press).
17. Belloc E, Méndez R. A deadenylation negative feedback mechanism governs meiotic metaphase arrest. Nature 2008; 452: 1017–21.
18. Xu Y, Zhang JJ, Grifo JA, et al. DNA methylation patterns in human tripronucleate zygotes. Mol Hum Reprod 2005; 11: 167–71.
19. Morgan HD, Santos F, Green K, et al. Epigenetic reprogramming in mammals. Hum Mol Genet 2005; 14: R47–58.
20. Duranthon V, Watson AJ, Lonergan P. Preimplantation embryo programming: transcription, epigenetics, and culture environment. Reproduction 2008; 135: 141–50.
21. Menezo Y Jr, Russo G, Tosti E, et al. Expression profile of genes coding for DNA repair in human oocytes using pangenomic microarrays, with a special focus on ROS linked decays. J Assist Reprod Genet 2007; 24: 513–20.
22. Derijck A, van der Heijden G, Giele M, et al. DNA double-strand break repair in parental chromatin of mouse zygotes, the first cell cycle as an origin of de novo mutation. Hum Mol Genet 2008; 17: 1922–37.
23. Barton TS, Robaire B, Hales BF. DNA damage recognition in the rat zygote following chronic paternal cyclophosphamide exposure. Toxicol Sci 2007; 100: 495–503.
24. Liu L, Bailey SM, Okuka M, et al. Telomere lengthening early in development. Nat Cell Biol 2007; 9: 1436–41.
25. Shen Y, Betzendahl I, Tinneberg HR, et al. Enhanced polarizing microscopy as a new tool in aneuploidy research in oocytes. Mutat Res 2008; 651: 131–40.
26. Van Blerkom J. Mitochondria in human oogenesis and preimplantation embryogenesis: engines of metabolism, ionic regulation and developmental competence. Reproduction 2004; 128: 269–80.
27. Dumollard R, Ward Z, Carroll J, et al. Regulation of redox metabolism in the mouse oocyte and embryo. Development 2007; 134: 455–65.
28. Dumollard R, Duchen M, Carroll J. The role of mitochondrial function in the oocyte and embryo. Curr Top Dev Biol 2007; 77: 21–49.
29. Tang CJ, Hu HM, Tang TK. NuMA expression and function in mouse oocytes and early embryos. J Biomed Sci 2004; 11: 370–6.
30. Oh B, Hwang SY, Solter D, et al. Spindlin, a major maternal transcript expressed in the mouse during the transition from oocyte to embryo. Development 1997; 124: 493–503.
31. Baluch DP, Capco DG. GSK3 beta mediates acentromeric spindle stabilization by activated PKC zeta. Dev Biol 2008; 317: 46–58.
32. Azoury J, Verlhac MH, Dumont J. Actin filaments: key players in the control of asymmetric divisions in mouse oocytes. Biol Cell 2009; 101: 69–76.
33. Zhang X, Ma C, Miller AL, et al. Polar body emission requires a RhoA contractile ring and Cdc42-mediated membrane protrusion. Dev Cell 2008; 15: 386–400.
34. Li H, Guo F, Rubinstein B, et al. Actin-driven chromosomal motility leads to symmetry breaking in mammalian meiotic oocytes. Nat Cell Biol 2008; 10: 1301–8.
35. Puppo A, Chun JT, Gragnaniello G, et al. Alteration of the cortical actin cytoskeleton deregulates Ca^{2+} signaling, monospermic fertilization, and sperm entry. PLoS ONE 2008; 3: e3588.
36. Abbott AL, Ducibella T. Calcium and the control of mammalian cortical granule exocytosis. Front Biosci 2001; 6: D792–806.
37. Ajduk A, Ma'agocki A, Maleszewski M. Cytoplasmic maturation of mammalian oocytes: development of a mechanism responsible for sperm-induced Ca^{2+} oscillations. Reprod Biol 2008; 8: 3–22.
38. Hansen DV, Tung JJ, Jackson PK. CaMKII and polo-like kinase 1 sequentially phosphorylate the cytostatic factor Emi2/XErp1 to trigger its destruction and meiotic exit. Proc Natl Acad Sci USA 2006; 103: 608–13.
39. Wu JQ, Kornbluth S. Across the meiotic divide—CSF activity in the post-Emi2/XErp1 era. J Cell Sci 2008; 121: 3509–14.
40. Terret ME, Wassmann K, Waizenegger I, et al. The meiosis I-to-meiosis II transition in mouse oocytes requires separase activity. Curr Biol 2003; 13: 1797–802.

41. Lee J, Okada K, Ogushi S, et al. Loss of Rec8 from chromosome arm and centromere region is required for homologous chromosome separation and sister chromatid separation, respectively, in mammalian meiosis. Cell Cycle 2006; 5: 1448–55.
42. Tam OH, Aravin AA, Stein P, et al. Pseudogene-derived small interfering RNAs regulate gene expression in mouse oocytes. Nature 2008; 453: 534–8.
43. Watanabe T, Totoki Y, Toyoda A, et al. Endogenous siRNAs from naturally formed dsRNAs regulate transcripts in mouse oocytes. Nature 2008; 453: 539–43.
44. Lykke-Andersen K, Gilchrist MJ, Grabarek JB, et al. Maternal Argonaute 2 is essential for early mouse development at the maternal-zygotic transition. Mol Biol Cell 2008; 19: 4383–92.
45. Wassarman PM, Litscher ES. Mammalian fertilization: the egg's multifunctional zona pellucida. Int J Dev Biol 2008; 52: 665–76.
46. Coy P, Cánovas S, Mondéjar I, et al. Oviduct-specific glycoprotein and heparin modulate sperm-zona pellucida interaction during fertilization and contribute to the control of polyspermy. Proc Natl Acad Sci USA 2008; 105: 15809–14.
47. Duncan FE, Stein P, Williams CJ, et al. The effect of blastomere biopsy on preimplantation mouse embryo development and global gene expression. Fertil Steril 2008; (in press).
48. Shen Y, Stalf T, Mehnert C, et al. High magnitude of light retardation by the zona pellucida is associated with conception cycles. Hum Reprod 2005; 20: 1596–606.
49. Schultz RM, Letourneau GE, Wassarman PM. Program of early development in the mammal: changes in patterns and absolute rates of tubulin and total protein synthesis during oogenesis and early embryogenesis in the mouse. Dev Biol 1979; 68: 341–59.
50. Tatone C, Carbone MC, Gallo R, et al. Age-associated changes in mouse oocytes during postovulatory in vitro culture: possible role for meiotic kinases and survival factor BCL2. Biol Reprod 2006; 74: 395–402.
51. Ma W, Zhang D, Hou Y, et al. Reduced expression of MAD2, BCL2, and MAP kinase activity in pig oocytes after in vitro aging are associated with defects in sister chromatid segregation during meiosis II and embryo fragmentation after activation. Biol Reprod 2005; 72: 373–83.
52. Hall VJ, Compton D, Stojkovic P, et al. Developmental competence of human in vitro aged oocytes as host cells for nuclear transfer. Hum Reprod 2007; 22: 52–62.
53. Hodges CA, Revenkova E, Jessberger R, et al. SMC1 beta-deficient female mice provide evidence that cohesins are a missing link in age-related nondisjunction. Nat Genet 2005; 37: 1351–5.
54. Liu L, Keefe DL. Defective cohesin is associated with age-dependent misaligned chromosomes in oocytes. Reprod Biomed Online 2008; 16: 103–12.
55. Subramanian VV, Bickel SE. Aging predisposes oocytes to meiotic nondisjunction when the cohesin subunit SMC1 is reduced. PLoS Genet 2008; 4: e1000263.
56. Steuerwald N, Cohen J, Herrera RJ, et al. Association between spindle assembly checkpoint expression and maternal age in human oocytes. Mol Hum Reprod 2001; 7: 49–55.
57. Hamatani T, Falco G, Carter MG, et al. Age-associated alteration of gene expression patterns in mouse oocytes. Hum Mol Genet 2004; 13: 2263–78.
58. Steuerwald NM, Steuerwald MD, Mailhes JB. Post-ovulatory aging of mouse oocytes leads to decreased MAD2 transcripts and increased frequencies of premature centromere separation and anaphase. Mol Hum Reprod 2005; 11: 623–30.
59. Pan H, Ma P, Zhu W, Schultz RM. Age-associated increase in aneuploidy and changes in gene expression in mouse eggs. Dev Biol 2008; 316: 397–407.
60. Huntriss J, Picton HM. Epigenetic consequences of assisted reproduction and infertility on the human preimplantation embryo. Hum Fertil (Camb) 2008; 11: 85–94.
61. Haaf T. Methylation dynamics in the early mammalian embryo: implications of genome reprogramming defects for development. Curr Top Microbiol Immunol 2006; 310: 13–22.
62. Horsthemke B, Ludwig M. Assisted reproduction: the epigenetic perspective. Hum Reprod Update 2005; 11: 473–82.
63. Liang XW, Zhu JQ, Miao YL, et al. Loss of methylation imprint of Snrpn in postovulatory aging mouse oocyte. Biochem Biophys Res Commun 2008; 371: 16–21.
64. Wan LB, Pan H, Hannenhalli S, et al. Maternal depletion of CTCF reveals multiple functions during oocyte and preimplantation embryo development. Development 2008; 135: 2729–38.
65. Chian RC, Huang JY, Tan SL, et al. Obstetric and perinatal outcome in 200 infants conceived from vitrified oocytes. Reprod Biomed Online 2008; 16: 608–10.
66. Shih W, Rushford DD, Bourne H, et al. Factors affecting low birthweight after assisted reproduction technology: difference between transfer of fresh and cryopreserved embryos suggests an adverse effect of oocyte collection. Hum Reprod 2008; 23: 1644–53.
67. Cobo A, Kuwayama M, Pérez S, et al. Comparison of concomitant outcome achieved with fresh and cryopreserved donor oocytes vitrified by the Cryotop method. Fertil Steril 2008; 89: 1657–64.

68. Larman MG, Minasi MG, Rienzi L, et al. Maintenance of the meiotic spindle during vitrification in human and mouse oocytes. Reprod Biomed Online 2007; 15: 692–700.
69. Cobo A, Pérez S, De los Santos MJ, et al. Effect of different cryopreservation protocols on the metaphase II spindle in human oocytes. Reprod Biomed Online 2008; 17: 350–9.
70. Katz-Jaffe MG, Larman MG, Sheehan CB, et al. Exposure of mouse oocytes to 1,2-propanediol during slow freezing alters the proteome. Fertil Steril 2008; 89(5 Suppl): 1441–7.
71. Eichenlaub-Ritter U, Shen Y, Tinneberg HR. Manipulation of the oocyte: possible damage to the spindle apparatus. Reprod Biomed Online 2002; 5: 117–24.
72. Montag M, Schimming T, van der Ven H. Spindle imaging in human oocytes: the impact of the meiotic cell cycle. Reprod Biomed Online 2006; 12: 442–6.
73. Fulka J Jr, Moor RM, Fulka J. Sister chromatid separation and the metaphase-anaphase transition in mouse oocytes. Dev Biol 1994; 165: 410–17.
74. Homer HA, McDougall A, Levasseur M, et al. Mad2 is required for inhibiting securin and cyclin B degradation following spindle depolymerisation in meiosis I mouse oocytes. Reproduction 2005; 130: 829–43.
75. Polan'ski Z. In-vivo and in-vitro maturation rate of oocytes from two strains of mice. J Reprod Fertil 1986; 78: 103–9.
76. Magli MC, Ferraretti AP, Crippa A, et al. First meiosis errors in immature oocytes generated by stimulated cycles. Fertil Steril 2006; 86: 629–35.
77. Li Q, Mckenzie LJ, Matzuk MM. Revisiting oocyte-somatic cell interactions: in search of novel intrafollicular predictors and regulators of oocyte developmental competence. Mol Hum Reprod 2008; 14: 673–8.
78. Richards JS. Genetics of ovulation. Semin Reprod Med 2007; 25: 235–42.
79. Hsieh M, Lee D, Panigone S, et al. Luteinizing hormone-dependent activation of the epidermal growth factor network is essential for ovulation. Mol Cell Biol 2007; 27: 1914–24.
80. Tani T, Kato Y, Tsunoda Y. Aberrant spindle assembly checkpoint in bovine somatic cell nuclear transfer oocytes. Front Biosci 2007; 12: 2693–705.
81. Ebner T, Moser M, Tews G. Is oocyte morphology prognostic of embryo developmental potential after ICSI? Reprod Biomed Online 2006; 12: 507–12.
82. Ubaldi F, Rienzi L. Morphological selection of gametes. Placenta 2008; 29(Suppl B): 115–20.
83. Wood JR, Dumesic DA, Abbott DH, et al. Molecular abnormalities in oocytes from women with polycystic ovary syndrome revealed by microarray analysis. J Clin Endocrinol Metab 2007; 92: 705–13.
84. Dumont J, Petri S, Pellegrin F, et al. A centriole- and RanGTP-independent spindle assembly pathway in meiosis I of vertebrate oocytes. J Cell Biol 2007; 176: 295–305.
85. Brunet S, Dumont J, Lee KW, et al. Meiotic regulation of TPX2 protein levels governs cell cycle progression in mouse oocytes. PLoS ONE 2008; 3: e3338.

8 Expression of Functional Aquaporins in Oocytes and Embryos and the Impact on Cryopreservation

François Chauvigné and Joan Cerdà

Laboratory of the Institut de Recerca i Tecnologia Agroalimentàries (IRTA)-Institut de, Ciències del Mar, Consejo Superior de Investigaciones Científicas (CSIC), Barcelona, Spain

INTRODUCTION

The development of successful methods for cryopreservation of oocytes and embryos must ensure that cells are not damaged by chilling, the formation of extracellular and intracellular ice, the chemical toxicity of cryoprotectants, fracturing, and osmotic swelling and shrinkage. Two main methods are currently employed for cryopreservation of oocytes and embryos: slow cooling rate with low concentration of cryoprotectants, and fast cooling after incubation in highly concentrated cryoprotectant solutions (vitrification) (1,2). In both techniques, however, one of the main challenges is to rapidly replace intracellular water with cryoprotectant. Vitrification protocols on specific developmental stages of oocytes or embryos result in acceptable survival rates (3,4), but identical protocols often fail when they are applied to other stages of the same species (5). These observations suggest that the cryobiological properties of oocytes and embryos may change during development. Investigation of the mechanisms regulating water and solute permeability in oocytes and embryos, for which current knowledge is still scarce, is essential to improve current cryopreservation protocols.

Understanding of the permeability properties of biological membranes was revolutionized by the discovery of the first water-specific membrane channel, aquaporin-1 (AQP1), by Peter Agre and collaborators (6). It is now recognized that water transport across membranes not only occurs by simple diffusion through the lipid bilayers, but also through aquaporins, particularly when rapid water permeability is required for physiological processes such as secretion, reabsorption, or osmotic stress. Aquaporins have been identified in many different tissues where water transport is important (7). The recent discovery of aquaporins in mammalian gametes and embryos will greatly contribute to uncover the mechanisms for water and solute movement in these complex systems. Recent reports also suggest that aquaporins could be the molecular pathway by which water and/or solutes move across the plasma membrane during cryopreservation. These advances are the focus of this chapter.

AQUAPORIN STRUCTURE AND FUNCTION

Aquaporins consist of six transmembrane domains, connected by five loops (A–E), and have their N- and C-termini located intracellularly (Fig. 1). One molecule consists of two repeats (hemipores) which are 180° mirror images. Each repeat contains the highly-conserved asparagine-proline-alanine (NPA) motif (in loops B and E), which is the hallmark of the major intrinsic protein superfamily of proteins to which aquaporins belong. The folding of the two loops, following the predicted "hourglass" model (8), is important for the formation of the water pore, as has been corroborated by cryo-electron microscopy and X-ray crystal structure studies (9,10). Aquaporins are present in the membrane as tetramers, but, unlike ion channels, each monomer contains its own pore. Most mammalian aquaporins are inhibited by mercurials. In AQP1, the site of mercurial inhibition was demonstrated at Cys-189 proximal to the NPA motif of loop E (11) (Fig. 1).

The 13 aquaporins that have been identified so far in mammals can be divided into two groups on the basis of their permeability characteristics, which generally coincide with specific amino acid sequence patterns (12) (Table 1). In the first group (aquaporins), there are the channels that are mainly permeated by water, such as AQP0, AQP1, AQP2, AQP4, AQP5, AQP6, and AQP8.

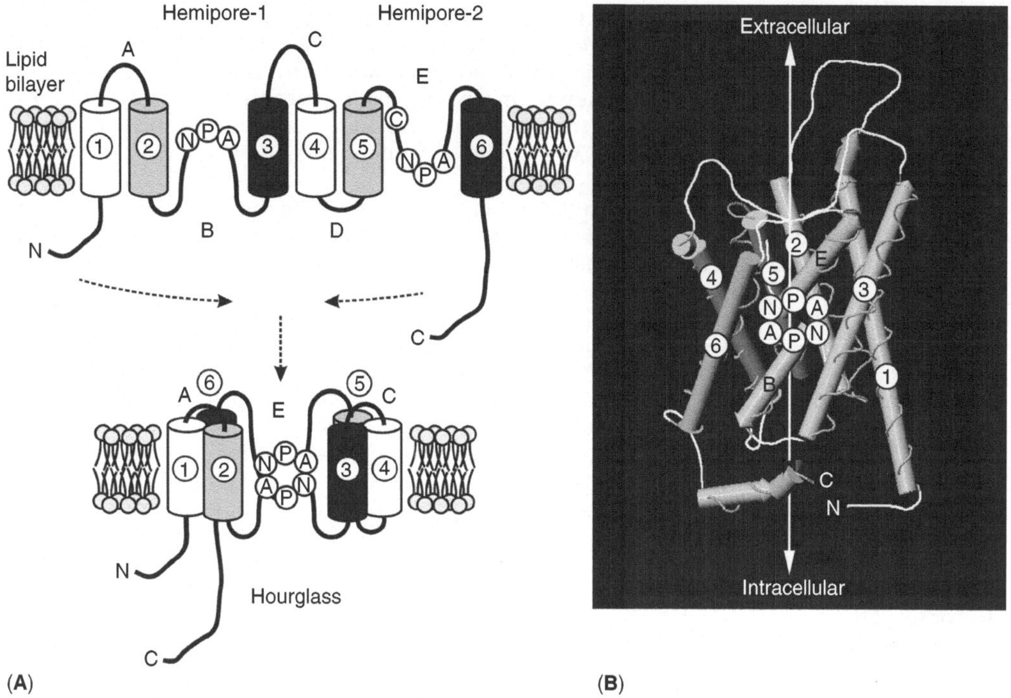

Figure 1 Topology and structure of the prototypical AQP1. (**A**) "'Hourglass" model for AQP1 topology in which arrangement of loops E and E with highly conserved NPA motifs forms a single aqueous pathway through the AQP1 subunit. Redrawn from Jung et al. (8) (**B**) Crystal structure of AQP1 monomer based on protein data bank (PDB) No. 1J4N viewed from the side. The arrow highlights the route taken by water, which can move in both directions through the channel. In both panels, the membrane-spanning helices are numbered (1—6). The Cys potentially responsible in the inhibition of water permeability by mercurial compounds of most AQPs is indicated in A. *Abbreviation*: NPA, asparagine-proline-alanine.

Table 1 Permeability Characteristics of Mammalian Aquaporins

Aquaporin	Functional permeability
AQP0	Water
AQP1	Water, nitric oxide, carbon dioxide
AQP2	Water
AQP3	Water, glycerol, urea
AQP4	Water
AQP5	Water
AQP6	Water (low), anions (NO_3^-, Cl^-)
AQP7	Water, glycerol, urea, arsenite
AQP8	Water, ammonia, hydrogen peroxide, urea
AQP9	Water, glycerol, urea, arsenite, large solutes (carbamides, polyols, purines, pyrimidines)
AQP10	Water, glycerol, urea
AQP11	Water
AQP12	Unknown

AQP6, however, is also permeated by anions (13), and AQP8, which, based on sequence analysis, might form a different subgroup within the family (14), could also be permeated by urea (15,16). The second group (aquaglyceroporins) includes AQP3, AQP7, AQP9, and AQP10 that are permeated by water to varying degrees and by small, non-charged solutes, in particular glycerol (17).

AQP11 and AQP12 are two aquaporins with poorly conserved NPA motifs (18), and although water permeation through AQP11 has been demonstrated (19), the permeability properties of AQP12 are as yet unknown. Recently, however, particular aquaporin isoforms (AQP1, AQP7, AQP8, AQP9) have been reported to conduct unconventional permeants, such as the nonpolar gases carbon dioxide and nitric oxide, the polar gas ammonia, the oxidative oxygen species hydrogen peroxide, and the metalloids antimonite, arsenite, and silicic acids (20), expanding the functional repertoire of aquaporins.

The most striking feature of the aquaporin channels is perhaps their high selectivity and efficiency with regard to water or glycerol transport, and the strict exclusion of ions including protons. Two main constriction sites have been identified in AQP1 and the *Escherichia coli* glycerol transporter (GlpF), as models for aquaporins and aquaglyceroporins, respectively, that underlie these features (21). The first constriction is located in the center of the pore in the NPA region, where the two asparagines located at the end of the two half helices (B and E in Fig. 1B) create an electrostatic barrier in the NPA region, which, together with desolvation, is essential for proton exclusion. The second constriction, located close to the extracellular exit of the channel and referred as the aromatic/arginine (ar/R) constriction, forms the narrowest region of the pore and is therefore generally assumed to be important for channel selectivity. In AQP1, the narrow ar/R site (2.8 Å) is formed by Phe, His, Cys, and Arg, which provides a hydrophilic environment creating a hydrophobicity and size filter (22). In contrast, in GlpF, and essentially in all other aquaglyceroporins, the ar/R region is wider (3.4 Å) and more hydrophobic due to the lack of His and substitution of the Cys by a second aromatic residue, which allows the passage of polyols and urea, and possibly of other small solutes such as NH_3, CO_2, or O_2 (22).

Aquaporins are widely distributed in different tissues and cell types with important water transport or secretory roles (23). However, the localization of aquaporins in the cell is not always seen at the plasma membrane. For instance, AQP6 is colocalized alongside H^+-ATPase in intracellular vesicles of acid-secreting α-intercalated cells from renal collecting duct, AQP11 is found in the endoplasmic reticulum of cells of the proximal tubule, and AQP12 appears in intracellular organelles of pancreatic acinar cells (13,24,25). Studies using genetically modified rodents and the identification of humans with altered aquaporin genes have demonstrated the physiological role of several aquaporins during water homeostasis in kidney, brain, and skin, as well as during the metabolism of adipocytes. Excellent reviews summarizing these findings have recently been published (26,27). Here we concentrate on the information available on the expression, localization, and role of aquaporins in reproductive tissues, oocytes and embryos.

AQUAPORINS IN THE FEMALE REPRODUCTIVE SYSTEM

Reproductive Tract

Metabolic processes during reproduction depend on fluid secretion and reabsorption, and therefore aquaporins may play a role in fluid exchange in the female reproductive tract. At least 10 aquaporin isoforms (AQP1–AQP9) have been shown to be expressed in female reproductive organs, and their specific expression patterns suggest that they might be involved in water movement between the intraluminal, interstitial, and capillary compartments (28).

In the uterus, steroid hormones induce water imbibition in the uterine endometrium. This water then crosses the epithelial cells into the lumen, leading to a decrease in viscosity of uterine luminal fluid (29). The expression of aquaporins has been demonstrated in the stroma (AQP8), myometrium (AQP1 and AQP8), endometrium (AQP1 and AQP2), and in glandular (AQP5 and AQP9) and luminal epithelial cells (AQP2, AQP3, and AQP5) of the uterus, and this expression seems to be regulated by estrogens and progesterone (29–32). The expression of human endometrial AQP2 is menstrual-cycle-dependant and reaches a high level at the midsecretory phase, the time of embryo implantation (30). These observations suggest that aquaporins may play important roles in hormone-mediated water and solute transport during uterine imbibition, as well as regulating luminal fluid volume at the time of blastocyst implantation (30,32). In addition, aquaporin isoforms (AQP3–AQP5 and AQP8) are also present in the mouse cervix, which may contribute to the changes in the organization of the collagen network and water content in the cervical connective tissue that occur during gestation, which allows cervical dilatation during labor (33).

Table 2 Expression Pattern of AQPs in Mammalian Ovarian Follicles and Embryos

	AQP1	AQP3	AQP5	AQP6	AQP7	AQP8	AQP9	Ref.
Ovarian Follicle								
Granulosa cells	N/D	N/D	+	N/D	+	+	+	36,37
Oocyte	N/D	•	N/D	N/D	•	N/D	•	38,39,40
Embryo								
Zygote	•	•/+	•	•	•	N/D	•/+	48,51
4–8 cells	•	•/+	•	•	•	N/D	•/+	38,48,51
Morula	•	•/+	•	•	•	+	•/+	38,48,51,52
Blastocyst	•	•/+	•	•	•	•/+	•/+	38,48–51

•, reported mRNA expression; +, reported protein expression; N/D, mRNA and/or protein not detected.

Aquaporins have also been found in epithelial cells of the oviduct (AQP5, AQP8, and AQP9) (34), and in the cell membrane of associated smooth muscle cells (AQP1) (35). The location of aquaporins in these areas may provide the mechanism for regulating ovum transport in the fallopian tube by altering the tubal luminal diameter during the estrus cycle (35). In addition, they may control the production of oviductal fluid for fertilization and early embryonic development (28).

Ovarian Follicles and Oocytes

Ovarian folliculogenesis is accompanied by the formation and expansion of the fluid-filled antrum. Development of this cavity requires water influx, which may occur by transcellular or pericellular transport mechanisms. Experiments testing the uptake of radiolabelled water and inulin (a complex sugar restricted to the extracellular compartment) by isolated rat antral follicles in vitro, in the presence or absence of $HgCl_2$, as well as the ability of granulosa cells to swell after hypotonic challenge, suggest that granulosa cell aquaporins may mediate transcellular water movement in antral follicles (36). Flow cytometry and immunocytochemical studies support this conclusion since they demonstrate the presence of AQP7–AQP9 in granulosa cells (36,37) (Table 2). Interestingly, AQP5 has also been recently found in flattened follicle cells of pig primordial follicles (37).

In contrast, in the mammalian oocyte only mRNAs encoding AQP3, AQP7, and AQP9 have been reported, whereas AQP1, AQP5, AQP6, and AQP8 were not detected (38,39) (Table 2). In rat oocytes, AQP9 mRNA is expressed in proestrus (immature) oocytes but not in estrous (mature) oocytes, in line with the decrease in water and solute permeability of oocytes during maturation (39). In naturally obtained mouse metaphase II (mature) oocytes, however, transcripts of AQP3 can be detected by a polymerase chain reaction after reverse transcription of RNA, but expression is reduced when oocytes are obtained by controlled ovarian hyperstimulation (40). These data suggest that permeability of immature oocytes is mediated by AQP9 (39), whereas that of mature oocytes may be mediated by AQP3 (40). However, this hypothesis is questionable because neither AQP3, AQP9, or AQP7 polypeptides have been demonstrated in mammalian oocytes, and the level of water permeability of oocytes, and its temperature dependency, is not necessarily consistent with the presence of water channels (41) (see next section).

In lower vertebrates, some aquaporin isoforms are highly expressed in the oocyte. In the amphibian *Xenopus laevis*, oocytes express AQP3 (42) as well as high levels of mRNA encoding a novel aquaglyceroporin unknown in mammals, termed AQPxlo, with similar permeability characteristics to rat AQP3 (43). However, water permeability of *X. laevis* oocytes is low, and it has been speculated that even though oocytes contain large amounts of stored AQPxlo mRNA, protein expression in oocytes is possibly low until embryogenesis (43). Most notably, a novel AQP1-like channel, originating from a teleost-specific gene duplication event, is found in marine teleosts that produce highly hydrated eggs, buoyant in seawater (44,45). This water-selective aquaporin, termed Aqp1b, is translocated into the oocyte plasma membrane during meiotic maturation where it mediates water influx into the oocyte following the osmotic gradient created by the hydrolysis of yolk proteins and the accumulation of ions (46,47). These findings suggest that comparative studies in lower vertebrates may be of great interest for understanding aquaporin regulation and function in mammalian oocytes.

Embryos

In mouse embryos, mRNAs encoding seven aquaporin isoforms, AQP1, AQP3, AQP5–AQP9 have been found from the one-cell stage up to the blastocyst stage, although AQP8 mRNA is not detected in early cleavage stages (38,48,49) (Table 2). In morula- and blastocyst-stage embryos, AQP8 levels are reduced when embryos are cultured in glycerol-based hyperosmolar media (50). AQP9 polypeptides have been demonstrated in cell margins from eight-cell embryos onward, whereas AQP3 and AQP8 are found only from compacted morula stages (51,52). In the trophectoderm, a squamous epithelium that surrounds the fluid-filled blastocyst cavity (blastocoele), AQP3 and AQP8 are detected in basolateral membranes, while AQP9 is predominantly observed within the apical membrane. AQP3 is also observed in the margins of the apolar cells of the inner cell mass (embryoblast) (49,51). Barcroft et al. (51) have demonstrated that these embryonic aquaporins are functional by measuring variations in water movement across the trophectoderm following exposure of blastocysts to a hyperosmotic gradient in the presence of mercurial inhibitors. These results support the current model of blastocyst formation in which aquaporins mediate trans-trophectodermal water movements during cavitation following the osmotic gradient created by Na^+/K^+-ATPase and ion channels (53).

ROLE OF AQUAPORINS IN WATER AND SOLUTE PERMEABILITY OF OOCYTES AND EMBRYOS

The significant involvement of water channels in the permeability of oocytes and embryos can be deduced from the permeability to water and its dependency on temperature, as the movement of water through channels is much less dependent on temperature than that through the lipid bilayer (5). The temperature dependency of permeability is expressed by the Arrhenius activation energy (E_a). Diffusion of water across the cell membrane requires a relatively large osmotic gradient, as the E_a of trans-cellular water transport is much higher ($E_a > 10$ kcal/mol) than that observed in the presence of aquaporins (54). In general, it appears that an osmotic water permeability coefficient (P_f) higher than 10 µm/sec and an E_a lower than 6 kcal/mol is suggestive of water transport through aquaporins, although there is not a priori theoretical basis for such values (54). Thus, this is not a strict criteria but a rough estimation for deducing predominant pathways for water and solute movement in the cell. Similarly, although such criteria have not been reported for the movement of small neutral solutes across the plasma membrane, it seems reasonable that high solute permeability (P_s) and low E_a values may indicate aquaporin-mediated solute transport (41).

Oocyte Permeability

As stated earlier, mammalian oocytes express AQP3, AQP7, and AQP9. However, mature oocytes have a P_f of 9.9 ± 0.7 µm/sec [hydraulic conductivity (L_p) of 0.4 ± 0.0 µm/min/atm] at 20°C, 0.7 ± 0.1 µm/min/atm at 25°C, or 0.5 ± 0.02 µm/min/atm at 19°C in mouse (41,52,55,56), 10.5 ± 4.7 µm/sec (0.5 ± 0.2 µm/min/atm) at 20°C in cow (57), 10.2 ± 2.9 µm/sec at 22°C in the golden hamster (58), and 9.9 ± 3.4 µm/sec (0.4 ± 0.1 µm/min/atm) at 20°C in human (59), whereas the E_a values are 12.3–14.5, 7.8, 8.0 ± 1.0, and 8.6 ± 5.1 kcal/mol, respectively. The permeabilities of mouse oocytes, at 25°C, to cryoprotectants such as glycerol (P_{Gly}), ethylene glycol (P_{EG}), and propylene glycol (P_{PG}), when tested at hypertonic concentrations (1.5 M), are 0.02 ± 0.02, 0.6 ± 0.0, and 1.7 ± 0.6 × 10^{-3} cm/min, with a remarkably high E_a, 41.6, 17.3, and 20.3 kcal/mol, respectively (41,52). In human oocytes, the P_{EG} is 0.6 × 10^{-3} cm/min (60). These data are slightly or clearly outside the suggested P_f and E_a values for aquaporin-mediated water and solute transport, and therefore they might indicate that water and solute permeation in oocytes occurs mainly by simple diffusion across the lipid bilayer rather than through aquaporins. This hypothesis is reinforced by the finding that water permeability of mouse oocytes is not sensitive to mercurial compounds or phloretin (52), known inhibitors of aquaglyceroporin function (61,62). However, it has also been reported that goat immature oocytes have a P_f of 22.3 ± 2.5 µm/sec (63), and that rat immature oocytes are permeable to water (P_f 12.8 ± 0.9 µm/sec) and mannitol in a mercury- and phloretin-sensitive manner (39). In addition, naturally obtained metaphase II mouse oocytes show a high and $HgCl_2$-sensitive water permeability (P_f 78 ± 3 µm/sec) which is associated with higher fertilization rates (40). Although these studies did not provide E_a estimations for water or solute permeation, it seems that the role of aquaporins during water and solute permeability in mammalian oocytes, and subsequent viability, is not yet clear. Future studies using targeted aquaporin gene knockdown strategies in oocytes might help to clarify this question.

Similar to mammals, fish mature oocytes are less permeable than immature oocytes to water and cryoprotectants (64,65). In freshwater teleosts, such as medaka (*Oryzias latipes*) or zebrafish (*Danio rerio*), the L_p of fully-grown immature oocytes at 25°C has been estimated to be 0.13 to 0.37 µm/min/atm, whereas that of metaphase II oocytes is 0.06 to 0.1 µm/min/atm. The permeability of mature oocytes to different cryoprotectants, including dimethylsulfoxide (DMSO), is close to zero, whereas that of immature oocytes is 1.5 to 3.0×10^{-3} cm/min. Zebrafish mature oocytes however are less susceptible to chilling injury than immature oocytes, probably because they are less water permeable (66). In contrast, there appears to be increased water permeability in oocytes of marine teleosts undergoing meiotic maturation, due to the transient insertion of Aqp1b in the oocyte plasma membrane, which mediates the hydration of the oocyte (46,47). Such hydration may account for a three- to eight-fold increase in volume of mature oocytes compared with immature oocytes (47).

Embryo Permeability

A different situation is seen in mammalian embryos, where considerably higher P_f and P_s and lower E_a values are found. It has been calculated that the L_p and P_{Gly} at 25°C of mouse morula embryos are 4.4 ± 1.8 µm/min/atm and 4.7 ± 1.5 × 10^{-3} cm/min, with E_a values of 6.3 and 10 kcal/mol, respectively (52). Mouse morulae are also more permeable to EG, with a P_{EG} of 10 ± 3.7×10^{-3} cm/min (E_a of 9.1 kcal/mol) at 25°C, than to other cryoprotectants such as PG or DMSO (41). Mouse blastocysts have an L_p of 3.6 ± 1.7 µm/min/atm and an E_a of 5.1 kcal/mol (52), and the L_p of rabbit trophectoderm epithelium has been reported to be 1118 µm/min/atm on day 5 (67). The increased water and solute permeability of embryos at the morula and blastocyst stages correlates with the appearance of AQP3, AQP8, and AQP9 polypeptides in the plasma membrane of blastomeres and trophectodermal cells, and of mercury-sensitive water transport (51,52). In contrast, in four-cell stage embryos, when low and exclusively cytoplasmic immunoreaction for AQP3 and AQP9 is seen, low L_p and high E_a values (0.63 µm/min/atm and 11.6 kcal/mol, respectively) have been reported (52,68). Altogether, these results support the role of aquaporins in water and solute permeability of mammalian embryos.

In the mouse morula stage, however, AQP3, AQP8, and AQP9 have all been detected at the cell-cell contact domains of blastomeres, so the specific contribution of each aquaporin isoform in the movement of water and various cryoprotectants is unclear. Edashige and collaborators (41) have recently clarified the role of AQP3 through the suppression of AQP3 expression by injecting double-stranded RNA (dsRNA) of AQP3 at the one-cell zygote stage. The results of this study demonstrate that AQP3 expressed in morulae may play an important role in permeability to water and neutral solutes, but this contribution depends on the concentration of external solutes. In this study, however, suppression of AQP3 expression did not completely prevent glycerol and EG permeation, and therefore, it is possible that AQP9 may also be involved in the movement of these solutes. In addition, it has been shown that AQP3 knockout mice embryos can develop to term (69), which suggests that embryonic aquaporins may compensate each other for water and neutral solute transport during embryogenesis (41). This, however, remains to be demonstrated.

Artificial Expression of Aquaporins in Oocytes

Effective vitrification of oocytes relies on the rapid permeation of cryoprotectants and water in the cell. The cryopreservation of mature oocytes faces an important obstacle because these cells appear to have low permeability to water and cryoprotectants. One proposed strategy to improve the permeability of mature mouse oocytes is the artificial expression of AQP3 (41,70) (Fig. 2). Using this approach, it has recently been reported that mouse oocytes injected with rat or mouse AQP3 cRNA, when exposed to a hyperosmotic glycerol-based solution (1.5 M), have increased water and glycerol permeability with respect to the noninjected oocytes. Most notably,

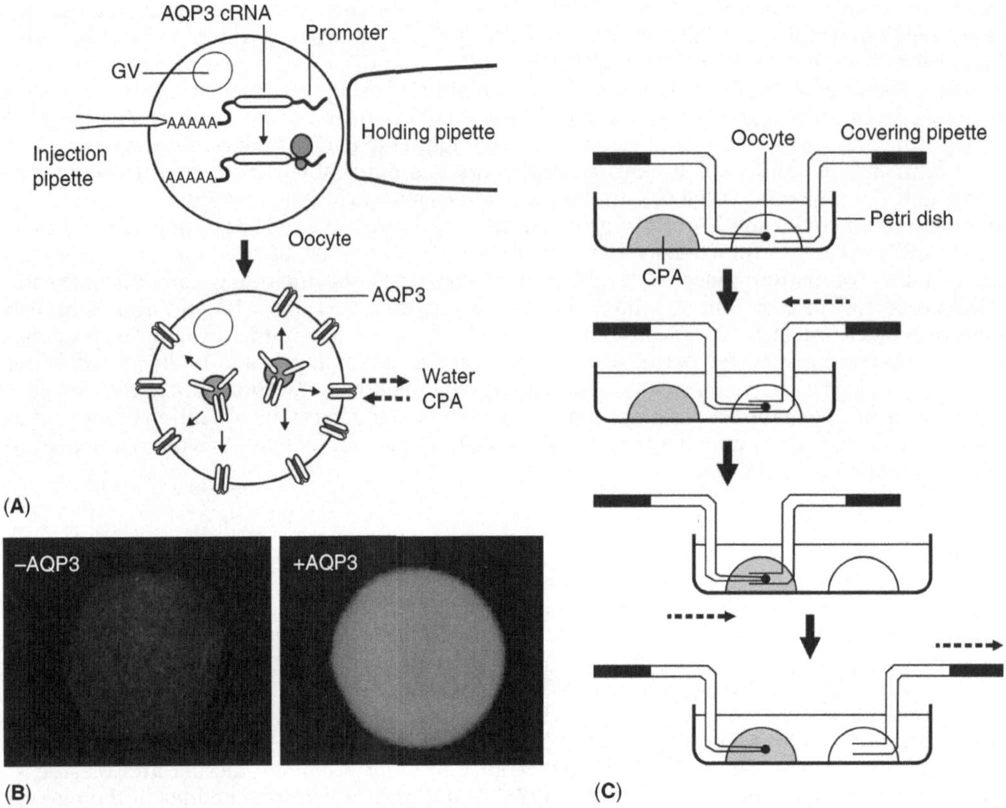

Figure 2 Artificial expression of AQP3 cRNA in oocytes and effect on their permeability to water and CPAs. (**A**) Drawing showing the microinjection of AQP3 cRNA into an oocyte, which drives translation and further translocation of the polypeptide into the oocyte plasma membrane. (**B**) Immunolocalization of rat AQP3 in an intact (noninjected, −AQP3) and a rat AQP3 cRNA-injected (+AQP3) mouse oocyte. *Source*: From Ref. 70. (**C**) Schematic representation of the device and the procedure for direct transfer of an oocyte (solid circle) to a CPA solution (grey hemisphere) from isotonic culture medium (open hemisphere) using a micromanipulator system. Redrawn from Ref. 70. *Abbreviations*: AQP, aquaporin; CPA, cryoprotectant agents; GV, germinal vesicle.

the survival of oocytes after vitrification correlated with the values of L_p and P_{Gly} for each oocyte. Thus, those oocytes with higher L_p and P_{Gly} values have higher survival rates and retain the ability to be fertilized after vitrification, although embryos cannot develop beyond the two-cell stage (70). These experiments demonstrate that over-expression of aquaporins in the oocyte improves survival during cryopreservation, although why subsequent development into blastocysts after fertilization is blocked is, as yet, unknown.

The over-expression of AQP3 cRNAs in mouse oocytes has also revealed unexpected phenomena regarding water and solute movements under hyperosmotic conditions in both oocytes and morula embryos (41). In glycerol solution, both the L_p and P_{Gly} of AQP3-injected oocytes and morulae are higher than in noninjected oocytes or in morulae injected with dsRNA of AQP3, indicating that, in modified oocytes and morulae, most water and glycerol molecules move through AQP3. However, in EG solution, although the P_{EG} values of both AQP3-injected oocytes and morulae are higher (and with a lower E_a) than in non-injected oocytes and dsRNA-injected morulae, the L_p values were similar. It has been hypothesized that, in the EG solution, the rapid permeation of EG into oocytes and morulae through AQP3 would decrease the movement of water to the outside through the same AQP3, because both EG and water would move through the same channel (41). This means that the osmotic difference inside and outside the plasma membrane would be lost by the rapid permeation of EG into the cell. While the mechanism

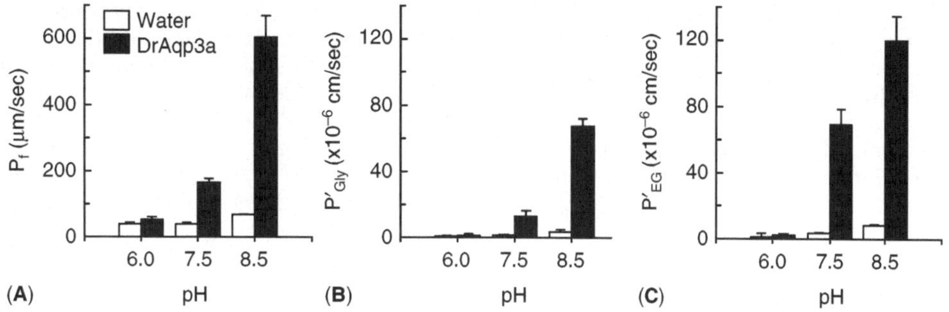

Figure 3 Osmotic water permeability (P_f), and apparent glycerol (P'_{Gly}) and ethylene glycol (P'_{EG}) permeability of *X. laevis* oocytes injected with water or 1 ng of zebrafish Aqp3a cRNA. For the calculation of P_f, oocytes were transferred to a 20 mOsm modified Barth's solution (MBS) for 20 sec (**A**), whereas for the calculation of P'_{Gly} and P'_{EG}, oocytes were exposed to isotonic MBS (200 mOsm) containing 160 mM of glycerol (**B**) or ethylene glycol (**C**) for 1 min. Data are the mean ± SEM (n = 8–10 oocytes).

underlying this phenomenon is not yet understood, it may be that oocytes artificially expressing AQP3, although an increased P_{EG} can reduce osmotic stress, still might suffer major cell damage produced by the formation of intracellular crystals due to the reduction in water efflux. This would be very unfortunate because EG seems to have relatively low toxicity as a cryoprotectant for mammalian oocytes, including human (60,71–74). The movement of water and EG in oocytes artificially expressing AQP3 needs to be studied in more detail before this strategy can be applied for successful cryopreservation of mammalian oocytes.

The artificial expression of rat AQP3 in fish and amphibian oocytes, as well as in fish embryos, has also been reported to improve the permeability to water and/or cryoprotectants (75–78). In zebrafish embryos, expression of AQP3 fused with a green fluorescent protein increases the permeability of the yolk sac to cryoprotectants due to the ubiquitous expression of AQP3 in the membranes, overcoming one of the main obstacles for the cryopreservation of fish embryos which is the low permeability of cryoprotectants into the yolk (75). These studies have also shown that AQP3-expressing embryos develop and reproduce normally (79). It has also been reported that frog and fish oocytes expressing mammalian AQP3 show increased permeability to DMSO, PG, and acetamide (76,77,80,81), which does not occur in mouse oocytes or embryos expressing AQP3 (41). However, in these studies fish and frog oocytes were injected with very high amounts of AQP3 cRNA (40 ng) when compared with mouse oocytes (2–10 pg), and it is possible that over-expression of AQP3 above physiological levels may cause abnormal permeation by solutes (41).

An additional strategy for the artificial expression of aquaglyceroporins in oocytes and embryos, not yet explored, is the use of nonmammalian aquaglyceroporin isoforms with different solute permeability characteristics. Recently, in zebrafish, which has seven aquaglyceroporins in the genome, we have isolated two different AQP3-like cDNAs (that share only 73% identity), Aqp3a, and Aqp3b, which have increased permeability to EG when expressed in *X. laevis* oocytes (82). Similar to mammalian AQP3 (83), water and solute permeation through zebrafish Aqp3a and Aqp3b is blocked by acidic pH, but maximum permeation is at pH 8.5 (Fig. 3), unlike mammalian AQP3, which seems to be more permeable at pH 7.5. Point mutations in the amino acid sequence of zebrafish Aqp3a increase water and EG permeability with respect to those of the wild type, or change the pH sensitivity. These findings suggest that zebrafish Aqp3a could be used as a model for investigating water and solute movement in aquaporin-expressing vertebrate oocytes, perhaps offering new possibilities for cryopreservation.

The osmotic flow of water across the cell membrane is another crucial aspect during cryopreservation. In baker's yeast (*Saccharomyces cerevisiae*), tolerance to rapid freezing conditions and expression of the water-selective aquaporins, Aqy1 and Aqy2, have been positively correlated based on microarray analysis of freeze resistant mutant strains and over-expression of Aqp1-1 or Aqy2-1 (84). It is believed that high levels of aquaporins in the plasma membrane of cells allow rapid water efflux, especially at freezing temperatures, resulting in the reduction in intracellular ice crystal formation and cell damage. Interestingly, a recent study reports that increased AQP2 translocation into the plasma membrane of rat kidney inner medullar collecting duct cells, by treatment with arginine vasopressin before freezing, results in a significant increase in cell viability on warming (85). Therefore, it is possible that the over-expression of water-selective aquaporins in mammalian oocytes, such as AQP1 or AQP4, which transport water more efficiently than aquaglyceroporins (86), might reduce osmotic stress and intracellular ice formation in oocytes, hence improving survival during cryopreservation. However, changes in oocyte water and/or solute permeability induced using this approach have not been reported.

CONCLUDING REMARKS

With the discovery of aquaporins and aquaglyceroporins in vertebrate oocytes and embryos, new questions on their role during oogenesis and early embryogenesis have arisen. However, published data can potentially provide the basis for a whole area of development on aquaporin-based techniques for oocyte cryopreservation. From the available reports, it appears that, although mammalian mature oocytes, in general, have low permeability in contrast to early embryos and might not express functional aquaporins in the plasma membrane, transient artificial expression of AQP3 in oocytes does affect their membrane permeability and the conditions suitable for cryopreservation. This finding opens up the possibility of testing other aquaglyceroporins, from mammalian or nonmammalian origin, which can be modified to enhance selected channel functions. Also, although not yet tested, artificial coexpression of aquaglyceroporins and water-selective aquaporins may optimize the movement in and out of cryoprotectants and water in oocytes, reducing the time of exposure to cryoprotectants during vitrification and the osmotic stress. However, before these methods can be successfully developed, additional research should be devoted to understanding the intracellular translational and post-translational mechanisms that regulate endogenous and exogenous aquaporin function both in immature and mature oocytes. This investigation will be important to optimize the permeability properties to water and neutral solutes of oocytes during cryopreservation.

ACKNOWLEDGEMENTS

The research conducted by the authors was financed by the Ministry of Education and Science (Spain), grants AGL2004-00316 and AGL2007-60262, and by the European Commission, Q5RS-2002-00784-CRYOCYTE and Marie Curie Research Training Network Aqua(glycero)porins, MRTN-CT-2006-035995.

REFERENCES

1. Fuller BJ, Paynter SJ. Cryopreservation of mammalian embryos. Methods Mol Biol 2007; 368: 325–39.
2. Gook DA, Edgar DH. Human oocyte cryopreservation. Hum Reprod Update 2007; 13: 591–605.
3. Whittingham DG, Leibo SP, Mazur P. Survival of mouse embryos frozen to –196 degrees and –269 degrees C. Science 1972; 178: 411–4.
4. Kasai M, Komi JH, Takakamo A, et al. A simple method for mouse embryo cryopreservation in a low toxicity vitrification solution, without appreciable loss of viability. J Reprod Fertil 1990; 89: 91–7.
5. Edashige K, Kasai M. The movement of water and cryoprotectants in mammalian oocytes and embryos and its relevance to cryopreservation. J Mamm Ova Res 2007; 24: 18–22.
6. Preston GM, Carroll TP, Guggino WB, et al. Appearance of water channels in Xenopus oocytes expressing red cell CHIP28 protein. Science 1992; 256: 385–7.
7. Agre P, King LS, Yasui M, et al. Aquaporin water channels-from atomic structure to clinical medicine. J Physiol 2002; 542: 3–16.
8. Jung JS, Preston GM, Smith BL, et al. Molecular structure of the water channel through aquaporin CHIP. The hourglass model. J Biol Chem 1994; 269: 14648–54.
9. Fu D, Libson A, Miercke LJ, et al. Structure of a glycerol-conducting channel and the basis for its selectivity. Science 2000; 290: 481–6.

10. Sui H, Han BG, Lee JK, et al. Structural basis of water-specific transport through the AQP1 water channel. Nature 2001; 414: 872–8.
11. Preston GM, Jung JS, Guggino WB, et al. The mercury-sensitive residue at cysteine 189 in the CHIP28 water channel. J Biol Chem 1993; 268: 17–20.
12. King LS, Kozono D, Agre P. From structure to disease: the evolving tale of aquaporin biology. Nat Rev Mol Cell Biol 2004; 5: 687–98.
13. Yasui M, Kwon TH, Knepper MA, et al. Aquaporin-6: an intracellular vesicle water channel protein in renal epithelia. Proc Natl Acad Sci USA 1999; 96: 5808–13.
14. Zardoya R. Phylogeny and evolution of the major intrinsic protein family. Biol Cell 2005; 97: 397–414.
15. Ishibashi K, Kuwahara M, Gu Y, et al. Cloning and functional expression of a new water channel abundantly expressed in the testis permeable to water, glycerol, and urea. J Biol Chem 1997; 272: 20782–6.
16. Ma T, Yang B, Verkman AS. Cloning of a novel water and urea-permeable aquaporin from mouse expressed strongly in colon, placenta, liver, and heart. Biochem Biophys Res Commun 1997; 240: 324–8.
17. Engel A, Stahlberg H. Aquaglyceroporins: channel proteins with a conserved core, multiple functions, and variable surfaces. Int Rev Cytol 2002; 215: 75–104.
18. Ishibashi K. Aquaporin subfamily with unusual NPA boxes. Biochim Biophys Acta 2006; 1758: 989–93.
19. Yakata K, Hiroaki Y, Ishibashi K, et al. Aquaporin-11 containing a divergent NPA motif has normal water channel activity. Biochim Biophys Acta 2007; 1768: 688–93.
20. Wu B, Beitz E. Aquaporins with selectivity for unconventional permeants. Cell Mol Life Sci 2007; 64: 2413–21.
21. De Groot BL, Grubmuller H. The dynamics and energetics of water permeation and proton exclusion in aquaporins. Curr Opin Struct Biol 2005; 15: 176–83.
22. Hub JS, de Groot BL. Mechanism of selectivity in aquaporins and aquaglyceroporins. Proc Natl Acad Sci USA 2008; 105: 1198–203.
23. Takata K, Matsuzaki T, Tajika Y. Aquaporins: water channel proteins of the cell membrane. Prog Histochem Cytochem 2004; 39: 1–83.
24. Morishita Y, Matsuzaki T, Hara-Chikuma M, et al. Disruption of aquaporin-11 produces polycystic kidneys following vacuolization of the proximal tubule. Mol Cell Biol 2005; 25: 7770–9.
25. Itoh T, Rai T, Kuwahara M, et al. Identification of a novel aquaporin, AQP12, expressed in pancreatic acinar cells. Biochem Biophys Res Commun 2005; 330: 832–8.
26. Rojek A, Praetorius J, Frøkiaer J, et al. A current view of the mammalian aquaglyceroporins. Annu Rev Physiol 2008; 70: 301–27.
27. Verkman AS. Mammalian aquaporins: diverse physiological roles and potential clinical significance. Expert Rev Mol Med 2008; 10: e13.
28. Huang HF, He RH, Sun CC, et al. Function of aquaporins in female and male reproductive systems. Hum Reprod Update 2006; 12: 785–95.
29. Jablonski EM, McConnell NA, Hughes FM, et al. Estrogen regulation of aquaporins in the mouse uterus: potential roles in uterine water movement. Biol Reprod 2003; 69: 1481–7.
30. He RH, Sheng JZ, Luo Q, et al. Aquaporin-2 expression in human endometrium correlates with serum ovarian steroid hormones. Life Sci 2006; 79: 423–9.
31. Lindsay LA, Murphy CR. Redistribution of aquaporins 1 and 5 in the rat uterus is dependent on progesterone: a study with light and electron microscopy. Reproduction 2006; 131: 369–78.
32. Lindsay LA, Murphy CR. Aquaporins are upregulated in glandular epithelium at the time of implantation in the rat. J Mol Histol 2007; 38: 87–95.
33. Anderson J, Brown N, Mahendroo MS, et al. Utilization of different aquaporin water channels in the mouse cervix during pregnancy and parturition and in models of preterm and delayed cervical ripening. Endocrinology 2006; 147: 130–40.
34. Brañes MC, Morales B, Ríos M, et al. Regulation of the immunoexpression of aquaporin 9 by ovarian hormones in the rat oviductal epithelium. Am J Physiol Cell Physiol 2005; 288: C1048–57.
35. Gannon BJ, Warnes GM, Carati CJ, et al. Aquaporin-1 expression in visceral smooth muscle cells of female rat reproductive tract. J Smooth Muscle Res 2000; 36: 155–67.
36. McConnell NA, Yunus RS, Gross SA, et al. Water permeability of an ovarian antral follicle is predominantly transcellular and mediated by aquaporins. Endocrinology 2002; 143: 2905–12.
37. Skowronski MT, Kwon TH, Nielsen S. Immunolocalization of Aquaporin-1, -5, and -9 in Female Pig Reproductive System. J Histochem Cytochem 2009; 57: 61–7.
38. Edashige K, Sakamoto M, Kasai M. Expression of mRNAs of the aquaporin family in mouse oocytes and embryos. Cryobiology 2000; 40: 171–5.

39. Ford P, Merot J, Jawerbaum A, et al. Water permeability in rat oocytes at different maturity stages: aquaporin-9 expression. J Membr Biol 2000; 176: 151–8.
40. Meng QX, Gao HJ, Xu CM, et al. Reduced expression and function of aquaporin-3 in mouse metaphase-II oocytes induced by controlled ovarian hyperstimulation were associated with subsequent low fertilization rate. Cell Physiol Biochem 2008; 21: 123–8.
41. Edashige K, Ohta S, Tanaka M, et al. The role of aquaporin 3 in the movement of water and cryoprotectants in mouse morulae. Biol Reprod 2007; 77: 365–75.
42. Schreiber R, Pavenstädt H, Greger R, et al. Aquaporin 3 cloned from Xenopus laevis is regulated by the cystic fibrosis transmembrane conductance regulator. FEBS Lett 2000; 475: 291–5.
43. Virkki LV, Franke C, Somieski P, et al. Cloning and functional characterization of a novel aquaporin from Xenopus laevis oocytes. J Biol Chem 2002; 277: 40610–6.
44. Fabra M, Raldúa D, Power DM, et al. Marine fish egg hydration is aquaporin-mediated. Science 2005; 307: 545.
45. Tingaud-Sequeira A, Chauvigné F, Fabra M, et al. Structural and functional divergence of two fish aquaporin-1 water channels following teleost-specific gene duplication. BMC Evol Biol 2008; 8: 259.
46. Fabra M, Raldúa D, Bozzo MG, et al. Yolk proteolysis and Aquaporin-1o play essential roles to regulate fish oocyte hydration during meiosis resumption. Dev Biol 2006; 295: 250–62.
47. Cerdà J, Fabra M, Raldúa D. In: Babin P, Cerdà J, Lubzens, eds. The Fish Oocyte: From Basic Studies to Biotechnological Applications. The Netherlands: Springer, 2007: 349–96.
48. Offenberg H, Barcroft LC, Caveney A, et al. mRNAs encoding aquaporins are present during murine preimplantation development. Mol Reprod Dev 2000; 57: 323–30.
49. Richard C, Gao J, Brown N, et al. Aquaporin water channel genes are differentially expressed and regulated by ovarian steroids during the periimplantation period in the mouse. Endocrinology 2003; 144: 1533–41.
50. Offenberg H, Thomsen PD. Functional challenge affects aquaporin mRNA abundance in mouse blastocysts. Mol Reprod Dev 2005; 71: 422–30.
51. Barcroft LC, Offenberg H, Thomsen P, et al. Aquaporin proteins in murine trophectoderm mediate transepithelial water movements during cavitation. Dev Biol 2003; 256: 342–54.
52. Edashige K, Tanaka M, Ichimaru N, et al. Channel-dependent permeation of water and glycerol in mouse morulae. Biol Reprod 2006; 74: 625–32.
53. Watson AJ, Natale DR, Barcroft LC. Molecular regulation of blastocyst formation. Anim Reprod Sci 2004; 82–83: 583–92.
54. Verkman AS, van Hoek AN, Ma T, et al. Water transport across mammalian cell membranes. Am J Physiol 1996; 270: C12–30.
55. Leibo SP. Water permeability and its activation energy of fertilized and unfertilized mouse ova. J Membr Biol 1980; 53: 179–88.
56. Paynter SJ, Fuller BJ, Shaw RW. Temperature dependence of Kedem-Katchalsky membrane transport coefficients for mature mouse oocytes in the presence of ethylene glycol. Cryobiology 1999; 39: 169–76.
57. Ruffing NA, Steponkus PL, Pitt RE, et al. Osmometric behavior, hydraulic conductivity, and incidence of intracellular ice formation in bovine oocytes at different developmental stages. Cryobiology 1993; 30: 562–80.
58. Benson CT, Critser JK. Variation of water permeability (Lp) and its activation energy (Ea) among unfertilized golden hamster and ICR murine oocytes. Cryobiology 1994; 31: 215–23.
59. Hunter J, Bernard A, Fuller B, et al. Plasma membrane water permeabilities of human oocytes: the temperature dependence of water movement in individual cells. J Cell Physiol 1992; 150: 175–9.
60. De Santis L, Coticchio G, Paynter S, et al. Permeability of human oocytes to ethylene glycol and their survival and spindle configurations after slow cooling cryopreservation. Hum Reprod 2007; 22: 2776–83.
61. Echevarría M, Windhager EE, Frindt G. Selectivity of the renal collecting duct water channel aquaporin-3. J Biol Chem 1996; 271: 25079–82.
62. Tsukaguchi H, Weremowicz S, Morton CC, et al. Functional and molecular characterization of the human neutral solute channel aquaporin-9. Am J Physiol 1999; 277: F685–96.
63. Le Gal F, Gasqui P, Renard JP. Differential osmotic behavior of mammalian oocytes before and after maturation: a quantitative analysis using goat oocytes as a model. Cryobiology 1994; 31: 154–70.
64. Valdez DM, Jr, Miyamoto A, Hara T, et al. Water- and cryoprotectant-permeability of mature and immature oocytes in the medaka (Oryzias latipes). Cryobiology 2005; 50: 93–102.
65. Seki S, Kouya T, Valdez DM, et al. The permeability to water and cryoprotectants of immature and mature oocytes in the zebrafish (Danio rerio). Cryobiology 2007; 54: 121–4.
66. Isayeva A, Zhang T, Rawson DM. Studies on chilling sensitivity of zebrafish (Danio rerio) oocytes. Cryobiology 2004; 49: 114–22.

67. Biggers JD, Bell JE, Benos DJ. Mammalian blastocyst: transport functions in a developing epithelium. Am J Physiol 1988; 255: C419–32.
68. Pfaff RT, Liu J, Gao D, et al. Water and DMSO membrane permeability characteristics of in-vivo- and in-vitro-derived and cultured murine oocytes and embryos. Mol Hum Reprod 1998; 4: 51–9.
69. Ma T, Song Y, Yang B, et al. Nephrogenic diabetes insipidus in mice lacking aquaporin-3 water channels. Proc Natl Acad Sci USA 2000; 97: 4386–91.
70. Edashige K, Yamaji Y, Kleinhans FW, et al. Artificial expression of aquaporin-3 improves the survival of mouse oocytes after cryopreservation. Biol Reprod 2003; 68: 87–94.
71. Miyake T, Kasai M, Zhu SE, et al. Vitrification of mouse oocytes and embryos at various stages of development in an ethylene glycol-based solution by a simple method. Theriogenology 1993; 40: 121–34.
72. Bautista JA, Kanagawa H. Current status of vitrification of embryos and oocytes in domestic animals: ethylene glycol as an emerging cryoprotectant of choice. Jpn J Vet Res 1998; 45: 183–91.
73. Santos RR, Tharasanit T, Van Haeften T, et al. Vitrification of goat preantral follicles enclosed in ovarian tissue by using conventional and solid-surface vitrification methods. Cell Tissue Res 2007; 327: 167–76.
74. Gautam SK, Verma V, Palta P, et al. Effect of type of cryoprotectant on morphology and developmental competence of in vitro-matured buffalo (Bubalus bubalis) oocytes subjected to slow freezing or vitrification. Reprod Fertil Dev 2008; 20: 490–6.
75. Hagedorn M, Lance SL, Fonseca DM, et al. Altering fish embryos with aquaporin-3: an essential step toward successful cryopreservation. Biol Reprod 2002; 67: 961–6.
76. Valdez DM, Jr, Hara T, Miyamoto A, et al. Expression of aquaporin-3 improves the permeability to water and cryoprotectants of immature oocytes in the medaka (Oryzias latipes). Cryobiology 2006; 53: 160–8.
77. Yamaji Y, Valdez DM, Jr, Seki S, et al. Cryoprotectant permeability of aquaporin-3 expressed in Xenopus oocytes. Cryobiology 2006; 53: 258–67.
78. Seki S, Kouya T, Hara T, et al. Exogenous expression of rat aquaporin-3 enhances permeability to water and cryoprotectants of immature oocytes in the zebrafish (Danio rerio). J Reprod Dev 2007; 53: 597–604.
79. Lance SL, Peterson AS, Hagedorn M. Developmental expression of aquaporin-3 in zebrafish embryos (Danio rerio). Comp Biochem Physiol C Toxicol Pharmacol 2004; 138: 251–8.
80. Meinild AK, Klaerke DA, Zeuthen T. Bidirectional water fluxes and specificity for small hydrophilic molecules in aquaporins 0-5. J Biol Chem 1998; 273: 32446–51.
81. Tsukaguchi H, Shayakul C, Berger UV, et al. Molecular characterization of a broad selectivity neutral solute channel. J Biol Chem 1998; 273: 24737–43.
82. Cerdà J, Chauvigné F. New aquaporin-based approaches for the cryopreservation of fish oocytes and embryos. European Aquaculture Society, Special Publication 2008; 37: 131–32.
83. Zeuthen T, Klaerke DA. Transport of water and glycerol in aquaporin 3 is gated by H(+). J Biol Chem 1999; 274: 21631–6.
84. Tanghe A, Kayingo G, Prior BA, et al. Heterologous aquaporin (AQY2-1) expression strongly enhances freeze tolerance of Schizosaccharomyces pombe. J Mol Microbiol Biotechnol 2005; 9: 52–6.
85. Wang W, Ben RN. Upregulation and protein trafficking of aquaporin-2 attenuate cold-induced osmotic damage during cryopreservation. In Vitro Cell Dev Biol Anim 2004; 40: 67–70.
86. Yang B, Verkman AS. Water and glycerol permeabilities of aquaporins 1–5 and MIP determined quantitatively by expression of epitope-tagged constructs in Xenopus oocytes. J Biol Chem 1997; 272: 16140–6.

9 Oocyte Storage at Different Developmental Stages

Mafalda Rato[1] and Carlos E. Plancha[2]

[1] *CEMEARE—Centro Médico de Assistência à Reprodução, Lisboa, Portugal and*

[2] *Biology of Reproduction Unit, Institute of Histology and Developmental Biology, Faculty of Medicine of Lisbon, Lisboa, Portugal*

FERTILITY STORAGE: DIVERSITY AS A RULE

Although embryo cryopreservation is a very advantageous technique for couples resorting to assisted reproduction therapies, its routine use has also resulted in a series of ethical, legal, and religious constrains related to embryo storage and/or discarding. Some countries have limited or even outlawed its use, stressing the need for alternative approaches. Despite these restrictive decisions for infertile couples, basic investigation at the cryobiology level has been continuously addressed, leading to new possibilities for female patients wishing to preserve their fertility potential. At the same time, oocyte banking has become a tangible option not only for human oocyte donation (1), but also to help maintain, in a safe and efficient manner, the diversity of organisms that are rare, endangered, or bred for agricultural or medicinal value (2).

From a clinical perspective, several situations could benefit from oocyte storage. Cases of premature ovarian failure, either due to intrinsic or iatrogenic causes (3,4), are a clear example, but a series of other situations raise equivalent concerns. These relate to legal/ethical constrains of surplus embryo storage, oocyte donation, or even the controversial trend of delaying parenthood (5). Other cases include ovarian tissue freezing (3), where concomitant antral follicle puncture is advised, or oocyte pickup with no accompanying sperm collection. Each particular case could benefit from oocyte storage at a particular developmental stage and, if technically optimized, one should not be considered more advantageous than the other, as the several possible situations impose a diversity of solutions.

OVERVIEW OF OOCYTE DEVELOPMENTAL STAGES

When considering storage of ovarian tissue, cumulus–oocyte complexes (COC), or oocytes, we cannot avoid being confronted with both cellular heterogeneity and distinct levels of complexity. From a cortical ovarian organization, rich in extracellular matrix and several cell types, to the cellular interactions between the immature oocyte and its surrounding cumulus cells, complexity is the key theme. At the intracellular level, the oocyte displays a cytoskeleton arrangement that further contributes to cellular vulnerability. In every case, the intrinsic physiological and biophysical characteristics of each oocyte's developmental stage add new variables to the cryopreservation equation.

Immature Oocytes

Fully grown, immature mammalian oocytes are arrested at prophase I of meiosis and are characterized by the presence of a prominent nucleus known as the germinal vesicle (GV). Although its size varies from one model to another, all oocytes must reach a species-specific predetermined volume that seems to correspond to the ability to support correct maturation and early embryo development (6). In every case, the oocyte is surrounded by and metabolically coupled to its companion granulosa cells through microtubule- and microfilament-rich trans-zonal processes (TZPs) (7,8) (Fig. 1). Decondensed chromosomes enclosed within the nuclear envelope support active transcription until the oocyte reaches its final diameter (9). Depending on the species, microtubules can be present in an interphase-like pattern, with microtubule reinforcement around the GV and at the cortex, or in a mitosis-like pattern (10,11). Dense microfilament bundles can also be observed at both the cortex and around the GV (12) and their presence can be related to the specific positioning of this organelle (13) (Fig. 2). Organelles such as the Golgi apparatus and endoplasmic reticulum locate in the cytoplasm along with mitochondria and cortical granules (14). A slight degree of vacuolization is commonly found in this state (15). The plasma membrane is characterized by low concentrations of cholesterol and polyunsaturated fatty acids (16), provided with numerous microvilli projecting into the perivitelline space (17), and its aquaporin content has been suggested to be higher than in maturing MI or mature MII oocytes (18). The oocyte is surrounded by a completely assembled *zona pellucida* in which glycoproteins are functionally linked giving it a characteristic spongy texture.

Quiescent and early growing immature oocytes are also worth considering in the context of female germ cell preservation. The preservation of both quiescent oocytes inside primordial follicles and early growing oocytes in primary or secondary follicles relates to the issue of ovarian cortical tissue freezing, which will not be extensively discussed in this chapter. In short, their potential for fertility preservation can be considered clinically relevant only when performed by a multidisciplinary team. This team should not only be able to guarantee proper cortical ovarian tissue recovery and freezing, but also to assure the technical expertise for subsequent function re-establishment (3,4,19). Oocytes at these developmental stages need to go through in vivo growth after successful ovarian tissue transplantation (20) or, theoretically, through prolonged culture before oocyte maturation, which we are still far from being able to manipulate in most large mammal models and particularly in humans (21).

The recovery and preservation of this pool of immature oocytes contained in the ovarian tissue is of significant interest, considering the potential for fertility recovery in the context of animal and human reproduction. However, they are associated with several difficulties that exclude them from current practical use. First, these cells have not acquired complete functionality like the fully grown immature oocyte. Second, in vitro approaches are still not effective. Third, freezing is particularly harmful to these follicles, as opposed to the earlier primordial stages. Nevertheless, in spite of all these difficulties, its use could prove highly beneficial, particularly in the context of avoiding primordial follicle exhaustion after ovarian tissue transplantation.

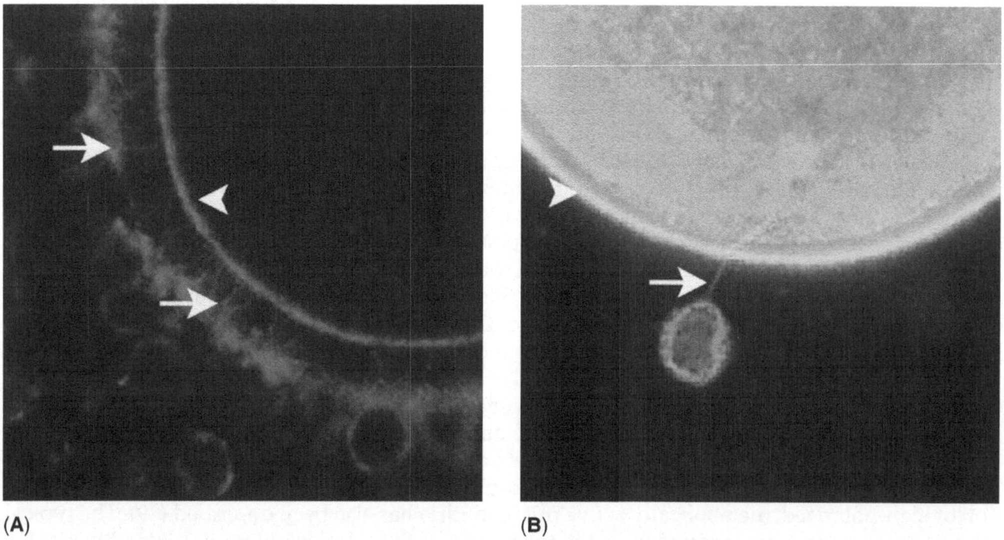

Figure 1 (**A**) Fully grown germinal vesicle of hamster oocyte surrounded by cumulus cells exhibiting microfilament-rich trans-zonal processes (arrows). (**B**) Fully grown mouse oocyte with attached cumulus cell projecting a microtubule-rich transzonal process across the *zona pellucida* (arrow). This oocyte was mechanically stripped before labeling. Note a strong microfilament staining at the oocyte cortex (arrow heads in **A** and **B**). *See Color Plates on Page xvi.*

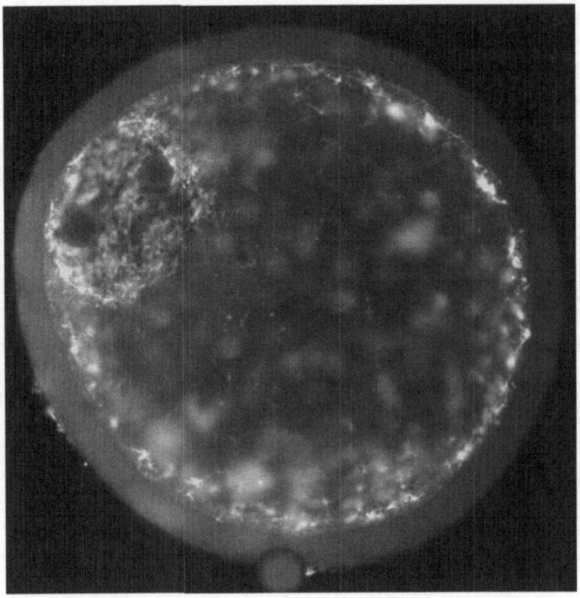

Figure 2 Fully grown germinal vesicle of hamster oocyte with microtubule network (green) around the germinal vesicle (blue) and at the cortical region. This oocyte was mechanically stripped before labeling. *See Color Plates on Page xvii.*

Mature Oocytes

Oocyte maturation is a short process but includes many important events occurring in the nuclear and cytoplasmic compartments. The main players in nuclear maturation are the chromosomes and the cytoskeleton, with meiotic spindles ensuring correct chromosome segregation and asymmetric cellular division. The key determinants in cytoplasmic maturation may well be the cumulus cells and their interactions with the oocyte, as they promote the availability of crucial metabolites and signaling molecules to the oocyte. These somatic cells have an essential role in maturation and, in some models such as the hamster, their presence is absolutely necessary until the MI stage (22). Recently it has been demonstrated in mice that immediately after luteinizing hormone stimulation, gap junctions between the oocyte and corona radiata cells are transiently closed, reopening soon afterwards (23), showing that somatic-germ cell communication maintains even at the latest stage of oogenesis.

The mature mammalian oocyte is promptly distinguished by the presence of the first polar body in the perivitelline space. Although it retains the large volume-to-surface ratio established during the growth phase, its condition is now of an M-phase (mitotic or meiotic) arrested cell, exhibiting the meiotic spindle as its most distinctive trait. The chromosomes are equatorially located at the microtubular spindle, which is associated with an enriched microfilament domain overlying the spindle at the oocyte cortex (24) (Fig. 3). As centrioles disappear at the pachytene stage of prophase I, spindle poles organize from the pericentriolar material containing microtubule organizing centers that include γ-tubulin and pericentrin (24,25). In contrast to the stable, well-aligned equatorial plate chromosomes, the oocyte spindle is a rather dynamic structure where most microtubules undergo rapid cycles of assembly and disassembly (26). Mitochondria, evenly dispersed through the cytoplasm (27), are often arranged in association with smooth endoplasmic reticulum elements (either tubules or vesicles) to form voluminous aggregates and, less frequently, small mitochondria–vesicle complexes (28). Particularly, a pericortical distribution of high-polarized, metabolically active mitochondria has also been described (29). The typical cortical granules appear stratified in one to three rows at the subplasmalemmal area (30). Finally, mature oocyte membranes are characterized by high (approximately 80%) saturated fatty acids and low (approximately 6%) polyunsaturated fatty acids content (31). This particular proportion of lipidic elements is believed to be responsible for its higher fluidity and resistance to chilling injury at low temperatures in comparison with immature stages (16).

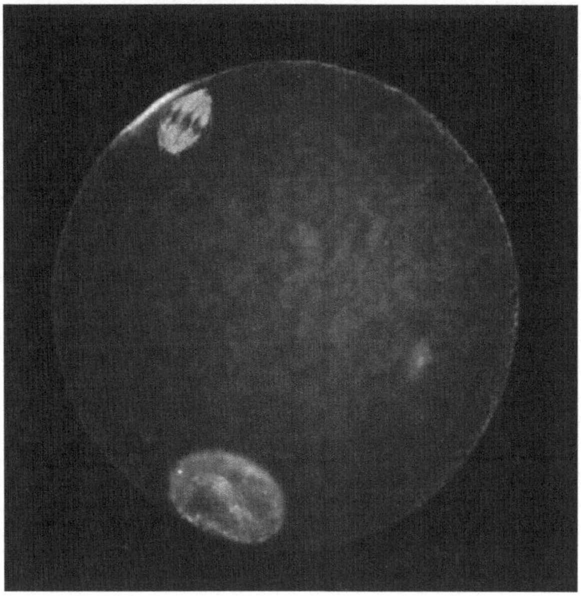

Figure 3 In vitro-matured hamster oocyte depicting a cortical second meiotic spindle with chromosomes equatorially aligned and the first polar body. Note the strong microfilament labeling overlying the spindle at the oocyte cortex. This oocyte was mechanically stripped before labeling. Microtubules (green), microfilaments (red), DNA (blue). *See Color Plates on Page xvii.*

A particularly relevant aspect pertaining to mature oocytes derives from in vitro maturation (IVM). Despite the relative success of oocyte IVM, differences between in vivo- and in vitro-matured oocytes emerge at both cytological and molecular levels (25). Beyond the normal appearance of a regular metaphase II oocyte characterized by the presence of a polar body, in vitro-matured oocytes may display some irregular features that reflect a diminished competence (32) and may lead to decreased resistance to cryopreservation. In the mouse, some of the features relate to abnormal centrosome number, loss of cell polarity, and absence of some morphofunctional characteristics of a normal spindle such as a dispersed pericentrin pattern (25,32). Other differences regarding protein synthesis (33), translation and transcription patterns (34), patterns of gene methylation (35), mitochondrial distribution (36), and ATP (37) and glutathione content (38) have also been reported. Aquaporin content has also been suggested to be lower in in vitro-matured human oocytes when compared to their in vivo-matured counterparts (39). Although this difference may reflect a more general culture effect upon cells, it can indicate that permeability characteristics of matured oocytes may be altered by culture conditions.

LESSONS FROM OOCYTE STORAGE AT DIFFERENT DEVELOPMENTAL STAGES

The current and recently reinforced belief that a limited primary oocyte pool is established at the time of birth (40) stresses the relevance of increasingly successful oocyte preservation attempts. Although much information concerning the cryopreservation of semen and embryos has been gathered, and turned into routine procedures, oocyte preservation at an efficient rate is still an elusive task. Efforts are continuously being made to circumvent such difficulties and several models have contributed to the understanding of the physiological and biophysical characteristics common to these cells. However, taking into account the distinctive features of these female gametes, diversity advises us caution in extrapolating results from one model to the other.

Over the past years, several promising results have been published regarding MII oocyte cryopreservation (41,42), but few have been translated to the in vitro fertilization laboratory in the form of a well-established protocol (43–45). The small, but increasing number of children born after mature oocyte preservation confirms that the right direction has been taken, but also reflects the need to pursue more efficient methodologies. As an attempt to avoid some problems related to this strategy, immature oocyte preservation has gathered some attention. Reports span a variety of models (46–48) but despite some encouraging outcomes, only one human live birth after immature oocyte cryopreservation has been reported (49).

Survival is the first challenge for oocyte freezing. Factors that influence oocyte survival relate to cryoprotectant agent (CPA) toxicity, osmotic cell shrinkage, intracellular ice formation, chilling temperatures, and osmotic swelling during CPA removal (50). However, one should bear in mind that success of oocyte cryopreservation cannot only be evaluated through cell survival, but more importantly through effective fertility preservation. In that sense, extending the evaluation of the freezing process to a functional level, both CPA type and concentration have been shown to influence post-thaw developmental potential (51–53). In particular, CPA influence depends on specific events such as the induced osmotic response, which in turn also depends on the oocyte's developmental stage (42,54).

Although data on slow freezing and vitrification point to increasingly higher oocyte survival rates, post-thaw events remain drastically compromised (55,56). Nevertheless, the aforementioned success rates need to be carefully interpreted as they vary within a wide range of values (1,57), with such disparities being probably related to the degree of cellular evaluation. For instance, morphological assessment of a polar body or correctly assembled spindle as a maturation marker represents a necessarily different evaluation level when compared to a functional evaluation, such as fertilization and developmental competence.

Immature Oocyte Storage

Immature oocytes rely on efficient cumulus–oocyte communication in order to correctly mature. Although some reports indicate post-thaw IVM as the main determinant for the technique's low success rates (58,59), others point to the preservation process itself as the most important factor (60). Great efforts have been made over the past years with the aim of elevating both survival and developmental rates (42) for immature oocytes. However, freezing immature oocytes while maintaining its maturation potential turned out to be one of the hardest challenges in this field as two main problems arise: (*i*) regarding TZPs and associated cytoskeleton preservation and (*ii*) relating to distinct cell volume between germ and somatic cells.

Accordingly, cryopreservation protocols imply a series of conformational changes either during CPA equilibration or during extracellular ice formation, which can drastically modify the physical interaction between the two cellular compartments. Such mechanical stress is, in most cases, responsible for either TZP rupture or oocyte–cumulus cell membrane damage (61). Furthermore, a deleterious CPA effect on the actin and microtubule cytoskeleton that support TZPs has also been reported, leading to subsequent cumulus cell loss during freezing (62–64) (Fig. 4). All these effects are potential threats to oocyte–cumulus bidirectional communication, possibly compromising oocytes function.

Another big obstacle to maturation potential maintenance is cumulus cell survival per se. The huge volume difference between these cells and the oocyte hinders the optimization of a freezing protocol that is able to sustain the survival of both cell types (65). Even when ice crystals are avoided, as with vitrification, volume disproportion dictates different reactions to CPA concentrations. In that sense, freezing seems to act on cumulus–oocyte communication in much the same way mechanical stripping does (i.e., negatively affecting meiotic progression). Most structural abnormalities detected following post-thaw IVM relate to spindle morphology, chromosome alignment (42,57,60), and modifications of cytoplasmic and nuclear organization (28) (Fig. 4). Although vitrification has been reported to support post-thaw meiotic resumption in a more efficient manner than slow cooling, spindle assembly is still highly compromised after IVM (66). However, all these argumentations may change soon as some teams are already successfully vitrifying immature oocytes, and successfully maturing them in vitro afterwards (Brambillasca F, personal results).

Regarding oocyte survival, immature oocyte resistance has been associated with membrane permeability and the presence of cumulus cells during CPA equilibration and freezing. GV oocytes have been reported to be less permeable than its mature counterparts, possibly due to its membrane lipidic composition (16) and cortical microfilament network (67). Furthermore,

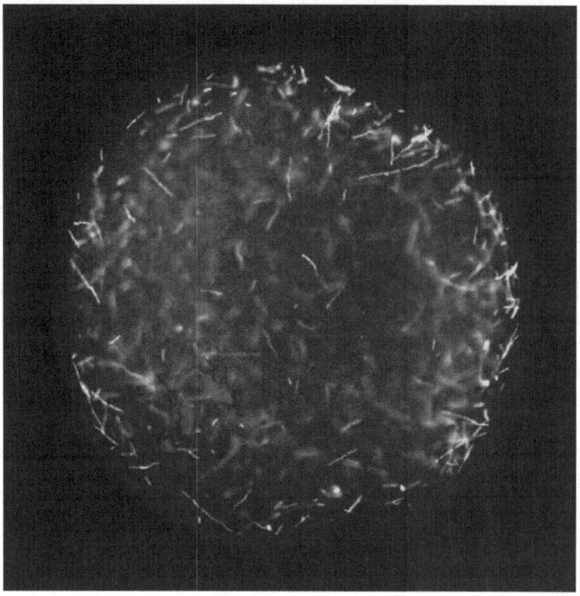

Figure 4 Cultured, fully grown hamster germinal vesicle oocyte after slow freezing and fast thawing. In this oocyte the absence of cumulus cells is a consequence of the cryopreservation process, negatively affecting meiotic progression. No mechanical stripping was done. Note the persistence of the germinal vesicle (blue) and the reappearance of a microtubule interphase-like network (green). *See Color Plates on Page xvii.*

cumulus cells have been shown to render considerable protection against CPA toxicity either by slowing down CPA penetration in the case of ethylene glycol or by conferring protection through osmotic swelling during CPA removal (50). Although contrasting data indicate that cumulus cells may act as an obstacle to oocyte CPA uptake or even as potential ice nucleation sites (68), its absolute requirement for subsequent maturation dictates its presence during freezing.

In order to improve not only oocyte and cumulus cell survival but also maturation rates after freezing, a rather simple approach focused on cellular membrane protection has been proposed. CPA equilibration at physiological temperature, associated with low toxicity CPA and a natural mixture of phospholipids and antioxidants (e.g., egg yolk) proved very efficient in minimizing the negative influence of the most common freezing protocols. As it is not desirable to use a poorly defined animal substance in clinical settings, the development of defined lipidic mixtures with a similar protective action is a very attractive possibility. Such a rationale led to a series of interesting ideas on the importance of preserving membrane organization rather than simply aiming for cell survival. Given the importance proper cytoskeleton organization has for correct maturation and fertilization, greater attention should be given to protecting the interplay between the cytoskeleton and the plasma membrane. In that sense, membrane protection would be expected to indirectly promote correct cytoskeleton functioning during and following maturation (65).

Another risk for membrane integrity is the cellular organization of the COC. Distinct attempts to preserve the three-dimensional COC structure have focused on cytoskeleton protection, either by depolymerizing or by stabilizing it. These approaches span several models and some results, mostly in association with vitrification, are starting to prove beneficial in terms of maturation (69) and developmental potential (70,71), possibly due to better TZP preservation (72). Following the same rationale, reports on oocyte osmotic properties point to an elevation of membrane permeability after actin depolymerization possibly due to a more uniform CPA penetration (67). Regarding functional studies on COC integrity following freezing, the hamster may prove to be an efficient model in terms of TZP integrity evaluation, as meiotic progression halts at the pre-metaphase I stage if that particular communication pathway is impaired (10). In that sense, a statistically significant improvement of the maturation potential has been observed following a slow cooling protocol with prior microtubule depolymerization (Rato M and Plancha CE, personal results).

Species-specific features also strongly contribute to diverse immature oocyte cryopreservation outcomes. One such example is the influence lipid content exerts over the chilling sensitivity, as oocytes with low lipid content such as those of the mouse, cat, rabbit, and human are more likely to be successfully cryopreserved than porcine, cattle, and buffalo oocytes (52). Specifically regarding the porcine model, the presence of a large number of intracytoplasmic lipid droplets has proven to exert a negative influence on the oocyte (73). Although the concept of delipation before cryopreservation has been introduced as a means to circumvent this problem (74) and even improve maturation rates (73), some recent modified vitrification protocols such as Solid Surface Vitrification have yielded slight improvements regarding survival, maturation potential, and blastocyst production without any micromanipulation or delipation procedures (75).

Despite all the different approaches, immature oocytes are still functionally compromised by freezing. Currently, efforts to determine and protect the major targets of COC disruption are mainly focused on chilling, CPA, and ice crystal-related injuries. On the other hand, IVM of freshly collected, fully grown immature oocytes is proving its efficiency by an increasing number of births. Regarding this immature stage, only the combination of these promising techniques will allow the elaboration of a freezing protocol able to sustain cell survival, post-thaw maturation, and, above all, fertility preservation in a safe and efficient manner.

Mature Oocyte Storage

Presently, mature oocyte storage will benefit from the establishment of a preservation protocol with minimal interference on cell viability and maximum safeguarding of developmental potential, as survival per se does not seem to be an issue anymore, in both slow freezing and vitrification protocols. Understandably, temperature-related spindle disassembly has been continuously addressed and although it is an ever present drawback of the current cryopreservation protocols, the encouraging number of babies born from frozen-thawed mature oocytes is a feedback that compels us to pursue higher efficiency.

For MII oocytes, osmotic stress and its associated volume changes are the main lines of injury during cryopreservation as it influences not only cell survival but also spindle organization (76). In order to define a theoretical base for the design of freezing solutions, some reports have been produced on the permeability characteristics of MII oocyte membranes to different CPAs (39,54). In that sense, the definition of dehydration strategies able to prevent excessive osmotic stress and minimize the exposure to high CPA concentration will surely prove beneficial for subsequent embryonic development (76). Furthermore, specific permeability characteristics of in vivo- and in vitro-matured oocytes should also be taken into account (39). The presence or absence of cumulus cells during mature oocyte freezing is another matter of discussion. While some studies report no protective effect of somatic cells (46), others have shown a positive effect upon post-thaw survival and fertilization rates (77). In that context, it has been hypothesized that cumulus cells could render some protection against sudden osmotic changes induced by rapid CPA passage through the ooplasm, but an effective protection upon oocyte developmental competence is still to be proven.

Despite these and other technical adjustments, equilibration protocols still aim for the same gold standard: to avoid ice crystal formation with minimum influence on oocyte morphology and physiology. Slow cooling strategies now favor higher sucrose concentrations and low toxic/rapid permeating CPAs (78). On the other hand, vitrification focuses on achieving high cooling rates by minimizing freezing volumes and CPA equilibration duration in order to alleviate CPA toxicity (79). Nevertheless, both slow freezing and vitrification should aim not only to improve survival but also to maintain post-thaw structural and developmental integrity.

Several reports point to structural modifications of the mature oocyte after slow freezing (28). Those include premature cortical granule exocitosis with subsequent zona hardening (14), loss of mitochondria polarity (29), decreased mitochondria number (27), increased microvacuolization (28), abnormal cytoskeleton (27,80) and spindle configuration, chromosome scattering, and chromatid non-disjunction (24) (Fig. 5). After vitrification, much less modifications have been reported. Fracture and detachment of microvilli have been observed, but regarding mitochondria and cortical granule disposition, no alterations seem to be induced (55). It appears that not one but the combination of some of the above-mentioned effects may be responsible for the reported low developmental competence after either slow freezing or vitrification.

While zona hardening is nowadays easily overcome by micromanipulation (intracytoplasmic sperm injection), other types of alterations can significantly impair further development. For instance, irreversible loss of mitochondria polarity is related to abnormal regulation of free intracellular Ca^{2+} (81). Another example is vacuolization, an important feature related to fertilization failure (82) that may be associated with membrane rupture of secondary lysosomes and of the oolemma itself (28). Regarding the cytoskeleton, the disruption of the oocyte architecture and related chromosomal and cytokinesis abnormalities is a widespread concern, as it may ultimately lead to development failure. Most profound cytoskeleton alterations include the disruption of the subplasmalemmal microfilament cortex, extensive appearance of cytoplasmic microtubular asters, and spindle disorganization with subsequent chromosome scattering (24). Mature oocyte exposure to specific CPAs was also reported as a disturbing agent for both the microfilament (62) and the microtubule networks (66), while oocyte handling below physiological temperatures (83,84), or freeze and thaw cycles (85,86), have been specifically related to microtubule depolymerization.

Nonetheless, most of these alterations seem partially reversible when appropriate manipulation after thawing is carried out. Although similar observations of post-thaw spindle and cytoskeleton reassembly seem to be a widespread phenomena across different animal models and in humans (62,85), a species-dependent response to cooling, CPA, and cryopreservation cannot be disregarded (55,84). Interestingly, human oocyte handling at room temperature during a pre-freezing equilibration series does not seem to have a significant influence on spindle morphology (1,85). This observation, which contrasts with previous data (87,88), suggests a protective action during the equilibration period attributable to the pre-freezing medium and possibly related to the documented ability of some CPAs, specifically propanediol, to stabilize the spindle structure (89).

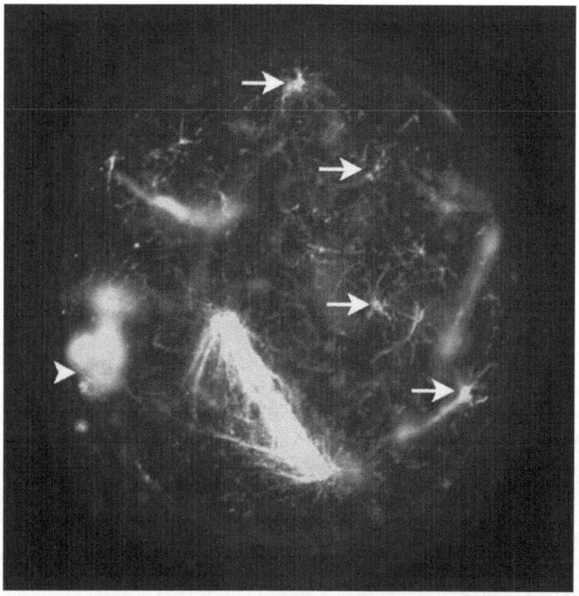

Figure 5 In vitro-matured hamster oocyte after slow freezing at the germinal vesicle stage. A distorted second meiotic spindle (green), with non-aligned chromosomes (blue), is clearly observed. Note the first polar body with a microfilament (red) rich region (arrow head), and several foci of premature cytoplasmic microtubule polymerization (arrows). *See Color Plates on Page xviii.*

Recent reports have highlighted that spindle and cytoskeleton injuries are not the sole factors responsible for post-thawing developmental impairment. Beyond these morphological traits, DNA integrity (59), the oocyte's proteome (90), and cell physiology (29,56) are reported to be modified after freezing. For instance, reports on DNA fragmentation after slow freezing (59) or vitrification (91) indicated DNA as a preferential target for cryoinjury and as a possible cause for low embryonic development. Strengthening this idea, the degree of injury seems to be protocol dependent and intracellular ice formation during slow cooling or devitrification is indicated as one of the probable causes for such damage (59). Moreover, events such as intracellular calcium elevation and proteome deregulation after CPA equilibration and freezing have also revealed protocol dependency (92). In that sense, the elimination of extracellular calcium from the equilibration media may alleviate such an effect (56).

In face of the series of alterations induced by freezing and its associated protocols, some modifications have been proposed in order to elevate both oocyte survival and competences. As mentioned previously, sucrose supplementation to the freezing media proved to be an advantageous strategy for increased oocyte survival (78) and spindle integrity (86). Calcium removal from the cryopreservation media has also yielded some interesting results in terms of oocyte physiology (56); also, post-thaw incubation seems an advantageous proposal regarding fertilization and early embryonic development (85,86). Regarding vitrification, attempts to improve its outcome include mainly the use of freezing devices (79,93) and cytoskeletal stabilizing agents (69,94).

FINAL REMARKS

The continuous search for the most efficient cryopreservation protocol has already been responsible for a series of successes. The increasing numbers of babies born from frozen oocytes, along with reassuring follow-up results, represent the main driving force to keep pursuing top efficiency. While there was a time when embryo preservation was the only available option to infertile couples, the demands of a progressively more encompassing management of human infertility has led to new attempts at tackling the complexity of one of the largest animal cells. Although an ongoing task, these challenges will continue to lead us to a better understanding of female gamete biology, further strengthening the clinical relevance of oocytes cryopreservation.

ACKNOWLEDGMENTS

We thank Paulo Navarro-Costa for providing Figure 1B, and for both critical reading of the chapter and helpful suggestions.

REFERENCES

1. Boiso I, Marti M, Santaló J, et al. A confocal analysis of the spindle and chromosome configurations of human oocytes cryopreserved at the germinal vesicle and metaphase II stage. Hum Reprod 2002; 17: 1885–91.
2. Rodrigues P, Limback D, McGinnis LK, et al. Oogenesis: prospects and challenges for the future. J Cell Physiol 2008; 216: 355–65.
3. Oktay K. Ovarian tissue cryopreservation and transplantation: preliminary findings and implication for cancer patients. Hum Reprod 2001; 66: 526–34.
4. Schmidt KLT, Andersen CY, Loft A, et al. Follow-up of ovarian function post-chemotherapy following ovarian cryopreservation and transplantation. Hum Reprod 2005; 20: 3539–46.
5. Heng BC. Delayed motherhood through oocyte and ovarian tissue cryopreservation–a perspective from Singapore. Reprod Biomed Online 2006; 12: 660–2.
6. Griffin J, Emery BR, Huang I, et al. Comparative analysis of follicle morphology and oocyte diameter in four mammalian species (mouse, hamster, pig and human). J Exp Clin Assist Reprod 2006; 3: 2.
7. Albertini DF. Regulation of meiotic maturation in the mammalian oocyte: interplay between exogenous cues and the microtubule cytoskeleton. Bioessays 1992; 14: 97–103.
8. Eppig JJ. Oocyte control of ovarian follicular development and function in mammals. Reprod 2001; 122: 829–38.

9. Moore GMP, Lintern-Moore S, Peters H, et al. RNA synthesis in the mouse oocyte. J Cell Biol 1974; 60: 416–22.
10. Plancha CE, Albertini DF. Hormonal regulation of meiotic maturation in the hamster oocyte involves a cytoskeleton-mediated process. Biol Reprod 1994; 51: 852–64.
11. Plancha CE. Oocyte maturation in mammals: experience from the hamster model. Ass Reprod Tech Androl 1995; 8: 77–90.
12. Kim NH, Chung HM, Cha KY, et al. Microtubule and microfilament organization in maturing human oocytes. Hum Reprod 1998; 13: 2217–22.
13. Alexandre H, Van Cauwenberge A, Mulnard J. Involvement of microtubules and microfilaments in the control of the nuclear movement during maturation of mouse oocyte. Dev Biol 1989; 136: 311–20.
14. Ghetler Y, Skutelsky E, Nun IB, et al. Human oocyte cryopreservation and the fate of cortical granules. Fertil Steril 2006; 86: 210–16.
15. Sundstrom P, Nilsson BO, Liedholm P, et al. Ultrastructural Characteristics of human oocytes fixed at a follicular puncture or after culture. J In Vitro Fert Embryo Transf 1985; 3: 232–42.
16. Ghetler Y, Yavin S, Shalgi R, et al. The effect of chilling on membrane lipid phase transition in human oocytes and zygotes. Hum Reprod 2005; 20: 3385–9.
17. Motta PM, Nottola SA, Micara G, et al. Ultrastucture of human unfertilized oocytes and polyspermic embryos in a IVF-ET program. Ann NY Acad Sci 1988; 541: 367–83.
18. Parisi M, Notrica J, Ford P, et al. Human oocyte cryopreservation and its relationship with water permeability at different maturity stages. Reprod Technol 2000; 10: 253–5.
19. Demeestere I, Simon P, Buxant F, et al. Ovarian function and spontaneous pregnancy after combined heterotopic and orthotopic cryopreserved ovarian tissue transplantation in a patient previously treated with bone marrow transplantation: case report. Hum Reprod 2006; 21: 2010–14.
20. Rosendahl M, Loft A, Byskov AG, et al. Biochemical pregnancy after fertilization of an oocyte aspirated from an heterotopic autotransplant of cryopreserved ovarian tissue: case report. Hum Reprod 2006; 21: 2006–9.
21. Smitz JE, Cortvrind RG. The earliest stages of folliculogenesis in vitro. Reprod 2002; 123: 185–202.
22. Plancha CE, Albertini DF. Protein synthesis requirements during resumption of meiosis in the hamster oocyte: early nuclear and microtubule configurations. Mol Reprod Dev 1992; 33: 324–32.
23. Norris RP, Freudzon M, Mehlmann LM, et al. Luteinizing hormone causes MAP kinase-dependent phosphorylation and closure of connexin 43 gap junctions in mouse ovarian follicles: one or two paths to meiotic resumption. Dev 2008; 135: 3229–38.
24. Mandelbauma J, Anastasiou O, Lévyd R, et al. Effects of cryopreservation on the meiotic spindle of human oocytes. Eur J Obst Gynecol Reprod Biol 2004; 113J: S17–23.
25. Sanfins A, Lee GY, Plancha CE, et al. Distinctions in meiotic spindle structure and assembly during in vitro and in vivo maturation of mouse oocytes. Biol Reprod 2003; 69: 2059–67.
26. Gorbsky G, Simerly C, Schatten G, et al. Microtubules in the metaphase-arrested mouse oocyte turn over rapidly. Proc Natl Acad Sci 1990; 87: 6049–53.
27. Rho GJ, Kim S, Yoo JG, et al. Microtubulin configuration and mitochondrial distribution after ultrarapid cooling of bovine oocytes. Mol Reprod Dev 2002; 63: 464–70.
28. Nottola SA, Macchiarelli G, Coticchio G, et al. Ultrastructure of human mature oocytes after slow cooling cryopreservation using different sucrose concentrations. Hum Reprod 2007; 22: 1123–33.
29. Jones A, Van Blerkom J, Davis P, et al. Cryopreservation of metaphase II human oocytes effects mitochondrial membrane potential: implications for developmental competence. Hum Reprod 2004; 19: 1861–6.
30. Sathananthan AH, Ng SC, Bongso A, et al. Visual Atlas of Early Human Development for Assisted Reproductive Technology. Singapore: National University of Singapore, 1993.
31. Matorras R, Ruiz JI, Mendoza R, et al. Fatty acid composition of fertilization-failed human oocytes. Hum Reprod 1998; 13: 2227–30.
32. Plancha CE, Sanfins A, Rodrigues P, et al. Cell polarity during folliculogenesis and oogenesis. Reprod Biomed Online 2005; 10: 478–84.
33. Schultz RM, Lamarca MJ, Wassarman PM. Absolute rates of protein synthesis during meiotic maturation of mammalian oocytes in vitro. Proc Natl Acad Sci 1978; 75: 4160–4.
34. Eichenlaub-Ritter U, Peschke M. Expression in in vivo and in vitro growing and maturing oocytes: focus on regulation of expression at the translational level. Hum Reprod Up 2002; 8: 21–41.
35. Lucifero D, Mertineit C, Clarke HJ, et al. Methylation dynamics of imprinted genes in mouse germ cells. Genomics 2002; 79: 530–8.
36. Nishi Y, Takeshita T, Sato K, et al. Change of the mitochondrial distribution in mouse ooplasm during in vitro maturation. Journal Nippon Med Sch 2003; 70: 408–15.
37. Combelles CMH, Albertini DF. Assessment of oocyte quality following repeated gonadotropin stimulation in the mouse. Biol Reprod 2003; 68: 812–21.

38. De Matos DG, Nogueira D, Cortvrindt R, et al. Capacity of adult and prepubertal mouse oocytes to undergo embryo development in the presence of cysteamine. Mol Reprod Dev 2003; 64: 214–18.
39. Van den Abbeel E, Schneider U, Liu J, et al. Osmotic responses and tolerance limits to changes in external osmolalities, and oolemma permeability characteristics, of human in vitro matured MII oocytes. Hum Reprod 2007; 22: 1959–72.
40. Begum S, Papaioannou VE, Gosden RG. The oocyte population is not renewed in transplanted or irradiated adult ovaries. Hum Reprod 2008; 23: 2326–30.
41. Gook DA, Schiewe MC, Osborn SM, et al. Intracytoplasmic sperm injection and embryo development of human oocytes cryopreserved using 1,2-propanediol. Hum Reprod 1995; 10: 2637–41.
42. Chen ZJ, Li M, Li Y, et al. Effects of sucrose concentration on the developmental potential of human frozen–thawed oocytes at different stages of maturity. Hum Reprod 2004; 19: 2345–9.
43. Porcu E, Fabbri R, Ciotti PM, et al. Ongoing pregnancy after intracytoplasmic sperm injection of epididymal spermatozoa into cryopreserved human oocytes. J Assist Reprod Gen 1999; 16: 183–5.
44. Fabbri R, Porcu E, Marsella T, et al. Technical aspects of oocyte cryopreservation. Mol Cell Endocrinol 2000; 169: 39–42.
45. Chian R-C, Gilbert L, Huang JYJ, et al. Live birth after vitrification of in vitro matured human oocytes. Fertil Steril 2008; in press.
46. Mandelbaum J, Junca AM, Plachot M, et al. Cryopreservation of human embryos and oocytes. Hum Reprod 1988; 3: 117–19.
47. Van der Elst J, Nerinckx S, Van Steirteghem AC. In vitro maturation of mouse germinal vesicle-stage oocytes following cooling, exposure to cryoprotectants and ultrarapid freezing: limited effect on the morphology of the second meiotic spindle. Hum Reprod 1992; 7: 1440–6.
48. Comizzoli P, Wildt DE, Pukazhenthi BS. Effect of 1,2-propanediol versus 1,2-ethanediol on subsequent oocyte maturation, spindle integrity, fertilization, and embryo development in vitro in the domestic cat. Biol Reprod 2004; 71: 598–604.
49. Tucker MJ, Wright G, Morton PC, et al. Birth after cryopreservation of immature oocytes with subsequent in vitro maturation. Fertil Steril 1998; 70: 578–9.
50. Miyake T, Kasai M, Zhu SE, et al. Vitrification of mouse oocytes and embryos at various stages of development in an ethylene glycol-based solution by a simple method. Theriogenol 1993; 40: 121–34.
51. Gook DA, Osborn SM, Johnston WIH. Cryopreservation of mouse and human oocytes using 1,2-propanediol and the configuration of the meiotic spindle. Hum Reprod 1993; 8: 1101–9.
52. Wani NA, Maurya SN, Misra AK, et al. Effect of cryoprotectants and their concentration on in vitro development of vitrified-warmed immature oocytes in buffalo (Bubalus bubalis). Theriogenol 2004; 61: 831–42.
53. Isachenko V, Montag M, Isachenko E, et al. Aseptic vitrification of human germinal vesicle oocytes using dimethyl sulfoxide as a cryoprotectant. Fertil Steril 2006; 85: 741–7.
54. Agca Y, Liu J, McGrath JJ, et al. Membrane permeability characteristics of metaphase II mouse oocytes at various temperatures in the presence of Me_2SO. Cryobiol 1998; 36: 287–300.
55. Wu C, Rui R, Dai J, et al. Effects of cryopreservation on the developmental competence, ultrastructure and cytoskeletal structure of porcine oocytes. Mol Reprod Dev 2006; 73: 1454–62.
56. Larman MG, Katz-Jaffe MG, Sheehan CB, et al. 1,2-propanediol and the type of cryopreservation procedure adversely affect mouse oocyte physiology. Hum Reprod 2007; 22: 250–9.
57. Park SE, Lee KA, Son WY, et al. Chromosome and spindle configurations of human oocytes matured in vitro after cryopreservation at the germinal vesicle stage. Fertil Steril 1997; 68: 920–6.
58. Candy CJ, Wood MJ, Whittingham DG, et al. Cryopreservation of immature mouse oocytes. Hum Reprod 1994; 9: 1738–42.
59. Men H, Monson RL, Parrish JJ, et al. Detection of DNA damage in bovine metaphase II oocytes resulting from cryopreservation. Mol Reprod Dev 2003; 64: 245–50.
60. Cooper A, Paynter SJ, Fuller BJ, et al. Differential effects of cryopreservation on nuclear or cytoplasmic maturation in vitro in immature mouse oocytes from stimulated ovaries. Hum Reprod 1998; 13: 971–8.
61. Ruppert-Lingham CJ, Paynter SJ, Godfrey J, et al. Developmental potential of murine germinal vesicle stage cumulus-oocyte complexes following exposure to dimethylsulphoxide or cryopreservation: loss of membrane integrity of cumulus cells after thawing. Hum Reprod 2003; 18: 392–8.
62. Vincent C, Pickering SJ, Johnson MH, et al. Dimethylsulphoxide affects the organization of microfilaments in the mouse oocyte. Mol Reprod Dev 1990; 26: 227–35.
63. Navarro-Costa P, Correia SC, Gouveia-Oliveira A, et al. Effects of mouse ovarian tissue cryopreservation on granulosa cell–oocyte interaction. Hum Reprod 2005; 20: 1607–14.
64. Vandevoort CA, Shirley CR, Hill DL, et al. Effects of cryoprotectants and cryopreservation on germinal vesicle-stage cumulus–oocyte complexes of rhesus monkeys. Fertil Steril 2007; 90: 805–16.
65. Isachenko EF, Nayudu PL. Vitrification of mouse germinal vesicle oocytes: effect of treatment temperature and egg yolk on chromatin and spindle normality and cumulus integrity. Hum Reprod 1999; 14: 400–8.

66. Rojas C, Palomo MJ, Albarracín JL, et al. Vitrification of immature and in vitro matured pig oocytes: study of distribution of chromosomes, microtubules, and actin microfilaments. Cryobiol 2004; 49: 211–20.
67. Legal F, Massip A. Cryopreservation of cattle oocytes: effects of meiotic stage, cycloheximide treatment, and vitrification procedure. Cryobiol 1994; 38: 290–300.
68. Acker JP, Larese A, Yang H, et al. Intracellular ice formation is affected by cell interactions. Cryobiol 1999; 38: 363–71.
69. Fujihira T, Kishida R, Fukui Y. Developmental capacity of vitrified immature porcine oocytes following ICSI: effects of cytochalasin B and cryoprotectants. Cryobiol 2004; 49: 286–90.
70. Park SE, Chung HM, Cha KY, et al. Cryopreservation of ICR mouse oocytes: improved post-thawed preimplantation development after vitrification using Taxol™, a cytoskeleton stabilizer. Fertil steril 2001; 75: 1177–84.
71. Fuchinoue K, Fukunaga N, Chiba S, et al. Freezing human immature oocytes using cryoloops with taxol in the vitrification solution. J Assist Reprod Gen 2004; 21: 307–9.
72. Vieira AD, Mezzalira A, Barbieri DP, et al. Calves born after open pulled straw vitrification of immature bovine oocytes. Cryobiol 2002; 45: 91–4.
73. Park KI, Kwon IK, Han MS, et al. Effects of partial removal of cytoplasmic lipid on survival of vitrified germinal vesicle stage pig oocytes. J Reprod Dev 2005; 51: 151–60.
74. Nagashima H, Kashiwazaki N, Ashman RK, et al. Removal of cytoplasmic lipid enhances the tolerance of porcine embryos to chilling. Biol Reprod 1994; 51: 618–22.
75. Gupta MK, Uhm SJ, Lee HT. Cryopreservation of immature and in vitro matured porcine oocytes by solid surface vitrification. Theriogenol 2007; 67: 238–48.
76. De Santis L, Coticchio G, Paynter S, et al. Permeability of human oocytes to ethylene glycol and their survival and spindle configurations after slow cooling cryopreservation. Hum Reprod 2007; 22: 2776–83.
77. Imoedemhe DG, Sigue AB. Survival of human oocytes cryopreserved with or without the cumulus in 1,2-propanediol. J Assist Reprod Gen 1992; 9: 323–7.
78. Fabbri R, Porcu E, Marsella T, et al. Human oocyte cryopreservation: new perspectives regarding oocyte survival. Hum Reprod 2001; 16: 411–16.
79. Kuwayama M, Vajta G, Kato O, et al. Highly efficient vitrification method for cryopreservation of human oocytes. Reprod Biomed Online 2005; 11: 300–8.
80. Eroglu A, Toth TL, Toner M. Alterations of the cytoskeleton and polyploidy induced by cryopreservation of metaphase II mouse oocytes. Fertil Steril 1998; 69: 944–57.
81. Van Blerkom J, Davis P, Alexander S. Inner mitochondrial membrane potential, cytoplasmic ATP content and free Ca2+ levels in metaphase II mouse oocytes. Hum Reprod 2003; 18: 2429–40.
82. Ebner T, Moser M, Sommergruber M, et al. Occurrence and developmental consequences of vacuoles throughout preimplantation development. Fertil Steril 2005; 83: 1635–40.
83. Pickering SJ, Johnson MH. The influence of cooling on the organization of the meiotic spindle of the mouse oocyte. Hum Reprod 1987; 2: 207–16.
84. Amann RR, Parks JE. Effects of cooling and rewarming on the meiotic spindle and chromosomes of in vitro-matured bovine oocytes. Biol Reprod 1994; 50: 103–10.
85. Rienzi L, Martinez F, Ubaldi F, et al. Polscope analysis of meiotic spindle changes in living metaphase II human oocytes during the freezing and thawing procedures. Hum Reprod 2004; 19: 656–59.
86. Bianchi V, Coticchio G, Fava L, et al. Meiotic spindle imaging in human oocytes frozen with a slow freezing procedure involving high sucrose concentration. Hum Reprod 2005; 20: 1078–83.
87. Sathananthan A, Trouson A, Freeman L, et al. The effects of cooling human oocytes. Hum Reprod 1988; 3: 968–77.
88. Van der Elst J, Van den Abbeel E, Jacobs R, et al. Effects of 1,2-porpanediol and dimethylsulphoxide on the meiotic spindle of the mouse oocyte. Hum Reprod 1988; 3: 960–7.
89. Joly C, Bchini O, Boulekbache H, et al. Effects of 1,2-propanediol on the cytoskeletal organization of the mouse oocyte. Hum Reprod 1992; 7: 374–8.
90. Katze-Jaffe MG, Larman MG, Sheehan CB, et al. Exposure of mouse oocytes to 1,2-propanediol during slow freezing alters the proteome. Fertil Steril 2008; 89: 1441–7.
91. Huang JYJ, Chen HY, Park JYS. Comparison of spindle and chromosome configuration in in vitro- and in vivo-matured mouse oocytes after vitrification. Fertil Steril 2008; 90: 1424–32.
92. Gardner DK, Sheehan CB, Rienzi L, et al. Analysis of oocyte physiology to improve cryopreservation procedures. Theriogenol 2007; 67: 64–72.
93. Succu S, Leoni GG, Bebbere D, et al. Vitrification devices affect Structural and molecular status of in vitro matured ovine oocytes. Mol Reprod Dev 2207; 74: 1337–44.
94. Schmidt DW, Nedambale TL, Kim C, et al. Effect of cytoskeleton stabilizing agents on bovine matured oocytes following vitrification. Fertil Steril 2004; 82: S26.

10 Oocyte Cryopreservation in the Mouse

Maureen J. Wood
Department of Obstetrics and Gynaecology, University of Aberdeen, Maternity Hospital, Foresterhill, Aberdeen, U.K.

INTRODUCTION

The low temperature storage of mammalian oocytes at all stages of development was pioneered in the mouse. Sherman and Lin (1,2) began to investigate the effects of cooling and exposure to cryoprotectants on mouse metaphase II oocytes half a century ago. At about the same time, the viability of immature oocytes frozen in ovarian tissue was demonstrated by the birth of live young after the tissue was transplanted (3). Interest in oocyte cryopreservation waned but was reignited when mouse embryos were frozen in 1972 (4,5). In 1976, Parkening et al. (6) reported the birth of three offspring from metaphase II oocytes frozen and held at –75°C before warming, fertilization in vitro, and transfer to surrogate mothers. Just a year later, the possibility of prolonged oocyte storage became a reality when Whittingham (7) obtained live births from frozen oocytes that had been stored at –196°C in liquid nitrogen. Mature oocytes were vitrified for the first time in 1988 (8). Isolated immature oocytes (9) and immature oocytes contained within isolated ovarian follicles (10,11) were frozen successfully in the 1990s.

As soon as embryo cryopreservation became possible, it was recognized by biomedical scientists as having a vital role in safeguarding genetically important mouse strains against accidental loss, facilitating the transport of stocks worldwide, and sparing laboratories the enormous cost of maintaining stocks not immediately required for research. Oocyte freezing was hailed as offering, in addition, the option of preserving the haploid mouse genome and thus allowing the subsequent creation of novel genetic combinations; it was soon superseded by sperm freezing. Currently, the major demand is for the storage of vast numbers of transgenic and mutant mouse stocks and for these freezing spermatozoa is often more efficient than freezing oocytes. Few laboratories regularly archive oocytes, although stocks of cryopreserved oocytes may be useful in specific projects (12–15). The mouse oocyte has served as a model for the development of basic cryobiological theory (16), but it is mainly the rapidly developing field of human assisted reproduction that has sustained interest in mouse oocyte cryopreservation.

Human oocytes are a scarce resource. The ethics of having donors undergo ovarian stimulation and oocyte collection purely to provide oocytes for research are questionable. The alternative of accepting oocytes from women requiring assisted conception or women donating primarily for clinical use, risks compromising the outcome of treatment. Furthermore, research in an animal model, before new procedures are adopted for clinical treatment or before a licence is granted for research on human oocytes, is not only sound practice but may also be a requirement of statutory bodies such as the U.K.'s Human Fertilization and Embryology Authority. The mouse has frequently been that model.

Mouse oocytes are available on demand in comparatively large numbers, thus enabling sound experimental design and valid analysis of results. Techniques for mouse superovulation, in vitro fertilization (IVF), embryo culture, and embryo transfer at all preimplantation stages are well established. In mouse, the ultimate test of oocyte viability, that is, fertilization and the proof of full developmental potential by the birth of normal mice, is readily available; results can be obtained within a month of oocyte cryopreservation. The bicornuate uterus of the mouse allows the transfer of control embryos and experimental embryos obtained from cryopreserved oocytes into the same foster mother (17), with a concomitant increase in statistical power.

Most of the problems, whether real or potential, associated with cryopreserving human oocytes were identified initially in the mouse. These include low and variable rates of survival (7,18–20), reduced rates of fertilization in vitro (7,18,21), and an increased occurrence of polyploidy in embryos obtained by IVF from frozen oocytes (18,21). The potential for disruption of the metaphase spindle and chromosome scattering during cooling (22) raised the spectre of an increased occurrence of chromosome nondisjunction after fertilization of cryopreserved oocytes, with a resultant production of aneuploid embryos (21). Some potentially deleterious effects of cryopreservation on oocyte physiology have been identified recently (23).

The aim of this chapter is to provide an historical perspective of developments in mouse oocyte cryopreservation and to indicate how research on the mouse metaphase II oocyte helped to lay a foundation for the introduction of oocyte freezing into clinical practice.

SURVIVAL, FERTILIZATION AND VIABILITY

Mouse oocytes are cryopreserved by slow controlled rate freezing or by vitrification. The outcome of cryopreservation, fertilization in vitro, and subsequent development in vitro and in vivo in some key studies is summarized in Tables 1–4. The lists are not comprehensive but other references are included in the text.

The survival and subsequent viability in vitro of frozen mouse oocytes (Table 1) is variable, ranging from just 4% of oocytes surviving and none being fertilized normally (data not shown in Table 1) (24,51) to rates of survival and fertilization of ~80% or more and subsequent development to blastocysts of >65% of the embryos obtained (25,26,28). This variability is perhaps not surprising given the number and interdependence of the factors, from mouse strain through all the stages of cryopreservation and handling in vitro, that contribute to success.

Most of the methods reported for oocyte freezing were derived empirically from those proven for mouse embryo storage, although, notably, a procedure identified as optimal in a theoretical exercise (30) corresponds closely with that previously followed by several groups (7,21,25,28,29). Further refinement of this procedure appeared to increase survival and fertilization rates (31), but the comparison was made retrospectively. Typically, cumulus enclosed or cumulus free oocytes incubated at ~0°C with 1.5 M dimethylsulfoxide (DMSO) are cooled slowly (0.3–0.5°C/min) to temperatures between –30°C and –40°C or below –65°C before rapid cooling and storage in liquid nitrogen. Oocytes are thawed slowly (~8–20°C/min) or rapidly (>200°C/min) before dilution of the cryoprotectant. The permutations of the procedures for equilibrating the oocytes with the cryoprotectant before cooling and returning the thawed oocytes to isotonic conditions (19,24,28,31,34,52 for example), are too numerous to list, but exert profound effects on survival and fertilization. The presence or absence of the cumulus cells (data not shown in Table 1) during freezing is usually considered to have no influence on survival (7,28), although one group (24) reported that their removal before cooling decreased survival from 72% to 5%.

DMSO has been the cryoprotectant of choice for mouse oocytes (Table 1) and in that respect the mouse is not a good model for humans, because after some initial success (53,54), DMSO was soon replaced by 1,2-propanediol (PrOH) in combination with sucrose for freezing human oocytes (51,55,56). The use of PrOH in mouse was limited by reports of poor survival (4%) after freezing (51) and an increased incidence of oocyte degeneration and parthenogenetic activation after exposure to PrOH at temperatures above 0°C (57–59). One group (19) reported similar, albeit low (49–63%), rates of survival after freezing in DMSO or PrOH, whereas 87% of oocytes survived in a mixture of DMSO with PrOH. The adverse effects of PrOH in mouse may be mediated by the protracted rise it induces in levels of intracellular calcium, similar to that seen at fertilization (60). Recently, however, Stachecki et al. (32–34) demonstrated that when the sodium content of the medium is reduced, PrOH is highly protective for mouse oocytes. The rates of survival (76–90%), fertilization (70–85%), and development in vivo (48%) were similar to the highest rates achieved with a conventional medium and DMSO (survival 80–91%; fertilization 75–86%; development of normal fetuses 56%) (28). A similar approach was successfully applied in human and several live births have resulted (61,62). Glycerol, although protective, is usually considered impractical for the cryopreservation of oocytes because equilibration and return to isotonic conditions, which is compatible with survival and development similar to oocytes in DMSO, takes a total of two hours at 37°C (26).

After initial enthusiasm in the 1990s few attempts have been made to develop mouse oocyte vitrification (Table 2), reflecting in part the lack of demand from the scientific community to archive mouse oocytes (12). The challenge of cryopreserving the temperature sensitive oocytes and embryos of some domestic species has led to significant advances in the approach to vitrification (63), but these have not been pursued systematically in mouse. Nonetheless, the mouse oocyte is still seen as a model in which to investigate the effects of vitrification on cell physiology (23).

Table 1 Survival, Fertilization, and Development In Vitro of Mouse Oocytes Cooled Slowly (0.3–0.5°C/min) Before Rapid Cooling to –196°C

	Freezing Medium					% Fertilized[c]		% Blastocysts[d]	
Research group	Macromolecular supplement	Cryoprotectant	Temperature of addition (°C)	Temperature at which slow cooling is terminated[a] (°C)	% Survived[b]	Frozen oocytes	Fresh oocytes	Frozen oocytes	Fresh oocytes
Parkening et al. (6)	BSA	DMSO	0	Held at –75°C	44	65	81	37	35
Whittingham (7)	BSA	DMSO	0	–80	55–58	48–62	61–71	86	80
Glenister et al. (21)	BSA	DMSO	0	–80	76	36	60	ND	ND
Ko and Threlfall (20)	BSA?	PrOH	RT	–40	51	14	54	ND	ND
Carroll et al. (18)	BSA	DMSO	0	–40	61	46	79	75	78
Trounson and Kirby (24)	BSA	DMSO	0	–80	72	59	91	ND	ND
Schroeder et al. (25)	FBS	DMSO	0	–80	83	88	80	67	77
Hunter et al. (26)	BSA	Glycerol	37	–35	85	79		74	
	BSA	DMSO	0	–65	88	60	95	88	~92
Bouquet et al. (27)	BSA	DMSO	0	–80	67	37	62	ND	ND
Carroll et al. (28)	BSA	DMSO	0	–40	80–86	28–79			
	FBS	DMSO	0	–40	80–91	75–86			
	PVA	DMSO	0	–40	85	42	85–93	ND	ND
George et al. (29)	FBS	DMSO	4	–60	ND	54	50	77	89
Karlsson et al. (30)	BSA?	DMSO	4	–70	75	65	88	50	81
Eroglu (31)	BSA	DMSO	4	–70	ND	95	92	ND	ND
Stachecki et al. (32–34)	FBS	PrOH + S[e]	RT	–33	76–97	71–89[f]	ND	58–84	ND
Lane and Gardner (35)	FBS	PrOH + S[e]	RT?	–33	81	49[f]	73	65	95
Valojerdi and Salehnia (36)	BSA	PrOH + S	RT?	–30	85	1	88	0	33
Huang et al. (37)	FBS?	PrOH + S[e]	RT?	–33	73	40	84	20	75

[a]Samples then cooled rapidly to –196°C.

[b]Number of morphologically normal oocytes after thawing expressed as a proportion of the number of oocytes frozen or the number of oocytes recovered from frozen-thawed cumulus masses.

[c]Number of two-cell embryos expressed as a proportion of the number of oocytes inseminated (assumes insemination of all normal oocytes unless authors state otherwise)

[d]Number of blastocysts expressed as a proportion of the number of two-cell embryos cultured in vitro after IVF (assumes all two-cell embryos were cultured unless authors state otherwise).

[e]Sodium-depleted freezing medium.

? Denotes that the macromolecular supplement (column 2) or the temperature of addition of the cryoprotective solution (column 4) was not plainly specified in the paper.

[f]Zonae of frozen oocytes were breached with a laser before insemination.

Abbreviations: BSA, bovine serum albumin (crystalline or fraction V); DMSO, dimethysulfoxide; FBS, fetal bovine serum; ND, no data; PrOH, 1,2-propanediol; RT, room temperature; S, sucrose.

Table 2 Survival, Fertilization, and Subsequent Development In Vitro of Vitrified Mouse Oocytes

	Vitrification solution					% Fertilized[b]		% Blastocysts[c] developed from	
Research group	Protein	Cryoprotectants	Temperature of exposure (°C)	Container	% Survived[a]	Vitrified oocytes	Fresh oocytes	Vitrified oocytes	Fresh oocytes
Kola et al. (8)	BSA?	D+A+P+PEG	4	Straw	71	27	87	ND	ND
Nakagata (38)	BSA	D+A+P+PEG	RT	Glass tube	84	82	98	ND	ND
Kono et al. (39)	BSA	D+A+P+PEG	4	Straw	83	71	80	80	96
Shaw et al. (40)	BSA	D+A+P+PEG	5–8	Straw	80	83	90	84	94
Shaw et al. (41)	BSA	D+A+P+PEG	RT	Straw	68	76	93	61	93
Wood et al. (42)	FBS	D	RT	Straw	81	80	94	ND	ND
Bos-Mikich et al. (43)	FBS	D	RT	Straw	93	85	92	ND	ND
Hotamisligil et al. (44)	BSA	E+S	RT	Straw	90	85	86	61	65
O'Neil et al. (45)	FBS	D+PEG	RT	Straw	95	91	93	73	77
O'Neil et al. (46)	FBS	D+AFGP	2–4	Straw	86	94	93	89	85
O'Neil et al. (47)	FBS	D+PEG	RT	Straw	94	90		71	
	FBS	D+DEX	RT	Straw	38	57	96	61	81
dela Peña et al. (48)	FBS	E+RAF	RT	Straw	~90	~94	ND	~87	ND
Valojerdi and Salehnia (36)	BSA	E+F+S	RT	Straw	80	85	88	32	33
Chen et al. (49)	HCS	E+S	RT	Straw	77	86		86	
	HCS	E+S	RT	OPS	63	88		88	
	HCS	E+S	RT	CPS	79	89		89	
	HCS	E+S	RT	Grid	39	81	92	81	92
Lane and Gardner (35)	HSA	D+E+F+S	37	Cryoloop	99	70[d]	73	96	95
Huang (37)	BSA?	E+P+S	RT	Cryoleaf	92	64	84	31	75
Endoh et al. (15)	BSA	D+E+F+S	RT	Cryotop	~75	>85[e]	ND	~87	ND

[a]Number of morphologically normal oocytes recovered after warming expressed as a proportion of the number of oocytes vitrified.
[b]Number of two-cell embryos expressed as a proportion of the number of oocytes inseminated (assumes insemination of all normal oocytes unless authors state otherwise).
[c]Number of blastocysts expressed as a proportion of the number of two-cell embryos cultured in vitro after IVF (assumes all two-cell embryos were cultured unless authors state otherwise).
[d]Zonae of vitrified oocytes breached before insemination.
[e]Sperm injected into cytoplasm.

? Denotes that the protein content of the vitrification solution was not specified plainly in the paper.

Abbreviations: A, acetamide; AFGP, antifreeze glycoprotein; BSA, bovine serum albumin; CPS, closed pulled straw; D, dimethylsulfoxide; DEX, dextran; E, ethylene glycol; F, Ficoll; FBS, fetal bovine serum; Grid, electron microscope grid; HCS, human cord serum; HSA, human serum albumin; ND, no data; OPS, open pulled straw; P, 1,2-propanediol; PEG, polyethylene glycol; RAF, raffinose; RT, room temperature; S, sucrose; Straw, 0.25 ml insemination straw.

Table 3 Outcome of Transfer of Embryos Produced by IVF from Frozen Oocytes

		Frozen				Control		
Research group	Stage transferred	No. of embryos	% Implanted	% Late-stage fetuses	% Live young	No. of embryos	% Implanted	% Late-stage fetuses
Whittingham (7)	2-cell	49	59	35		20	45	45
	2-cell	24			46			
	Blastocyst	21	76	24		12	58	25
	Blastocyst	12			58			
Glenister et al. (21)	2-cell	149	67	50		222	67	53
Kola et al. (8)	2-cell	154	54	30		242	73	55
Trounson and Kirby (24)	2-cell	154	55	30		152	77	64
Carroll et al. (28)	2-cell	74[a]	65	57		26[a]	69	61
	2-cell	42[b]	60	55				
George et al. (29)	2-cell	115	70	36		144	71	47
	Blastocyst	112	46	30		133	68	51
Lane and Gardner (35)	Blastocyst	42	52	26		60	87	68
Stachecki et al. (50)	2-cell[c]	185	73	48		175[d]	62	40
	Blastocyst[c]	163	62	12		163[d]	76	47

[a]Oocytes frozen and handled up to the stage of embryo culture in medium containing fraction V bovine serum albumin.
[b]Oocytes frozen and handled up to the stage of embryo culture in medium containing fetal bovine serum.
[c]Embryos developed from oocytes frozen in sodium-depleted medium.
[d]Control embryos produced in vivo.

Table 4 Outcome of Transfer of Embryos Produced by IVF from Vitrified Oocytes

		Vitrified				Control			
Research group	Stage transferred	No. of embryos	% Implanted	% Late-stage fetuses	% Live young	No. of embryos	% Implanted	% Late-stage fetuses	% Live young
Kola et al. (8)	2-cell	70	41	19		242	73	55	
	2-cell	94	63	26					
Nakagata (38)	2-cell	120			46	78			60
Wood et al. (42)	2-cell	240	79	40		160	84	61	
Bos-Mikich et al. (43)	2-cell	136	77	49		145	68	38	
		13			38				
Lane and Gardner (35)	Blastocyst	92	88	57		60	87	68	
Endoh et al. (15)	2-cell	58[a]	57		26[c]	80[a]	76		56[c]
		48–78[b]	14–31		5–15[c]	71–116[b]	32–71		15–30[c]

[a]Embryos from B6D2F1 hybrid × B6D2F1 hybrid.
[b]Embryes from inbred strains: C57Bl/6J; DBA/2J; C3H/HeN; BALB/cA.
[c]Delivered by caesarean section.

Mouse oocytes were first vitrified (8) in solution VS1 (a mixture of DMSO, PrOH, acetamide, and polyethylene glycol) introduced by Rall and Fahy (64) for embryo vitrification, but the outcome was very discouraging: survival (71%) was similar to that of frozen oocytes, but fertilization (27% vs. 58%) and subsequent embryo development were low. An increased number of abnormal fetuses was seen after the transfer of embryos derived by IVF from oocytes exposed to the solution, with or without vitrification. In addition, the vitrification and warming procedures were cumbersome, involving prolonged stepwise addition and removal of cryoprotectants, partly at room temperature and partly at 4°C. For more than a decade, research focused on developing safer solutions and minimizing the toxic and osmotic damage sustained when oocytes were equilibrated with the cryoprotectants before cooling and when the cryoprotectants were removed after warming (38–42,44–48). Increased rates of survival (80–95%) and fertilization (>80%) resulted from some of the modifications (38,40,42,45–48) but the variability between samples (41,42,47) was a major drawback.

The newer methods of vitrification, which combine increased rates of cooling and warming with lower concentrations of cryoprotectants (63), have been applied sporadically in mouse (15,35,37,49,60,65). Given the variety of solutions and containers used, it is meaningless to compare the relative success of the different approaches, particularly as fertilization was in some cases assisted either by breaching the zona pellucida before insemination (35) or by injecting sperm into the oocyte cytoplasm (ICSI) (15).

There is no consensus about the capacity for development in vitro of embryos derived from oocytes frozen in DMSO, although it is clear that those obtained from oocytes frozen in PrOH have a reduced capacity to form blastocysts in vitro (Table 1). When embryos were transferred at the two-cell stage to the oviducts of foster mothers, their potential for development to late-stage fetuses (7,21,28,50) or live mice (6,7) was unimpaired, although sometimes the number of embryos transferred was low. In contrast, when the embryos were cultured to the blastocyst stage before transfer, their capacity for full fetal development was significantly reduced compared with controls obtained from fresh oocytes fertilized in vitro (29,35) or blastocysts developed in vivo (50). It is pertinent for application in human assisted conception to speculate that embryos derived from frozen oocytes have a reduced ability to withstand the stress of prolonged culture.

Most groups report similar development in vitro among embryos obtained from vitrified and fresh oocytes (Table 2). The high postimplantation losses and recovery of abnormal fetuses in the first report of oocyte vitrification (8) raised alarm about the risky nature of vitrification. Subsequent studies (Table 4) on the first generation of vitrification solutions with cryoprotectant concentrations >6 M confirmed the reduced developmental potential in vivo (38,42). After the transfer of embryos of hybrid and inbred mouse strains obtained from oocytes vitrified with lower concentrations of cryoprotectants, development to late stage fetuses (35) and the live birth rate (15) were significantly impaired. Wood et al. (42) reported similar rates of implantation for two-cell embryos from oocytes vitrified in 6 M DMSO and controls; but the postimplantation losses were almost doubled in the vitrified group (49% vs. 27%). In contrast, two years later in the same laboratory with a similar vitrification solution, postimplantation losses (37%) were lower than previously reported (49%) and did not differ from unfrozen controls (44%) (43). An inspection of laboratory records showed that rates of postimplantation loss among controls in several studies varied from 17% to 44%, suggesting that caution must be exercised when interpreting viability data as factors unrelated to cryopreservation may influence the outcome of the embryo transfers.

Comparisons between slow controlled rate freezing and vitrification have claimed the superiority of vitrification according to several criteria, including oocyte survival (35,37,66), fertilization (35,37), zona hardening (66), development in vitro (35–37) and in vivo (35), blastocyst cell number and the allocation of cells to the inner cell mass (35), post-warming ultrastructure (35–37), and post-cryopreservation metabolism (35) and protein expression (60). In only two of these comparisons (35,37) were the freezing procedures suitable for *mouse* oocytes but in neither was the outcome of freezing equal to that achieved previously with conventional (28) or sodium depleted (34) medium. Vitrification may well offer a more reliable approach than freezing for oocyte cryopreservation, but until the findings are confirmed for oocytes frozen according to procedures that are appropriate and *reliable* for *mouse* oocytes the conclusions must be treated cautiously.

FAILED FERTILIZATION AND ZONA HARDENING

Conventional freezing in DMSO reduces the rate of fertilization by up to 50% among mouse oocytes inseminated in vitro (7,18,21,26,27,30,67,68). The cause of this reduction lies primarily at the level of the zona pellucida: DNA labelling showed that sperm had not penetrated the zona in about 80% of oocytes in which fertilization failed (68); fertilization was restored to control levels by zona-drilling (67) or complete removal of the zona (68) before fertilization in vitro. The "hardening" of the zona of frozen oocytes that inhibits sperm penetration is generally attributed to exocytosis of the cortical granules, similar to that triggered by the entry of the fertilizing spermatozoon (69,70), although evidence of premature cortical granule release was not always found (68). Alternative mechanisms may be involved in the inhibition of sperm entry, or it is possible that only a subpopulation of cortical granules is released during cryopreservation (36,68,71,72). Whatever the mechanism, some oocytes are plainly less susceptible than others as fertilization is rarely a total failure.

The effect of vitrification on fertilization is more difficult to dissect. The only study specifically addressing the question (65) reported a significant decrease in fertilization in oocytes vitrified in solutions containing DMSO and ethylene glycol (EG). Early attempts to vitrify mouse oocytes in solutions containing ≥6 M penetrating cryoprotectants resulted in comparatively high rates of fertilization (Table 2), although these were often significantly lower than in controls (8,38,41,42) and with extreme variability between individual samples of oocytes (41,42,45–47). With lower concentrations of cryoprotectants (35,36,49), fertilization was similar in vitrified and control oocytes, whether (35) or not (36,49) the zona was breached before insemination.

Different stages of the cryopreservation procedure have been implicated in zona hardening, including cooling to temperatures close to 0°C (73) and exposure to cryoprotectants (DMSO, PrOH, EG) at temperatures above 0°C (60,65,69,70,74). Attention usually focuses on the effect of exposure to cryoprotectants before cooling, but the period after thawing appears to be equally important for zona hardening (74) and the maintenance of high rates of fertilization in frozen oocytes (28,31,32,52). The time course of zona hardening as a result of cryoprotectant exposure at or above room temperature warrants further investigation, as it has implications for the development of vitrification procedures.

The release of cortical granules after sperm entry is mediated by a rise in levels of intracellular calcium and it is tempting to speculate that a similar mechanism is effective during cryopreservation: Exposure of mouse oocytes to PrOH and EG at room temperature (60,75,76) and to PrOH, DMSO or EG at 37°C (60,65) increased the level of intracellular calcium and hardened the zona. The effect on calcium levels of equilibrating oocytes with DMSO at 0°C, the most common procedure for mouse oocyte freezing (Table 1), has not been tested. Removing calcium from the medium reduced the rise in intracellular calcium in response to PrOH and EG but not to DMSO (60,65), indicating that in the response to DMSO calcium is released from internal stores. Vitrification in calcium-free medium containing EG, DMSO and sucrose, or freezing in calcium-free medium with PrOH and sucrose, significantly reduced zona hardening and increased the rate of fertilization (60,65). The development in vitro of the vitrified embryos was similar to controls (65) but the full potential of embryos cryopreserved in the absence of calcium remains to be demonstrated in vivo.

The problem of failed fertilization is easily overcome in mouse. Simply breaching the zona pellucida (32,35,67) removes the barrier to sperm penetration, but the risk of polyspermic fertilization precludes this approach in human. Zona hardening is prevented by freezing and recovering oocytes in medium containing a protective protein supplement (25,28,32). The rates of fertilization in frozen oocytes were increased almost to that in freshly collected oocytes when the medium was supplemented with fetal bovine serum (FBS) instead of bovine serum albumin (BSA) (28). The risk of disease transmission from animal-derived proteins prohibits the addition of FBS to media for human use; a possible alternative, human serum albumin (HSA), appears to offer no protection (77) but the effect of rHSA is not known. Seemingly minor modifications of the cryopreservation procedure may be sufficient to maintain high rates of fertilization in mouse without the addition of FBS (31,52). Cryopreserved human oocytes have been fertilized almost exclusively by ICSI since the mid 1990s (56,78,79) but freezing protocols have changed considerably since then. In the light of a recent report of similar rates of fertilization (82–83%) in frozen oocytes achieved with conventional IVF and ICSI (80), it is perhaps timely to question the need for universal ICSI.

CYTOSKELETON AND CHROMOSOMES

Events in the oocyte during fertilization and the first cleavage division depend upon a functional cytoskeleton. Species differences in cytoskeletal organization make it unwise to extrapolate the effects of cryopreservation in mouse directly to human (81,82). Nevertheless, the possibility that cryopreservation may compromise the cytoskeleton prompted extensive research in the mouse, with attention focused primarily on the resilience of the metaphase spindle.

There is extensive evidence that cooling and exposure to cryoprotectants have profound effects on spindle morphology and chromosome location (22,58,59,83–89). Stabilization of the spindle with paclitaxel (Taxol™) before vitrification had no influence on survival (>80%) but promoted blastocyst development (59% vs. 24%) (13), although others (35) have reported higher rates of blastocyst development (>90%) with better developed inner cell masses in the absence of a stabilizer. In practice, "Does cryopreservation disrupt the spindle?" is not the relevant question. It is more important to know whether the spindle can reorganize and function normally after warming. A remarkable number of cryopreserved oocytes have apparently normal spindles with correctly aligned chromosomes immediately after warming or after a period of recovery at 37°C (13,31,37,49,66,88,90–93), although, in a significant proportion the spindle remains disorganized (31,37,49,66,90). However, spindle morphology does not necessarily correspond with function and, with few exceptions (93), the cryopreservation protocols tested have not yet been established as compatible with high rates of post-fertilization embryonic development in mouse. Non-invasive examination of the oocyte spindle (87) will provide an opportunity to associate cryopreservation induced changes with subsequent viability in vitro and even in vivo.

Cytogenetic analysis of zygotes obtained by IVF from cryopreserved oocytes has provided convincing evidence for the normality of spindle function in those oocytes that are able to complete meiosis. The occurrence of aneuploidy at the first cleavage division is similar in embryos derived from frozen or vitrified oocytes and untreated oocytes (21,27,31,37,43,94,95). An early report (8) that vitrification and freezing increased the occurrence of aneuploidy has not been confirmed. There is little doubt that the fear of chromatid nondisjunction in cryopreserved mouse oocytes is unfounded, and the same appears to be true in human (96).

The microfilament network of the oocyte has a vital role in its structure and function, influencing, for example, the viscoelasticity of the cytoplasm, cortical granule exocytosis, chromosome movements, extrusion of the second polar body, movement of the pronuclei, and the first cleavage division. The effect of cryopreservation on this network has received little attention, although it has been known for almost 20 years that its organization is disturbed by cooling and exposure to cryoprotectants (82,97). Freezing and vitrification disrupt the intermediate filaments (36) and the network of cortical actin (31), although the latter recovers after incubation at 37°C. In a novel approach (98), proactive disassembly of the F-actin network before freezing preserved oocyte viscoelasticity, increased post-thaw survival, had no influence on fertilization, and promoted development in vitro. The potential of the embryos to develop after transfer in vivo remains to be evaluated. The authors suggest that this treatment prevents the stresses of osmotic shrinking and swelling from being transmitted to the plasma membrane and intracellular organelles.

The increased incidence of polyploidy, usually triploidy, in embryos derived from cryopreserved mouse oocytes is undisputed (8,21,27,31,43,94,95), although the source of the extra set of chromosomes in the triploid zygotes is more contentious. Morphological and cytogenetic analyses implicated retention of the second polar body (18,27), and this was presumed to be the result of a damaged microfilament network. Triple staining of microfilaments, microtubules, and DNA in combination with interference contrast microscopy to locate sperm tails also identified the extra chromosomes as maternal. However, the abnormal spindle remnants in all the digynic zygotes and the observation that 40% (5 out of 12) had extruded a second polar body lacking chromatin led the authors (31) to suggest that a failure of spindle rather than microfilament function was the cause of digyny. In contrast, when frozen (95) and vitrified(43) oocytes were parthenogenetically activated to allow cytogenetic examination of the maternal chromosomes in isolation, the incidence of digyny was low in cryopreserved and control oocytes alike, leading to the conclusion that the increased polyploidy is of paternal origin. The incidence of polyspermic fertilization in cryopreserved oocytes has not been examined directly but could be clarified by counting decondensing sperm heads in the oocyte cytoplasm. The possibility of producing polyploid embryos has not deterred the use of cryopreserved oocytes in assisted conception, as it is usual practice to examine all oocytes for evidence of normal fertilization while the pronuclei are still visible. Nonetheless, it is possible that a small proportion of triploid zygotes, in which the second polar body is extruded but both sets of maternal chromosomes are retained within one pronucleus (31), may be scored as normal.

CRYOPRESERVATION AND OOCYTE PHYSIOLOGY

Understanding the influence of cryopreservation on oocyte physiology will provide more effective endpoints than those currently used, that is, survival, fertilization and, to a lesser extent, development in vitro, for the development and refinement of new protocols (23,99). It has already been shown that cryopreservation alters oocyte metabolism (35), calcium signalling (60,65,76), and mRNA and protein expression (60,99). Pyruvate uptake was significantly reduced in frozen and vitrified oocytes compared with controls (35). Exposure to common cryoprotectants as well as the complete sequence of cryopreservation (60,65,75,76) causes a rise in levels of intracellular calcium, and thus may modify the downstream events of fertilization and embryo development. Exposure to PrOH and freezing (100,60) resulted in both up and down regulation of proteins but vitrification in a solution without PrOH had a less marked effect (60). Freezing decreased expression of the gene *CD9*, which is associated with female infertility in mice, possibly as a result of its role in sperm–oocyte fusion (99). The value of some of these data for frozen oocytes is limited by the use of protocols appropriate for freezing human rather than mouse oocytes (60,99,100). DMSO, the cryoprotectant preferred for mouse oocytes, modifies calcium signalling at 37°C (65) but its effects under conditions that would be used prior to freezing have not been tested. The effect of cryoprotectants at 37°C is more relevant for oocyte vitrification than freezing, and it is encouraging that vitrification in medium containing DMSO, EG, Ficoll, and sucrose appears to have less impact on oocyte physiology than freezing in PrOH and sucrose (35,60,65). It is unclear, however, whether the difference reflects an advantage of the vitrification procedure per se or the choice of a solution appropriate for the cryopreservation of *mouse* oocytes.

CONCLUSION

The value of the mouse oocyte in basic cryobiological research was recognized half a century ago. It offers the ultimate test of cell viability, that is, the birth of normal offspring and, because of its ready availability, has been used widely in the development of procedures for freezing mammalian oocytes. The ability of embryos resulting from the in vitro fertilization of frozen mouse oocytes to develop normally after transfer in vivo, provided reassurance and paved the way for the introduction of oocyte freezing in human assisted conception. The problems associated with oocyte freezing were first identified in the mouse; however, the extent to which the mouse has been a good model in which to investigate them has varied. In contrast with freezing, the mouse oocyte has contributed little as a model for human oocyte vitrification, except perhaps to discourage vitrification in the "first generation" solutions in which survival was extremely variable and had potentially deleterious effects on embryonic development. It seems likely that the mouse oocyte will continue to be seen as a model in which to assess the effects of cryopreservation on oocyte physiology, but this raises the fundamental issue of its suitability for testing cryopreservation procedures that are appropriate for the human rather than the *mouse* oocyte.

REFERENCES

1. Sherman JK, Lin TP. Survival of unfertilized mouse eggs during freezing and thawing. Proc Soc Exp Biol Med 1958; 98: 902–5.
2. Sherman JK, Lin TP. Temperature shock and cold storage of unfertilized mouse eggs. Fertil Steril 1959; 10: 384–96.
3. Parrott DMV. The fertility of mice with orthotopic ovarian grafts derived from frozen tissue. J Reprod Fert 1960; 1: 230–41.

4. Whittingham DG, Leibo SP, Mazur P. Survival of mouse embryos frozen to –196°C and –269°C. Science 1972; 178: 411–14.
5. Wilmut I. The effect of cooling rate, cryoprotective agent, and stage of development on survival of mouse embryos during freezing and thawing. Life Sci 1972; 11: 1071–9.
6. Parkening TA, Tsunoda Y, Chang MC. Effects of various low temperatures, cryoprotective agents and cooling rates on the survival, fertilizability and development of frozen-thawed mouse eggs. J Exp Zoolog 1976; 197: 369–74.
7. Whittingham DG. Fertilization in vitro and development of unfertilized mouse oocytes previously stored at –196°C. J Reprod Fert 1977; 49: 89–94.
8. Kola I, Kirby C, Shaw J, et al. Vitrification of mouse oocytes results in aneuploid zygotes and malformed foetuses. Teratology 1988; 38: 467–74.
9. Candy CJ, Wood MJ, Whittingham DG, et al. Cryopreservation of immature mouse oocytes. Hum Reprod 1994; 9: 1738–42.
10. Carroll J, Whittingham DG, Wood MJ, et al. Extra-ovarian production of mature viable mouse oocytes from frozen primary follicles. J Reprod Fert 1990; 90: 321–7.
11. Carroll J, Gosden RG. Transplantation of frozen-thawed mouse primordial follicles. Hum Reprod 1993; 8: 1163–7.
12. Glenister PH, Thornton CE. Cryoconservation – archiving for the future. Mammalian Genome 2000; 11: 565–71.
13. Park SE, Chung HM, Cha KY, et al. Cryopreservation of ICR mouse oocytes: improved post–thawed preimplantation development after vitrification using Taxol™, a cytoskeleton stabilizer. Fertil Steril 2001; 75: 1177–84.
14. Anzai M, Nishiwaki M, Yanagi M, et al. Application of laser-assisted zona drilling to in vitro fertilization of cryopreserved mouse oocytes with spermatozoa from a subfertile transgenic mouse. J Reprod Dev 2006; 52: 601–6.
15. Endoh K, Mochida K, Ogonuki N, et al. The developmental ability of vitrified oocytes from different mouse strains assessed by parthenogenetic activation and intracytoplasmic sperm injection. J Reprod Dev 2007; 53: 1199–206.
16. Leibo SP. Cryopreservation of oocytes and embryos: optimization by theoretical versus empirical analysis. Theriogenology 2008; 69: 37–47.
17. Rall WF, Wood MJ. High in vitro and in vivo survival of day 3 mouse embryos vitrified or frozen in a non–toxic solution of glycerol and albumin. J Reprod Fert 1994; 101: 681–8.
18. Carroll J, Warnes GM, Matthews CD. Increase in digyny explains polyploidy after in-vitro fertilization of frozen-thawed mouse oocytes. J Reprod Fert 1989; 85: 489–94.
19. Todorow SJ, Siebzehnrübl ER, Koch R, et al. Comparative results on survival of human and animal eggs using different cryoprotectants and freeze-thawing regimens. I Mouse and hamster. Hum Reprod 1989; 4: 805–11.
20. Ko Y, Threlfall WR. The effects of 1,2-propanediol as a cryoprotectant on the freezing of mouse oocytes. Theriogenology 1988; 29: 987–95.
21. Glenister PH, Wood MJ, Kirby C, et al. Incidence of chromosome anomalies in first-cleavage mouse embryos obtained from frozen-thawed oocytes fertilized in vitro. Gamete Res 1987; 16: 205–16.
22. Magistrini M, Szöllösi D. Effects of cold and isopropyl-N-phenylcarbamate on the second meiotic spindle of mouse oocytes. Eur J Cell Biol 1980; 22: 699–707.
23. Gardner DK, Sheehan CB, Rienzi L, et al. Analysis of oocyte physiology to improve cryopreservation procedures. Theriogenology 2007; 67: 64–72.
24. Trounson A, Kirby C. Problems in the cryopreservation of unfertilized eggs by slow cooling in dimethylsulfoxide. Fertil Steril 1989; 52: 778–86.
25. Schroeder AC, Champlin AK, Mobraaten LE, et al. Developmental capacity of mouse oocytes cryopreserved before and after maturation in vitro. J Reprod Fert 1990; 89: 43–50.
26. Hunter JE, Bernard A, Fuller B, et al. Fertilization and development of the human oocyte following exposure to cryoprotectants, low temperatures and cryopreservation: a comparison of two techniques. Hum Reprod 1991; 6: 1460–5.
27. Bouquet M, Selva J, Auroux M. The incidence of chromosomal abnormalities in frozen–thawed mouse oocytes after in-vitro fertilization. Hum Reprod 1992; 7: 76–80.
28. Carroll J, Wood MJ, Whittingham DG. Normal fertilization and development of frozen thawed mouse oocytes: protective action of certain macromolecules. Biol Reprod 1993; 48: 606–12.
29. George MA, Johnson MH, Howlett SK. Assessment of the developmental potential of frozen-thawed mouse oocytes. Hum Reprod 1994; 9: 130–6.
30. Karlsson JOM, Eroglu A, Toth TL, et al. Fertilization and development of mouse oocytes cryopreserved using a theoretically optimized protocol. Hum Reprod 1996; 11: 1296–305.

31. Eroglu A, Toth TL, Toner M. Alterations of the cytoskeleton and polyploidy induced by cryopreservation of metaphase II mouse oocytes. Fertil Steril 1998; 69: 944–57.
32. Stachecki JJ, Cohen J, Willadsen S. Detrimental effects of sodium during mouse oocyte cryopreservation. Biol Reprod 1998; 59; 395–400.
33. Stachecki JJ, Cohen J, Willadsen S. Cryopreservation of unfertilized mouse oocytes: the effect of replacing sodium with choline in the freezing medium. Cryobiology 1998; 37; 346–54.
34. Stachecki JJ, Willadsen S. Cryopreservation of mouse oocytes using a medium with low sodium content: effect of plunge temperature. Cryobiology 2000; 40; 4–12.
35. Lane M, Gardner DK. Vitrification of mouse oocytes using a nylon loop. Mol Reprod Dev 2001; 58: 342–7.
36. Valojerdi MR, Salehnia M. Developmental potential and ultrastructural injuries of metaphase II (MII) mouse oocytes after slow freezing or vitrification. J Assist Reprod Genet 2005; 22: 119–27.
37. Huang JYJ, Chen H-Y, Tan S-L, et al. Effect of choline-supplemented sodium-depleted slow freezing versus vitrification on mouse oocyte meiotic spindles and chromosome abnormalities. Fertil Steril 2007; 88: 1093–100.
38. Nakagata N. High survival rate of unfertilized mouse oocytes after vitrification. J Reprod Fert 1989; 87: 479–83.
39. Kono T, Kwon OY, Nakahara T. Development of vitrified oocytes after in vitro fertilization. Cryobiology 1991; 28: 50–4.
40. Shaw PW, Fuller BJ, Bernard A, et al. Vitrification of mouse oocytes: improved rates of survival, fertilization, and development to blastocysts. Mol Reprod Dev 1991; 29: 373–8.
41. Shaw PW, Bernard A, Fuller BJ, et al. Vitrification of mouse oocytes using short cryoprotectant exposure: effect of varying exposure times on survival. Mol Reprod Dev 1992; 33: 210–4.
42. Wood MJ, Barros C, Candy CJ, et al. High rates of survival and fertilization of mouse and hamster oocytes after vitrification in dimethylsulphoxide. Biol Reprod 1993; 49: 489–95.
43. Bos-Mikich A, Wood MJ, Candy CJ, et al. Cytogenetical analysis and developmental potential of vitrified mouse oocytes. Biol Reprod 1995; 53: 780–5.
44. Hotamisligil S, Toner M, Powers RD. Changes in membrane integrity, cytoskeletal structure, and developmental potential of murine oocytes after vitrification in ethylene glycol. Biol Reprod 1996; 55: 161–8.
45. O'Neil L, Paynter SJ, Fuller BJ, et al. Vitrification of mature mouse oocytes: improved results following addition of polyethylene glycol to a dimethyl sulfoxide solution. Cryobiology 1997; 34: 295–301.
46. O'Neil L, Paynter SJ, Fuller BJ, et al. Vitrification of mature mouse oocytes in a 6 M Me_2SO solution supplemented with antifreeze glycoproteins: the effect of temperature. Cryobiology 1998; 37: 59–66.
47. O'Neil L, Paynter SJ, Fuller BJ, et al. Vitrification of mature mouse oocytes in dimethylsulphoxide: improved results following the addition of polyethylene glycol but not dextran. Cryo letters 1998; 19: 141–6.
48. dela Peña EC, Takahashi Y, Atabay EC, et al. Vitrification of mouse oocytes in ethylene glycol-raffinose solution: effects of preexposure to ethylene glycol or raffinose on oocyte viability. Cryobiology 2001; 42: 103–11.
49. Chen S-U, Lien Y-R, Cheng Y-Y, et al. Vitrification of mouse oocytes using closed pulled straws (CPS) achieves a high survival and preserves good patterns of meiotic spindles, compared with conventional straws, open pulled straws (OPS) and grids. Hum Reprod 2001; 16: 2350–6.
50. Stachecki JJ, Cohen J, Schimmel T, et al. Fetal development of mouse oocytes and zygotes cryopreserved in a nonconventional freezing medium. Cryobiology 2002; 44: 5–13.
51. Gook DA, Osborn SM, Johnston WIH. Cryopreservation of mouse and human oocytes using 1,2-propanediol and the configuration of the meiotic spindle. Hum Reprod 1993; 8: 1101–9.
52. Wood MJ, Candy CJ. Reducing the time elapsed between thawing and insemination of frozen mouse oocytes increases the rate of fertilization. J Reprod Fert Abstract Series 1993; 11: 9.
53. Chen C. Pregnancy after human oocyte cryopreservation. Lancet 1986; 327: 884–6.
54. Van Uem JFHM, Siebzehnrübl ER, Schuh B, et al. Birth after cryopreservation of unfertilised oocytes. Lancet 1987; 329: 752–3.
55. Tucker M, Wright G, Morton P, et al. Preliminary experience with human oocyte cryopreservation using 1,2-propanediol and sucrose. Hum Reprod 1996; 11: 1513–5.
56. Porcu E, Fabbri R, Seracchioli R, et al. Birth of a healthy female after intracytoplasmic sperm injection of cryopreserved human oocytes. Fertil Steril 1997; 68: 724–6.
57. Shaw JM, Trounson AO. Parthenogenetic activation of unfertilized mouse oocytes by exposure to 1,2-propanediol is influenced by temperature, oocyte age, and cumulus removal. Gamete Research 1989; 24: 269–79.

58. Van der Elst J, Van den Abbeel E, Nerinckx S, et al. Parthenogenetic activation pattern and microtubular organization of the mouse oocyte after exposure to 1,2-propanediol. Cryobiology 1992; 29: 549–62.
59. Joly C, Bchini O, Boulekbache H, et al. Effects of 1,2-propanediol on the cytoskeletal organization of the mouse. Hum Reprod 1992; 7: 374–8.
60. Larman MG, Katz-Jaffe MG, Sheehan CB, et al. 1,2-propanediol and the type of cryopreservation procedure adversely affects mouse oocyte physiology. Hum Reprod 2007; 22: 250–9.
61. Quintans CJ, Donaldson MJ, Bertolino MV, et al. Birth of two babies using oocytes that were cryopreserved in a choline-based freezing medium. Hum Reprod 2002; 17: 3149–52.
62. Boldt J, Tidswell N, Sayers A, et al. Human oocyte cryopreservation: 5-year experience with a sodium-depleted slow freezing method. Reprod Biomed Online 2006; 13: 96–100.
63. Vajta G, Kuwayama M. Improving cryopreservation systems. Theriogenology 2006; 65: 236–44.
64. Rall WF, Fahy GM. Ice-free cryopreservation of mouse embryos at –196°C by vitrification. Nature 1985; 313: 573–5.
65. Larman MG, Sheehan CB, Gardner DK. Calcium-free vitrification reduces cryoprotectant-induced zona pellucida hardening and increases fertilization rates in mouse oocytes. Reproduction 2006; 131: 53–61.
66. Ko C-S, Ding D-C, Chu T-W, et al. Changes to the meiotic spindle and zona pellucida of mature mouse oocytes following different cryopreservation methods. Anim Reprod Sci 2008; 105: 272–82.
67. Carroll J, Depypere H, Matthews CD. Freeze-thaw-induced changes of the zona pellucida explains decreased rates of fertilization in frozen-thawed mouse oocytes. J Reprod Fert 1990; 90: 547–53.
68. Wood MJ, Whittingham DG, Lee S-H. Fertilization failure of frozen mouse oocytes is not due to premature cortical granule release. Biol Reprod 1992; 46: 1187–95.
69. Vincent C, Pickering SJ, Johnson MH. The hardening effect of dimethylsulphoxide on the mouse zona pellucida requires the presence of an oocyte and is associated with a reduction in the number of cortical granules present. J Reprod Fert 1990; 89: 253–9.
70. Schalkoff ME, Oskowitz SP, Powers RD. Ultrastructural observations of human and mouse oocytes treated with cryopreservatives. Biol Reprod 1989; 40: 379–93.
71. George MA, Johnson MH, Vincent C. Use of fetal bovine serum to protect against zona hardening during preparation of mouse oocytes for cryopreservation. Hum Reprod 1992; 7: 408–12.
72. Van Blerkom J, Davis PW. Cytogenetic, cellular and developmental consequences of cryopreservation of immature and mature mouse and human oocytes. Microsc Res Tech 1994; 27: 165–93.
73. Johnson MH, Pickering SJ, George MA. The influence of cooling on the properties of the zona pellucida of the mouse oocyte. Hum Reprod 1988; 3: 383–7.
74. Johnson MH. The effect on fertilization of exposure of mouse oocytes to dimethyl sulfoxide: an optimal protocol. J in vitro fertilisation and embryo transfer 1989; 6: 168–75.
75. Litkouhi B, Winlow W, Gosden RG. Impact of cryoprotective agent exposure on intracellular calcium in mouse oocytes at metaphase II. Cryo letters 1999; 20: 353–62.
76. Takahashi T, Igarashi H, Doshida M, et al. Lowering intracellular and extracellular calcium contents prevents cytotoxic effects of ethylene glycol-based vitrification solution in unfertilized mouse oocytes. Mol Reprod Dev 2004; 68: 250–8.
77. George MA, Johnson MH. Use of fetal bovine serum substitutes for the protection of the mouse zona pellucida against hardening during cryoprotectant addition. Hum Reprod 1993; 8: 1898–900.
78. Gook DA, Schiewe MC, Osborn SM, et al. Intracytoplasmic sperm injection and embryo development of human oocytes cryopreserved using 1,2-propanediol. Hum Reprod 1995; 10: 2637–41.
79. Kazem R, Thompson LA, Srikantharajah A, et al. Cryopreservation of human oocytes and fertilization by two techniques: in-vitro fertilization and intracytoplasmic sperm injection. Hum Reprod 1995; 10: 2650–4.
80. Li X-H, Chen S-U, Zhang X, et al. Cryopreserved oocytes of infertile couples undergoing assisted reproductive technology could be an important source of oocyte donation: a clinical report of successful pregnancies. Hum Reprod 2005; 20: 3390–4.
81. Vincent C, Johnson MH. Cooling, cryoprotectants, and the cytoskeleton of the mammalian oocyte. Oxford Reviews Reproductive Biology 1992; 14: 72–100.
82. Pickering SJ, Braude PR, Johnson MH, et al. Transient cooling to room temperature can cause irreversible disruption of the meiotic spindle in the human oocyte. Fertil Steril 1990; 54: 102–8.
83. Pickering SJ, Johnson MH. The influence of cooling on the organisation of the meiotic spindle of the mouse oocyte. Hum Reprod 1987; 2: 207–16.
84. Van der Elst J, Van den Abbeel E, Jacobs R, et al. Effect of 1,2-propanediol and dimethylsulphoxide on the meiotic spindle of the mouse oocyte. Hum Reprod 1988; 3: 960–7.
85. Sathananthan AH, Kirby C, Trounson AO, et al. The effects of cooling mouse oocytes. J Assist Reprod Genet 1992; 9: 139–48.

86. Johnson MH, Pickering SJ. The effect of dimethylsulphoxide on the microtubular system of the mouse oocyte. Development 1987; 100: 313–24.
87. Larman MG, Minasi MG, Rienzi L, et al. Maintenance of the meiotic spindle during vitrification in human and mouse oocytes. Reprod Biomed Online 2007; 15: 692–700.
88. Cooper A, O'Neil L, Bernard A, et al. The effect of protein source and temperature on the damage to mouse oocyte cytoskeleton after exposure to a vitrification solution. Cryo letters 1996; 17: 149–56.
89. O'Neil L, Paynter SJ, Fuller BJ, et al. Murine oocyte cytoskeletal changes, fertilisation and embryonic development following exposure to a vitrification solution. Cryo letters 1997; 18: 17–26.
90. Aigner S, Van der Elst J, Siebzehnrübl, et al. The influence of slow and ultra-rapid freezing on the organization of the meiotic spindle of the mouse oocyte. Hum Reprod 1992; 7: 857–64.
91. Chen S-U, Lien Y-R, Chen H-F, et al. Open pulled straws for vitrification of mature mouse oocytes preserve patterns of meiotic spindles and chromosomes better than conventional straws. Hum Reprod 2000; 15: 2598–603.
92. Gomes CM, Silva CASE, Acevedo N, et al. Influence of vitrification on mouse metaphase II oocyte spindle dynamics and chromatin alignment. Fertil Steril 2008; 90: 1396–404.
93. Stachecki JJ, Munné S, Cohen J. Spindle organization after cryopreservation of mouse, human, and bovine oocytes. Reprod Biomed Online 2004; 8; 664–72.
94. Bouquet M, Selva J, Auroux M. Effects of cooling and equilibration in DMSO, and cryopreservation of mouse oocytes, on the rates of in vitro fertilization, development, and chromosomal abnormalities. Mol Reprod Dev 1995; 40: 110–5.
95. Bos-Mikich A, Whittingham DG. Analysis of the chromosome complement of frozen-thawed mouse oocytes after parthenogenetic activation. Mol Reprod Dev 1995; 42: 254–60.
96. Coticchio G, Bonu MA, Sciajno R, et al. Truths and myths of oocyte sensitivity to controlled rate freezing. Reprod Biomed Online 2007; 15: 24–30.
97. Vincent C, Pickering SJ, Johnson MH, et al. Dimethylsulphoxide affects the organisation of microfilaments in the mouse oocyte. Mol Reprod Dev 1990; 26: 227–35.
98. Hosu BG, Mullen SF, Critser JK, et al. Reversible disassembly of the actin cytoskeleton improves the survival rate and developmental competence of cryopreserved mouse oocytes. PloS ONE 2008; 3: e2787.
99. Wen Y, Quintero R, Chen B, et al. Expression of CD9 in frozen-thawed mouse oocytes: preliminary experience. Fertil Steril 2007; 88: 526–9.
100. Katz-Jaffe MG, Larman MG, Sheehan CB, et al. Exposure of mouse oocytes to 1,2-propanediol during slow freezing alters the proteome. Fertil Steril 2008; 89: 1441–7.

11 Oocyte Storage in Domestic Species

Gábor Vajta,[1,2] Jason Conceicao,[1] and John Yovich[1,2]

[1]PIVET Medical Centre, Perth, Western Australia, Australia
[2]Cairns Fertility Centre, Cairns, Queensland, Australia

INTRODUCTION: THE PRIVILEGED ROLE OF DOMESTIC SPECIES IN MAMMALIAN EMBRYOLOGY RESEARCH

In contrast to many other fields of science, domestic animal research has pioneered advancements in some fields of reproductive biology including embryology and cryopreservation. Besides the well-know achievements in somatic cell nuclear transfer (1,2), considerable successes have been achieved in cryopreservation of spermatozoa (3) and preimplantation stage embryos (4,5). These discoveries were often made in parallel with or even preceding those in experimental animals. These results have had enormous consequences both on the practical application (e.g., commercial artificial insemination and frozen embryo transfer in cattle) and on the advancement of research in other mammalian species including humans. Recent achievements in the establishment of new cryopreservation, embryo culture, and gender determination methods, as well as the impressive advancement in the field of transgenic research, indicate that this positive trend will continue in the future.

There are three reasons that may explain the privileged role domestic species play in embryology research. First, there is easy and inexpensive access to almost unlimited numbers of oocytes in many domestic species including cattle, pig, and sheep. These animals are slaughtered in huge numbers and their ovaries are routinely discarded. Logistics is the only problem in obtaining these ovaries. (The relatively handicapped situation in equine embryology may at least be partially attributed to the less amenable access to horse oocytes.) Second, there is considerable pressure from the animal breeding industry for rapid genetic advancement using the most advanced technologies available. Practical ways need to be discovered for widespread on-field application. This pressure is especially strong in species with small litters, long gestation, and large size (to gain more from investment). This may explain the long-existing differences in the overall efficiency of embryo technologies in cattle versus pigs. (This gap, however, has closed rapidly during the past few years with pig embryology needed for highly profitable industrial biomedical application). The third factor that may explain the advanced status of domestic animal embryology is that there are many similarities between the features of oocytes and embryos from both domestic species and humans, in contrast to those from experimental animals. Although the mouse is still the preferred animal model for human embryology, some techniques (including cryopreservation of oocytes) can be performed more easily in the mouse than in the human. This makes the mouse species suboptimal for the development of more-efficient, less harmful oocyte cryopreservation technologies. The opposite problem occurs in rats. To obtain large quantities of oocytes from other experimental animals seems to be too complicated and expensive for most laboratories compared to domestic species.

CRYOPRESERVATION OF MAMMALIAN OOCYTES: GENERAL FEATURES

In contrast to male gametes, where freeze-drying and room-temperature storage may maintain fertilizing ability [although not viability (6)], at present the only available way for long-term preservation of mammalian oocytes is storage below –150°C, in practice at or around –196°C. The oocytes are immersed in or kept slightly above the level of liquid nitrogen. In other fields of cryobiology, storage itself is not the demanding part of the cryopreservation process, most problems occur at cooling and warming. Various kinds of injuries may occur during these processes including ice crystal formation, chilling injury, osmotic and toxic damage and fracture of the samples (7–9). The challenge is to minimize these injuries as well as to increase the resistance and regeneration ability of the sample. The two major cryopreservation techniques that have emerged over the past decades for oocytes and embryos use different approaches to achieve these goals. While traditional slow-rate freezing attempts to establish a balance between various sources of injuries by minimizing all of them to a tolerable level, vitrification focuses on the total elimination of ice crystal formation. Vitrification reduces chilling injury due to the rapid passage through the dangerous temperature zones above and slightly below 0°C (10,11). This was initially observed to be a by-product of vitrification; however, it is now seen as the integral element of the technique. In mammalian embryology, these strategies have become increasingly important for cryopreserving samples from many different species at various developmental stages.

Unfortunately oocytes from almost all mammalian species are very sensitive to cryoinjuries, and to store them successfully is one of the ultimate challenges in reproductive cryobiology. Reasons for this sensitivity have been described in detail elsewhere (12) and are also discussed in other chapters of this book. Briefly, major factors may include the *extreme size and round shape*, making the equal distribution of cryoprotectant solution within the oocyte difficult to achieve without side effects such as toxic and osmotic injuries; sensitivity to *chilling injury* that may cause irreversible damage to cytoplasmic lipid droplets, and serious, although potentially reversible damage to the meiotic spindle and membranous structures (13–17). The shock caused by the cryopreservation process may also induce *premature release of cortical granules*, consequently hardening the zona pellucida and reducing sperm penetration.

It should be noted, however, that there are some reservations regarding the above factors and their effects. Zygotes have at least the same size and exactly the same shape as oocytes; however, in many species they are much less sensitive to cryoinjuries. Pig oocytes and early embryos suffer irreversible chilling damage during slow-rate freezing, supposedly due to the extreme amount of cytoplasmic lipid droplets (Fig. 1), while cat embryos with similar amount of lipid droplets have been successfully cryopreserved by both traditional freezing and vitrification (18). Damage to the meiotic spindle does not occur in immature [germinal vesicle (GV) stage] oocytes theoretically, making these earlier, immature stage oocytes less prone to cryoinjuries. However, in most species, the cryopreservation of GV stage oocytes is much more demanding than that of matured ones (17,19,20). Explanations for this phenomenon include the changing permeability of membranes as observed in goat oocytes by Le Gal et al. (21). Recently, concerns were also raised regarding the commonly believed dramatic effect of zona hardening (8).

Other, practical factors have also hampered advancement in this field. Paradoxically, the need for oocyte cryopreservation may be far less important than that of preimplantation stage embryos in domestic animals. As mentioned earlier, fresh oocytes can be obtained anytime in large quantities for experimental purposes. When the genetics is important, a much safer option

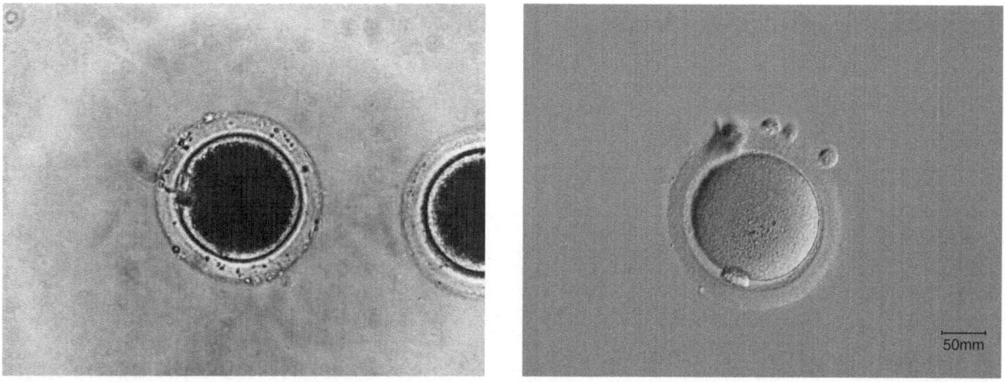

(A) (B)

Figure 1 Mature pig (**A**) and human oocytes (**B**). Pig oocytes contain lipid droplets that are believed to increase the sensitivity of these cells to cryodamage. *See Color Plates on Page xviii.*

is to perform either in vitro or in vivo fertilization with commercially available semen and cryopreserve these embryos before transfer for logistic reasons. Long-term storage of oocytes would be required only in special situations and does not attract much research money. On the other hand, as mentioned earlier, methods for mouse oocyte cryopreservation were available decades ago, but this approach was efficient for that species only. Among other factors the lack of pressure to advance domestic and experimental animal embryology contributed considerably to the prolonged inefficiency of mammalian oocyte cryopreservation in many species including humans.

However the rapidly increasing, almost imperative need to establish efficient protocols for human oocyte cryopreservation has generated a considerable demand for an appropriate animal model, and the most feasible choice was to test new methods in cattle and pigs. During the past decade this approach has become quite successful, especially in the field of vitrification. Based on established techniques, new areas of application were discovered in domestic animal embryology. These include the preservation of oocytes from rare and endangered species or from animals with special genetic value for potential future breeding programs.

ACHIEVEMENTS WITH DOMESTIC ANIMALS

As can be concluded from the previous chapter, the initial advancement in the cryopreservation of oocytes in domestic species has been far less impressive than that of preimplantation stage embryos (Table 1). The first calf born from traditionally frozen oocytes was reported by Fuku et al. only in 1992 (22), 15 years after the first mouse pup (30) and 6 years after the first human baby (31). On the other hand, oocyte vitrification was rapidly adapted to cattle, with the birth of the first offspring in the same year as that from traditional freezing (23). This occurred just three years after the birth of the first mouse pup (32), preceding by seven years the first publication reporting human success (33). In contrast to humans and the mouse, since the very first successes, vitrification has become the dominant approach for oocyte cryopreservation in domestic animals.

Oocyte vitrification has become less demanding and more efficient with the introduction of purpose-designed tools to increase cooling and warming rates. This has resulted in offspring born after double-vitrification, using both matured oocyte and blastocyst stage embryos, after in vitro fertilization (34). A similar technical approach was followed to produce calves after vitrification of immature oocytes. These immature oocytes underwent subsequent in vitro maturation, fertilization, and transfer with or without a second round of blastocyst vitrification (24,35). Vitrification of enucleated cattle oocytes resulted in full-term development of calves after embryonic cell nuclear transfer (25). The same method when applied to matured oocytes before enucleation eventually led to the birth of a healthy calf after somatic cell nuclear transfer (26).

Table 1 Offspring Born after Oocyte Cryopreservation in Domestic Animals

Species	Reported by	Method, sample, and further treatment
Cattle	Fuku et al. (22)	Slow freezing, IVM oocytes, IVF
Cattle	Hamano et al. (23)	Vitrification, IVM oocytes, IVF
Cattle	Vieria et al. (24)	Vitrification, immature oocytes, IVF
Cattle	Booth et al. (25)	Vitrification, IVM oocytes, embryonic cell NT
Cattle	Hou et al. (26)	Vitrification, IVM oocytes, somatic cell NT
Pig	Li et al. (27)	Vitrification, delipated IVM oocytes, somatic cell NT
Horse	Maclellan et al. (28)	Vitrification, in vivo-matured oocytes, IVF
Buffalo	Neglia et al. (29)	Vitrification, IVM oocytes, IVF

Abbreviations: IVF, in vitro fertilization; IVM, in vitro maturation; NT, nuclear transfer.

For a long period of time pigs were regarded as the most challenging domestic species to work with from an embryologist's point of view due to extreme sensitivity to any in vitro procedures. Additionally, for animal breeders the large litter and short gestation and generation interval makes porcine embryology far less attractive commercially than that in cattle. However, disadvantages from the breeders' point of view may become advantageous when biomedical application is considered. Similar features (size, genetics, organ structure, etc.) to humans make pigs excellent candidate animals for various applications including xenotransplantation and disease models. To achieve these goals genetic modifications are required, and a well-established embryology background seems to be indispensable. Accordingly, during the first decade of the 21st century, research in this direction has accelerated dramatically and resulted in rapid advancement predominantly in the field of somatic cell nuclear transfer (36). It should be acknowledged retrospectively that many proposed difficulties could be eliminated relatively easily with the application of a simple, proper approach.

With cryopreservation however, the extreme chilling sensitivity has made the advancement very challenging, even though vitrification of in vivo-derived pig embryos, with high cooling and warming rates, has eventually resulted in piglets being produced without any additional treatment (37). Zygotes, early cleavage stage embryos, and especially oocytes did not survive the procedure. It was discovered that the only possible way to obtain survival was to physically remove the most sensitive structures, namely the lipid droplets, from the cytoplasm. A method was originally introduced to delipate embryos by high-speed centrifugation with subsequent micromanipulation (38), however, this was rather complicated for routine application. A modification of the technique [partial zona digestion followed by centrifugation, creating a larger perivitelline space permitting complete separation of lipid droplets from the cytoplasm during and after centrifugation (39)] has made the delipation technique more practical and compatible with some new in vitro procedures including handmade cloning. Eventually these delipation techniques resulted in live piglets after somatic cell nuclear transfer using either zona-included or zona-free vitrified oocytes (27,40).

The third domestic species where offspring were obtained after cryopreservation of oocytes is the horse, and the successful method was again vitrification (28). Finally, a birth was also reported after the vitrification of buffalo (*Bubalus bubalis*) oocytes (29). So far, there has been no report published regarding the birth of live offspring from sheep and goat, although vitrification of ovine oocytes has already resulted in late-stage fetuses (Ledda, personal communication).

The initial, sporadic successes have demonstrated the possibility of live births from sheep and goats; however, the work was abandoned after the first achievements were published. The first three calves born after double-vitrification (from oocyte and blastocyst stage embryos, respectively) were slaughtered a month after their birth (Vajta, unpublished), and the laboratory that produced the first foals from vitrified oocytes had to discontinue the transfer program (Seidel, personal communication). Both decisions were made for financial reasons. Even the experimental use of cryopreserved domestic animal oocytes (suggested earlier to avoid logistic problems, for example, for somatic cell nuclear transfer) has remained unexploited: Researchers, prefer to use fresh oocytes instead of cryopreserved ones with their intrinsic handicap. Accordingly, even the most optimistic among us may not see the number of domestic animals born alive after oocyte vitrification exceed much more than twenty worldwide. These numbers barely justify the invested effort.

Due to the rapidly increasing need for animal models to establish an efficient oocyte cryopreservation method in humans, research in this field has been markedly intensified, including research with domestic animals. Next we will summarize the achievements from two major areas of research, the use of cytoskeleton relaxants and the application of a stress to induce stress tolerance.

CYTOSKELETON RELAXANTS AND STABILIZERS

The idea to use cytoskeleton-modifying agents to minimize the effect of deformations during cryopreservation—due to the rather drastic osmotic effects that may occur, especially at vitrification—is not new. A cytoskeleton relaxant, cytochalasin B, has been successfully used to improve survival rates of vitrified pig embryos (41). In MII-phase oocytes it may also prevent damage to the meiotic spindle. So far, the achievements with cytochalasin B, a commonly used cytoskeleton relaxant are controversial. According to some recent publications, treatment with cytochalasin B before vitrification improved in vitro development of mature ovine oocytes (42), but decreased the survival of immature ones (43). It did not have any effect on porcine oocytes regardless of their maturation status (44). On the other hand, according to earlier publications, cytochalasin B treatment improved survival rates and developmental competence of porcine in vitro-matured oocytes (45). It also improved maturation rates of immature porcine oocytes (46,47). It is worthwhile to mention that, although the species has not yet been domesticated, there was no improvement in maturation rates when immature mink whale oocytes were treated with cytochalasin B before vitrification (48).

The apparent confusion is hard to interpret: one may propose that there are some beneficial effects from cytochalasin B treatment under certain circumstances. However, considering the potential risks of cytochalasin B treatment, it does not seem to be a very promising approach to improve the survival and developmental competence of human oocytes.

There is another agent with a microtubule stabilizing effect that may become a more appropriate candidate. Although the initial findings were disappointing (49), recent results unanimously support the positive effect of the chemotherapeutic drug Taxol in preventing cryodamage to domestic animal oocytes. Pretreatment of cow oocytes with Taxol prevented spindle and cortical granule abnormalities after vitrification and resulted in higher cleavage and blastocyst rates (50). Blastocyst development was also improved when porcine and ovine oocytes were treated with Taxol before vitrification (42,51). A recent ultra-structural study has also confirmed that Taxol treatment was very efficient in stabilizing the metaphase plate and the spindle morphology of cow oocytes subjected to vitrification (52). These, almost unanimous, observations in domestic animals have confirmed the initially reported beneficial effect of Taxol pretreatment on in vitro developmental rates of mouse oocytes (53). On the other hand, the ultimate goal of application in humans will be a difficult task, because Taxol at present is regarded as a very potent anticancer drug that attacks mitosis through microtubules (54). Accordingly any suggestion for its embryological application in humans will possibly create a strong and a partially justified aversion. It should also be noted that there is almost no data regarding the in vivo long-term effect of this treatment. Extensive studies of fetal and postnatal development are required, and apart from the traditional experimental animals, pigs may be excellent candidates for this purpose.

STRESS FOR STRESS TOLERANCE

Another possible way to improve survival and developmental rates after cryostorage of mammalian zygotes and embryos has been discovered recently, first in experimental animals, then more widely in domestic animals. The fact that in humans and other mammals the appropriate physical (or even mental) stress may induce subsequent stress tolerance has been a widely acknowledged phenomenon for years. The physiological principles, however, were discovered by János (Hans) Selye and colleagues only in the middle of the last century (55). This research has also proven that the response is not limited to the initial impulse, as expressed by the name: general adaptation syndrome. Later, a similar phenomenon was also discovered at the cellular level in eukaryotes and even in bacteria. Many recent publications have dealt with the biochemical and molecular biological basis of these processes (56–58). The factors involved may include stress proteins (heat, cold, and other shock proteins, most of them with chaperone activity) and different pathways, including those mediated with glutathione (59,60).

The reason why stress-induced stress tolerance has not been exploited in zygotes and early embryos is (among other reasons) that their cells are regarded as being transcriptionally inactive. However, according to recent investigations this inactivity is not absolute and transcription as well as post-translational stability can be modified by different impacts. For example, the heat-induced increase in the transcription of the HSP70 gene has been observed in bovine embryos as early as at the two-cell stage, much earlier than the activation of the whole genome at the 8- to 16-cell stage (61).

Earlier, high hydrostatic pressure (HHP) was demonstrated to have a profound effect on cell metabolism: it changes various intracellular biochemical reactions including passive and active transport through the membranes as well as the protein profile of cells (62–64). HHP also induces production of heat shock proteins both in prokaryotes and in eukaryotes (65–67). As the intensity and length of HHP can be measured exactly, its impact is consistent and repeatable, affecting all samples uniformly, so it seems to be an appropriate way to induce the initial stress.

To achieve maximum protection the initial stress should be sublethal, which means a rather high impact of 200 to 800 bars that equates to the pressure at 8000 to 10,000 meters below sea level. A special device is required to induce this. This pressure should be applied for one to two hours, and then a recovery period of one to two hours is required before cryopreservation.

The first experiments were performed with in vivo-derived murine and in vitro-produced bovine blastocysts. HHP treatment resulted in a considerable increase in survival and hatching rates after cryopreservation (68–70). HHP was also found to be surprisingly efficient for porcine sperm cryopreservation, improving motility, changing the protein profile, and increasing the litter number by 50% after artificial insemination (71–74).

Eventually the work became focused on porcine oocytes, where HHP did not result in morphological alterations detectable by light microscopy, but induced higher in vitro shock resistance and developmental competence after vitrification and parthenogenetic activation (75,76).

The impressive results encouraged researchers to test another sublethal injury to induce tolerance. When applied to in vitro-matured porcine oocytes, a 1% (w/v) increase in the sodium chloride concentration of the media over the normal physiological range for one hour increased blastocyst rates after vitrification (77). The resistance could also be induced by elevated concentrations of sucrose and trehalose (78); thus, the most feasible explanation for this result is a common osmotic effect.

The possibility of stress-induced stress tolerance is far from being extensively exploited, and may have considerable applications in mammalian reproductive cryobiology including oocyte cryopreservation. These experiments focus on improving survival rates and developmental competence, and to also prove the safety of the entire procedure, for potential human application. Domestic animals are indispensable for this work, and eventually efficiency may reach a level where commercial and other practical applications (including rescue of eggs from endangered species and breeds) become a realistic goal.

CONCLUSION

At present, cryostorage of oocytes from domestic species is a very limited part of mammalian reproductive cryobiology, largely due to the relative inefficiency of available procedures and the lack of need from animal breeders for cryopreserved oocytes. However, techniques developed with the use of easily accessible domestic animal oocytes can be applied to many species including humans. On the basis of recent encouraging results, the efficient cryostorage of oocytes will eventually reach a level appropriate for practical application as well as for breeding and preservation of genetics in domestic species.

ACKNOWLEDGMENTS

The authors thank Sergio Ledda and George Seidel, Jr., for providing data for this review.

REFERENCES

1. Willadsen SM. Nuclear transplantation in sheep embryos. Nature 1986; 320: 63–5.
2. Wilmut I, Schnieke AE, McWhir J, Kind AJ, Campbell KH. Viable offspring derived from fetal and adult mammalian cells. Nature 1997; 385: 810–13.
3. Polge C. Fertilizing capacity of bull spermatozoa after freezing at 79°C. Nature 1952; 169: 626–7.
4. Wilmut I, Rowson LEA. Experiments on the low temperature preservation of cow embryos. Vet Rec 1973; 92: 686–90.
5. Willadsen SM, Polge C, Rowson LE, et al. Deep freezing of sheep embryos. J Reprod Fertil 1976; 46: 151–4.
6. Wakayama T, Yanagimachi R. Development of normal mice from oocytes injected with freeze-dried spermatozoa Nat Biotechnol 1998; 26: 639–41.

7. Mazur P. Cryobiology: the freezing of biological systems. Science 1970; 168: 939–49.
8. Coticchio G, Bonu MA, Sciajno R, et al. Truths and myths of oocyte sensitivity to controlled rate freezing. Reprod Biomed Online 2007; 15: 24–30.
9. Leibo SP. Cryopreservation of oocytes and embryos: optimization by theoretical versus empirical analysis. Theriogenology 2008; 69: 2–47.
10. Shaw JM, Jones GM. Terminology associated with vitrification and other cryopreservation procedures for oocytes and embryos. Hum Reprod Update 2003; 9: 583–605.
11. Vajta G, Nagy ZP. Are traditional freezers still needed in the embryo laboratory? Review on vitrification. Reprod Biomed Online 2006; 12: 779–96.
12. Stachecki JJ, Cohen J. An overview of oocyte cryopreservation. Reprod Biomed Online 2004; 9: 152–63.
13. Martino A. Pollard JA, Leibo SP. Effect of chilling bovine oocytes on their developmental competence. Mol Reprod Dev 1996; 45: 503–12.
14. Pickering S, Braude P, Hohnson M, et al. Transient cooling to room temperature can cause irreversible disruption of the meiotic spindle in the human oocyte. Fertil Steril 1990; 54: 102–8.
15. Sathananthan AH, Ng SC, Trounson A, et al. The effects of ultrarapid freezing on meiotic and mitotic spindles of oocytes and embryos. Gamete Res 1998; 21: 385–401.
16. Rienzi L, Martinez F, Ubaldi F, et al. Polscope analysis of meiotic spindle changes in living metaphase II human oocytes during the freezing and thawing procedures. Hum Reprod 2004; 19: 655–9.
17. Ghetler Y, Yavin S, Shalgi R, et al. The effect of chilling on membrane lipid phase transition in human oocytes and zygotes. Hum Reprod 2005; 20: 3385–9.
18. Dresser BL, Gelwicks EJ, Wachs KB, Keller GL. First successful transfer of cryopreserved feline (*Felis catus*) embryos resulting in live offspring. J Exp Zool 1988; 246: 180–6.
19. Leibo SP, Martion A, Kobayashi S, Pollard JW. Stage-dependent sensitivity of oocytes and embryos to low temperatures. Anim Reprod Sci 1996; 42: 45–53.
20. Men H, Monson RL, Rutledge JJ. Effect of meiotic stage and maturation protocols on bovine oocyte's resistance to cryopreservation. Theriogenology 2002; 57: 1095–103.
21. Le Gal F, Gasqui P, Renard JP. Differential osmotic behaviour of mammalian oocytes before and after maturation: a quantitative analysis using goat oocytes as a model. Cryobiology 1994; 31: 154–70.
22. Fuku E, Kojima T, Shioya Y, Marcus GJ, Downey BR. In vitro fertilization and development of frozen-thawed bovine oocytes. Cryobiology 1992; 29: 485–92.
23. Hamano S, Koikeda A, Kuwayama M, Nagai T. Full-term development of in vitro-matured, vitrified and fertilized bovine oocytes. Theriogenology 1992; 38: 1085–90.
24. Vieria AD, Mezzalira A, Barieri DP, et al. Calves born after open pulled straw vitrification of immature bovine oocytes. Cryobiology 2002; 45: 91–4
25. Booth PJ, Vajta G, Høj A, et al. Full-term development of nuclear transfer calves produced from open pulled straw (OPS) vitrified cytoplasts. Theriogenology 1999; 51: 99–1006.
26. Hou YP, Dai YP, Zhu SE, et al. Bovine oocytes vitrified by the open pulled straw method and used for somatic cell cloning supported development to term. Theriogenology 2005; 64: 1389–91.
27. Li R, Lai L, Wax D, et al. Cloned transgenic swine via in vitro production and cryopreservation. Biol Reprod 2006; 75: 226–30.
28. Maclellan LJ, Carnevale EM, Coutinho DA, et al. Pregnancies from vitrified equine oocytes collected from super-stimulated and non-stimulated mares. Theriogenology 2002; 58: 911–19.
29. Neglia G, Gasparrini B, Caracciolo DI, et al. First pregnancies to term after transfer of buffalo vitrified embryos entirely produced in vitro. Vet Res Commun (Special Issue) 2004; 28: 233–6.
30. Whittingham DG. Fertilization in vitro and development to term of unfertilized mouse oocytes previously stored at –196°C. J Reprod Fertil 1977; 49: 89–94.
31. Chen C. Pregnancy after human oocyte cryopreservation. Lancet 1986; 1: 884–6.
32. Nakagata N. High survival rate of unfertilized mouse oocytes after vitrification. J Reprod Fertil 1989; 87: 479–83.
33. Kuleshova L, Gianaroli L, Magli C, Trounson A. Birth following vitrification of a small number of human oocytes. Hum Reprod 1999; 14: 3077–9.
34. Vajta G, Kuwayama M, Holm P, et al. Open pulled straw vitrification: a new way to reduce cryoinjuries of bovine ova and embryos. Mol Reprod Dev 1998; 51: 53–8.
35. Vieria AD, Forell F, Feltrin C, Rodrigues JL. Calves born after direct transfer of vitrified bovine in vitro-produced blastocysts derived from vitrified immature oocytes. Reprod Domest Anim 2008; 43: 314–18.
36. Vajta G, Zhang Y, Machaty Z. Somatic cell nuclear transfer in pigs: recent achievements and future possibilities. Reprod Fertil Dev 2007; 19: 403–23.
37. Berthelot, et al. 2000.

38. Nagashima H, Kashiwazaki N, Ashman RJ, et al. Removal of cytoplasmic lipid enhances the tolerance of porcine embryos to chilling. Biol Reprod 1994; 51: 618–22.
39. Esaki R, Ueda H, Kurome M, et al. Cryopreservation of porcine embryos derived from in vitro-matured oocytes. Biol Reprod 2004; 71: 432–7.
40. Du Y, Li J, Kragh PM, et al. Piglets born from vitrified cloned blastocysts produced with a simplified method of delipation and nuclear transfer. Cloning Stem Cells 2007; 9: 470–6.
41. Dobrinsky JR, Pursel VG, Long CR, Johnson LA. Birth of piglets after transfer of embryos cryopreserved by cytoskeletal stabilization and vitrification. Biol Reprod 2000; 62: 546–70.
42. Zhang J, Nedambale TL, Yang M, Li J. Improved development of ovine matured oocytes following solid surface vitrification (SSV). Effect of cumulus cells and cytoskeleton stabilizer. Anim Reprod Sci 2008; (in press).
43. Bogliolo L, Ariu F, Fois S, et al. Morphological and biochemical analysis of immature ovine oocytes vitrified with or without cumulus cells. Theriogenology 2007; 68: 1138–49.
44. Gupta MK, Uhm SJ, Lee HT. Cryopreservation of immature and in vitro matured porcine oocytes by solid surface vitrification. Theriogenology 2007; 67: 238–48.
45. Somfai T, Dinnyés A, Sage D, et al. Development to the blastocyst stage of parthenogenetically activated in vitro matured porcine oocytes after solid surface vitrification. Theriogenology 2006; 66: 415–22.
46. Isachenko V, Soler C, Isachenko E, Perez-Sanchez F, Grishchenko V. Vitrification of immature porcine oocytes: effects of lipid droplets, temperature, cytoskeleton, and addition and removal of cryoprotectant. Cryobiology 1998; 36: 250–3.
47. Fujihira T, Kishida R, Fukui Y. Developmental capacity of vitrified immature porcine oocytes following ICSI: effects of cytochalasin B and cryoprotectants. Cryobiology 2004; 49: 286–90.
48. Fujihira T, Kobayashi M, Hochi S, et al. Developmental capacity of Antarctic mink whale (*Balaenoptera bonaerensis*) vitrified oocytes following in vitro maturation, and parthenogenetic activation or intracytoplasmic sperm injection. Zygote 2006; 14: 89–95.
49. Fujihira T, Nagai H, Fukui Y. Relationship between equilibration times and the presence of cumulus cells, and effect of Taxol treatment for vitrification of in vitro matured porcine oocytes. Cryobiology 2005; 51: 339–43.
50. Morató R, Izquierdo D, Albarracín JL, et al. Effects of pre-treating in vitro matured bovine oocytes with the cytoskeleton stabilizing agent taxol prior to vitrification. Mol Reprod Dev 2008; 75: 191–201.
51. Shi WQ, Zhu SE, Zhang D, et al. Improved development by taxol pre-treatment after vitrification of in vitro matured porcine oocytes. Reproduction 2006; 131: 795–804.
52. Morató R, Mogas T, Maddox-Hyttel P. Ultrastructure of bovine oocytes exposed to Taxol prior to OPS vitrification. Mol Reprod Dev 2008; 75: 1318–26.
53. Park SE, Chung HM, Cha KY, et al. Cryopreservation of ICR mouse oocytes: improved post-thawed preimplantation development after vitrification using taxol, a cytoskeleton stabilizer. Fertil Steril 2001; 75: 1177–84.
54. Jordan MA, Kamath K. How do microtubule-targeted drugs work? An overview. Curr Cancer Drug Targets 2007; 7: 730–42.
55. Perdrizet GA. Hans Selye and beyond: responses to stress. Cell Stress Chaperones 1997; 2: 214–19.
56. Akerfelt M, Trouillet D, Mezger V, Sistonen L. Heat shock factors at a crossroad between stress and development. Ann NY Acad Sci 2007; 1113: 15–27.
57. Arya R, Mallik M, Lakhotia SC. Heat shock genes—integrating survival and death. J Biosci 2007; 32L: 595–610.
58. Söti C, Csermely P. Protein stress and stress proteins: implications in aging and disease. J Biosci 2007; 32: 511–15.
59. Edwards JL, King WA, Kawarsky SJ, Ealy AD. Responsiveness of early embryos to environmental insults: potential protective roles of HSP70 and gluthatione. Theriogenology 2001; 55: 209–23.
60. Hansen PJ. To be or not to be—determinants of embryonic survival following heat shock. Theriogenology 2007; 68S: S40–S48.
61. Hansen, et al. 2001.
62. Aldridge BE, Bruner LJ. Pressure effects on mechanisms of charge transport across bilayer membranes. Biochim Biophys Acta 1985; 817: 343–54.
63. Abe F, Kato C, Horikoshi K. Pressure-regulated metabolism in microorganisms. Trends Microbiol 1999; 7: 447–53.
64. Lammi MJ, Elo MA, Sironen RK, et al. High hydrostatic pressure-induced changes in cellular protein synthesis. Biorheology 2004; 41: 309–13.
65. Welch TJ, Farewell A, Neidhardt FC, Bartlett DH. Stress response of Escherichia coli to elevated hydrostatic pressure. J Bacteriol 1993; 175: 7170–7.

66. Kaarniranta K, Elo M, Sironen R, et al. Hsp70 accumulation in chondrocytic cells exposed to high continuous hydrostatic pressure coincides with mRNA stabilization rather than transcriptional activation. Proc Natl Acad Sci USA 1998; 95: 2319–24.
67. Wemekamp-Kamphuis HH, Karatzas AK, Wouters JA, Abee T. Enhanced levels of cold-shock proteins in *Listeria monocytogenes* LO28 upon exposure to low temperature and high hydrostatic pressure. Appl Environ Microbiol 2002; 68: 456–463.
68. Pribenszky C, Molnár M, Cseh S, Solti L. Improving post-thaw survival of cryopreserved mouse blastocysts by hydrostatic pressure challenge. Anim Reprod Sci 2005; 87: 143–50.
69. Pribenszky C, Molnar M, Horváth A, Harnos A, Szenci O. Hydrostatic pressure induced increase in post-thaw motility of frozen boar spermatozoa. Reprod Fertil Dev 2005; 18: 162–3.
70. Pribenszky C, Siquiera E, Molnár M, Rumpf R. Substantial increase in the post-warm survival, developmental dynamics and hatching rate of vitrified IVP bovine blastocysts by hydrostatic pressure pre-treatment, Reprod Fertil Dev 2007; (in press).
71. Pribenszky C, Molnár M, Ulrich P, et al. Pressure assisted cryopreservation: a novel possibility for IVP bovine blastocyst cryopreservation. Reprod Domest Anim 2005; 40: 338–44.
72. Pribenszky C, Molnár M, Horváth A, et al. Improved post-thaw motility, viability and fertility are achieved by hydrostatic pressure treated bull semen. Reprod Fertil Dev 2007; 19: 181–2.
73. Kuo YH, Pribenszky C, Huang SY. Higher litter size is achieved by the insemination of high hydrostatic pressure-treated frozen-thawed boar semen. Proceedings of the 6th International Conference on Boar Semen Preservation, Alliston, Ontario, Canada, 2007: III.-22.
74. Huang SY, Pribenszky C, Kuo YH, et al. The effect of hydrostatic pressure treatment on the protein profile of boar spermatozoa before and after freezing. Proceedings of the 6th International Conference on Boar Semen Preservation, Alliston, Ontario, Canada, 2007: I.-34.
75. Pribenszky C, Du Y, Molnár M, Harnos A, Vajta G. Increased stress tolerance of matured pig oocytes after high hydrostatic pressure treatment. Anim Reprod Sci 2008; 106: 200–7.
76. Du Y, Pribenszky C, Molnár M, et al. High hydrostatic pressure (HHP): a new way to improve in vitro developmental competence of porcine matured oocytes after vitrification. Reproduction 2008; 135: 13–17.
77. Lin L, Du Y, Liu Y, et al. Elevated NaCl concentration improves cryotolerance and developmental competence of porcine oocytes. Reprod Biomed Online (in press).
78. Lin L, Kragh PM, Purup S, et al. Osmotic stress induced by sodium chloride, sucrose or trehalose improves cryotolerance and developmental competence of porcine oocytes. Reprod Fertil Dev 2008; (in press).
79. Aman RR, Parks JE. Effects of cooling and rewarming of the meiotic spindle and chromosomes of in vitro matured bovine oocytes. Biol Reprod 1994; 50: 103–10.
80. Rojas C, Palomo MJ, Albarracín JL, Mogas T. Vitrification of immature and in vitro matured pig oocytes: study of distribution of chromosomes, microtubules, and actin microfilaments. Cryobiology 2004; 49: 211–20.

12 Control of the Solution Effect During Controlled-Rate Cooling: Principles and Practical Application

James J. Stachecki
Tyho-Galileo Research Laboratories, Livingston, New Jersey, U.S.A.

Much of the information we know today about solution effects during controlled-rate freezing comes from previously published works. Specifically, those papers published between 1950 and 1980. Therefore, the topic of solution effects is not new, but actually very old, especially if one considers the works of Luyet, (1,2), for example. These early works have laid the foundation for modern cryobiology, at least as it applies to the storage of mammalian cells, including oocytes and embryos. The information in the collective works of Lovelock, Meryman, Mazur, Polge, Smith, Levitt, Farrant, Willadsen, Luyet and company provide us an invaluable resource from which we can continue to learn about the effects of cryopreservation on cells. It is therefore beneficial to review some of these manuscripts in order to comment on the principles of controlled-rate cooling. The idea of solution effects or how the intracellular environment as well as the extracellular environment will impact the cells during cooling comes into play. These effects can also be described, if only in part, due to our incomplete understanding of exactly what happens during any form of cellular cooling and/or freezing. The importance of having a general understanding of solution effects, and for that matter, of controlled-rate cooling, is so that we are better able to understand how to freeze cells in such a way that they can remain viable upon rewarming. This chapter will therefore discuss the principles and the practical applications of controlled-rate cooling and solution effects. For the purposes of this book, I will focus on mammalian oocytes. What is confusing, or can be, is that the principles or theoretical ideas about solution effects and cooling, and how to cool to obtain viability, sometimes do not work when applied to a dynamic and complex system such as a living cell. Discussions of this type may best serve as a starting point for evaluating the complexities of cellular cryopreservation. Practical application of principles and/or theories leads to interesting and often unexpected results. The unexpected results we obtain then have us thinking of ways to explain them in a scientifically sound manner.

Controlled-rate cooling theories have been applied successfully, and sometimes not so successfully, to a variety of cell types including yeast, plants, red blood cells, stem cells, and other assorted mammalian cells. Controlled-rate cooling refers to the decrease in temperature in a controlled manner, typically with a computer controlled freezing machine. However, any cooling in which the rate is known and can be adjusted could be called controlled-rate cooling. In its simplest form, this could involve placing a beaker of alcohol in a freezer or other cold environment and monitoring the temperature decline over time (3). Controlled-rate cooling also refers to the rate at which the cell adapts to its changing environment. In other words, controlling the rate of dehydration in a manner that brings about an equilibrium between the outside environment and the intracellular environment, also known as equilibrium cooling. Solution effects refer to those effects that can occur during cooling, whether it be in equilibrium or not. The effects caused by solutes, penetrating and nonpenetrating cryoprotectants, and the osmotic environment are collectively known as solution effects. The potential problem of solution effects involves the change(s) in a solution or cytoplasm that result from dehydration, increased solute concentration, pH changes, and precipitation of solutes (4,5).

To begin our discussion of solution effects we can first examine what occurs in solutions of known composition. Phase diagrams are used to describe equilibrium situations in which two or more phases of matter exist together in pure substances or in solutions. The manner in

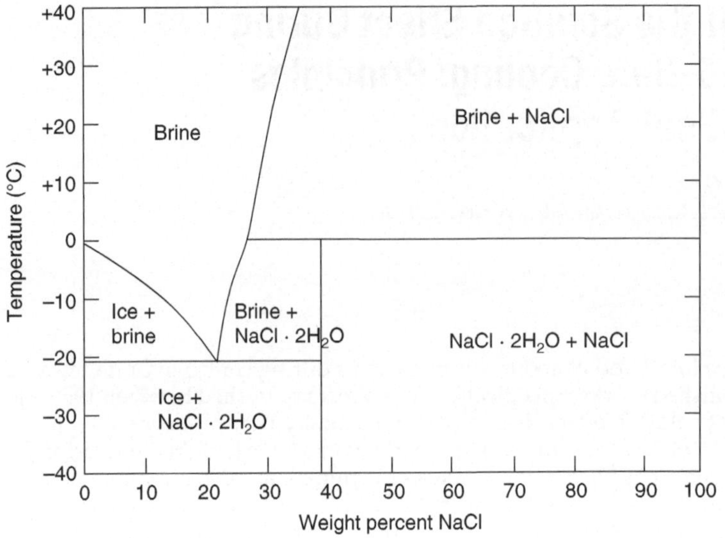

Figure 1 Phase diagram for sodium chloride and water.

which the substances are cooled determines the mixture of phases that exist when they become solid. Phase diagrams have been widely used in cryobiology since Cocks and Brower published an article showing their utility (6). In biological systems, the primary component is water; the entire system is a collection of compartments filled with an aqueous solution. As aqueous solutions are cooled, the water forms a crystalline solid (ice) which has almost no solubility for the solutes that were in the aqueous solution. As pure ice forms, the solutes will be confined to the remaining liquid phase, becoming more concentrated. Since this lowers the freezing point of the aqueous liquid, the system can remain in equilibrium with a substantial unfrozen fraction. As cooling continues, the solubility limit of the solution will also be reached, leading to the precipitation of solutes. These events are succinctly described by a phase diagram. The simplest type of phase diagram is for binary systems, systems in which there are only two phases present. Figure 1 shows the phase diagram for sodium chloride and water.

Starting at the left hand side of the diagram, if the temperature of a solution with 0% salt is lowered, the freezing point occurs at 0°C. If the solution has salt dissolved in it (i.e., the concentration of salt is below the solubility limit), then the mixture will exist in the brine compartment. As the temperature is lowered, the weight percent of NaCl does not change until the thick line in Figure 1 is reached. This line defines the freezing point of the solution. Further cooling will take the solution along the curve defined by the thick line until the eutectic point is reached at –21.2°C. At this point, the unfrozen compartment of the mixture is saturated with NaCl; any further cooling will cause salt to precipitate out of the mixture. For freezing biological systems, this left side of the phase diagram is the most important as it describes the osmolality of the solution in which the cells exist. The osmolality will follow this curve down to –21.2°C, where it will hold constant until the temperature is raised once again.

Although the binary phase diagram for sodium chloride and water is useful for understanding injury to cells, most cryopreservation protocols use at least one cryoprotective additive to reduce the freeze-thaw injury. In such cases, there are three compounds that must be considered: water, NaCl, and the cryoprotectant. A system with three components is described by a ternary phase diagram. Diagrams have been published for the ternary systems DMSO-NaCl-H_2O and glycerol-NaCl-H_2O. From these diagrams, it is clear that the solubility and eutectic (solidification point) behavior of a single solute can be altered significantly by the amount and type of additional solutes introduced into the system [e.g., dimethylsulfoxide (DMSO) serves as a solvent for NaCl as well as being a solute in water]. It is also clear that equilibrium between solids and liquid becomes increasingly complex as the number of components is increased.

In a recent paper, Fahy reviews several ways of estimating or calculating the amount of ice formed in an aqueous cryoprotectant solution at different temperatures based on phase diagrams (7). One of the simplest methods is to make estimates based on the behavior of ideal solutions. Equations that describe such relationships are not new and have been published elsewhere (7). These equations are important in gaining an understanding of how ideal solutions behave, however, keep in mind that an oocyte or cell is not an ideal solution. Fahy discusses the correlation between freezing injury and osmotic stress and goes on to explain the concept of colligative cryoprotection.

If we examine the cellular effects of controlled-rate cooling, based on the discussions above, there are a number of phenomena that occur, namely (*i*) an increase in osmolarity of the extra- and intracellular solute concentration and (*ii*) a decrease in cell volume (initially, but the cell may re-expand somewhat from uptake of solute and/or penetrating cryoprotectant). Both of these have been shown to cause cell damage. Lovelock applied the binary phase diagram in Figure 1 and the above equations to cryobiology by explaining the freezing injury suffered by red blood cells, as well as providing an explanation of the cryopreservative effects of such compounds as glycerol and DMSO. Lovelock (8) showed that the increasing extracellular solute concentration that occurs during cooling when water freezes out as ice was responsible for red blood cell lysis (>0.8 mol/L NaCl) and that the addition of a cryoprotectant (glycerol) reduced the amount of ice formed at any temperature, thereby effectively reducing the concentration of electrolytes produced. This was among the first experiments (mainly done in human erythrocytes) to describe solution effects, namely osmotic stress and cell injury. The collective works of many investigators (8–14) have supported a link between observed injury after thawing and a failure of membrane semipermeability during freezing that allows solute uptake during freezing and entrapment of the solute upon thawing. Similarly, Meryman showed a consistent relationship between salt (specifically, NaCl) osmolality and injury during freezing in the presence of a wide variety of cryoprotectants (15). Meryman proposed that injury is linked to an inability of the cell to shrink below a critical minimum volume, causing hydrostatic pressure across the cell membrane thereby increasing its permeability to solute (9,10,16,17).

Therefore, cell shrinkage, increased osmolarity (intracellular), and solute uptake can be detrimental to cell survival. The rate at which cells are cooled will thus affect the rate of cell shrinkage (via water removal) and solute concentration and these in turn relate to cell survival (18). Mazur's inverted U curve demonstrated this, showing cell survival with increasing cooling rate until an optimum is reached and then declining with further increases in the cooling rate. Figure 2, redrawn from Leibo et al. (19), shows a modified version of this curve. The descending portion of the curve is linked to the formation of intracellular ice, and the cooling rate at which this descent takes place has been linked to the permeability of the cell to water (5,20).

Let us now examine what happens in the cell when we try to cool it in a solution containing cryoprotectants. The theoretical basis for cryopreservation of cells proposed by Mazur has been applied to many cell types including mouse (5) and human oocytes (21). According to Mazur's two-factor hypothesis for cellular damage when cooling, the growth of ice crystals

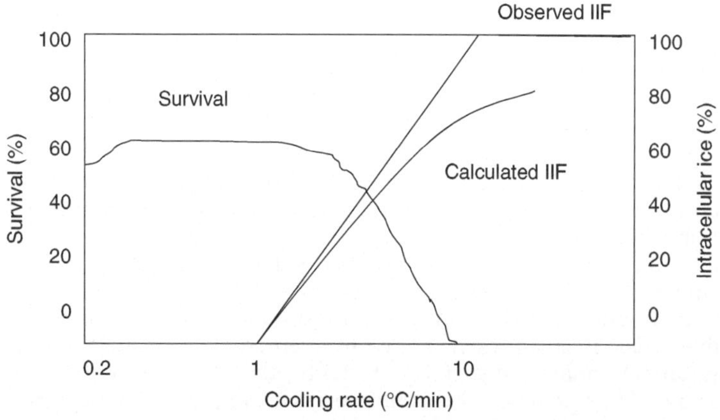

Figure 2 Relationship between cell survival, cooling rate, and IIF. Redrawn from Ref. 19. *Abbreviation*: IIF, intracellular ice formation.

intracellularly (IIF) and high concentrations of solutes resulting from the freezing of water (solution effects) were major causes of cell damage during freezing and thawing. Despite its incompleteness and imperfections, this hypothesis has been, and continues to be, very useful for getting to grips with mammalian-cell cryopreservation. Conventional freezing methods aim to minimize both IIF and solution effects by the use of penetrating cryoprotectants that interact with water to abolish eutectic freezing (6) and reduce the amount of ice formed at any temperature (8,10). It is a characteristic of these cryoprotectants that at a certain concentration [for 1,2-propanediol (PrOH), slightly above 70%; [see Figure 1 in Ref. 14] they prevent ice formation altogether (22). This concentration is well in excess of the starting concentration employed in conventional cell freezing methods (1–2 M). Here ice formation is initially induced extracellularly by seeding and, as a result of the solute gradient thus created, freezable water flows out of the cells to restore equilibrium in water activity across the cell membrane, minimizing the chance of IIF during cooling. As the temperature is gradually lowered, the concentration of cryoprotectant in the liquid phase, which includes the intracellular fluid, increases correspondingly until a level is reached at which additional formation and growth of ice crystals, although possible, are unlikely, even if the temperature drops further (22). Rather, the liquid phase turns into a glassy substance that solidifies without further crystal formation as the temperature continues to decrease. The unfrozen liquid phase remaining within the cells when they are plunged into liquid nitrogen (LN_2) should ideally, at least according to the two-factor hypothesis referred to above, consist of this glassy substance with all the original cell solutes remaining in solution (22). This theory suggests that when we slow-cool cells using a penetrating cryoprotectant such as PrOH and standard slow-cooling protocols, we are actually vitrifying the cells. Indeed, when we slow-cool human embryos, for example, typical survival rates range between 80% and 100%, for many in vitro fertilization centers. These survival rates would not be possible, at least according to Mazur and company, if IIF were occurring. This correlates well with the idea that slow-cooling is vitrification. Of course, this does not mean that IIF does not or cannot occur, it simply suggests that in conventional embryo freezing protocols, IIF is not a major source of cell damage.

In order to avoid IIF, the cooling rate down to the temperature at which no further ice formation is possible must be adjusted to allow time for most, if not all, freezable water to diffuse out of the cells. The flow of water and cryoprotectant through the cell membrane, on which dehydration and prevention of IIF depends, is related to membrane permeability. More specifically, dehydration and rehydration are influenced by the composition and the permeability characteristics of the cell membrane, the surface-to-volume ratio of the cell, the temperature, and the difference in osmotic pressure between the intracellular and extracellular environment. All these factors will effect cell survival and should be taken into account when dehydrating and rehydrating cells (23–25). Many investigators have thus looked at membrane permeability characteristics of a variety of cells including human oocytes (26–30). Paynter (26) suggests that variability in success rates with oocyte freezing is due in part to the different protocols that are in use today, and concludes that enough time is needed to dehydrate the cell and allow cryoprotectant to flow inside the cell in a manner that is sufficient but that does not dehydrate the cells too much and cause too large a volume change, as these have been related to cell demise, most likely originating from Meryman's minimum volume hypothesis and supported by others. Dehydration is complicated by the fact that different cell types have different permeability characteristics, and this may also differ within cell types, between cell stages. For example, dynamic changes were observed in the permeability of mouse oocytes during fertilization (23), and these most likely occur during human oocyte maturation as well. During controlled rate cooling, surface-to-volume ratio is an important factor, influencing the rate of dehydration and therefore the risk of IIF. This has been deemed responsible for the failures of the early attempts of freezing mature oocytes, due to their spherical shape and very large size. However, zygotes which have the same surface-to-volume ratio freeze relatively easily (31,32), as do activated unfertilized oocytes (Stachecki et al., unpublished data).

According to our equilibrium model (cooling in the presence of a cryoprotectant, PrOH), slow cooling must be continued at least to a temperature below –50°C before the cells are plunged into LN_2, in principle around –110°C when the cryoprotectant is PrOH, since in a water–PrOH system freezable water is still present above this temperature at atmospheric pressure. A similar line of argument applies when cryoprotectants other than PrOH are used. However, as amply demonstrated by experience, this is not necessary in practice where cells, including embryos, have been shown to be able to survive plunging from much higher temperatures even when conventional freezing procedures were employed. Admittedly, it is not difficult to think of reasons why the argument above is inaccurate, the main one being the assumption that phase equilibrium is maintained during cooling and rewarming. Even though the cell, theoretically, should remain highly permeable to water, even at reduced temperature below –20°C, calculating or observing this at such a low temperature is difficult at best. Phase equilibrium is rarely, if ever, the case especially in the lower temperature range where increasingly high viscosity of the liquid phase as the temperature is lowered interferes with the mobility of water molecules (22,33). This would suggest then that the majority of the water must be removed from the cell at a relatively high temperature, to avoid IIF, despite calculations of membrane permeability.

The conclusion above would help explain Fabbri's observations. Fabbri showed that addition of 0.3 M sucrose was better, in terms of oocyte survival, than 0.2 M and 0.1 M sucrose, and later studies have confirmed this (34). In terms of removal of water from the cell, the 0.3 M sucrose will dehydrate the cell more than 0.2 M sucrose. However, if we consider that after seeding the osmolarity increases dramatically far beyond what is necessary for adequate dehydration for cell survival, and if we believe that the cell can continue to lose water at and below the seeding temperature, based on the theories surrounding equilibrium cooling; then the effect of increasing the sucrose concentration from 0.1 to 0.2 M or 0.2 to 0.3 M (a paltry 100 mOsm) should, by fact, be negligible. Yet it is not. So how could this be possible? To start with, sucrose may have different effects. The presence of more sucrose would also reduce the total amount of ice formed and the concentration of salts during freezing (11,12), thus reducing osmotic effects from solutes, although the sucrose concentration itself would increase, thus balancing out the overall osmotic effect; so this line of reasoning does not go far in explaining the increase in survival. Sucrose will also raise the glass transition temperature and may therefore protect by arresting cell volume reduction at a higher temperature (35). Meryman has also presented evidence that sucrose can stimulate potassium efflux during cell swelling, reducing the chances of cell lysis due to post-hypertonic swelling (10). In addition, the added sucrose would provide osmotic support by its presence after thawing. These ideas do not seem to fully explain the observed increase in survival. However, if water permeability was severely inhibited (not in equilibrium), past a certain temperature (somewhere near the seeding temperature), this would increase the possibility of IIF occurring, despite slow cooling and the sufficiently high solute concentrations generated below the seeding temperature. This would help explain why increasing the sucrose concentration by 100 mOsm, at the start of the procedure, where the cell would be more likely to act as a perfect osmometer (36), could have such an impact on survival. It would then stand to reason that further increases in sucrose concentration beyond 0.3 M would lead to a further increase in survival rate, but only up to a point that the sucrose itself would become detrimental or have toxic effects. Thus, it is intriguing that there are no publications about this in the literature.

Although this chapter is about solution effects, let us take a minute and discuss IIF, as it has a potential role in cell death. We can see that, according to Figure 2 [derived from Leibo et al. 1975, and again in 1977 and 1978 (19,33,37)] survival decreases proportionally to increasing IIF, and is directly related to the cooling rate. From the figure, IIF is essentially zero, at least for mouse eggs in a solution of 1 M DMSO from which this graph is derived, when the cooling rate is below 1°C/min. This means that essentially no IIF can or does occur below this rate. Furthermore, Toner (38,39) showed that IIF by either homo- or heterogeneous nucleation is rendered ineffective by the addition of cryoprotectants. Standard slow-cooling protocols for mouse and human oocytes and embryos use 1.5 M cryoprotectant and cool at –0.3°C/min, well below this critical rate. Therefore, there should be virtually no IIF formed, and thus cellular demise must be explained by something other than IIF. However, as suggested above, if water permeability is inhibited and does not occur in equilibrium, as originally thought, and despite a slow cooling rate, IIF could pose a real problem, and removal of this excess water would, most likely, have to occur at a relatively high temperature.

If IIF does not occur using modern slow-cooling protocols containing cryoprotectants in excess of 1 M and cooling rates of less than 1°C/min, as suggested above, then, for arguments sake, solution effects would be responsible for most of the damage that occurs during equilibrium cooling. However, despite calculations and many experimental results, this too can be disputed. Most, if not all, cooling studies have used saline as their base solution, with sodium being the most abundant ion. The works of Lovelock, Meryman, Mazur, and company have used saline/sodium as their base solution in the majority, if not all, of their experiments. During cooling, with the addition of cryoprotectant(s) and after seeding, the solute concentration increases significantly causing the cell to shrink and dehydrate and the solute concentration to increase. Both of which, according to Meryman's minimal volume hypothesis and Mazur's work, could be detrimental to cell survival. An increased solute concentration could also lead to membrane alterations and solute uptake, as described above. The most likely and abundant candidate for uptake would be sodium. In his first paper on the subject, Stachecki theorized that because the energy-dependent Na^+/K^+ pump may be expected to become increasingly disabled as the temperature decreased, it is likely that the intracellular sodium load will have increased by the time the cell is transferred to LN_2. If so, this situation will still exist immediately after thawing and could be at least one reason for cellular demise. In the same paper, Stachecki et al. showed that replacing sodium chloride with choline chloride in the cryopreservation media, mouse oocytes survived at a significantly higher rate. They concluded that the high concentration of sodium ions, that is an inevitable consequence of using conventional cell handling media for cryopreservation, poses a specific threat to the cells, and that choline, which does not diffuse through the plasma membrane, is a promising candidate to replace sodium. By contrast, embryos have been successfully cryopreserved in sodium containing solutions for years, although survival rates rarely reach over 90%. Different membrane characteristics between oocytes and embryos (including activated eggs) can have a profound effect on their ability to survive freezing in a sodium environment. Thus, it would appear that sodium-loading either does not affect embryos or does not occur at a detrimental rate, as in oocytes.

In a previous paper, Toner et al. (40) showed that mouse zygotes exposed to high osmolar solutions (1200–2000 mOsm) of phosphate-buffered saline (PBS) supplemented with choline chloride remained intact after cooling to –40°C and rewarming (and some were able to continue developing in vitro), whereas PBS supplemented with sodium chloride, at any concentration, was lethal to the embryos subjected to the same cooling and rewarming regime. The osmotic effects of the choline, especially at a starting concentration of 1200 mOsm, would be increased significantly after seeding and by the time the temperature reached –40°C. By way of comparison, the salinity of an isotonic saline solution (around 300 mOsm), with no cryoprotectant, is about 5 molal at a temperature of –25°C (8). This would suggest that with a starting concentration of 1200 mOsm the salinity would be significantly higher than 5 molal. In Toner's experiment, zygotes were directly placed into the solutions causing a rapid exodus of water from the cell and cell shrinkage over 30% of their original volume. Additionally, there were no cryoprotectants to dilute out the osmotic effects and aid in preventing cell shrinkage, as are in conventional slow-cooling protocols. Even in a conventional protocol (26), after around three minutes of exposure to 1.5 mol/l PrOH in the presence of 0.3 M sucrose, oocyte volume shrinks drastically exceeding rapidly the 30% threshold excursion believed to be detrimental to cell viability. Collectively, these studies demonstrate that oocytes can shrink beyond what is thought to be detrimental to survival. It has also been shown that, in the absence of cooling, oocytes and embryos can withstand quite a lot of hypertonic exposure. Agca et al. (41) exposed bovine oocytes to PBS with increasing concentrations of sodium chloride to increase the osmolarity to supra-physiological concentrations (up to 4800 mOsm). They found that oocytes exposed to 2400 mOsm or lower, developed to the blastocyst stage, albeit at slightly reduced rates as compared with untreated controls. van Os and Zeilmaker (42) exposed mouse zygotes to solutions up to 3100 mOsm (by addition of NaCl) and rinsed them out without detriment to blastocyst formation. Although there are other studies, previously mentioned, that indicate hypertonic exposure is detrimental to survival following cooling, it would indeed be interesting to repeat those studies using choline-based and choline-supplemented media. It could well be that sodium toxicity during cooling was responsible for much of the observed cellular damage in many experiments and, to a much lesser extent, the solution effects of cell shrinkage and hypertonicity.

According to the theory of equilibrium cooling, cooling to lower temperatures will lead to the further dehydration of the cell, a lower chance for IIF, and an increase in solution effects and vice versa. In a subsequent study, Stachecki showed that the majority of mouse oocytes frozen in a choline-based medium remained intact after plunging into LN_2 from –20°C and that 32% were intact after plunging from –10°C. In the first successful mammalian embryo freezing experiments, plunge temperatures between –60°C and –120°C were employed and considered essential (43–45). Subsequently, it was shown that embryos could survive plunging from temperatures of –30°C to –36°C provided that rapid thawing was used (46). With plunge temperatures above –30°C, survival is greatly reduced when such protocols are used, particularly in the absence of extracellular dehydrators such as sucrose. Stachecki et al. showed that a conventional sodium-based freezing medium did not allow oocytes to survive and develop when a plunge temperature of –20°C was employed, whereas a choline-based medium did. There have been few previous studies that have examined embryo survival following plunging at a temperature greater than –30°C (47,48). Stachecki found that with choline the ability of mouse oocytes to survive cryopreservation was far greater than expected at plunge temperatures above –30°C, especially in the light of the literature on survival at these relatively high temperatures (48–50). Indeed Stachecki observed time and again that there was nearly 100% survival of mouse oocytes plunged at –20°C versus only 90% survival (on average with hundreds of oocytes frozen) when plunged at –33°C. At a higher plunge temperature, less dehydration would be expected and thus there would be less solution effects, as well as a greater chance for IIF. Besides the interesting observation that choline offered more protection than sodium was that lethal IIF was negligible even after cooling to only –20°C. This is in marked contrast to what has been expected or calculated in an equilibrium cooling system and helps support Figure 2 whereby cooling at a rate below 1°C/min and in the presence of a cryoprotectant IIF is virtually eliminated. By contrast, Stachecki found that cooling to –10°C led to only 60% survival at best, indicating the possibility of IIF damage to those oocytes. IIF will of course occur above a certain dehydration point, which, at least for the mouse egg, seems to be somewhere above –20°C. In an earlier study by Wood and Farrant (48), they froze mouse eight-cell embryos by plunging from –20°C to –25°C and found a similar result. Of the embryos plunged after cooling to only –20°C, 66% survived, whereas only 23% survived after being plunged from –25°C, suggesting that the further demise of embryos plunged at –25°C was due to something other than IIF. Of course, one must keep in mind that improper warming regimes can cause IIF to form, and the studies mentioned above used the most optimal warming protocol at their disposal.

Another interesting observation by Stachecki (51) was that although mouse eggs could survive plunging from –20°C, their development to the blastocyst stage was poorer (35% loss from two-cell to blastocyst, despite less solution effects) than those eggs plunged at –33°C where their loss in development was only 20% from two-cell to blastocyst stage. Here, at –33°C, one would expect greater solution effects, which could be the reason for the 10% reduction in survival, but development was 15% better overall. These results are difficult to explain in a complicated biological system, but it seems that IIF did not occur and solution effects were minimal, at least compared to the marked effect that choline-substitution had on mouse oocyte survival. The same was found for human oocytes (52).

In addition to the solution effects mentioned, the effect of penetrating cryoprotectants is also critical. As water is frozen out as ice, the cryoprotectant concentration will also dramatically increase. This will help chelate the remaining intracellular water, but at the same time could have detrimental toxic effects (53–55). The PrOH concentration of the extracellular liquid phase of media containing 1.5 M PrOH and 0.1 M sucrose at –33°C was approximately 47% to 48% (w/w) PrOH (51) which, according to Fahy (56), is high enough to suppress ice nucleation and achieve vitrification in a small sample and further supports the idea of negligible IIF in a standard equilibrium cooling system employing a cryoprotectant and cooling rate below 1°C/min. Cryoprotectant toxicity has been linked to the type of cryoprotectant, concentration, temperature, and duration of exposure. Cells can also be sensitized to injury in the presence of a cryoprotectant, in that more damage can occur once injury begins (7). Nonpenetrating cryoprotectants such as sucrose will reduce the concentration of penetrating cryoprotectant at any temperature, thereby reducing their toxic effects. This may, in part, help explain the added benefit of increasing the sucrose concentration to 0.3 M for freezing human oocytes (34,57), in addition to the benefit of added dehydration at a higher temperature, as previously discussed. Toxicity from the most commonly used penetrating cryoprotectants, although not thought to have significant effects on oocyte or embryo viability under conditions of conventional equilibrium cooling regimes, can have profound effects during rapid vitrification from room temperature (55,58,59). Of great interest is a recent paper by Wusteman et al. (60) that, for the first time, showed that choline actually reduces cryoprotectant toxicity (PrOH and DMSO). Although this study was done in smooth muscle cells and not oocytes, the findings are nonetheless relevant to this discussion.

Although this chapter is about solution effects during cooling, we must not overlook the fact that the solution effects during rewarming and rehydration are likely to be as important as other steps of the entire procedure, or perhaps even more so (3,7). Removal of the intracellular cryoprotectant and re-establishment of the original water content is an obvious source of osmotic stress. Therefore, it is possible to have 100% survival after cooling and then effectively kill the cells during the rewarming and rehydration phases. What complicates all the theories around successful storage of cells is that it is difficult to precisely identify when and where damage occurs. What happens, more often than not, is that cellular demise is associated with IIF or solution effects, ignoring the critical steps of rewarming and other potential damaging effects such as ion loading.

Solution effects, from all their different sources, although potentially detrimental to cell survival during and after equilibrium cooling, for the most part, do not fully explain the poor overall survival (typically under 80%) of mammalian oocytes. It seems logical, however, that other factors are responsible for cellular demise, and that solution effects from sodium ions, specifically, have led many into thinking that solution effects, in general, are problematic. Despite theoretical calculations of solute concentration and water permeability, experimental manipulations and practical application often give markedly different results. Based on the discussions herein, it was found that dehydration at a relatively high temperature in order to avoid IIF, in a medium with minimal sodium, while assuming that solution effects, although present, had a negligible negative impact, led to around 90% survival of human unfertilized metaphase II oocytes (52). Others have reported successful cryopreservation of human oocytes using a sodium-reduced medium (61–63). However, these studies used an earlier version of the sodium-depleted choline medium, and not the exact formulation Stachecki used to obtain 90% survival. For example, Boldt reported suboptimal survival rates (61% and 59%) using a sodium-depleted media that was either PBS or HEPES based (63). In the PBS group, fertilization after intracytoplasmic sperm injection was also inadequate (55.8%). Despite the poor survival rates and implantation rates, the authors concluded that sodium-depleted media is beneficial to the cryopreservation of human oocytes. Not only did Boldt et al. (2006) use a different version of the choline media reported by Stachecki et al. (52), they also modified the "optimized" protocol that Stachecki used. Boldt et al. changed the dehydration protocol as well as the freezing container used, which would alter the temperature of the oocytes at seeding as well as the warming rate obtained upon thawing, and they also changed the thawing protocol used by Stachecki, which was carefully derived to deliver an optimal warming and rehydration rate to aid in reducing cellular damage. These protocol modifications alone could well have produced the observed outcome, despite the benefits of using choline in the medium.

Stachecki, as well as others [Dr. Wilding, personal communication; (61)], using this media and protocol have obtained high survival rates. However, survival after rewarming and rehydration is only the first step in obtaining a viable pregnancy. It has been shown, not only in a choline-based system of slow-cooling human oocytes but in most others as well that embryo development and pregnancy rates are much lower than with nonfrozen oocytes (64,65). Solution effects, if we believe are of minimal impact on initial survival, may certainly be manifested later in the further development of the cells even if they are not immediately apparent upon rewarming. It is the solution effects and all the other effects of phase changes, pH modifications, solute precipitation, etc., which can occur during slow cooling and solidification, that combined could impact the overall survival of oocytes and cells.

With the current trend towards vitrification via plunging from room temperature, as opposed to vitrification after equilibrium cooling to –30°C or below, solution effects are markedly different. Here they come mainly from high cryoprotectant concentrations, rather than from a combination of high solute and cryoprotectant concentration. In a rapid cooled vitrification scenario concentrating solutes may only occur to a small degree and thus impact overall survival and development differently. Dehydration comes from the increased penetrating and nonpenetrating cryoprotectant concentrations used. These methods have their own hazards to be avoided, such as cryoprotectant toxicity, but at the same time minimize solution effects, at least those from solute concentration.

As mentioned in the introductory paragraph, the importance of having a general understanding of solution effects, and controlled-rate cooling, is so that we are better able to understand how to freeze cells in such a way that they can remain viable upon rewarming. Experimental results often differ from theory, as shown. When trying to rationalize our results, the complexity of a biological cell, such as the human oocyte, becomes apparent. It has been suggested herein that oocytes, as well as embryos, have a far greater tolerance to solution effects than originally thought and that different types of solution effects (i.e., sodium vs. choline ions) will produce vastly different results. Additional experiments geared towards identifying the limits of a cell's tolerance to various solution effects, where survival is not necessarily the goal, should help further our understanding of the cell's potential to survive storage at subzero temperatures.

ACKNOWLEDGEMENTS

I would like to kindly thank my coworkers who have helped me through the years including Steen Willadsen, Tim Schimmel, and Jacques Cohen. In addition I would like to thank our collaborators and their staff who have helped and supported my research: Santiago Munne, John Garrisi, Klaus Wiemer, Ricardo Azambuja, and Sandro Sabino.

REFERENCES

1. Luyet B. The vitrification of organic colloids and protoplasm. Biodynamica 1937; 1: 1–14.
2. Luyet B, Gehenio P. Life and death at low temperatures. In: Life and death at low temperatures. Normandy, MO: Biodynamica, 1940: 261.
3. Stachecki JJ, Cohen J, Schimmel T, Willadsen SM. Fetal development of mouse oocytes and zygotes cryopreserved in a nonconventional freezing medium. Cryobiology 2002; 44: 5–13.
4. Mazur P. Causes of injury in frozen and thawed cells. Fed Proc 1965; 24(Suppl 15): 175–82.
5. Mazur P. Freezing of living cells: mechanisms and implications. Am J of Physiol 1984; 247: C125–42.
6. Cocks F, Brower W. Phase diagram relationships in cryobiology. Cryobiology 1974; 11: 340–58.
7. Fahy GM. Theoretical considerations for oocyte cryopreservation by freezing. Reprod Biomed Online 2007; 14: 709–14.
8. Lovelock J. The haemolysis of human red blood cells by freezing and thawing. Biochem Biophys Acta 1953; 10: 414–26.
9. Meryman HT. Modified model for the mechanism of freezing injury in erythrocytes. Nature 1968; 218: 333–6.
10. Meryman H. The exceeding of a minimum tolerable cell volume in hypertonic suspension as a case of freezing injury. In: Wolstenholme G, O'Connor M, eds. The Frozen Cell. London: J & A Churchill, 1970: 51–67.
11. Farrant J, Woolgar AE. Cryoprotective additives and hypertonic hemolysis. Cryobiology 1970; 7: 56–60.
12. Farrant J, Woolgar AE. Human red cells under hypertonic conditions; a model system for investigating freezing damage. 2. Sucrose. Cryobiology 1972; 9: 16–21.
13. Mazur P. Cryobiology: the freezing of biological systems. Science 1970; 168: 939–49.
14. Stachecki JJ, Cohen J, Willadsen SM. Cryopreservation of unfertilized mouse oocytes: the effect of replacing sodium with choline in the freezing medium. Cryobiology 1998; 37: 346–54.
15. Meryman HT, Williams RT, Douglas MSJ. Freezing injury from "solution effects" and its preservation by natural or artificial cryopreservation. Cryobiology 1977; 14: 287–302.
16. Meryman HT. Osmotic stress as a mechanism of freezing injury. Cryobiology 1971; 8: 489–500.
17. Meryman HT. Freezing injury and its prevention in living cells. Annu Rev Biophys Bioeng 1974; 3: 341–63.
18. Mazur P. Kinetics of water loss from cells at subzero temperatures and the likelihood of intracellular freezing. Journal of general physiology 1963; 47: 347–69.
19. Leibo SP, McGrath JJ, Cravalho EG. Microscopic observation of intracellular ice formation in unfertilized mouse ova as a function of cooling rate. Cryobiology 1978; 15: 257–71.
20. Mazur P. Stopping biological time. The freezing of living cells. Annals of the New York Academy of Sciences 1988; 541: 514–31.
21. Chen C. Pregnancy after human oocyte cryopreservation. Lancet 1986; 1: 884–6.
22. Luyet B. Physical changes occurring in frozen solutions during rewarming and melting. In: Wolstenholme G, M OC, eds. The Frozen Cell. London: J & A Churchill, 1970: 27–50.
23. Jackowski S, Leibo SP, Mazur P. Glycerol permeabilities of fertilized and unfertilized mouse ova. J of Exp Zool 1980; 212: 329–41.

24. Leibo SP. Water permeability and its activation energy of fertilized and unfertilized mouse ova. J Membr Biol 1980; 53: 179–88.
25. Zawlodzka S, Takamatsu H. Osmotic injury of PC-3 cells by hypertonic NaCl solutions at temperatures above 0 degrees C. Cryobiology 2005; 50: 58–70. Epub 2004 Dec 15.
26. Paynter SJ, Borini A, Bianchi V, et al. Volume changes of mature human oocytes on exposure to cryoprotectant solutions used in slow cooling procedures. Hum Reprod 2005; 20: 1194–9. Epub 2005 Jan 21.
27. Kleinhans FW. Membrane permeability modeling: Kedem-Katchalsky vs a two-parameter formalism. Cryobiology 1998; 37: 271–89.
28. Newton H, Pegg DE, Barrass R, Gosden RG. Osmotically inactive volume, hydraulic conductivity, and permeability to dimethyl sulphoxide of human mature oocytes. J Reprod Fertil 1999; 117: 27–33.
29. Edashige K, Yamaji Y, Kleinhans FW, Kasai M. Artificial expression of aquaporin-3 improves the survival of mouse oocytes after cryopreservation. Biol Reprod 2003; 68: 87–94.
30. Yang D, Blohm P, Winslow L, Cramer L. A twin pregnancy after microinjection of human cryopreserved oocyte with a specially developed oocyte cryopreservation regime. Fertility and Sterility 1998; 70(Suppl 1):S239.
31. Senn A, Vozzi C, Chanson A, De Grandi P, Germond M. Prospective randomized study of two cryopreservation policies avoiding embryo selection: the pronucleate stage leads to a higher cumulative delivery rate than the early cleavage stage. Fertil Steril 2000; 74: 946–52.
32. Bianchi V, Coticchio G, Distratis V, et al. Differential sucrose concentration during dehydration (0.2 mol/l) and rehydration (0.3 mol/l) increases the implantation rate of frozen human oocytes. Reprod Biomed Online 2007; 14: 64–71.
33. Leibo SP. Fundamental cryobiology of mouse ova and embryos. In: The Freezing of mammalian embryos. Amsterdam: Elsevier Scientific Publishing Co, 1977: 69–96.
34. Fabbri R, Porcu E, Marsella T, et al. Human oocyte cryopreservation: new perspectives regarding oocyte survival. Hum Reprod 2001; 16: 411–16.
35. Takahashi T, Hirsh A, Erbe E, Williams RJ. Mechanism of cryoprotection by extracellular polymeric solutes. Biophys J 1988; 54: 509–18.
36. Mazur P, Schneider U. Osmotic responses of preimplantation mouse and bovine embryos and their cryobiological implications. Cell Biophysics 1986; 8: 259–85.
37. Leibo SP, McGrath JJ, Cravalho EG. Microscopic observation of intracellular ice formation in mouse ova as a function of cooling rate. Cryobiology 1975; 12: 579.
38. Toner M, Cravalho EG, Armant DR. Water transport and estimated transmembrane potential during freezing of mouse oocytes. Journal of Membrane Biology 1990; 115: 261–72.
39. Toner M, Cravalho EG, Karel M, Armant DR. Cryomicroscopic analysis of intracellular ice formation during freezing of mouse oocytes without cryoadditives. Cryobiology 1991; 28: 55–71.
40. Toner M, Cravalho EG, Stachecki J, et al. Nonequilibrium freezing of one-cell mouse embryos. Membrane integrity and developmental potential. Biophys J 1993; 64: 1908–21.
41. Agca Y, Liu J, Rutledge JJ, Critser ES, Critser JK. Effect of osmotic stress on the developmental competence of germinal vesicle and metaphase II stage bovine cumulus oocyte complexes and its relevance to cryopreservation. Mol Reprod Dev 2000; 55: 212–9.
42. van Os HC, Zeilmaker GH. Volumetric behaviour and survival of mouse zygotes and embryos in hyperosmotic media. Hum Reprod 1986; 1: 95–8.
43. Whittingham DG, Leibo SP, Mazur P. Survival of mouse embryos frozen to –196 degrees and –269 degrees C. Science 1972; 178: 411–4.
44. Willadsen SM, Polge C, Rowson LE, Moor RM. Deep freezing of sheep embryos. J Reprod Fertil 1976; 46: 151–4.
45. Wilmut I, Rowson LE. The successful low-temperature preservation of mouse and cow embryos. J Reprod Fertil 1973; 33: 352–3.
46. Willadsen SM. Factors affecting the survival of sheep embryos during-freezing and thawing. In: The freezing of mammalian embryos. Amsterdam: North-Holland Publishing Co; 1977: 175–201.
47. Van den Abbeel E, Van der Elst j, Van Steirteghem AC. The effect of temperature at which slow cooling is terminated and of thawing rate on the survival of one-cell mouse embryos frozen in dimethyl sulfoxide or 1,2-propanediol solutions. Cryobiology 1994; 31: 423–33.
48. Wood MJ, Farrant J. Preservation of mouse embryos by two-step freezing. Cryobiology 1980; 17: 178–80.
49. Friedler S, Giudice LC, Lamb EJ. Cryopreservation of embryos and ova. Fertil Steril 1988; 49: 743–64.
50. Whittingham DG, Wood M, Farrant J, Lee H, Halsey JA. Survival of frozen mouse embryos after rapid thawing from –196 degrees C. J Reprod Fertil 1979; 56: 11–21.

51. Stachecki JJ, Willadsen SM. Cryopreservation of mouse oocytes using a medium with low sodium content: effect of plunge temperature. Cryobiology 2000; 40: 4–12.
52. Stachecki JJ, Cohen J. An overview of oocyte cryopreservation. Reprod Biomed Online 2004; 9: 152–63.
53. Rall WF. Factors affecting the survival of mouse embryos cryopreserved by vitrification. Cryobiology 1987; 24: 387–402.
54. Wood MJ, Barros C, Candy CJ, et al. High rates of survival and fertilization of mouse and hamster oocytes after vitrification in dimethylsulphoxide. Biology of Reproduction 1993; 49: 489–95.
55. Hotamisligil S, Toner M, Powers RD. Changes in membrane integrity, cytoskeletal structure, and developmental potential of murine oocytes after vitrification in ethylene glycol. Biol of Reprod 1996; 55: 161–8.
56. Fahy G. Vitrification. In: McGrath J, Diller K, eds. Low temperature biotechnology: Emerging applications and engineering contributions. New York: American Society of Mechanical Engineering, 1988: 113–146.
57. Coticchio G, De Santis L, Rossi G, et al. Sucrose concentration influences the rate of human oocytes with normal spindle and chromosome configurations after slow-cooling cryopreservation*. Hum Reprod 2006; 21: 1771–6. Epub 2006 Mar 20.
58. Mukaida T, Wada S, Takahashi K, et al. Vitrification of human embryos based on the assessment of suitable conditions for 8-cell mouse embryos. Hum Reprod 1998; 13: 2874–9.
59. Fahy GM, Wowk B, Wu J, Paynter S. Improved vitrification solutions based on the predictability of vitrification solution toxicity. Cryobiology 2004; 48: 22–35.
60. Wusteman M, Rauen U, Simmonds J, Hunds N, Pegg DE. Reduction of cryoprotectant toxicity in cells in suspension by use of a sodium-free vehicle solution. Cryobiology 2008; 56: 72–9. Epub 2007 Nov 17.
61. Goud A, Goud P, Qian C, et al. Cryopreservation of human germinal vesicle stage and in vitro matured M II oocytes: influence of cryopreservation media on the survival, fertilization, and early cleavage divisions. Fertil Steril 2000; 74: 487–94.
62. Quintans CJ, Donaldson MJ, Bertolino MV, Pasqualini RS. Birth of two babies using oocytes that were cryopreserved in a choline-based freezing medium. Hum Reprod 2002; 17: 3149–52.
63. Boldt J, Tidswell N, Sayers A, Kilani R, Cline D. Human oocyte cryopreservation: 5-year experience with a sodium-depleted slow freezing method. Reprod Biomed Online 2006; 13: 96–100.
64. Coticchio G, Bonu MA, Bianchi V, Flamigni C, Borini A. Criteria to assess human oocyte quality after cryopreservation. Reprod Biomed Online 2005; 11: 421–7.
65. Petracco A, Azambuja R, Okada L, et al. Comparison of embryo quality between sibling embryos originating from frozen or fresh oocytes. Reprod Biomed Online 2006; 13: 497–503.

13 Cryopreservation and the Cytoskeleton of the Human Oocyte

Giovanni Coticchio and Maria Antonietta Bonu

Tecnobios Procreazione, Centre for Reproductive Health, Bologna, Italy

INTRODUCTION

In eukaryotic cells, variations in shape, active movement, and mechanics of cell division depend on the cytoskeleton, a sophisticated array of protein filaments that extends throughout the cytoplasm. The cytoskeleton also directs and provides the machinery for the intracellular transport of organelles. These functions are assured by three types of filaments, namely microtubules, actin filaments, and intermediate filaments. Each filament, formed from a different protein monomer, can be assembled in different structures whose functions, such as coordinated polymerization and depolymerization in time and space or interaction to one another and other cell components, are finely regulated by a wide range of associated proteins. In the context of oocyte physiology, the microtubules and actin filaments assist the cytoplasmic choreography of the meiotic process, and, in the course of oogenesis, contributes to an endowment of maternal-derived molecules and spatial clues that are essential for the development of the embryo (1,2). An example of the importance of the cytoskeleton for the function of the oocyte is represented by the involvement of microtubules—in the form of the metaphase I (MI) and metaphase II (MII) spindles—and actin filaments in the sequential segregation of homologous chromosomes and sister chromatids through two successive and highly asymmetric cell divisions that coincide with the emission of the first (PBI) and second (PBII) polar body. This process ensures that the zygote receives a haploid set of maternal chromosomes. The organization and dynamics of microtubules and actin filaments are known be affected by a variety of intrinsic and extrinsic factors (3–5). In fact, it is paradoxical that much of current understanding of microtubule dynamics during cell division in somatic and germ cells is a consequence of studies on the cold sensitivity of these structures. Damage to the MII spindle has emerged from studies that have tested the effects of physical–chemical conditions which are imposed, often unknowingly, during the return journey to and from physiological temperatures to cryogenic storage in liquid nitrogen (–196°C). This has generated the credence that oocytes cannot be safely cryopreserved. In reality, the question as to whether the oocyte cytoskeleton is subjected to damage after cryostorage does not imply a simple answer. Many factors, such as specific conditions imposed by the diverse cryopreservation protocols or nature of the biological material (differences in species, maturation stage, and intrinsic quality), can act independently or through complex interactions, influencing the response of the cytoskeleton to cryopreservation conditions, in a fashion that is not always understood or even recognized. For example, relatively minor differences in the degree of dehydration of the cytoplasm generated by exposure to cryoprotective agents (CPAs) before controlled rate slow cooling (CRSC) cryopreservation can influence the proportion of frozen-thawed oocyte with a normal MII spindle (6). Current evidence, which has expanded considerably over the last few years, collectively does not appear to be entirely consistent, probably as a result of differences in the protocols subjected to scrutiny and/or criteria and methods of analysis. It is perceived, however, that specific sets of cryopreservation conditions may generate rather unique influences that may be not necessarily reproduced under other conditions, making arduous the attempt to draw general conclusions. In this chapter, evidence on the effects of cryopreservation on the actin and microtubular structures of the fully grown human oocytes will be discussed, without taking in consideration the countless number of regulatory factors on which virtually no information is presently available.

MICROFILAMENTS

Organization and Function of Microfilaments in the Mammalian Oocyte

Microfilament fibers are formed from only one type of structure, the polymerized helical form of actin monomers. Actin filaments can acquire multiple shapes and functions in different cell types. Various nucleation and assembly factors, such a Arp2/3 and formins (7,8), can in fact establish whether actin filaments will organize into bundles and networks interacting with membranes or other cell components, or will provide the scaffolding for lamellipodia, microvilli, and other specialized cell protrusions. In the mature oocyte, actin filaments form a sub-oolemmal network, which is profoundly involved in the process of PB extrusion (see below), similar to what occurs in other cell types for less specialized forms of cytokinesis. Actin filaments also appear to play a role in other rather different functional contexts, during oocyte maturation, and the determination of oocyte asymmetry, which is established by the cortical localization of the meiotic spindle. In the mouse, during meiotic maturation in vitro, the centrally positioned germinal vesicle (GV) breaks down and the chromosomes, which in the meantime have acquired a condensed conformation, migrate to the periphery. Here, the newly formed MI spindle, which has accompanied chromosomes in their journey to the cortex, will dictate a highly asymmetric cell division and the extrusion of the PBI. Conditions which hinder the migration of the chromosomes toward the cortex have important developmental implications, because the MI spindle forms and remains in a more central position, determining the loss of a large mass of cytoplasm with the emission of the PBI (9). In the mouse, MI spindle migration and anchoring to the cortex relies on the action of actin microfilaments. The fine details of this process are not known, but recently it has emerged that myosin-II is not implicated, while a chromosome–actin interaction is required (10). In particular, relocation of chromosome to the cortex is initially associated to a cloud of actin filaments whose nucleation is mediated by the activity of Fmn2, a formin-family protein. Active chromosome movements appear to depend on actin turnover and an asymmetric distribution of actin, with a polarized accumulation of microfilaments behind the chromosomes/spindle during their relocation to the cortex. The universality of the importance of actin-mediated chromosome/spindle translocation in the unfolding of meiosis is challenged, however, by other lines of evidence. In fact, it seems that in preovulatory oocytes of other mammals (including humans), the GV acquires a peripheral location before it breaks down and therefore the chromosome/spindle complex does not require repositioning (2). It is possible that in mouse oocytes the central position of the GV, and the consequent necessity to move the chromosomes/spindle complex to the periphery after GV break down, may derive from a disanchoring from the cortex as an effect of a particular sensitivity to culture conditions (7,11). Irrespective of whether the chromosomes/MI spindle complex forms in the periphery or is vehicled from the center, in mouse oocytes, its presence induced a polarization of the cortex, which is critical for the correct emission of PBI and PBII. Actin filaments, initially distributed uniformly throughout the cortex, accumulate beneath the oolemma in correspondence with the chromosomes/MI spindle position (12,13). This is believed to limit the extension of the cleavage furrow, which determines the extrusion of PBI to the differentiated cortical area overlying the chromosomes, thereby reducing the loss of cytoplasm. A similar mechanism is thought to operate in coincidence with the emission of PBII.

Does Cryopreservation Affect the Oocyte Actin Organization and Function?

From the above scenario, it emerges that actin filaments are actively involved in a multiplicity of phases of the meiotic process and therefore represent a possible cytoskeletal target of cryopreservation-induced damage. Studies on this subject in human oocytes are rare, but data of in vitro maturation in frozen-thawed oocytes are compatible with a possible involvement of microfilaments in cell alterations found after freezing and thawing. For example, in comparison to fresh controls, in cryopreserved GV-stage oocytes recovered from unstimulated ovaries, the maturation rate (i.e., the proportion of MII oocytes after 48 hours of culture) may be reduced (56.9% vs. 76.3%, respectively) (14). Clearly, an inhibitory effect on the process of meiotic maturation may involve different aspects or stages of the cell cycle machinery, but it is possible that, under certain cryopreservation conditions, disarrangement of cortical actin prevents extrusion of PBI. Consistent with this hypothesis is the evidence that mature mouse oocytes frozen-thawed under suboptimal conditions may show an increased tendency to retain the PBII after

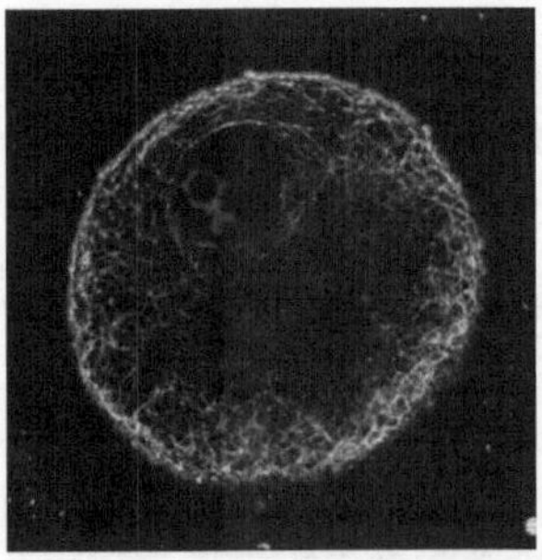

Figure 1 Human germinal vesicle (GV)-stage oocyte showing an interphase-like three-dimensional network of microtubules (in green) spanning across the cytoplasm. Within the GV, an uninterrupted ring of heterochromatin is visible around the nucleolus. The image is derived from a collaborative study between Tecnobios Procreazione and the laboratory of Prof. D. Albertini. *See Color Plates on Page xviii.*

insemination (11.8% vs. 1.3% of controls) (15). In this species, microfilaments appear actually sensitive to conditions that may occur during cryopreservation, as indicated by the disruption of cortical microfilaments upon treatment with 1.5 M dimethylsulfoxide (DMSO) at 37°C (16). However, a reduction in the effect of the CPA was observed when the treatment was performed at lower temperatures. Alterations in the cortical arrangement of microfilaments were also reported following exposure of mouse oocyte for a prolonged period (>5 minutes) at high concentrations (>4 M) of ethylene glycol (EG) at room temperature. In view of this, perhaps it is not incidental that the most successful CRSC protocols developed for mouse oocytes require treatment with 1.5 M DMSO a 4°C (17). Temperature- and concentration-dependent influence of CPA exposure on actin filaments have also been demonstrated in the rhesus monkey. GV-stage oocytes of this species treated with a medium supplemented with 1.0 or 2.0 M glycerol at room temperature lose the typical cortical layer of actin filaments, while this effect is prevented when CPA concentration is lowered to 0.5 M or temperature is reduced to 0°C (18). In human oocytes, the condition of cortical filaments after cryopreservation has been described only rarely. Nevertheless, in comparison to fresh controls, no alterations in frequency (around 80%) and quality of actin staining was found in oocytes frozen with CRSC protocols based on exposure to 1.5 M EG or 1,2-propanediol (PrOH) at room temperature (19). Another sign that in mature human oocytes cryopreserved by CRSC or vitrification microfilaments are not significantly affected is provided by the fact that no increase in the tendency to retain the PBII and give rise to three-pronuclear fertilization is found after intracytoplasmic sperm microinjection (20,21). Along the same lines, the shape and size of PBII, features that depend on the functioning of the cortical actin network, have never been reported to be abnormal in cryopreserved oocytes. It is not clear why, in comparison to the mouse, human oocytes appear less sensitive to disruption of the mechanism of PB extrusion on exposure to freezing conditions. Perhaps, this may reflect differences in organization of actin filaments in the oocyte cortex, as suggested by the fact that in humans the cortical actin is found as an uninterrupted and regular layer beneath the oolemma (19), and the process of polarization, which occurs in the mouse in coincidence with the subcortical positioning of the meiotic spindle, is not observed (Fig. 1). Another element that is suggestive of the preservation of the function of actin filaments in cryopreserved oocytes is the observation that, after sperm microinjection and fertilization, the first mitotic divisions generate cleaving embryos with blastomeres of normal size and shape in a proportion comparable to unfrozen control (21,22). This requires that the endowment of actin present in the unfertilized oocyte be passed on to the ensuing embryo unaltered in quantity and functional ability.

MICROTUBULES

Microtubule Dynamics During Oocyte Maturation

Microtubules are composed of heterodimers formed by α- and β-tubulin [see Ref. (23) for a review]. Individual units of this heterodimer are connected in a head-to-tail fashion to form protofilaments. In their turn, 13 of these protofilaments can juxtapose side-by-side and slightly out of phase to each other, to form the typical cylindrical and helical microtubular structure. Numerous functions fulfilled by microtubules require dynamic assembly (polymerization) and disassembly (depolymerization). This is driven by binding, hydrolysis, and exchange of guanosine triphosphate (GTP) on the β-tubulin monomer. Polymerization is initiated from free heterodimers loaded with GTP. Organization of heterodimers into protofilaments and cylinders of the microtubular wall is followed by hydrolysis of GTP into guanosine diphosphate and inorganic phosphate. The conformational change that accompanies GTP hydrolysis makes heterodimer interactions at the growing end of the microtubule rather unstable. Stabilization, however, can intervene as an effect of the creation of a terminal "cap" of tubulin-GTP. This gives rise to a metastable, blunt-ended microtubule that may remain temporarily stable, disassemble, or progress to further growth. This general machinery is finely controlled by a myriad of regulatory factors that promote microtubule polymerization, stabilization, or destabilization.

The dynamic character of microtubules can be seen at work in the maturing oocyte. Shortly before meiotic maturation is resumed at ovulation, in the GV-stage oocyte microtubules are organized in a dense and relatively static network of fibers which spans across the whole cytoplasm (24) (Fig. 2). As soon as the GV breaks down (as discussed, an event that occurs peripherally in the human oocyte), the microtubular array becomes more dynamic, establishes contacts with the chromosomes, which in the meantime have condensed, and organizes into the MI spindle. This is a relatively short-lived structure that expedites chromosome segregation during the first meiotic division and is afterwards remodeled to give origin to the MII spindle in the mature oocyte. While being highly dynamic in its basic components (the microtubules), to assure chromosome congression and continued alignment on the metaphase plate, the MII spindle as a whole is required to persist undisturbed in a bipolar configuration for a few hours (25), that is, during the temporal window in which the oocyte expresses its highest ability to develop into a viable embryo (26). Once fertilization is commenced, prompted by penetration of the spermatozoon, the MII spindle guides chromatid segregation concomitantly with the extrusion of the PBII, thereby ensuring completion of the meiotic process. The MII spindle plays other important roles in oocyte physiology. In a fashion similar to the first meiotic division (1,9), its sub-oolemmal position dictates a highly asymmetric second meiotic division and the extrusion of a small PBII, thereby minimizing the loss of organelles and molecules needed for the earliest stages of development. Spindle position also appears to play a critical developmental role, acting as a landmark for the establishment of the animal–vegetal axis and thereby contributing to set the geometrical coordinates of the embryo (2). Also, the MII spindle appears to be involved in another process not closely related to chromosome segregation and PBII extrusion, as suggested by the fact that it represents a preferential localization site of the src-family protein tyrosine kinases, proteins essential for late stages of fertilization (27).

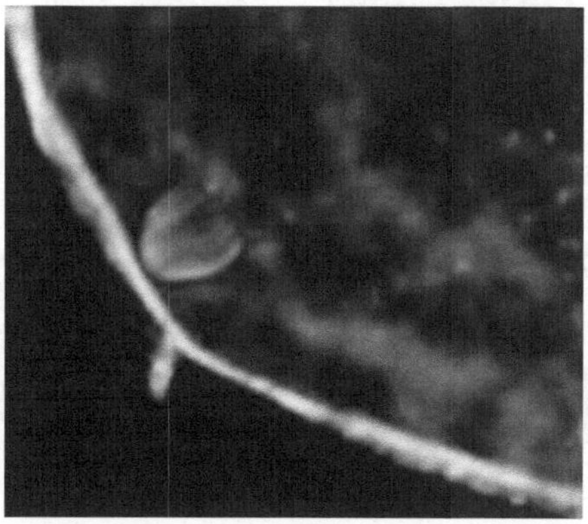

Figure 2 Detail of a human MII oocyte. Actin filaments (in red) are organized in a thin uninterrupted layer beneath the oolemma and do not exhibit a restricted localization coincident with the position of the MII spindle that instead is found in mouse oocytes. The end of the MII spindle oriented toward the center of the oocyte is clearly disorganized, having lost the typical polar convergence of microtubule fibers. The image is derived from a collaborative study between Tecnobios Procreazione and the laboratory of Prof. D. Albertini. *See Color Plates on Page xix.*

Effects of Cryopreservation on Microtubule and Chromosome Organization

As a highly dynamic and sensitive structure that depends on the finely regulated process of tubulin polymerization and depolymerization, it is not surprising that the MII spindle may be affected by diverse factors. Disparate spindle configurations may be found in fresh mature oocytes or in oocytes derived from various forms of manipulation (Fig. 3). The classical bipolar organization with chromosomes aligned on the metaphase plate is the one that is believed to assure the highest chances of flawless segregation of sister chromatids, while structures that diverge from this scheme are clearly at higher risk of meiotic errors (5). Spindle perturbances

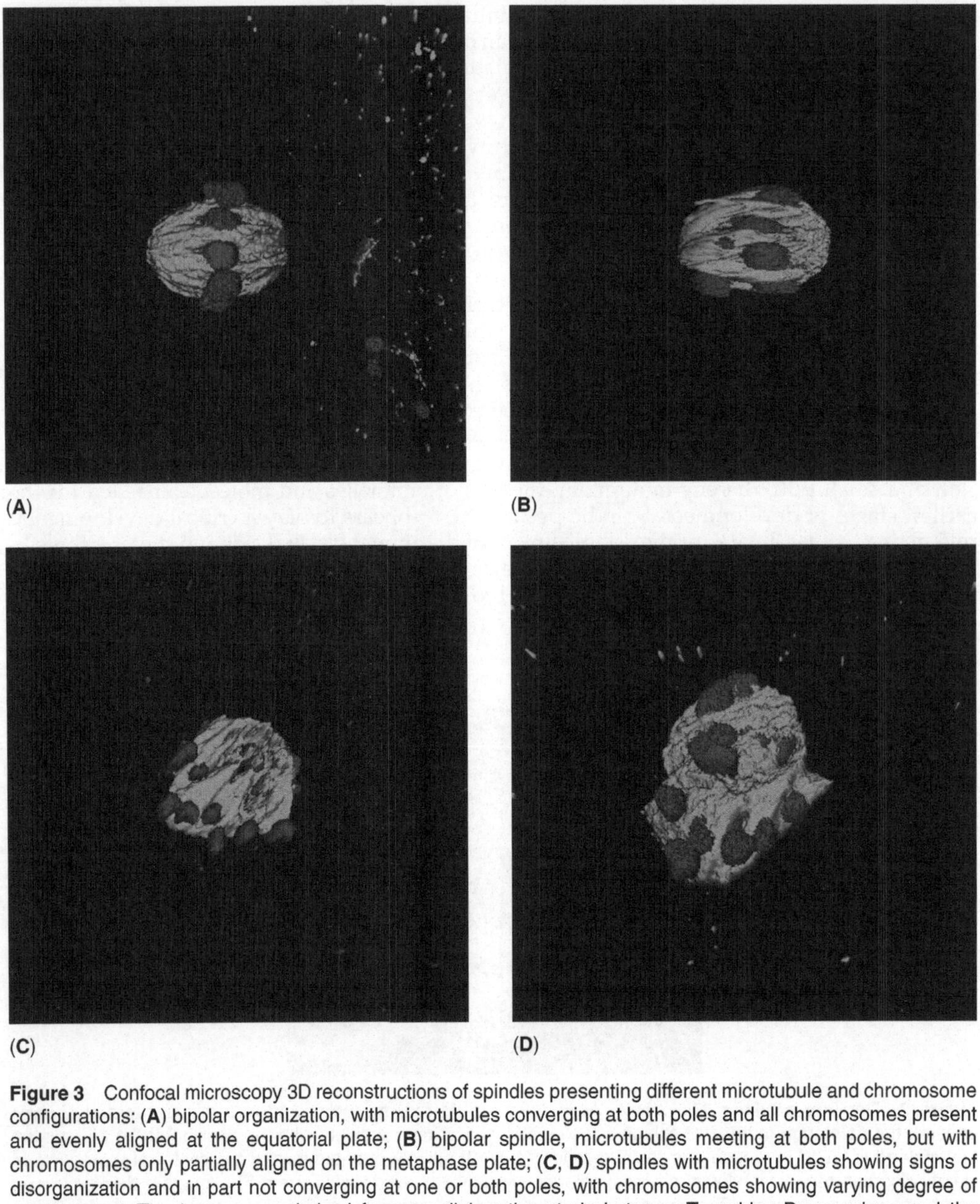

(A) (B) (C) (D)

Figure 3 Confocal microscopy 3D reconstructions of spindles presenting different microtubule and chromosome configurations: (**A**) bipolar organization, with microtubules converging at both poles and all chromosomes present and evenly aligned at the equatorial plate; (**B**) bipolar spindle, microtubules meeting at both poles, but with chromosomes only partially aligned on the metaphase plate; (**C**, **D**) spindles with microtubules showing signs of disorganization and in part not converging at one or both poles, with chromosomes showing varying degree of misalignment. The images are derived from a collaborative study between Tecnobios Procreazione and the laboratory of Prof. D. Albertini. *See Color Plates on Page xix.*

can have intrinsic origins, as indicated by the fact that the large majority of oocytes from older women (40–45 years) display microtubular abnormalities and chromosome displacement, as opposed to oocytes of younger women (20–25 years) in which those anomalies are only sporadically represented (5). These anomalies are believed to be the cellular basis of the increase in aneuploidies that affects the oocytes (and the resulting embryos) of older women. Inadequate temperatures may represent an extrinsic source of spindle damage and, likewise the previous case, meiotic errors. In a fundamental study conducted two decades ago, using epifluorescence microscopy, Pickering et al. (3) observed that in mature human oocytes cooled at room temperature for 10 to 30 minutes, the MII spindle suffers major structural alterations, including reduction in size, microtubular disorganization or disappearance, and chromosome dispersal. Upon rewarming to physiological temperature (37°C), some of these spindle anomalies may disappear, but others persist, predisposing oocytes to developmental failure. This bears obvious implications for the cryopreservation of these cells, a process that involves thermal transitions to and from liquid nitrogen temperature. In addition, low temperature preservation techniques, CRSC and vitrification, require the use of cryoprotectants whose action may affect spindle and chromosome configuration, as a consequence of acute osmotic stress and/or direct interference with the dynamics of tubulin polymerization. Therefore, analysis of the MII spindle has represented an intensely studied subject in both animal models and humans since the dawn of oocyte cryopreservation history.

Classically, the meiotic spindle may be visualized through immunofluorescence microscopy, which offers reliable and detailed information on microtubular structures. Chromosomes may also be visualized by using fluorescent DNA-specific dyes. Such a technique requires methods of preparation (fixation and staining) which are incompatible with the preservation of cell viability. Therefore, it cannot be adopted for oocyte selection in an in vitro fertilization (IVF) context, but is crucial for an objective assessment of oocyte quality after cryopreservation. Using the mouse model, Johnson and Pickering were among the first to investigate the effects of cryopreservation conditions on the oocyte MII spindle (28). They reported that exposure to DMSO initially generates the formation of microtubular asters around the multiple microtubule organizing centers normally found in the oocytes of this species. They also found that a prolonged treatment with the same CPA causes spindle disassembly and chromosome disarrangement. Together with the observation of the same authors that described how lower suboptimal temperature cause irreversible spindle depolymerization in human oocytes (3), this evidence contributed to generate the belief that cryopreservation was incompatible with the maintenance of an intact MII spindle. This appeared as a confirmation of the impression that oocytes were altogether particularly susceptible to cryopreservation, as suggested by the low survival rates experienced by those who first attempted to cryopreserve human oocytes in a clinical context using the CRSC approach (29). Data that emerged a few years later from the work of Gook et al. came as a surprise. By adopting a CRSC protocol involving 1.5 M PrOH and 0.1 M sucrose as a freezing mixture, these authors showed that, after freezing and thawing, 60% of human oocytes could be found with normal spindle and chromosome configuration (30). A 60% rate may appear relatively low, but it should be noted that in fresh controls it is not infrequent that the proportion of oocytes with a normal spindle is around 70% (6,31). From the studies of Gook et al. it was also evident that in frozen-thawed oocytes in which the spindle is absent or aberrant no stray chromosomes are observed (32). In other words, disassembly of the spindle, when it occurs, does not lead to chromosome dispersal throughout the cytoplasm. More recently, similar findings were reported by Zenzes et al. (33). This has important implications because, in the hypothesis of a possible spindle repolymerization after freezing-thawing, if chromosomes remain closely adjacent to each other, reestablishment of physical interactions with the spindle fibers and organization on the equatorial plate would seem more feasible in comparison to a situation in which they would need to be recaptured from mutually distant cytoplasmic locations and redeployed in a compact array. Gook et al. also provided evidence of normal sets of chromosomes in frozen-thawed oocytes undergoing fertilization, a sign that the second meiotic division had occurred undisturbed under the conditions tested by these authors (32). Confirmation that cryopreservation may be compatible with the maintenance of an intact spindle came from experiments in the mouse, in which it was described that, after CRSC or ultra-rapid freezing, a large number of oocytes (72% and 84%, respectively) with a normal, barrel-shaped MII spindle could be recovered after two hours of incubation from thawing (34). However, these findings were not considered conclusive to rule out damage to the MII spindle after cryopreservation. Some investigators have tested the hypothesis that cryopreservation-induced damage to the cytoskeletal apparatus may be circumvented if oocytes are stored at the GV stage, when microtubules are not organized in the MII spindle, and matured in vitro after thawing, therefore avoiding direct exposure of the MII spindle to cryopreservation conditions. Using supernumerary GV-stage oocytes from IVF patients, Baka et al. (35) found that, after CRSC cryopreservation with a protocol based on 1.5 M PrOH and 0.1 M sucrose as CPAs, in vitro maturation generated mature oocytes whose frequency of formation of normally organized spindles was not statistically different from in vitro or in vivo matured fresh controls (81%, 83%, and 89%, respectively). These results show that cryopreservation at the GV stage has some potential. However, this strategy—while attempting to resolve a problem, the preservation of the MII spindle—in fact, introduces another perhaps more important complication. In fact, it is recognized that in vitro maturation systems have not yet been developed to a stage capable to generate oocytes with a developmental potential comparable to fresh material. Other studies on GV stage oocytes have not confirmed the findings of Baka et al. In an investigation performed with oocytes retrieved from unstimulated patients, Park et al. (36) observed that cryopreservation at the GV stage has a detrimental effect on the rate of in vitro matured oocytes with normal chromosome and spindle configuration, in comparison with nonfrozen in vitro matured material. Similar results were obtained by Boiso et al. (37) also. As discussed above, microtubules are involved in highly dynamic processes during the whole process of meiotic maturation, shifting from an interphase-like intricate and diffuse array to the delicate and compact organization of the MI and MII spindles. Changes of the same magnitude are also likely to involve a countless number of cytoskeletal regulatory factors. Therefore, it is plausible that cryopreservation conditions, when applied at the GV stage, may alter some cytoskeletal elements and compromise indirectly, but significantly, phases of microtubular dynamics occurring at later stages of the maturation process. Another difficulty involved in the in vitro maturation approach derives from the fact that cryopreservation, apart from potentially affecting the oocyte proper, has major implications for the cumulus cells that are physically and functionally associated to the oocyte and play a crucial role for the acquisition of developmental potential (see chap. 9). Paradoxically, storage conditions that are compatible with oocyte survival after thawing, nevertheless, can destroy the intercellular communication between the germinal and somatic components within the cumulus cell–oocyte complex and even be lethal to the cumulus cells, compromising the oocyte ability to develop to the blastocyst stage (38). In addition, in vitro maturation technique per se can represent a possible source of perturbation to the normal process of formation of the MII spindle. For example, Cecklenak et al. (39) proved that the adoption of an inappropriate culture medium for in vitro maturation can severely affect the proportion of in vitro matured oocytes displaying a normal chromosome and spindle configuration. In conclusion, it seems that the option to cryopreserve oocytes at the GV stage to avoid damage to the MII spindle generates more problems than those that is hoped to solve, at least for the time being.

Probably for this reasons, more recent studies based on the hypothesis of spindle disruption after cryopreservation have again focused on mature oocytes. Mullen et al. (40) transiently exposed mature human oocytes to anisosmotic conditions generated by buffer solutions including different concentrations of sucrose, with the objective to simulate the osmotic stress occurring during dehydration and rehydration and ascertain possible consequences to the MII spindle structure. The authors found that both hypo- and hyperosmotic conditions caused a decrease in the proportion of oocytes with a normal spindle, in extreme cases with a total loss of a recognizable microtubular structure. However, such anisosmotic conditions were obtained in the absence of penetrating CPAs (DMSO, EG, and PrOH), molecules which interact with tubulin influencing microtubule dynamics and under certain circumstances inducing microtubule polymerization or stabilization (28,41). Our group has been directly involved in investigations aiming to ascertain the conditions of the MII spindle in human oocytes after cryopreservation. In particular, in a study published in 2006 (6), we compared the effect of two distinct CRSC protocols differing in the concentration of sucrose in the freezing solution. In control oocytes, we ascertained tubulin staining in about 90% of oocytes, although only in a smaller fraction (73.1% of the total) we observed an orderly bipolar array of microtubules and a regular chromosome distribution. These figures are comparable to those reported by Stachecki et al. (31). The fraction of oocytes with abnormal spindles varies depending on age (5), a circumstance that suggested us to limit our observations to oocytes donated by patients younger than 36 years. In our experience, cryopreservation with a method in which sucrose concentration in the loading solution was 0.3 M did not essentially affect the regularity of spindle and chromosome arrangement. By contrast, the application of a protocol involving a lower sucrose concentration (0.1 M) was associated to a limited but statistically significant reduction (50.8%) in the rate of oocytes with normal spindle and chromosome configurations. Cobo et al. (42) have provided evidence that the chromosome complement of embryos derived from cryopreserved oocytes is unaffected, a circumstance that requires as a prerequisite that the MII spindle organization is unaltered after cryopreservation. In particular, using probes for chromosomes 13, 18, 21, X, and Y, these authors conducted a preimplantation genetic diagnosis assessment, ascertaining that embryos developed from oocytes frozen with a 1.5 M PrOH and 0.2 M sucrose protocol had a rate of aneuploidy similar to the one found in embryos from fresh oocytes (28% and 26%, respectively). However, it should be noticed that only 21 embryos from frozen oocytes were examined in this study. Evidence converging with our experience has been described also by Stachecki et al. (31), whose data suggest that the proportion of oocytes with barrel-shaped spindles and chromosomes regularly aligned on the spindle equatorial plane is unchanged after cryopreservation (76.7% vs. 69.7% in fresh and frozen oocytes, respectively). It is interesting to note that these authors used a CRSC protocol involving the replacement of sodium with the less toxic cation choline, to reduce the so-called solution effect, consisting of the increase in solute concentration occurring during extracellular ice formation. It appears then that the preservation of an intact spindle after CRSC is a goal that may be achieved under certain conditions. However, as a confirmation of the impression that the effects of cryopreservation on the microtubular structure depend strictly on the protocol employed, another study conducted by our group has come to different conclusions. In particular, after cryopreservation with a CRSC protocol including 1.5 M EG and 0.2 M sucrose, we observed that the percentage of oocytes with a bipolar spindle and chromosomes aligned on the equatorial plane was moderately but significantly lower in comparison to fresh oocytes (53.8% and 70%, respectively). Recently, the trend of evidence which indicates that the oocyte spindle may endure the process of cryopreservation has been confirmed by a study in which four different protocols were compared (CRSC 1.5 M PrOH + 0.2 M sucrose; CRSC 1.5 M PrOH + 0.3 M sucrose; CRSC 1.5 M PrOH + 0.3 mol/L sucrose with Na^+-depleted, choline-replaced media and cryotip vitrification) (43). The proportion of oocytes showing bipolar spindle and equatorial chromosome alignment ranged from 73.9% to 88.9% in the different groups, an incidence that in all cases was not statistically different in comparison with fresh controls. These data confirm the results generated previously and concerning the CRSC 1.5 M PrOH + 0.3 M sucrose (44) and the Na^+ depleted–choline replaced (31) protocols. Other more comprehensive analyses will need to be performed especially on the vitrification approach, in consideration that current applications are based in most cases on the use of the cryotop as a storage device. This may not be without consequences, because the design of the storage device influences the volume in which oocytes are vitrified, with possible consequences on the rate of cooling and warming and in final analysis on the efficiency by which oocytes are vitrified. Overall, this study is nevertheless rather reassuring, confirming that maintenance of spindle organization may be achieved by a number of cryopreservation conditions. The fact that the vitrification approach can also preserve the MII spindle is particularly important, in the view of the increasing number of IVF programs that are introducing oocyte vitrification as a standard cryopreservation procedure.

Dynamic Observation of the MII Spindle in Cryopreserved Oocytes

In the last several years, technical advances in polarized light microscopy have made possible the observation of the oocyte MII spindle in a dynamic and noninvasive fashion, thereby allowing repeated observations of the same specimen over time. Polarized light microscopy is based on the phenomenon of birefringence by which highly orderly structures, such as the spindle microtubules, generate the decomposition of a single incident beam of polarized light into two orthogonal rays. This creates a difference in contrast between the spindle and the rest of the cell, which may be detected by imaging devices (e.g., the Polscope or the Oosight) that digitally amplify birefringence signals and, after computational manipulations, make quantifiable the degree of microtubule orientation. Polarized light microscopy does not require any preparative technique of the specimen and does not affect cell viability, conditions that have strongly motivated its use as a tool to study spindle presence, organization, and dynamics after cryopreservation and, therefore, to select oocytes of higher developmental potential. A detailed account of the evidence generated by the use of polarized light microscopy for the analysis of cryopreserved oocytes may be found elsewhere in this volume (see chap. 14). In general, as an effect of its absolute noninvasiveness, polarized light microscopy has confirmed its validity to study spindle dynamics, or more precisely disappearance and reappearance, during the phases of dehydration and rehydration in both CRSC and vitrification protocols (45–47). However, recent evidence has questioned the ability of the Polscope, a particular application of polarized light microscopy adopted by several human IVF laboratories, to provide detailed information on the actual constitution of the MII spindle. In particular, after comparison of data generated from the same material by the Polscope and high-performance confocal microscopy, it has emerged that, after cryopreservation, quantitative Polscope analysis, that is, measurement of retardance, is unable to discriminate among spindles with different configurations or different chromosome distributions (44). This suggests that the Polscope may be a rather inefficient method for assessing the precise structure of the MII spindle and, as a result, for noninvasive oocyte assessment and selection after cryopreservation or other IVF manipulations. Another recent study (48), not based on polarized light microscopy but on high-performance confocal microscopy, has addressed the question of the oocyte MII spindle dynamics after cryopreservation. In the section Effects of Cryopreservation on Microtubule and Chromosome Organization of this chapter it has been discussed how several lines of evidence based on confocal microscopy indicate that different cryopreservation conditions are compatible with the maintenance of an apparently normal MII spindle. These data, however, were produced by considering only one time point, usually chosen between one and three hours, after the end of the thawing (or warming in the case of vitrification) and rehydration procedures. While it is possible that, as an effect of CPAs or other unrecognized factors, the spindle may endure the entire cryopreservation procedure without undergoing depolymerization, in fact, it is perhaps more plausible that the spindle may depolymerize and repolymerize with different dynamics in dependence of the specific conditions imposed by a given protocol, as suggested by at least one polarized light microscopy study (47). For this reason, using high-resolution confocal microscopy, the dynamics of the meiotic spindle reassembly in human oocytes cryopreserved by a CRSC protocol based on 1.5 M PrOH + 0.3 M sucrose has been investigated over a period of three hours. It has been found that, immediately following thawing and rehydration, cryopreserved oocytes display a variety of spindle and chromosome abnormalities. However, recovery of a bipolar spindle with aligned chromosomes occurs within one hour from rehydration. Extending the post-thaw/rehydration culture for longer intervals (2–3 hours) results in progressive loss of bipolar spindle structure coincident with chromosome displacement. Therefore, an active process of spindle repair and chromosome alignment appears to be acting within one hour from the end of the thawing/rehydration procedure, followed by a deterioration in the forces that maintain chromosomes in a metaphase configuration. The causes of this biphasic spindle dynamics are not currently known, but it may be hypothesized that energy sources for spindle recovery (i.e., tubulin polymerization) and for the activity of microtubule motors, which keep the spindle poles focused and assure appropriate attachment of the chromosomes at the equatorial plate, may be compromised by cryopreservation. This hypothesis is consistent with the observation that mitochondrial function may be affected in frozen-thawed oocytes (49). Irrespective of the specific merit of these data, which are relevant to a particular set of freezing and thawing conditions, and despite the reassuring evidence provided by several studies, it appears that the question of a possible influence of cryopreservation on the oocyte MII spindle still demands a substantial investigation effort. In future studies, the analysis of MII spindle in cryopreserved oocytes will require a methodology that contemplates the possibility to detect morpho-functional changes within a time period of interest. It will also be very important to include novel or more accurate criteria to establish spindle normalcy, in the light of evidence which suggests that spindle size, microtubule density, and other parameters may indicate a condition of abnormality, and yet coexist with bipolarity and equatorial chromosome alignment (48,50,51).

CONCLUSIONS

Because of its inherent dynamic properties and sensitivity to a variety of intrinsic and extrinsic influences, the cytoskeleton of the human oocyte has been suspected to be a potential target of adverse cryopreservation conditions, with possible implications for the health of children developed from cryopreserved oocytes. In reality, the complement of polymerized actin of the mature oocyte appears rather insensitive to cryopreservation, but the paucity of observations conducted so far suggests that this matter would benefit from a more extensive analysis. The issue of the MII spindle constitution after cryopreservation has attracted more interest, leading to several studies performed under a variety of cryopreservation conditions. Currently, a consensus seems to have converged around the position that preservation of spindle integrity is compatible with at least some CRSC and vitrification approaches. However, it appears difficult to draw definite conclusions because even relatively minor protocol modifications, such as diverse dehydration conditions or type of storage devices, can have different and unpredictable downstream effects on microtubule organization. Therefore, each cryopreservation protocol, especially those newly developed, should be assessed separately. Novel insights of the MII spindle apparatus could derive from technical and methodological advances in confocal microscopy that have integrated previously unrecognized morphometric parameters of normalcy with the traditional assessments of spindle bipolarity and chromosome alignment. This will increase our understanding of the cytoskeleton, irrespective of the concerns raised by the possible effects of cryopreservation.

REFERENCES

1. Barrett SL, Albertini DF. Allocation of gamma-tubulin between oocyte cortex and meiotic spindle influences asymmetric cytokinesis in the mouse oocyte. Biol Reprod 2007; 76: 949–57.
2. Albertini DF, Barrett SL. The developmental origins of mammalian oocyte polarity. Semin Cell Dev Biol 2004; 15: 599–606.
3. Pickering SJ, Braude PR, Johnson MH, et al. Transient cooling to room temperature can cause irreversible disruption of the meiotic spindle in the human oocyte. Fertil Steril 1990; 54: 102–8.
4. Pickering SJ, Johnson MH, Braude PR, et al. Cytoskeletal organization in fresh, aged and spontaneously activated human oocytes. Hum Reprod 1988; 3: 978–89.
5. Battaglia DE, Goodwin P, Klein NA, et al. Influence of maternal age on meiotic spindle assembly in oocytes from naturally cycling women. Hum Reprod 1996; 11: 2217–22.
6. Coticchio G, De Santis L, Rossi G, et al. Sucrose concentration influences the rate of human oocytes with normal spindle and chromosome configurations after slow-cooling cryopreservation. Hum Reprod 2006; 21: 1771–6.
7. Leader B, Lim H, Carabatsos MJ, et al. Formin-2, polyploidy, hypofertility and positioning of the meiotic spindle in mouse oocytes. Nat Cell Biol 2002; 4: 921–8.
8. Hall A. Rho GTPases and the control of cell behaviour. Biochem Soc Trans 2005; 33: 891–5.
9. Choi T, Fukasawa K, Zhou R, et al. The Mos/mitogen-activated protein kinase (MAPK) pathway regulates the size and degradation of the first polar body in maturing mouse oocytes. Proc Natl Acad Sci U S A 1996; 93: 7032–5.
10. Li H, Guo F, Rubinstein B, et al. Actin-driven chromosomal motility leads to symmetry breaking in mammalian meiotic oocytes. Nat Cell Biol 2008; 10: 1301–8.
11. Maro B, Verlhac MH. Polar body formation: new rules for asymmetric divisions. Nat Cell Biol 2002; 4: E281–3.
12. Verlhac MH, Lefebvre C, Guillaud P, et al. Asymmetric division in mouse oocytes: with or without Mos. Curr Biol 2000; 10: 1303–6.
13. Halet G, Carroll J. Rac activity is polarized and regulates meiotic spindle stability and anchoring in mammalian oocytes. Dev Cell 2007; 12: 309–17.
14. Son WY, Park SE, Lee KA, et al. Effects of 1,2-propanediol and freezing-thawing on the in vitro developmental capacity of human immature oocytes. Fertil Steril 1996; 66: 995–9.
15. Carroll J, Warnes GM, Matthews CD. Increase in digyny explains polyploidy after in-vitro fertilization of frozen-thawed mouse oocytes. J Reprod Fertil 1989; 85: 489–94.
16. Vincent C, Pickering SJ, Johnson MH, et al. Dimethylsulphoxide affects the organisation of microfilaments in the mouse oocyte. Mol Reprod Dev 1990; 26: 227–35.
17. Carroll J, Wood MJ, Whittingham DG. Normal fertilization and development of frozen-thawed mouse oocytes: protective action of certain macromolecules. Biol Reprod 1993; 48: 606–12.
18. Younis AI, Toner M, Albertini DF, et al. Cryobiology of non-human primate oocytes. Hum Reprod 1996; 11: 156–65.
19. De Santis L, Coticchio G, Paynter, et al. Permeability of human oocytes to ethylene glycol and their survival and spindle configurations after slow cooling cryopreservation. Hum Reprod 2007; 22: 2776–83.

20. Borini A, Sciajno R, Bianchi V, et al. Clinical outcome of oocyte cryopreservation after slow cooling with a protocol utilizing a high sucrose concentration. Hum Reprod 2006; 21: 512–17.
21. Cobo A, Kuwayama M, Perez S, et al. Comparison of concomitant outcome achieved with fresh and cryopreserved donor oocytes vitrified by the cryotop method. Fertil Steril 2008; 89: 1657–64.
22. Bianchi V, Coticchio G, Distratis V, et al. Differential sucrose concentration during dehydration (0.2 mol/L) and rehydration (0.3 mol/L) increases the implantation rate of frozen human oocytes. Reprod Biomed Online 2007; 14: 64–71.
23. Howard J, Hyman AA. Dynamics and mechanics of the microtubule plus end. Nature 2003; 422: 753–8.
24. Combelles CM, Cekleniak NA, Racowsky C, et al. Assessment of nuclear and cytoplasmic maturation in in-vitro matured human oocytes. Hum Reprod 2002; 17: 1006–16.
25. Brunet S, Maro B. Cytoskeleton and cell cycle control during meiotic maturation of the mouse oocyte: integrating time and space. Reproduction 2005; 130: 801–11.
26. Ducibella T. Biochemical and cellular insights into the temporal window of normal fertilization. Theriogenology 1998; 49: 53–65.
27. McGinnis LK, Albertini DF, Kinsey WH. Localized activation of Src-family protein kinases in the mouse egg. Dev Biol 2007; 306: 241–54.
28. Johnson MH, Pickering SJ. The effect of dimethylsulphoxide on the microtubular system of the mouse oocyte. Development 1987; 100: 313–24.
29. Al-Hasani S, Diedrich K, van der Ven, et al. [Initial results of the cryopreservation of human oocytes]. Geburtshilfe Frauenheilkd 1986; 46: 643–4.
30. Gook DA, Osborn SM, Johnston WI. Cryopreservation of mouse and human oocytes using 1,2-propanediol and the configuration of the meiotic spindle. Hum Reprod 1993; 8: 1101–9.
31. Stachecki JJ, Munne S, Cohen J. Spindle organization after cryopreservation of mouse, human, and bovine oocytes. Reprod Biomed Online 2004; 8: 664–72.
32. Gook DA, Osborn SM, Bourne H, et al. Fertilization of human oocytes following cryopreservation: normal karyotypes and absence of stray chromosomes. Hum Reprod 1994; 9: 684–91.
33. Zenzes MT, Bielecki R, Casper RF, et al. Effects of chilling to 0 degrees C on the morphology of meiotic spindles in human metaphase II oocytes. Fertil Steril 2001; 75: 769–77.
34. Aigner S, Van der Elst J, Siebzehnrubl E, et al. The influence of slow and ultra-rapid freezing on the organization of the meiotic spindle of the mouse oocyte. Hum Reprod 1992; 7: 857–64.
35. Baka SG, Toth TL, Veeck LL, et al. Evaluation of the spindle apparatus of in-vitro matured human oocytes following cryopreservation. Hum Reprod 1995; 10: 1816–20.
36. Park SE, Son WY, Lee SH, et al. Chromosome and spindle configurations of human oocytes matured in vitro after cryopreservation at the germinal vesicle stage. Fertil Steril 1997; 68: 920–6.
37. Boiso I, Marti M, Santalo J, et al. A confocal microscopy analysis of the spindle and chromosome configurations of human oocytes cryopreserved at the germinal vesicle and metaphase II stage. Hum Reprod 2002; 17: 1885–91.
38. Ruppert-Lingham CJ, Paynter SJ, Godfrey J, et al. Developmental potential of murine germinal vesicle stage cumulus-oocyte complexes following exposure to dimethylsulphoxide or cryopreservation: loss of membrane integrity of cumulus cells after thawing. Hum Reprod 2003; 18: 392–8.
39. Cekleniak NA, Combelles CM, Ganz DA, et al. A novel system for in vitro maturation of human oocytes. Fertil Steril 2001; 75: 1185–93.
40. Mullen SF, Agca Y, Broermann DC, et al. The effect of osmotic stress on the metaphase II spindle of human oocytes, and the relevance to cryopreservation. Hum Reprod 2004; 19: 1148–54.
41. Joly C, Bchini O, Boulekbache H, et al. Effects of 1,2-propanediol on the cytoskeletal organization of the mouse oocyte. Hum Reprod 1992; 7: 374–8.
42. Cobo A, Rubio C, Gerli S, et al. Use of fluorescence in situ hybridization to assess the chromosomal status of embryos obtained from cryopreserved oocytes. Fertil Steril 2001; 75: 354–60.
43. Cobo A, Perez S, De los Santos MJ, et al. Effect of different cryopreservation protocols on the metaphase II spindle in human oocytes. Reprod Biomed Online 2008; 17: 350–9.
44. Coticchio G, Sciajno R, Hutt KJ, et al. Comparative analysis of the metaphase II spindle of human oocytes through polarized light and high performance confocal microscopy. Fertil Steril Accepted for publication.
45. Rienzi L, Martinez F, Ubaldi F, et al. Polscope analysis of meiotic spindle changes in living metaphase II human oocytes during the freezing and thawing procedures. Hum Reprod 2004; 19: 655–9.
46. Bianchi V, Coticchio G, Fava L, et al. Meiotic spindle imaging in human oocytes frozen with a slow freezing procedure involving high sucrose concentration. Hum Reprod 2005; 20: 1078–83.
47. Ciotti PM, Porcu E, Notarangelo L, et al. Meiotic spindle recovery is faster in vitrification of human oocytes compared to slow freezing. Fertil Steril 2008; DOI: 10.1016/j.fertnstert.2008.03.013.

48. Bromfield J, Coticchio G, Hutt K, et al. Meiotic spindle dynamics in human oocytes following slow-cooling cryopreservation. Submitted for publication.
49. Jones A, Van Blerkom J, Davis P, et al. Cryopreservation of metaphase II human oocytes effects mitochondrial membrane potential: implications for developmental competence. Hum Reprod 2004; 19: 1861–6.
50. Sanfins A, Lee GY, Plancha CE, et al. Distinctions in meiotic spindle structure and assembly during in vitro and in vivo maturation of mouse oocytes. Biol Reprod 2003; 69: 2059–67.
51. Shen Y, Betzendahl I, Sun F, et al. Non-invasive method to assess genotoxicity of nocodazole interfering with spindle formation in mammalian oocytes. Reprod Toxicol 2005; 19: 459–71.

14 Non-Invasive Assessment of Cryopreserved Oocytes Through Polarized Light Microscopy

Markus Montag, Maria Köster, Benjamin Rösing, Katrin van der Ven, and Hans van der Ven
Department of Gynecological Endocrinology & Reproductive Medicine, University of Bonn, Bonn, Germany

INTRODUCTION

The interest in cryopreservation of oocytes in human assisted reproduction is increasing. There are numerous reasons for this, and some of the more prominent ones are fertility preservation in young cancer patients, cryobanking of oocytes for future use to delay motherhood, legal restrictions regarding the number of oocytes that can be used for insemination in a fresh cycle after puncture (as in Italy), and oocyte donation programs. Although oocyte cryopreservation using a slow freezing protocol was successful in achieving pregnancies as early as 1986/1987 (1,2), the field has recently attracted much interest due to encouraging data obtained with oocyte vitrification (3,4). Another important stimulus was the change in legislation in Italy a few years ago, where nowadays only a restricted number of oocyte can be used for insemination (5). Legal restrictions seem to be a driving force for exploring new technologies and techniques and especially publications from Italian groups on cryopreservation of metaphase II oocytes have recently increased.

The main factor that affects the success rate of an oocyte freezing program is the cryopreservation procedure by itself. The reason is that cryopreservation may directly affect the oocyte or have an indirect impact on subsequent embryo development. This is influenced by the methodological approach as well as by the quality and properties of the solutions applied during the freeze-thaw procedure. Another important factor is the quality of the oocyte prior to freezing, which may be influenced by the stimulation protocol and patient driven factors (e.g., maternal age).

Therefore, investigations on oocyte quality are of utmost interest in order to reveal the state of the oocyte at the very beginning and to allow detecting changes during the whole process, which may be due to the intervention made during the freeze/thaw process.

As these oocytes will be used for therapeutic purposes, all investigations on oocyte quality should be non-invasive in order to guarantee that the oocytes remain healthy and intact.

PARAMETERS OF OOCYTE QUALITY PRIOR TO FREEZING THAT CAN BE ASSESSED NON-INVASIVELY

Assessing oocyte quality prior to the freezing process will not only help in identifying the current quality status of an oocyte, but it will also enable to visualize and document any changes that may have occurred during freezing and thawing.

Morphological Parameters

Obviously, morphological parameters of the oocyte are primarily assessed. Certain parameters in fresh oocytes such as polar body morphology or cytoplasmic features (6–9) and position and shape of the metaphase II spindle (10–15) were shown to correlate with the success of assisted reproductive technology (ART). Although these parameters allow for a qualitative grading of all oocytes from one patient, they do not allow for a definite and conclusive all or none decision about whether an oocyte is completely devoid of any implantation potential. Therefore, assessment of oocyte quality prior to cryopreservation is not a means to exclude oocytes from the freezing process, but to assist in identifying oocytes that after freezing and thawing may be preferentially used for further therapeutic treatment. Morphological parameters that seem to have a negative influence on the outcome of cryopreservation are a large perivitelline space, presence of vacuoles, and/or central granules in the ooplasm.

Non-Morphological Parameters

Besides morphological parameters other parameters exist, which to some extent give an indirect picture of the physiological state of an oocyte. Among these are appearance and birefringence of the zona pellucida (16–19), respiration rate measurements (20), and in the near future probably also tests on the expression of genes related to oocyte quality (21). As for the morphological parameters, assessing zona birefringence, oocyte respiration, or gene expression profiles will not enable the identification of those oocytes that will implant at a rate of 100%.

PARAMETERS OF CRYOPRESERVED OOCYTE QUALITY THAT CAN BE ASSESSED NON-INVASIVELY

Most of the parameters mentioned above could be used for assessing oocyte quality after cryopreservation; however, so far only few of them have been investigated toward their potential for that purpose.

With regard to morphology, the overall morphological appearance of a frozen-thawed oocyte is one parameter that can be assessed. Major detrimental changes during freezing or thawing either due to oocyte quality or due to technical problems will result in degeneration of the oocyte, which can be easily detected.

Respiration Rate Measurements

The most promising possibilities for assessing oocyte quality are those non-invasive methods that provide some information about oocyte physiology. One method that can detect oocyte health is the respiration rate, which correlates with the ability of an oocyte to produce ATP—a key factor for subsequent fertilization and embryo development (22). This topic is an open field for research, as so far only data on fresh oocytes are available (20). The data from Ref. (20) show that the baseline respiration rate of an oocyte cultured over three hours is a good means to detect oocytes that are atretic, arrested, properly matured, or abnormal. It would be interesting to assess the baseline respiration rate of oocytes before freezing and after thawing and to compare the values in order to detect those oocytes that, in a physiological way, did survive the whole procedure without any compromise and thus may have the best potential for further fertilization and embryo development.

Polarized Light Microscopy

Another approach to non-invasive quality assessment is polarized light microscopy. Polarized light microscopy was already introduced a century ago in cell biology to investigate the structure and development of skeletal and cellular components in animal cells (23). Further, it has been used to visualize the microtubule-dependent birefringence of the mitotic spindle (24), whose filamentous nature was revealed in 1953 in living cells by Inoué using time-lapse movies recorded with the polarizing microscope (25). It was demonstrated that microtubules are responsible for spindle birefringence and that spindle retardance is related to microtubule density (26). With the improvements in computer technology, a new type of polarization microscope system enabled real-time visualization of birefringent structures (27). This truly non-invasive method was used in the field of ART to investigate human oocytes and to visualize the two birefringent structures that are present in oocytes at the same time: the spindle and the inner layer of the zona pellucida (Fig. 1).

Spindle Imaging of Cryopreserved Metaphase II Oocytes

Spindle imaging was initially applied to investigate the effect of the presence and location of the spindle on fertilization rates and embryo viability in fresh oocytes. It was shown that oocytes presenting a spindle showed significantly higher fertilization rates (10,12–15,17,28,29), higher blastocyst formation rates (10,17), and higher pregnancy and implantation rates (15), although few studies reported the opposite (11,30). Therefore, the spindle and especially the functions mediated by an intact spindle seem to be an important part of the human oocyte and the fertilization process.

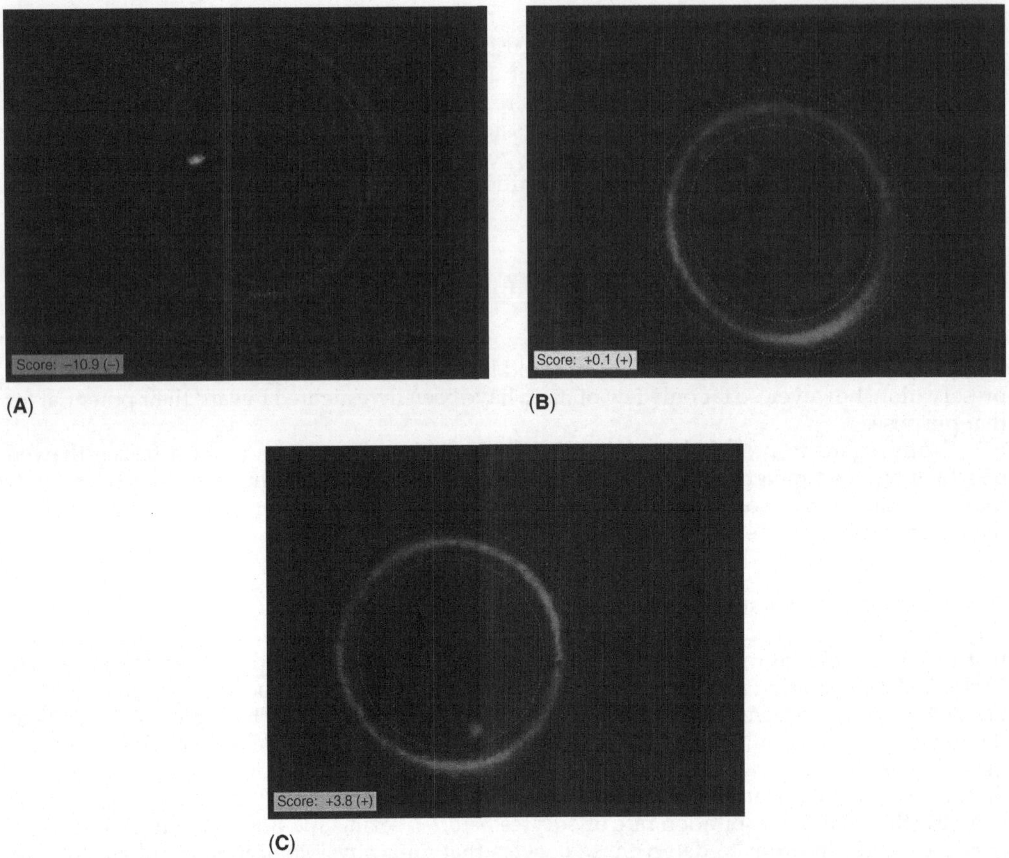

Figure 1 Zona and spindle imaging in metaphase II oocytes. The presence and location of a spindle within a metaphase II oocyte can be easily assessed by polarization microscopy. Due to its birefringent nature, the spindle will appear as an intense birefringent structure in the cytoplasm and close to the first polar body. Besides the spindle, the inner ring of the zona pellucida has some birefringent properties and can be visualized. An automatic zona evaluation system (Octax PolarAide™) enables a qualitative analysis of the zona birefringence and thus a subclassification of the quality of the corresponding oocytes, which is displayed as a score. The oocytes shown are characterized by a low score/low quality (**A**), intermediate score/intermediate quality (**B**), and high score/high quality (**C**). For details on zona imaging see Refs. (18) and (19). *See Color Plates on Page xx.*

Although the presence or absence of a spindle is discussed in the literature, one must be aware that spindle visibility may also depend on physiological characteristics of oocytes. The dynamics of spindle formation during oocyte maturation, in particular during the transition from metaphase I to metaphase II, was assessed by video cinematography (31). During this transitional step the human spindle behaves in a highly dynamic manner and is even absent for a certain time period. This phenomenon has been already mentioned by others (14,32,33). Therefore, the timing of spindle imaging seems to be of utmost clinical importance, and oocyte classification cannot rely on the presence or absence of the first polar body. Some oocytes classified as metaphase II based on the presence of a first polar body by conventional light microscopy can actually be in early telophase I with spindle remnants linking the polar body and the ooplasm, as seen by polarization microscopy. From the point of oocyte maturation, these oocytes can be classified as being immature. Quite a large number of these "immature" oocytes will acquire a metaphase II spindle within two to three hours in culture (31). In view of cryopreservation of oocytes, spindle imaging may help to start the freezing process at a time when the oocyte has reached a defined maturational state and this may be an important point for later evaluation and comparison of the results achieved.

Visualizing a spindle is dependent on external factors also. The use of glass-bottom dishes for spindle imaging is mandatory, as polarized light cannot pass through plastic. Another characteristic feature of the spindle is the dynamic instability of the microtubules involved in spindle formation (34) and, hence, the sensitivity of the spindle to temperature and pH. Consequently, the most important aspect with regard to spindle visualization is the proper control of these essential factors.

Besides pH, temperature is another important factor that has an impact on the spindle. Pickering and colleagues showed almost 20 years ago (35) that the meiotic spindle in mature human oocytes undergoes major structural changes after cooling at room temperature for 10 to 30 minutes. Upon re-warming to physiological temperature (37°C), not all spindles will return to their normal state. Similar studies were performed with polarization microscopy and it was reported that in human oocytes spindles start to disintegrate at a temperature of 33°C (36,37). Below a certain threshold of 25°C the spindle will never form again (36) and even partial spindle recovery may result in the formation of a deficient spindle, which may not separate chromosomes properly in the following meiotic division and give rise to aneuploid oocytes. Therefore, temperature-induced spindle disassembly may have a dramatic impact on fertilization and embryo development.

Changes in temperature are part of the cryopreservation procedure, and in combination with the use of cryoprotectant, temperature change is a major intervention in cryopreservation. Therefore, cryopreservation techniques such as slow cooling and vitrification may affect spindle and chromosome configuration.

Consequently, polarized light microscopy has been applied in the cryopreservation of spindle-positive metaphase II oocytes. Following cryopreservation of human oocytes by a slow freezing/rapid thawing protocol (33), spindles were present in 37% of all oocytes after thawing; however, they all disappeared in the following washing steps and after further incubation a spindle finally reappeared within three hours in 57% of all thawed oocytes. Another study on slow freezing of human metaphase II oocytes also confirmed that following thawing a re-organization of the spindles occurs within three to five hours (38). Since the report of these studies, the solutions used for cryopreservation of human oocytes by slow cooling have been improved (39,40).

Whether polarization microscopy is a good tool for studies on fine changes in spindle assembly has been questioned recently by a study on thawed oocytes, where polymerized microtubules were visualized by fluorescence confocal scanning microscopy although the spindle was not organized in a normal way (41). The same authors reported in more detail about the qualitative use of polarization microscopy in fresh as well as frozen oocytes (42). They concluded that the assumption that polarization microscopy-based qualitative detection of spindle birefringence is indicative of normal spindle organization and chromosome organization is not correct. Their results also show that retardance measurements are not informative on the degree of order or polymerization of the meiotic spindle, something that has already been questioned by others also (19). Altogether this allows the conclusion that polarization microscopy cannot allow an objective morphometric evaluation of the spindle. However, polarization microscopy remains a valid approach to study spindle dynamics in a non-invasive way.

Besides slow freezing, vitrification as an alternate technique for cryopreservation of immature and mature human oocytes has aroused much interest due to the success that had been reported by various authors (3,4,43).

As with the slow freezing protocol, polarization microscopy studies were undertaken to investigate the presence of the metaphase II spindle during the vitrification process. In contrast to the slow freezing protocol, Chen and co-workers reported that after vitrification and warming of mouse metaphase II oocytes, spindles were found in 50% of the warmed oocytes and in another 25% the spindle appeared within the following two hours (44). In addition, in contrast to the slow freezing/rapid thawing protocol, spindles remained present and did not disappear. Larman and colleagues (45) also reported that processing oocytes during the vitrification procedure at temperatures below 37°C resulted in spindle depolymerization, whereas maintaining the temperature at a physiological point left the spindle intact and unaffected. When they used a slow freezing protocol instead, they found that the meiotic spindle was not preserved.

It is very difficult to draw any final conclusions from the studies published until now. Further, one should be aware that the term slow cooling as well as vitrification is indicative for a certain technique, although the details of how the technique is actually performed may differ from laboratory to laboratory (39,40,46). Besides the variations in temperature that are used during the equilibration steps in the freezing or thawing solutions in both methods, there are numerous variations in terms of the composition of these solutions.

In our studies we found a very good visibility of the spindle in metaphase II oocytes during freezing and thawing using a slow freezing protocol (Fig. 2). However, in contrast to the findings by Larman and colleagues (45), we also noticed a loss of the spindle in some oocytes during vitrification and a re-formation within a certain time period, similar to the findings in slow freezing. This can probably be attributed to the way we perform vitrification (47), which may be different in other studies or compared with other techniques.

Based on our observations and in view of the profound data presented by Coticchio et al. (41,42), the visualization of a spindle during or shortly after freezing and thawing may be compromised due to unknown factors. This was extensively discussed in a recent review (48) especially in view of the role of the cryoprotectants, whose presence in the cytoplasm of the oocyte may have altered the optical characteristics in a way that, under certain circumstances, does not allow visualizing a still existing spindle.

Besides the presence of the spindle, spindle location may also be an important factor. The location of the spindle in relation to the first polar body, as a prognostic tool, was initially investigated by Rienzi et al. (12). These authors reported that oocytes with a deviation of the spindle location from the position of the polar body of more than 90° showed lower fertilization rates. However, the discussion on the relevance of spindle deviation is still unclear, as the daily routine work in the laboratory indicates that this phenomenon could be due to manipulation and stress caused by oocyte denudation (49). In addition, Fang et al. (50) showed that the use of polarized light microscopy in detecting spindle deviation in oocytes did not lead to higher fertilization rates compared to a control group without spindle imaging, although numbers were low in this study. We observed that after freezing and thawing of a human oocyte, the location of the polar body may have altered probably due to the osmotic stress and/or other mechanic interventions (data not shown). Therefore, the location of the spindle in human metaphase II oocytes in relation to the first polar body seems to be of no prognostic value with regard to oocyte quality after the freeze/thaw process.

Zona Imaging

Besides the visualization of spindle birefringence, polarized light microscopy also enables the visualization of structures within the zona pellucida. The multilaminar structure of the zona pellucida was first identified in hamster oocytes (51). The inner zona layer exhibits the highest amount of birefringence, and these findings were verified in human oocytes (52) and stimulated further studies in human ART on the value of zona imaging. As for respiration rate measurements, zona imaging using polarization microscopy is a prognostic tool for oocyte quality and subsequent embryo development (16–19). The ratio is that unfertilized metaphase II oocytes are classified by a non-invasive single investigation using an automatic scoring module that allows for user-independent zona imaging. The measuring device is based on the automatic detection of the birefringence of the inner zona layer. Once detected, a software module automatically starts to calculate and display in real time a zona score based on the intensity and distribution of the birefringence at 180 measuring points. Zona birefringence as revealed by polarization microscopy is directly linked to the paracrystalline network structure of the zona (53,54), which is formed during follicular maturation by the oocyte (56) and to a lower extent by the granulosa cells (55–57). A high birefringence of the inner zona layer appears to be primarily an indication of an optimal formation of this ordered structure of the inner zona layer during oocyte maturation. Therefore, zona imaging by polarization light microscopy is an indirect measure of oocyte quality and also of embryo quality, as embryo development on day 3—but not on day 2—is superior in embryos derived from oocytes with high and uniform birefringence of the inner zona layer (18,19).

All studies on the prognostic value of zona imaging were performed with fresh metaphase II oocytes. We used polarization microscopy to investigate in a preliminary study zona imaging of metaphase II oocytes during freezing and thawing with a slow cryopreservation

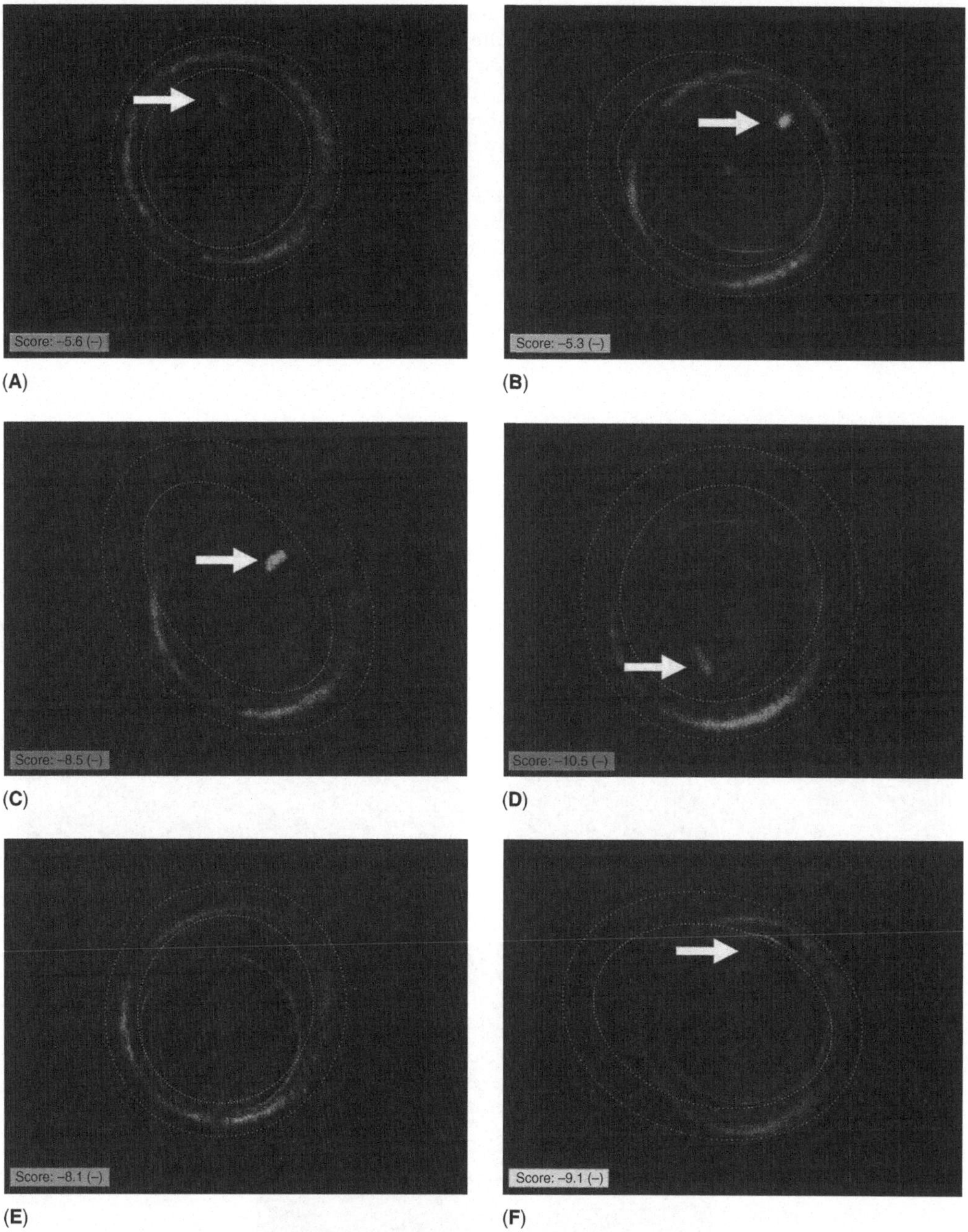

Figure 2 Spindle imaging in metaphase II oocytes during slow cooling cryopreservation. The presence and location of a spindle within a metaphase II oocyte was assessed prior to cryopreservation (**A**) and in equilibration solutions 2 (**B**) and 4 (**C**) by polarization microscopy. Following thawing of the same oocyte, the spindle was clearly visible in the first thawing solution (**D**) but disappeared in the following solutions (**E**). After incubation for another 2 hr in culture medium, the spindle did form again (**F**). For slow freezing/rapid thawing of denuded human oocytes we used a commercial kit (OC-Freeze/OC-Thaw; Cook, Limerick, Ireland) and essentially followed the instructions given by the manufacturer. The final freezing solution contains 1.5 M propanediol and 0.2 M sucrose in a HEPES buffered salt solution, while the first thawing solution contains 1.0 M propanediol and 0.3 M sucrose in a HEPES buffered salt solution. A detailed protocol is given in Ref. (46). *See Color Plates on Page xxi.*

protocol (Fig. 3). As shown in Figure 4, the initial zona score of the oocytes, assessed prior to freezing, did change during the freezing and the thawing process. Notably, three to four hours after thawing, which is the time point at which intracytoplasmic sperm injection may be performed, the score was higher compared to the initial value but with time it did return to the initial score. These data show that exposure to the cryoprotectant has an impact on the zona and in particular the inner birefringent ring of the zona is affected. Our investigations do not allow drawing a conclusion on the exact biophysical nature of these changes. However, a change in the zona score indicates a re-arrangement of the filamentous structures within the zona and a rising score indicates a shift toward a state with a higher order. Probably this change does reflect a hardening of the zona. Interestingly, Nottola and colleagues (58) reported that three to four hours after thawing human metaphase II oocytes showed signs of zona hardening in an electron microscopic analysis, which is in accordance with our observations using polarization microscopy. However, our preliminary results allow us to conclude that the zona may undergo further changes and that the state present three to four hours after thawing is not fixed.

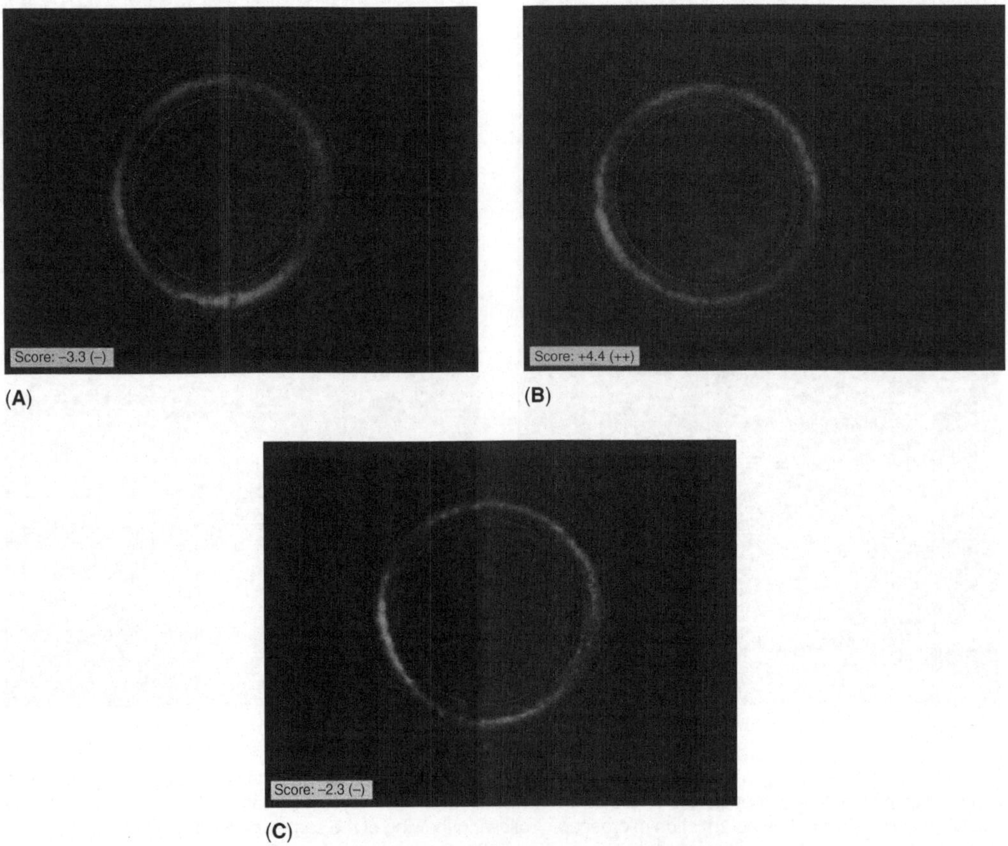

Figure 3 Zona imaging in metaphase II oocytes during slow cooling cryopreservation. Zona imaging of a metaphase II oocytes revealed that 3.5 hr after thawing the zona score (**B**) was different from the value prior to freezing (**A**). Following culture for another 18 hr, the zona score was almost similar to that in the beginning (**C**). This indicates that the changes induced by the cryopreservation procedure do have an impact on the architecture of the zona pellucida. The higher value 3.5 hr after thawing may indicate a temporary hardening at least of the inner ring of the zona pellucida. The freeze/thaw procedure was performed as described in the legend to Figure 2. *See Color Plates on Page xxii.*

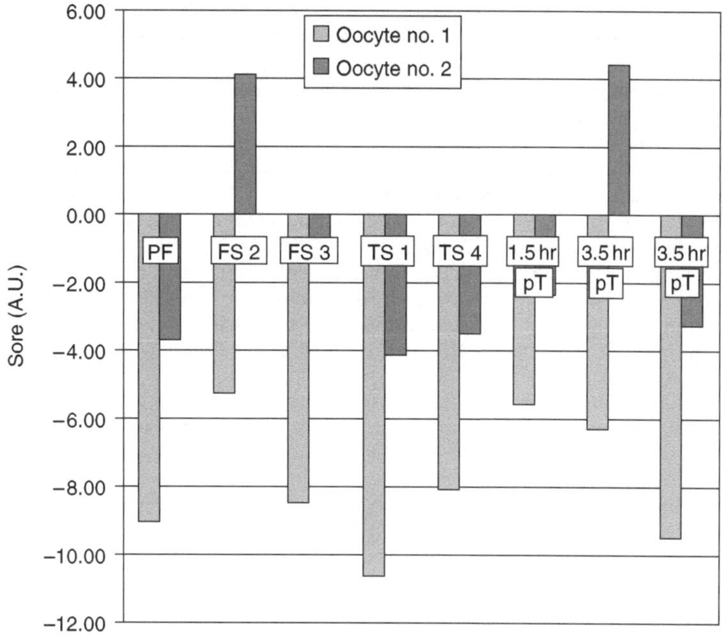

Figure 4 Change of the zona score in two metaphase II oocytes during slow cooling cryopreservation. The freeze/thaw procedure was performed as described in the legend to Figure 2. *Abbreviations*: PF, prior freezing; FS 2, freezing solution 2; FS 3, freezing solution 3; TS 1, thawing solution 1; TS 4, thawing solution 4; pT, post thawing.

In view of assessing oocyte quality, we may conclude that the cryopreservation procedure does also impact the zona structure. Therefore, the prognostic benefit of the zona can be relied on only if the imaging is done at proper time intervals. Further studies in this direction will be undertaken to investigate whether the changes in the zona structure can be used as another indicator for the quality of the oocyte or for the quality of a freezing protocol.

CONCLUSIONS

Assessing cryopreserved oocytes by polarization light microscopy has the potential to identify oocytes with good quality and high potential for further embryo development. The key factors that can be investigated after cryopreservation and thawing are the re-formation of the spindle (spindle imaging) and the texture of the zona pellucida (zona imaging). Polarized light microscopy of cryopreserved oocytes will generate data that can be used as a prognostic criterion to identify the embryo with the highest implantation potential. However, the technique must be properly applied in order to avoid artifacts, which may have an impact on the final results.

REFERENCES

1. Chen C. Pregnancy after human oocyte cryopreservation. Lancet 1986; 1: 884–6.
2. Van Uem JFHM, Siebzenhrübel ER, Schun B, et al. Birth after cryopreservation of unfertilized oocytes. Lancet 1987; 1: 752–3.
3. Kuleshova L, Gianaroli L, Magli L, et al. Birth following vitrification of small number of human oocytes. Hum Reprod 1999; 14: 3077–9.
4. Yoon TK, Chung HM, Lim JM, et al. Pregnancy and delivery of healthy infants developed from vitrified oocytes in a stimulated *in vitro* fertilization-embryo transfer program. Fertil Steril 2000; 74: 180–1.

5. Benagiano G, Gianaroli L. The new Italian IVF legislation. Reprod Biomed Online 2004; 9: 117–25.
6. Serhal PF, Ranieri DM, Kinis A, et al. Oocyte morphology predicts outcome of intracytoplasmic sperm injection. Hum Reprod 1997; 12: 1267–70.
7. Xia P. Intracytoplasmic sperm injection: correlation of oocyte grade based on polar body, perivitelline space and cytoplasmic inclusions with fertilization rate and embryo quality. Hum Reprod 1997; 12: 1750–5.
8. Ebner T, Yaman C, Moser M, et al. Prognostic value of first polar body morphology on fertilization rate and embryo quality in intracytoplasmic sperm injection. Hum Reprod 2000; 15: 427–30.
9. Ebner T, Moser M, Tews G. Is oocyte morphology prognostic of embryo developmental potential after ICSI? Reprod Biomed Online 2006; 12: 507–12.
10. Wang WH, Meng L, Hackett RJ, Keefe DL. Developmental ability of human oocytes with or without birefringent spindles imaged by Polscope before insemination. Hum Reprod 2001; 16: 1464–8.
11. Moon JH, Hyun CS, Lee SW, et al. Visualization of the metaphase II meiotic spindle in living human oocytes using the Polscope enables the prediction of embryonic developmental competence after ICSI. Hum Reprod 2003; 18: 817–20.
12. Rienzi L, Ubaldi F, Martinez F, et al. Relationship between meiotic spindle location with regard to polar body position and oocyte developmental potential after ICSI. Hum Reprod 2003; 18: 1289–93.
13. Cohen Y, Malcov M, Schwartz T, et al. Spindle imaging: a new marker for optimal timing of ICSI? Hum Reprod 2004; 19: 649–54.
14. De Santis L, Cino I, Rabellotti E, et al. Polar body morphology and spindle imaging as predictors of oocyte quality. Reprod Biomed Online 2005; 11: 36–42.
15. Madaschi C, Carvalho de Souza Bonetti T, Paes de Almeida Ferreira Braga D, et al. Spindle imaging: a marker for embryo development and implantation. Fertil Steril 2008; 90: 194–8.
16. Shen Y, Stalf T, Mehnert C, et al. High magnitude of light retardation by the zona pellucida is associated with conception cycles. Hum Reprod 2005; 20: 1596–606.
17. Rama Raju GA, Prakash GJ, Krishna KM, Madan K. Meiotic spindle and zona pellucida characteristics as predictors of embryonic development: a preliminary study using polscope imaging. Reprod Biomed Online 2007; 14: 166–74.
18. Montag M, Schimming T, Köster M, et al. Oocyte zona birefringence intensity is associated with embryonic implantation potential in ICSI cycles. Reprod Biomed Online 2008; 16: 239–44.
19. Montag M, van der Ven H. Oocyte assessment and embryo viability prediction: birefringence imaging. Reprod Biomed Online 2008; 17: 454–60.
20. Scott L, Berntsen J, Davies D, et al. Human oocyte respiration-rate measurement—potential to improve oocyte and embryo selection? Reprod Biomed Online 2008; 17: 461–9.
21. Patrizio P, Fragouli E, Bianchi V, et al. Molecular methods for selection of the ideal oocyte. Reprod Biomed Online 2007; 15: 346–53.
22. Dumollard R, Marangos P, Fitzharris G, et al. Sperm-triggered [Ca^{2+}] oscillations and Ca^{2+} homeostasis in the mouse egg have an absolute requirement for mitochondrial ATP production. Development 2004; 131: 3057–67.
23. Schmidt WJ. Die Bausteine des Tierkörpers in polarisiertem Lichte. Bonn: Cohen, 1924.
24. Swann MM, Mitchison JM. Refinements in, polarized light microscopy. J Exp Biol 1950; 27: 226–37.
25. Inoué S. Polarization optical studies of the mitotic spindle. I. The demonstration of spindle fibers in living cells. Chromosoma 1953; 5: 487–500.
26. Sato H, Ellis GW, Inoué S. Microtubular origin of mitotic spindle form birefringence: demonstration of the applicability of Wiener's equation. J Cell Biol 1975; 67: 501–17.
27. Oldenbourg R, Mei G. New polarized light microscope with precision universal compensator. J Microsc 1995; 180: 140–7.
28. Cooke S, Tyler JP, Driscoll GL. Meiotic spindle location and identification and its effect on embryonic cleavage plane and early development. Hum Reprod 2003; 18: 2397–405.
29. Rienzi L, Ubaldi F, Iacobelli M, et al. Meiotic spindle visualization in living human oocytes. Reprod Biomed Online 2005; 10: 192–8.
30. Chamayou S, Ragolia A, Alecci C, et al. Meiotic spindle presence and oocyte morphology do not predict clinical ICSI outcomes: a study of 967 transferred embryos. Reprod Biomed Online 2006; 13: 661–7.
31. Montag M, Schimming T, van der Ven H. Spindle imaging in human oocytes: the impact of the meiotic cell cycle. Reprod Biomed Online 2006; 12: 442–6.
32. Eichenlaub-Ritter U, Shen Y, Tinneberg HR. Manipulation of the oocyte: possible damage to the spindle apparatus. Reprod Biomed Online 2002; 5: 117–24.
33. Rienzi L, Martinez F, Ubaldi F, et al. Polscope analysis of meiotic spindle changes in living metaphase II human oocytes during the freezing and thawing procedures. Hum Reprod 2004; 19: 655–9.
34. Kirschner M, Mitchison T. Beyond self-assembly: from microtubules to morphogenesis. Cell 1986; 45: 329–42.

35. Pickering SJ, Braude PR, Johnson MH, et al. Transient cooling to room temperature can cause irreversible disruption of the meiotic spindle in the human oocyte. Fertil Steril 1990; 54: 102–8.
36. Wang WH, Meng L, Hackett RJ, et al. Limited recovery of meiotic spindles in living human oocytes after cooling-rewarming observed using polarized light microscopy. Hum Reprod 2001; 16: 2374–8.
37. Wang WH, Meng L, Hackett RJ, et al. Rigorous control during intracytoplasmic sperm injection stabilizes the meiotic spindle and improves fertilization and pregnancy rates. Fertil Steril 2002; 77: 1274–7.
38. Bianchi V, Coticchio G, Fava L, et al. Meiotic spindle imaging in human oocytes frozen with a slow freezing procedure involving high sucrose concentration. Hum Reprod 2005; 20: 1078–83.
39. Borini A, Sciajno R, Bianchi V, et al. Clinical outcome of oocyte cryopreservation after slow cooling with a protocol utilizing a high sucrose concentration. Hum Reprod 2006; 21: 512–17.
40. De Santis L, Coticchio G, Paynter S, et al. Permeability of human oocytes to ethylene glycol and their survival and spindle configurations after slow cooling cryopreservation. Hum Reprod 2007; 22: 2776–83.
41. Coticchio G, De Santis L, Rossi G, et al. Sucrose concentration influences the rate of human oocytes with normal spindle and chromosome configurations after slow-cooling cryopreservation. Hum Reprod 2006; 21: 1771–6.
42. Coticchio G, Sciajno R, Hutt K, et al. Comparative analysis of the metaphase II spindle of human oocytes through polarized light and high performance confocal microscopy. Fertil Steril 2009; in press.
43. Isachenko V, Montag M, Isachenko E, et al. Aseptic vitrification of human germinal vesicle oocytes using dimethyl sulfoxide as a cryoprotectant. Fertil Steril 2006; 85: 741–7.
44. Chen CK, Wang CW, Tsai WJ, et al. Evaluation of meiotic spindles in thawed oocytes after vitrification using polarized light microscopy. Fertil Steril 2004; 82: 666–72.
45. Larman MG, Minasi MG, Rienzi L, Gardner DK. Maintenance of the meiotic spindle during vitrification in human and mouse oocytes. Reprod Biomed Online 2007; 15: 692–700.
46. Bianchi V, Coticchio G, Diatris V, et al. Differential sucrose concentration during dehydration (0.2 mol/L) and re-hydration (0.3 mol/L) increases the implantation rate of frozen human oocytes. Reprod Biomed Online 2007; 14: 64–71.
47. Isachenko V, Montag M, Isachenko E, et al. Aseptic technology of vitrification of human pronuclear oocytes using open-pulled straws. Hum Reprod 2005; 20: 492–6.
48. De Santis L, Cino I, Coticchio G, et al. Objective evaluation of the viability of cryopreserved oocytes. Reprod Biomed Online 2007; 15: 338–45.
49. Taylor TH, Chang CC, Elliott T, et al. Effect of denuding on polar body position in in-vitro matured oocytes. Reprod Biomed Online 2008; 17: 515–19.
50. Fang C, Tang M, Li T, et al. Visualization of meiotic spindle and subsequent embryo development in *in vitro* and *in vivo* matured human oocytes. J Assist Reprod Genet 2007; 24: 547–51.
51. Keefe D, Tran P, Pellegrini C, Oldenbourg R. Polarized light microscopy and digital image processing identify a multilaminar structure of the hamster zona pellucida. Hum Reprod 1997; 12: 1250–2.
52. Pelletier C, Keefe DL, Trimarchi JR. Noninvasive polarized light microscopy quantitatively distinguishes the multilaminar structure of the zona pellucida of living human eggs and embryos. Fertil Steril 2004; 81: 850–6.
53. Wassarman PM, Jovine L, Litscher ES. Mouse zona pellucida genes and glycoproteins. Cytogenet Genome Res 2004; 105: 228–34.
54. Nikas G, Paraschos T, Psychoyos A, Handyside AH. The zona reaction in human oocytes as seen with scanning electron microscopy. Hum Reprod 1994; 9: 2135–8.
55. Sinowatz F, Toepfer-Petersen E, Kolle S, Palma G. Functional morphology of the zona pellucida. Anat Histol Embryol 2001; 30: 257–63.
56. Bogner K, Hinsch KD, Nayudu P, et al. Localization and synthesis of zona pellucida proteins in the marmoset monkey (*Callithrix jacchus*) ovary. Molecular Hum Reprod 2004; 10: 481–8.
57. Gook D, Martic M, Borg J, Edgar DH. Identification of zona pellucida proteins during human folliculogenesis. Hum Reprod 2004; 19(Suppl 1): 140.
58. Nottola SA, Machiarelli G, Coticchio G, et al. Ultrastructure of human mature oocytes after slow cooling cryopreservation using different sucrose concentrations. Hum Reprod 2007; 22: 1123–33.

15 Impact of Cryopreservation on Oocyte Physiology, Metabolism, and the Proteome

Mark G. Larman and David K. Gardner
Department of Zoology, The University of Melbourne, Victoria, Australia

INTRODUCTION

Following the first birth from a cryopreserved human oocyte over 20 years ago (1), very few births were reported during the next decade. The initial success of Chen's technique, which was based on a slow freezing method developed using the mouse oocyte as a model, could not be repeated. Furthermore, subsequent laboratory studies revealed the potential negative effects of cooling and exposure to cryoprotectants on oocyte physiology, which raised concerns regarding the safety of such a procedure. Meiotic spindle disruption, chromosome abnormalities, zona hardening, and reduced fertilization were the initial indicators that suggested oocyte cryopreservation protocols were suboptimal and consequently had a negative impact on the oocyte. The unique physiology and membrane composition of the metaphase-II-ovulated oocyte evidently provided greater challenges to scientists than that of the embryo. The report of the first birth from a cryopreserved human embryo (2) was around the same time as that of the oocyte (1). The efficacy of human embryo cryopreservation, however, has established the process as a routine assisted reproductive technique. Frozen embryos were transferred in around 15% of the total number of in vitro fertilization (IVF) cycles performed in the United States in 2005 (Assisted Reproductive Technology Report by Centers for Disease Control and Prevention). Almost 30% of these frozen embryo transfers resulted in a live birth, which equates to around 6000 births in 2005. This is also the case in Europe, where the number of live births derived from frozen embryo transfers is around 8000/yr (3). In comparison, oocyte cryopreservation is still considered an experimental procedure by the American Society of Reproductive Medicine and has resulted in approximately 500 live births worldwide, in two decades. However, encouraging outcomes for clinical oocyte cryopreservation are now being reported (4,5).

What has made the oocyte so difficult to cryopreserve compared with the embryo? The pronuclear oocyte is routinely cryopreserved, so it does not appear an issue of the surface area to volume ratio. The plasma membrane and zona pellucida of the oocyte undergo a dynamic change at fertilization. Thus, differences in permeability to water and cryoprotectants would mean that successful cryopreservation protocols for embryos may not directly translate to oocytes. A unique and evident difference between the oocyte and embryo is the presence of the meiotic spindle. This highly dynamic structure is critical for appropriate chromosome segregation and is extremely temperature sensitive. Therefore, it is imperative that cryopreservation protocols maintain the meiotic spindle and chromosome separation capabilities. Another specific characteristic of the oocyte is that it is primed for activation by the sperm. This means that the zona pellucida of a metaphase II (MII) oocyte is in its unfertilized configuration. It has not undergone the physical hardening and proteolytic modification that occurs following the sperm induced release of the oocyte's cortical granules, which facilitates part of the block to polyspermy. Due to the ability to bypass the zona pellucida and directly place the sperm in the cytoplasm through intracytoplasmic sperm injection (ICSI), cryopreservation induced zona hardening has not been considered as a major technical issue. However, zona hardening is but one of many early-activation events, triggered through a common messenger, an increase in intracellular calcium. Therefore, it is plausible that other cellular events that are normally activated by the fertilizing sperm are also precociously initiated by cryopreservation, leading to an altered physiological state.

This chapter will consider the effects of cryopreservation on oocyte physiology that have been observed using various forms of microscopy (electron, polarized light, epifluorescent, and confocal), and analyzed through metabolomics and proteomics. Furthermore, such effects will be discussed with regard to the cryopreservation technique (slow freezing or vitrification) and cryoprotectants that have the best potential to maintain oocyte physiology.

ZONA PELLUCIDA MODIFICATION AND REDUCED FERTILIZATION

Due to the difficulties in obtaining fresh human oocytes, most of the research into the effects of cryopreservation has been performed on the mouse. These studies will be discussed initially to provide a background on how cryopreservation affects the mammalian zona pellucida, which in turn has the potential to reduce fertilization rates.

The first report of mammalian (mouse) oocyte cryopreservation noted a reduction in fertilization rate (6). This was subsequently confirmed by numerous other studies in the proceeding decade (7–10). In mouse oocytes, both cooling and exposure to cryoprotectants was found to induce physical hardening of the zona pellucida (11,12). Breaching the zona or removing it completely prior to insemination, however, permitted fertilization (13,14). This suggested that the process of cryopreservation leads to zona hardening, explaining the reduction in fertilization rates. The normal process of zona hardening, which acts as a block to polyspermy, is triggered by the penetrating sperm (see Ref. (15) for a review). The sperm introduces a protein (phospholipase Cζ) that causes an initial large increase in intracellular calcium, followed by oscillations that last several hours (16). One of the downstream events of the initial increase is the calcium-dependent process of cortical granule exocytosis (17,18). The cortical granules release their contents into the zona pellucida, leading to a physical hardening and loss of sperm-binding capacity through the proteolytic cleavage of sperm-binding proteins (15).

The exact mechanism of zona hardening through cooling and cryoprotectant exposure is unclear and may involve different processes. Dimethylsulfoxide (DMSO) and 1,2-propanediol (PrOH) were shown to cause cortical granule exocytosis of up to 80% following a stepwise increase to 1.5 M (19). The cryoprotectant does not appear to have a direct effect on the zona pellucida, as exposure to DMSO failed to cause zona hardening unless the oocyte was present (20). Furthermore, studying electron micrographs of these oocytes revealed that only the treatments that led to zona hardening corresponded to a significant reduction in cortical granules. Additional evidence that DMSO triggers cortical granule exocytosis and reduces fertilization was provided by the observation that a sperm-binding protein (ZP2), present in the zona pellucida, was modified following DMSO exposure (20). Proteolysis of ZP2 inhibited sperm binding and occurred only in the presence of the oocyte. These investigations indicate that cryoprotectants (DMSO and PrOH) cause cortical granule release, which in turn leads to zona hardening, modification of the sperm-binding protein ZP2, and a significant reduction in fertilization.

Interestingly, adding fetal bovine serum (FBS) to the medium before and during exposure to DMSO significantly reduces zona hardening and proteolysis of ZP2 despite cortical granule release (21). A component of FBS, fetuin, is thought to be responsible for reducing zona hardening, most likely by acting as a competitive inhibitor for the enzyme(s) released by the cortical granules (22). Using FBS or fetal calf serum (FCS) to prevent zona hardening, following exposure to DMSO, allowed subsequent fertilization to be performed (23,24). It appears that only FBS and FCS have enough fetuin to reliably reduce zona hardening sufficiently to allow fertilization, although some fractions of bovine serum albumin and sources of human cord serum are capable of improving fertilization rates (22,24). FBS was used to demonstrate that cryoprotectant exposure leads to only partial cortical granule exocytosis (23). Although cortical granules had been released following exposure to DMSO (in the presence of FBS), sufficient cortical granules were still available to induce zona hardening following a second DMSO alone treatment. Partial exocytosis may help explain why other reports have failed to see a significant reduction in cortical granules using a fluorescent assay (14,25). As a number of studies, using electron microscopy, have revealed that both cooling and cryoprotectant exposure result in reduced cortical granules, it is plausible that inadequate sensitivity or difficulty in quantitative measurement may not permit the fluorescent assay to detect partial cortical granule exocytosis during cryopreservation.

DMSO is not the only cryoprotectant to cause zona hardening. Matson et al. (26) used PrOH in a slow freezing protocol and demonstrated significant zona hardening of mouse oocytes. Furthermore, using a vitrification protocol that exposed mouse oocytes to 8% DMSO

(1.13 M) and 8% ethylene glycol (EG) (1.43 M) for one minute followed by 16% DMSO (2.25 M) and 16% EG (2.86 M) for 30 seconds at 37°C (27) induced significant zona hardening and reduction in fertilization (28) (Fig. 1). As mentioned above, zona hardening triggered by the sperm is a calcium-dependent event, and recent studies have demonstrated that exposure of mouse oocytes to three commonly used cryoprotectants (DMSO, PrOH, and EG) cause an increase in intracellular calcium (28–31).

Undertaking analysis of MII oocytes using a fluorescent intracellular calcium indicator revealed that the individual cryoprotectants, at concentrations comparable to the initial exposure, were capable of causing a large transient increase in calcium similar to the initial calcium transient caused by a fertilizing sperm (Fig. 2). Surprisingly, the source of calcium increase was different for cryoprotectants (i.e., intracellular or extracellular). Performing the cryoprotectant exposure in calcium-free medium demonstrated that DMSO causes an increase by releasing calcium from intracellular stores, whereas EG and PrOH cause an influx of calcium across the plasma membrane. Subsequently, performing vitrification with calcium-free medium significantly reduced zona hardening and increased fertilization (28), indicating that these changes in oocyte physiology are the consequence of a cryoprotectant induced intracellular calcium increase. To support the findings that cryoprotectants cause zona hardening through an increase in calcium, which in turn leads to a reduction in fertilization, two additional treatments were performed. First, introducing a calcium chelator into the oocytes, which would sequester any increase in calcium before it could cause cortical granule exocytosis, significantly reduced zona hardening to a level comparable to non-vitrified oocytes. Second, using a laser to breach a small section of the zona pellucida following standard vitrification/warming, which would permit direct access for the sperm, significantly increased fertilization rates similar to non-vitrified oocytes.

Importantly, the release of calcium caused by the cryoprotectants can cause more than zona hardening. Increases in intracellular calcium act as important cell signaling second messengers. Therefore, cortical granule release is not the only activation event triggered by an increase in calcium at fertilization. A precocious increase in calcium could also affect

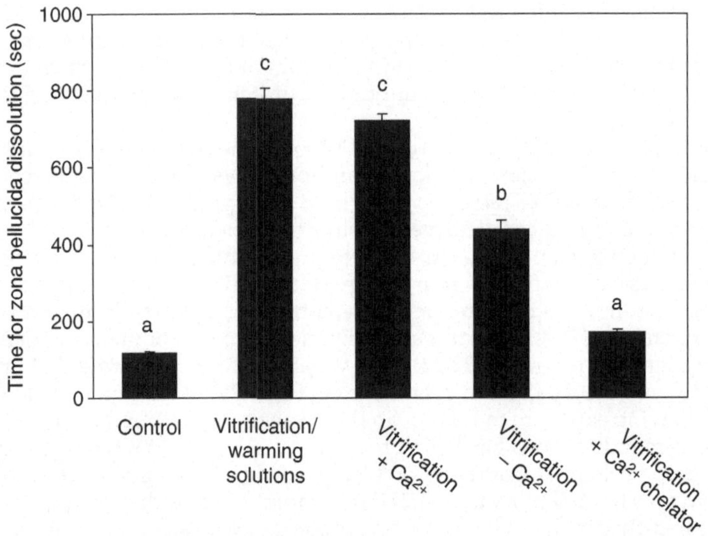

Figure 1 Cryoprotectants and vitrification cause zona hardening of mouse MII oocytes. To assess zona hardening, oocytes were incubated in a 1% chymotrypsin solution. The time taken for the zona pellucida to disappear completely was recorded for each treatment. Exposing the oocytes to the vitrification/warming solutions alone increased zona hardening significantly, as did vitrification with media containing calcium. In contrast, reducing the calcium by vitrifying in calcium-free media or introducing a calcium chelator into the oocytes, significantly reduced the level of zona hardening. Different letters are significantly different ($p < 0.01$) Data from Ref. (28).

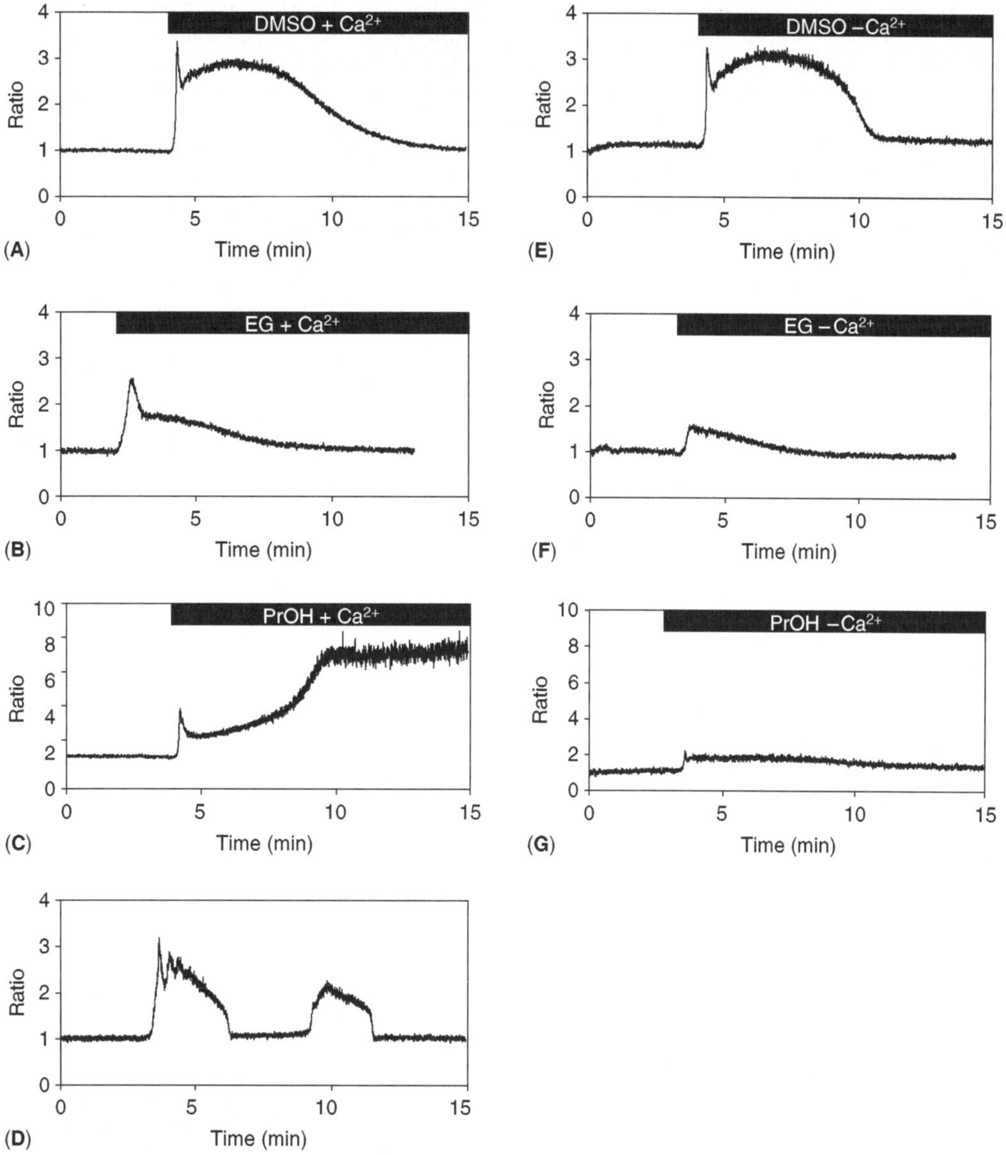

Figure 2 The effect of cryoprotectants on intracellular calcium of mouse MII oocytes. Mouse oocytes were loaded with a fluorescent calcium indicator and monitored with a photomultiplier-based system. Oocytes were treated with 1.5M DMSO (**A**, **E**), EG (**B**, **F**), or PrOH (**C**, **G**) in the presence or absence, respectively, of extracellular calcium. DMSO and EG caused a transient increase in intracellular calcium similar to that at fertilization (**D**). PrOH increased intracellular calcium for the duration of the exposure. Removing calcium from the medium significantly reduced the increase in intracellular calcium in response to EG and PrOH. The response to DMSO was unaffected, indicating that the increase was from intracellular calcium release and not extracellular calcium influx. Data from Refs. (31) and (106). *Abbreviations*: DMSO, dimethylsulfoxide; EG, ethylene glycol; PrOH, 1,2-propanediol.

levels/activity of cell cycle proteins such as M-phase promoting factor (MPF). Downregulation of MPF is required for meiosis resumption and is normally triggered by the sperm-induced calcium oscillations (32–34). Reduced levels of MPF at the time of ICSI would affect sperm decondensation and formation of the pronucleus from the sperm. The levels of intracellular calcium experienced by the oocyte have physiological sequelae. For example, activated oocytes that show a different calcium release profile develop at different rates and may even undergo apoptosis (35–38). So, although ICSI can circumvent the physical block to fertilization caused by zona hardening and proteolytic modification, it should be noted that zona hardening also reflects additional effects on oocyte physiology.

Due to limited studies, the degree of zona hardening and fertilization reduction in human oocytes is somewhat unclear. Initial reports of human oocyte cryopreservation (1986–1991) were made before ICSI had been developed as a form of fertilization that bypasses sperm interactions with the zona pellucida and plasma membrane (39). Therefore, from the mouse data discussed above it seems that the high fertilization rate (83%) reported by Chen (1986) (1) may have been due to the exposure of oocytes to cryoprotectant at a low temperature and in the presence of FBS. It is difficult to compare subsequent studies, not only due to the rather limited numbers of oocytes used and the variability in oocyte quality (i.e., fresh, aged, in vitro matured) but also the variation in cryopreservation protocol. These early reports on oocyte cryopreservation do, however, suggest a 10% to 50% decrease in fertilization rates for cryopreserved oocytes compared with fresh (40–44) oocytes. An interesting side note is that the membrane properties of human MII oocytes matured in vitro appear different to those of freshly ovulated oocytes, which in turn affects their response to cryopreservation (Larman and Gardner, unpublished observations).

With concerns of reduced fertilization rates, the advent of ICSI provided an opportunity to overcome any possible effects that cooling and cryoprotectants may have on the zona pellucida (45,46). Two initial publications indicated that ICSI may improve fertilization rates over IVF. In the first reported study of using ICSI on cryopreserved human oocytes, Gook and colleagues (45) demonstrated an improvement, albeit statistically nonsignificant, in fertilization rates using ICSI over IVF (71% and 55%, respectively). Unfortunately, due to regulatory issues, this study was carried out in two different clinics, which means that the sperm and oocytes used in the two groups were different, making direct comparison difficult. Kazem et al. (47) reported a significant ($p < 0.005$) increase in fertilization rates using ICSI over IVF (45.9% and 13.5%, respectively). The total number of cryopreserved oocytes used for comparing fertilization in these two studies, however, is small (56 for ICSI and 55 for IVF). Another factor to consider is that in 1995 ICSI was still relatively new, and the success of this procedure was lower compared with what is now expected. In contrast, a more recent comparison of the fertilization methods concluded that there was no difference in fertilization rates between ICSI and IVF (83% and 85%, respectively) (48). The reason for this outcome is unclear, especially considering that the cryopreservation protocol used (49) has been shown to induce cortical granule release and apparent zona hardening (50).

Following the study of electron micrographs involving the cooling of human oocytes to 0°C in the presence of DMSO, Sathananthan and colleagues (51) noted that the cortical granules were well preserved even though cooled oocytes did not appear to have the two to three rows of cortical granules underlying the plasma membrane seen in control oocytes. In contrast, stepwise increases in DMSO or PrOH released 70% to 80% of cortical granules in all oocytes (19). Li et al. (2005) (48) reported that no cortical granule exocytosis took place following cryopreservation using a fluorescent probe to stain the granules. There were, however, no control oocytes available, making it difficult to determine if there was indeed a change in fluorescence and therefore the number of cortical granules. These apparently conflicting studies may be explained by the observation that cryoprotectant exposure induces only partial cortical granule exocytosis (23), as discussed above. There are, however, no extensive studies on the effect of cryopreservation on fresh human oocytes to indicate if zona hardening occurs to the extent to warrant ICSI, but recent reports (using electron microscopy) do indicate the loss of cortical granules following cryopreservation (50,52,53).

MEIOTIC SPINDLE DEPOLYMERIZATION

At MII the oocyte is arrested with the maternal chromosomes suspended on an organized, bipolar, microtubule structure that is the meiotic spindle. The spindle must be formed and positioned correctly, with aligned chromosomes, to ensure proper chromosome segregation during second polar body extrusion. The intact spindle is also required for downregulation of MPF (54).

It is, therefore, imperative to ensure, for meiotic resumption and appropriate separation of the maternal chromosomes following fertilization, that the meiotic spindle is complete. This is of significance considering the numerous reports detailing the disappearance of the meiotic spindle during cryopreservation (51,55–67).

Interestingly, the meiotic spindle is not a static structure and is under constant flux. Microtubule dimers (composed of α- and β-tubulin) are preferentially lost at the microtubule organizing center (centrosome) and added at the end that is associated at the chromosome kinetochore. The rate at which the microtubule dimers are lost/added determines the state of the meiotic spindle. The meiotic spindle will rapidly depolymerize if oocytes are cooled below 37°C (68–71). Significantly, the spindle in mouse oocytes can recover from cooling. Following a short exposure to 4°C, the spindle in mouse oocytes were found to take around one hour to repolymerize (72). However, the human oocyte appears to be more sensitive to temperature than that of the mouse (71). Allowing human oocytes to cool from 37°C, spindle disassembly had occurred within five minutes, at which point the temperature was 27°C. The spindle could repolymerize within 20 minutes if the oocytes were immediately returned to 37°C. Unfortunately, if oocytes were maintained at a cooled temperature and then returned to 37°C, the spindle failed to reform in the same time period. The effect of cooling may even be permanent as the majority of human oocytes cooled to room temperature for 10 minutes failed to repolymerize their spindles within four hours of being returned to 37°C (73,74). As the human oocyte spindle is exquisitely sensitive to cooling, significant concerns were raised for the ability of cryopreserved oocytes to support appropriate chromosome separation and meiotic resumption.

To determine the effect of cryopreservation on the MII oocyte, it is possible to fix and stain the spindle and chromosomes. The advantage of this technique is the ability to obtain high-resolution images of the microtubules and chromosomes. The obvious disadvantage is that it only provides a snap shot, for example, the recovery of individual oocytes over time cannot be monitored. As the meiotic spindle is composed of highly organized microtubules, it retards polarized light (75). To enable polarized light microscopy to be performed more easily and provide a quantifiable image of retardance, liquid crystal optics, employing circularly polarized light and novel image processing algorithms were combined [Polscope; (76)]. Thus, the birefringent retardation of the polarized light (i.e., retardance) provides a measure of the order and density of the microtubules comprising the meiotic spindle (77). As this technique provides images of the spindle using transmitted light, it is noninvasive and does not affect embryo development and viability (71).

The combination of the two techniques outlined above has provided invaluable insight into the effect of the cryopreservation process on the meiotic spindle. It is difficult to compare directly the different studies using human oocytes due to the variability in oocyte quality and cryopreservation methods employed. However, when oocytes were fixed and stained immediately after thawing or at subsequent timed intervals, it was noted that in many cases the spindle was highly disorganized or depolymerized and completely absent. Over the course of one to five hours post-thaw recovery, the majority of spindles would repolymerize to some degree (56,61–63,65,66). However, the concern remained that even if the meiotic spindle does repolymerize after cryopreservation, are the chromosomes correctly aligned? In the mouse, scattered chromosomes associated with multiple microtubule asters were observed in the cytoplasm following cryoprotectant exposure and cryopreservation (70,72,78). Fortunately, unlike the mouse, human chromosomes appear to remain clustered in the cytoplasm (51,56,73). Subsequently, during depolymerization in the human oocyte, an association between the kinetochore and microtubule organizing center is maintained so that the chromosomes are effectively tethered and await meiotic spindle reformation. Despite the potential for chromosome missegregation (8,57), it appears that oocytes are actually at no greater risk of aneuploidy or polyploidy than noncryopreserved oocytes (56,61,63,65,79–82).

Using polarized light microscopy has enabled a more dynamic investigation of meiotic spindle depolymerization and repolymerization following cryopreservation. One of the most interesting discoveries was that the spindle actually increases in birefringence during the equilibration phase, which occurs at room temperature, and in fact depolymerizes during the thawing procedure, when the cryoprotectant is being removed (60). It appears that cryoprotectants provide a protective effect against spindle depolymerization at temperatures below 37°C (60,83). The mechanism of protection is unclear. It was suggested that tubulin dimer concentration in the cytosol would increase during dehydration, which would drive the reaction towards

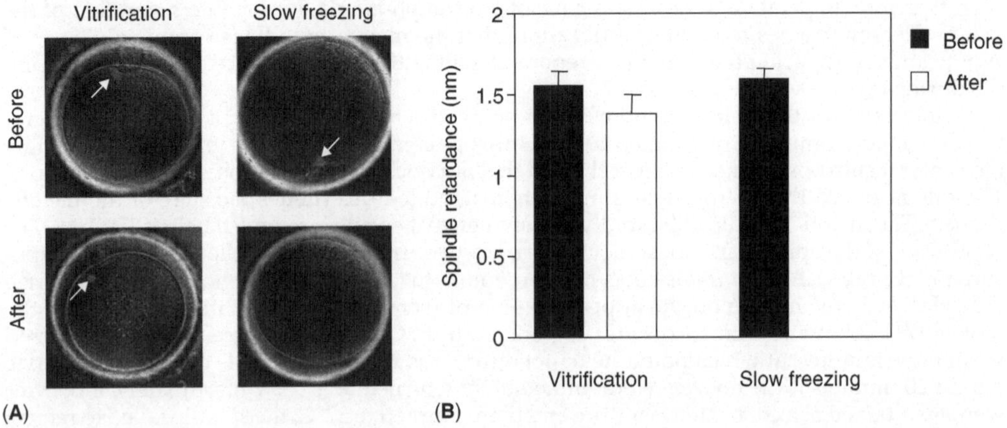

Figure 3 The effect of vitrification and slow freezing on spindle retardance of sibling human MII oocytes. (**A**) Polarized light microscopy (Oosight™) was used to image the meiotic spindle before and immediately after vitrification/warming at 37°C or slow freezing/thawing. (**B**) The degree of spindle retardance indicates that the spindle was not significantly altered following vitrification. There was no spindle to measure in any of the slow frozen oocytes. Arrows indicate the position of the meiotic spindle. Data replicated from Ref. (64).

increased polymerization. This explanation, however, now seems unlikely given the observation that human oocytes exposed to hypo- and hyperosmotic conditions can exhibit depolymerized meiotic spindles (84).

Recent studies have indicated that the meiotic spindle is maintained or recovers more rapidly following vitrification and warming (59,64,85). Each of these studies used polarized light microscopy to examine the oocytes before vitrification and immediately after warming. Using a protocol that exposed the mouse oocytes to cryoprotectants at room temperature and warmed them at 37°C (imaging every 15 minutes for three hours), Chen et al. (59) revealed that 48% and 72% of the oocytes possessed spindles within 15 and 60 minutes of warming, respectively. Larman and colleagues (64) demonstrated that a vitrification protocol that exposed individual mouse and human oocytes to vitrification and warming solutions at 37°C maintained the spindle, which was in stark contrast to human oocytes with depolymerized spindles following slow freezing (Fig. 3). A comparison of each of these three types of cryopreservation methods plus vitrification and warming at room temperature was performed on human oocytes (85). In agreement with Larman and colleagues (2007) (64), keeping the oocyte at a physiological temperature (37°C) during exposure to the solutions for vitrification and warming was significantly better for maintaining the meiotic spindle. Almost half of the oocytes had a spindle in the final warming solution and this increased to 100% within 15 minutes of culture.

COMPROMISED METABOLIC ACTIVITY

Surprisingly, only a few groups have undertaken research to investigate the effect of cryopreservation on oocyte and embryo metabolism. Considering that embryo developmental competence is tightly correlated with metabolism, which can be measured noninvasively, makes metabolism assessment a powerful assay for cryopreservation efficacy. The metabolic activities of bovine, mouse, and human embryos have been measured after cryopreservation and shown to be related to subsequent development (86–90). Lane et al. (2002) (88) also demonstrated that analysis of metabolism can be used to improve cryopreservation conditions. Analysis of pyruvate uptake, by individual two-cell mouse embryos demonstrated that slow freezing significantly reduced metabolism. In contrast, pyruvate uptake by vitrified embryos was twice that of slow frozen embryos. Furthermore, embryo viability of the subsequent blastocysts was assessed by cell number, trophectoderm/inner cell mass outgrowth, and glucose uptake. Each of these parameters indicated that the embryos cryopreserved using vitrification were significantly

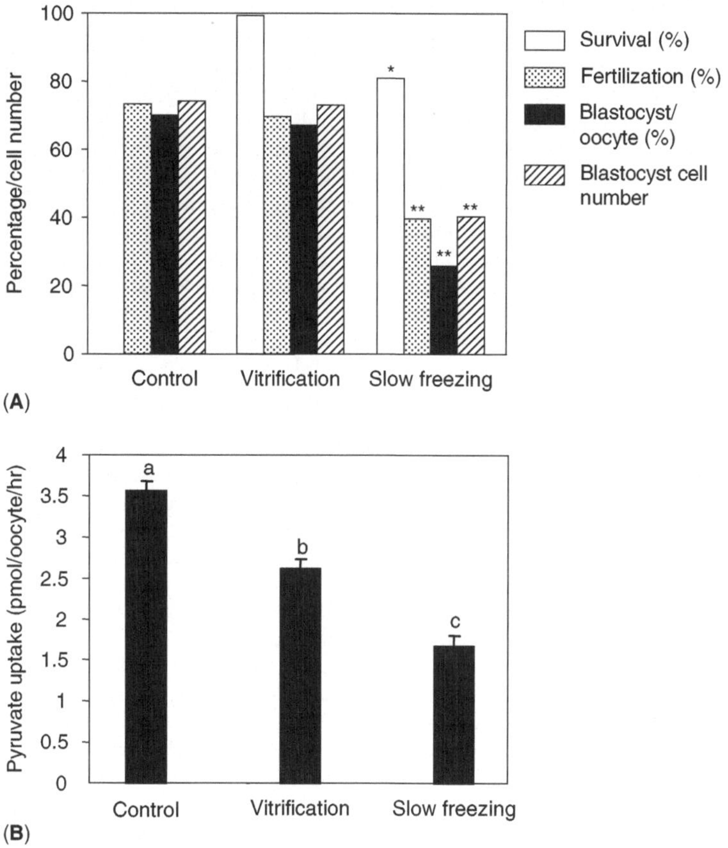

Figure 4 Developmental capacity of mouse MII oocytes following vitrification and slow freezing. Mouse oocytes were cryopreserved using vitrification or slow freezing. (**A**) Subsequent survival, fertilization, and embryo development were monitored. Oocytes that were slow frozen were significantly impaired in all parameters. Embryos derived from vitrified oocytes were comparable to the control. $^{*}p < 0.05$; $^{**}p < 0.01$ (**B**) Pyruvate uptake by the oocytes was measured, as an assessment of metabolic activity (using ultramicrofluorometry) during the next three hours after cryopreservation. The metabolism of slow frozen oocytes was significantly less than those vitrified. Both cryopreservation treatments led to significantly different pyruvate uptake compared to noncryopreserved control oocytes. Data from Ref. (27). Different letters denote significant differences ($p < 0.05$).

more viable than those slow frozen. The uptake and oxidation of pyruvate is the major energy source for the early embryo (91) and is dependent on pyruvate transfer across the inner mitochondrial membrane. Disruption of this process through reactive oxygen species (ROS) production appears to account for the difference in metabolism between slow frozen and vitrified early embryos, as the addition of a ROS scavenger (ascorbate) to the slow freezing medium restored pyruvate uptake to the level achieved with vitrification and increased viability (88).

Metabolism measurement has also been used to assess the impact of cryopreservation on oocytes (27). Mouse oocytes were slow frozen or vitrified and subsequent survival and metabolism of the oocytes were recorded. Due to zona hardening, a small breach in the zona pellucida was made using a laser (see section "Zona Pellucida Modification and Reduced Fertilization") to permit fertilization. Subsequent fertilization, embryo development, and viability were also assessed. Figure 4 clearly demonstrates that in all the parameters measured, oocytes and subsequent embryos are affected less by vitrification and thus, generate more viable embryos.

ABERRANT PROTEIN EXPRESSION

Changes in gene expression profiles following cryopreservation of mouse and human embryos have been reported (92–94). These studies are valuable in the overall assessment of cryopreservation, but proteins are the functional moiety. The ability to acquire protein profiles (proteome) from small groups of oocytes (n = 5) has recently enabled the comparison of cryopreservation techniques (31,95). Surface-enhanced laser desorption/ionization time-of-flight mass spectrometry (SELDI-TOF MS) is a form of chromatography that utilizes chemically modified or receptor/ligand bound surfaces to select for different proteins. The proteins captured on the surface can then be isolated by TOF MS, due to differences in mass/charge ratio, following ionization (96).

Proteomic analysis of mouse MII oocytes undergoing slow freezing or vitrification revealed significant alterations in the proteins expressed by oocytes that had been slow frozen (31). Using positively/negatively charged surfaces and TOF MS capable of detecting peptides and proteins <20 kDa, numerous proteins were found to be up- or downregulated in slow frozen oocytes compared to control (noncryopreserved) and vitrified MII oocytes. Seventeen positively charged proteins showed a significant alteration in expression levels following slow freezing. Twelve proteins were upregulated and the remainder downregulated. Twenty-one negatively charged proteins showed a significant alteration in expression levels following slow freezing. Fourteen proteins were downregulated and the remainder upregulated. Figure 5 shows three negatively charged proteins that are upregulated and four downregulated by slow freezing, but not vitrification. Hierarchical cluster analysis, which groups samples with similar protein expression profiles together, revealed the proteome of vitrified oocytes to be closer to the control than those slow frozen. The identity of the affected proteins has not been established, nor has the mechanism of up- or downregulation.

Regarding the etiology of cryo-induced trauma, SELDI-TOF MS has also been used to discern which segment of the slow freezing protocol has the most significant effect on the proteome (95). Analysis of mouse oocytes at the different stages of the slow freezing protocol indicated that aberrations in protein expression are principally caused by the prolonged exposure to PrOH during the original dehydration phase. One potential protein that is upregulated by

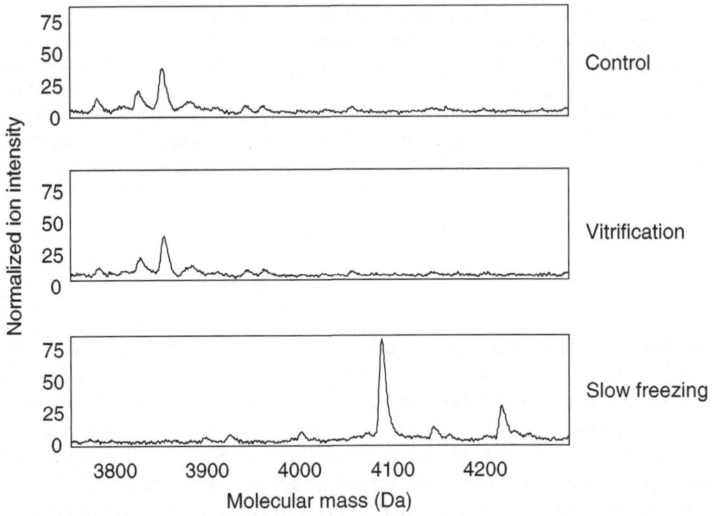

Figure 5 Expression of negatively charged proteins in mouse MII oocytes following vitrification and slow freezing. SELDI-TOF MS was used to investigate the effect of cryopreservation on the proteome of mouse oocytes. Protein profiles enhanced around the 3800–4200 Da range demonstrate that slow freezing significantly ($p < 0.05$) downregulates four proteins and upregulates three proteins compared to control (noncryopreserved) and vitrified oocytes. Data from Ref. (31). Abbreviation: SELDI-TOF MS, surface-enhanced laser desorption/ionization time-of-flight mass spectrometry.

PrOH is E2 ubiquitin-conjugating enzyme. This protein tags other proteins for degradation, so it could be involved in removing damaged or denatured proteins following slow freezing. Analyzing the effects of each stage of the cryopreservation procedure on the proteome should allow a more directed approach to improve oocyte and embryo cryopreservation methods.

SUMMARY

Cryopreservation methods have generally been improved empirically through a focus on survival, fertilization, and subsequent development. To optimize a procedure, it is critical to be able to assess and understand the effect it has on the physiology of the cell. Measurement of the effect of oocyte cryopreservation on the zona pellucida, meiotic spindle, metabolism, and protein expression facilitates justification and modification of the various requirements of the cryopreservation procedure. As discussed in this chapter and in agreement with others, a number of these analyses have demonstrated that vitrification has less of a detrimental impact on oocyte physiology than slow freezing (65,97–100).

It is clear that cryopreservation leads to cortical granule exocytosis, zona hardening, and reduced fertilization in the mouse oocyte. Whether the human oocyte is affected to the same extent remains unclear, but research has led to cryopreservation media that contain fetuin or are calcium-free, which have both been shown to reduce zona hardening and increase fertilization rates. Of some concern is the observation that DMSO causes an increase in calcium through intracellular release, probably by disrupting intracellular organelles. Although DMSO is an effective cryoprotectant (due to its high membrane permeability characteristic), there is growing evidence, albeit in other cell types, that DMSO has the potential to cause apoptosis and differentiation (101,102). It may also be the cause of toxicity observed from frozen bone marrow and stem cell transplants (103) (NIH funded study NCT00631787).

It is worth noting that all cryoprotectants are toxic depending on their concentration and exposure time. Analysis of the effect of a slow freezing protocol (using PrOH) indicated that the aberrant protein expression observed by slow freezing is due to the chronic exposure of the oocytes to PrOH. The ability to assess the effect of cryopreservation on the proteins being expressed by the oocyte and perturbations in metabolism will allow the development of protocols with minimized impact on oocyte physiology.

The ability to monitor the meiotic spindle after cryopreservation using various techniques has demonstrated that vitrification has a higher capacity to maintain or reform the dynamic microtubule structure than slow freezing. Previously, it was necessary to delay fertilization one to three hours following cryopreservation to allow the spindle to reform, thereby increasing oocyte aging. A vitrification protocol that maintains the oocytes at a physiological temperature during exposure to the vitrification and warming solutions has provided the opportunity to proceed with fertilization immediately following warming.

Of course, there are caveats with any given method of analysis. Cryopreservation is a rapidly growing field with new protocols, media, and devices constantly being reported, making it very difficult to compare the latest techniques. Live births have resulted from a plethora of cryopreservation techniques and protocols, so it is clear that multiple methods have the potential to succeed. Perhaps some of the effects of cryopreservation are insignificant or only transient and the oocyte is capable of recovery. However, stresses have been shown to accumulate in the embryo, resulting in compromised physiology and viability (104,105). Therefore, it is prudent to minimize stress on the oocytes to maximize developmental potential.

REFERENCES

1. Chen C. Pregnancy after human oocyte cryopreservation. Lancet 1986; 1: 884–6.
2. Zeilmaker GH, Alberda AT, van Gent I, Rijkmans CM, Drogendijk AC. Two pregnancies following transfer of intact frozen-thawed embryos. Fertil Steril 1984; 42: 293–6.
3. Andersen AN, Gianaroli L, Felberbaum R, de Mouzon J, Nygren KG. Assisted reproductive technology in Europe, 2001. Results generated from European registers by ESHRE. Hum Reprod 2005; 20: 1158–76.
4. Chian RC, Huang JY, Tan SL, et al. Obstetric and perinatal outcome in 200 infants conceived from vitrified oocytes. Reprod Biomed Online 2008; 16: 608–10.

5. Kuwayama M, Vajta G, Kato O, Leibo SP. Highly efficient vitrification method for cryopreservation of human oocytes. Reprod Biomed Online 2005; 11: 300–8.
6. Whittingham DG. Fertilization in vitro and development to term of unfertilized mouse oocytes previously stored at –196 degrees C. J Reprod Fertil 1977; 49: 89–94.
7. Glenister PH, Wood MJ, Kirby C, Whittingham DG. Incidence of chromosome abnormalities in first-cleavage mouse embryos obtained from frozen-thawed oocytes fertilized in vitro. Gamete Res 1987; 16: 205–16.
8. Carroll J, Warnes GM, Matthews CD. Increase in digyny explains polyploidy after in-vitro fertilization of frozen-thawed mouse oocytes. J Reprod Fertil 1989; 85: 489–94.
9. Trounson A, Kirby C. Problems in the cryopreservation of unfertilized eggs by slow cooling in dimethyl sulfoxide. Fertil Steril 1989; 52: 778–86.
10. Todorow SJ, Siebzehnrubl ER, Spitzer M, et al. Comparative results on survival of human and animal eggs using different cryoprotectants and freeze-thawing regimens. II. Human. Hum Reprod 1989; 4: 812–6.
11. Johnson MH, Pickering SJ, George MA. The influence of cooling on the properties of the zona pellucida of the mouse oocyte. Hum Reprod 1988; 3: 383–7.
12. Johnson MH. The effect on fertilization of exposure of mouse oocytes to dimethyl sulfoxide: an optimal protocol. J Assis Reprod Gen 1989; 6: 168–75.
13. Carroll J, Depypere H, Matthews CD. Freeze-thaw-induced changes of the zona pellucida explains decreased rates of fertilization in frozen-thawed mouse oocytes. J Reprod Fertil 1990; 90: 547–53.
14. Wood MJ, Whittingham DG, Lee SH. Fertilization failure of frozen mouse oocytes is not due to premature cortical granule release. Biol Reprod 1992; 46: 1187–95.
15. Sun Q-Y. Cellular and molecular mechanisms leading to cortical reaction and polyspermy block in mammalian eggs. Microsc Res and Techniq 2003; 61: 342–8.
16. Saunders CM, Larman MG, Parrington J, et al. PLC zeta: a sperm-specific trigger of Ca(2+) oscillations in eggs and embryo development. Development 2002; 129: 3533–44.
17. Kline D, Kline JT. Repetitive calcium transients and the role of calcium in exocytosis and cell cycle activation in the mouse egg. Dev Biol 1992; 149: 80–9.
18. Tahara M, Tasaka K, Masumoto N, et al. Dynamics of cortical granule exocytosis at fertilization in living mouse eggs. Am J Physiol 1996; 270: C1354–61.
19. Schalkoff ME, Oskowitz SP, Powers RD. Ultrastructural observations of human and mouse oocytes treated with cryopreservatives. Biol Reprod 1989; 40: 379–93.
20. Vincent C, Pickering SJ, Johnson MH. The hardening effect of dimethylsulphoxide on the mouse zona pellucida requires the presence of an oocyte and is associated with a reduction in the number of cortical granules present. J Reprod Fertil 1990; 89: 253–9.
21. Vincent C, Turner K, Pickering SJ, Johnson MH. Zona pellucida modifications in the mouse in the absence of oocyte activation. Mol Reprod Dev 1991; 28: 394–404.
22. George MA, Johnson MH. Use of fetal bovine serum substitutes for the protection of the mouse zona pellucida against hardening during cryoprotectant addition. Hum Reprod 1993; 8: 1898–900.
23. George MA, Johnson MH, Vincent C. Use of fetal bovine serum to protect against zona hardening during preparation of mouse oocytes for cryopreservation. Hum Reprod 1992; 7: 408–12.
24. Carroll J, Wood MJ, Whittingham DG. Normal fertilization and development of frozen-thawed mouse oocytes: protective action of certain macromolecules. Biol Reprod 1993; 48: 606–12.
25. Van Blerkom J, Davis PW. Cytogenetic, cellular, and developmental consequences of cryopreservation of immature and mature mouse and human oocytes. Microsc Res Tech 1994; 27: 165–93.
26. Matson PL, Graefling J, Junk SM, Yovich JL, Edirisinghe WR. Cryopreservation of oocytes and embryos: use of a mouse model to investigate effects upon zona hardness and formulate treatment strategies in an in-vitro fertilization programme. Hum Reprod 1997; 12: 1550–3.
27. Lane M, Gardner DK. Vitrification of mouse oocytes using a nylon loop. Mol Reprod Dev 2001; 58: 342–7.
28. Larman MG, Sheehan CB, Gardner DK. Calcium-free vitrification reduces cryoprotectant-induced zona pellucida hardening and increases fertilization rates in mouse oocytes. Reproduction 2006; 131: 53–61.
29. Litkouhi B, Winlow W, Gosden RG. Impact of cryoprotective agent exposure on intracellular calcium in mouse oocytes at metaphase II. Cryoletters 1999; 20: 353–62.
30. Takahashi T, Igarashi H, Doshida M, et al. Lowering intracellular and extracellular calcium contents prevents cytotoxic effects of ethylene glycol-based vitrification solution in unfertilized mouse oocytes. Mol Reprod Dev 2004; 68: 250–8.
31. Larman MG, Katz-Jaffe MG, Sheehan CB, Gardner DK. 1,2-propanediol and the type of cryopreservation procedure adversely affect mouse oocyte physiology. Hum Reprod 2007; 22: 250–9.

32. Lorca T, Galas S, Fesquet D, et al. Degradation of the proto-oncogene product p39mos is not necessary for cyclin proteolysis and exit from meiotic metaphase: requirement for a Ca(2+)-calmodulin dependent event. Embo J 1991; 10: 2087–93.
33. Nixon VL, Levasseur M, McDougall A, Jones KT. Ca(2+) oscillations promote APC/C-dependent cyclin B1 degradation during metaphase arrest and completion of meiosis in fertilizing mouse eggs. Curr Biol 2002; 12: 746–50.
34. Marangos P, Carroll J. Fertilization and $InsP_3$-induced Ca^{2+} release stimulate a persistent increase in the rate of degradation of cyclin B1 specifically in mature mouse oocytes. Dev Biol 2004; 272: 26–38.
35. Bos-Mikich A, Whittingham DG, Jones KT. Meiotic and mitotic Ca^{2+} oscillations affect cell composition in resulting blastocysts. Dev Biol 1997; 182: 172–9.
36. Gordo AC, Wu H, He CL, Fissore RA. Injection of sperm cytosolic factor into mouse metaphase II oocytes induces different developmental fates according to the frequency of [Ca(2+)](i) oscillations and oocyte age. Biol Reprod 2000; 62: 1370–9.
37. Gordo AC, Rodrigues P, Kurokawa M, et al. Intracellular calcium oscillations signal apoptosis rather than activation in in vitro aged mouse eggs. Biol Reprod 2002; 66: 1828–37.
38. Ducibella T, Huneau D, Angelichio E, et al. Egg-to-embryo transition is driven by differential responses to Ca(2+) oscillation number. Dev Biol 2002; 250: 280–91.
39. Palermo P, Joris H, Devroey P, Van Steirteghem AC. Pregnancies after intracytoplasmic injection of a single spermatazoon into an oocyte. Lancet 1992; 340: 17–18.
40. Mandelbaum J, Junca AM, Tibi C, et al. Cryopreservation of immature and mature hamster and human oocytes. Ann N Y Acad Sci 1988; 541: 550–61.
41. AL-Hasani S, Kirsch J, Diedrich K, Blanke S, et al. Successful embryo transfer of cryopreserved and in-vitro fertilized rabbit oocytes. Hum Reprod 1989; 4: 77–9.
42. Siebzehnruebl ER, Todorow S, van Uem J, et al. Cryopreservation of human and rabbit oocytes and one-cell embryos: a comparison of DMSO and propanediol. Hum Reprod 1989; 4: 312–7.
43. Hunter JE, Bernard A, Fuller B, Amso N, Shaw RW. Fertilization and development of the human oocyte following exposure to cryoprotectants, low temperatures and cryopreservation: a comparison of two techniques. Hum Reprod 1991; 6: 1460–5.
44. Gook DA, Osborn SM, Bourne H, Johnston WI. Fertilization of human oocytes following cryopreservation: normal karyotypes and absence of stray chromosomes. Hum Reprod 1994; 9: 684–91.
45. Gook DA, Cschiewe M, Osborn SM, et al. Intracytoplasmic sperm injection and embryo development of human oocytes cryopreserved using 1,2-propanediol. Hum Reprod 1995; 10: 2637–41.
46. Porcu E, Fabbri R, Seracchioli R, et al. Birth of a healthy female after intracytoplasmic sperm injection of cryopreserved human oocytes. Fertil Steril 1997; 68: 724–6.
47. Kazem R, Thompson LA, Srikantharajah A, et al. Cryopreservation of human oocytes and fertilization by two techniques: in-vitro fertilization and intracytoplasmic sperm injection. Hum Reprod 1995; 10: 2650–4.
48. Li X-H, Chen S-U, Zhang X, et al. Cryopreserved oocytes of infertile couples undergoing assisted reproductive technology could be an important source of oocyte donation: a clinical report of successful pregnancies. Hum Reprod 2005; 20: 3390–4.
49. Fabbri R, Porcu E, Marsella T, et al. Human oocyte cryopreservation: new perspectives regarding oocyte survival. Hum Reprod 2001; 16: 411–16.
50. Nottola SA, Coticchio G, De Santis L, et al. Ultrastructure of human mature oocytes after slow cooling cryopreservation using different sucrose concentrations. Hum Reprod 2007; 22: 1123–33.
51. Sathananthan AH, Trounson A, Freemann L, Brady T. The effects of cooling human oocytes. Hum Reprod 1988; 3: 968–77.
52. Nottola SA, Coticchio G, De Santis L, et al. Ultrastructure of human mature oocytes after slow cooling cryopreservation with ethylene glycol. Reprod Biomed Online 2008; 17: 368–77.
53. Ghetler Y, Skutelsky E, Ben Nun, et al. Human oocyte cryopreservation and the fate of cortical granules. Fertil Steril 2006; 86: 210–6.
54. Winston NJ, Mcguinness O, Johnson MH, Maro B. The exit of mouse oocytes from meiotic M-phase requires an intact spindle during intracellular calcium release. J Cell Sci 1995; 108 (Pt 1): 143–51.
55. Aigner S, Van der Elst J, Siebzehnrubl E, et al. The influence of slow and ultra-rapid freezing on the organization of the meiotic spindle of the mouse oocyte. Hum Reprod 1992; 7: 857–64.
56. Gook DA, Osborn SM, Johnston WI. Cryopreservation of mouse and human oocytes using 1, 2-propanediol and the configuration of the meiotic spindle. Hum Reprod 1993; 8: 1101–9.
57. Eroglu A, Toner M, Leykin L, Toth T. Cytoskeleton and polyploidy after maturation and fertilization of cryopreserved germinal vesicle—stage mouse oocytes. J Assis Reprod Gen 1998; 15: 447–54.

58. Boiso I, Marti M, Santalo J, et al. A confocal microscopy analysis of the spindle and chromosome configurations of human oocytes cryopreserved at the germinal vesicle and metaphase II stage. Hum Reprod 2002; 17: 1885–91.
59. Chen CK, Wang CW, Tsai WJ, et al. Evaluation of meiotic spindles in thawed oocytes after vitrification using polarized light microscopy. Fertil Steril 2004; 82: 666–72.
60. Rienzi L, Martinez F, Ubaldi F, et al. Polscope analysis of meiotic spindle changes in living metaphase II human oocytes during the freezing and thawing procedures. Hum Reprod 2004; 19: 655–9.
61. Stachecki JJ, Munne S, Cohen J. Spindle organization after cryopreservation of mouse, human, and bovine oocytes. Reprod Biomed Online 2004; 8: 664–72.
62. Bianchi V, Coticchio G, Fava L, Flamigni C, Borini A. Meiotic spindle imaging in human oocytes frozen with a slow freezing procedure involving high sucrose concentration. Hum Reprod 2005; 20: 1078–83.
63. Coticchio G, De Santis L, Rossi G, et al. Sucrose concentration influences the rate of human oocytes with normal spindle and chromosome configurations after slow-cooling cryopreservation. Hum Reprod 2006; 21: 1771–6.
64. Larman MG, Minasi MG, Rienzi L, Gardner DK. Maintenance of the meiotic spindle during vitrification in human and mouse oocytes. Reprod Biomed Online 2007; 15: 692–700.
65. Huang JY, Chen HY, Tan SL, Chian RC. Effect of choline-supplemented sodium-depleted slow freezing versus vitrification on mouse oocyte meiotic spindles and chromosome abnormalities. Fertil Steril 2007; 88: 1093–100.
66. Cobo A, Perez S, De los Santos MJ, et al. Effect of different cryopreservation protocols on the metaphase II spindle in human oocytes. Reprod Biomed Online 2008; 17: 350–9.
67. De Santis L, Coticchio G, Paynter S, et al. Permeability of human oocytes to ethylene glycol and their survival and spindle configurations after slow cooling cryopreservation. Hum Reprod 2007; 22: 2776–83.
68. Magistrini M, Szollosi D. Effects of cold and of isopropyl-N-phenylcarbamate on the second meiotic spindle of mouse oocytes. Eur J Cell Biol 1980; 22: 699–707.
69. Van der Elst J, Van den Abbeel E, Nerinckx S, Van Steirteghem A. Parthenogenetic activation pattern and microtubular organization of the mouse oocyte after exposure to 1,2-propanediol. Cryobiology 1992; 29: 549–62.
70. Sathananthan AH, Kirby C, Trounson A, Philipatos D, Shaw J. The effects of cooling mouse oocytes. J Assist Reprod Gen 1992; 9: 139–48.
71. Wang WH, Meng L, Hackett RJ, Keefe DL. Developmental ability of human oocytes with or without birefringent spindles imaged by Polscope before insemination. Hum Reprod 2001; 16: 1464–8.
72. Pickering SJ, Johnson MH. The influence of cooling on the organization of the meiotic spindle of the mouse oocyte. Hum Reprod 1987; 2: 207–16.
73. Pickering SJ, Braude PR, Johnson MH, Cant A, Currie J. Transient cooling to room temperature can cause irreversible disruption of the meiotic spindle in the human oocyte. Fertil Steril 1990; 54: 102–8.
74. Almeida PA, Bolton VN. The effect of temperature fluctuations on the cytoskeletal organisation and chromosomal constitution of the human oocyte. Zygote 1995; 3: 357–65.
75. Inoue S, Oldenbourg R. Microtubule dynamics in mitotic spindle displayed by polarized light microscopy. Mol Biol Cell 1998; 9: 1603–7.
76. Keefe D, Liu L, Wang W, Silva C. Imaging meiotic spindles by polarization light microscopy: principles and applications to IVF. Reprod Biomed Online 2003; 7: 24–9.
77. Oldenbourg R. Polarized light microscopy of spindles. Method Cell Biol 1999; 61: 175–208.
78. Johnson MH, Pickering SJ. The effect of dimethylsulphoxide on the microtubular system of the mouse oocyte. Development 1987; 100: 313–24.
79. Van Blerkom J. Developmental failure in human reproduction associated with chromosomal abnormalities and cytoplasmic pathologies in meiotically mature oocytes. In: van Blerkom J, ed. The Biological Basis of Early Human Reproductive Failure. New York: Oxford University Press, 1994: 283–326.
80. Bos-Mikich A, Wood MJ, Candy CJ, Whittingham DG. Cytogenetical analysis and developmental potential of vitrified mouse oocytes. Biol Reprod 1995; 53: 780–5.
81. Chen SU, Lien YR, Chen HF, et al. Open pulled straws for vitrification of mature mouse oocytes preserve patterns of meiotic spindles and chromosomes better than conventional straws. Hum Reprod 2000; 15: 2598–603.
82. Cobo A, Rubio C, Gerli S, et al. Use of fluorescence in situ hybridization to assess the chromosomal status of embryos obtained from cryopreserved oocytes. Fertil Steril 2001; 75: 354–60.

83. George MA, Pickering SJ, Braude PR, Johnson MH. The distribution of alpha- and gamma-tubulin in fresh and aged human and mouse oocytes exposed to cryoprotectant. Mol Hum Reprod 1996; 2: 445–56.
84. Mullen SF, Agca Y, Broermann DC, et al. The effect of osmotic stress on the metaphase II spindle of human oocytes, and the relevance to cryopreservation. Hum Reprod 2004; 19: 1148–54.
85. Ciotti PM, Porcu E, Notarangelo L, et al. Meiotic spindle recovery is faster in vitrification of human oocytes compared to slow freezing. Fertil Steril 2008; DOI: 10.1016/j.fertnstert.2008.03.013.
86. Gardner DK, Pawelczynski M, Trounson AO. Nutrient uptake and utilization can be used to select viable day 7 bovine blastocysts after cryopreservation. Mol Reprod Dev 1996; 44: 472–5.
87. Kaidi S, Bernard S, Lambert P, et al. Effect of conventional controlled-rate freezing and vitrification on morphology and metabolism of bovine blastocysts produced in vitro. Biol Reprod 2001; 65: 1127–34.
88. Lane M, Maybach JM, Gardner DK. Addition of ascorbate during cryopreservation stimulates subsequent embryo development. Hum Reprod 2002; 17: 2686–93.
89. Walker DL, Gardner DK, Lane M, et al. Cryosystem assessment by glucose uptake of murine blastocysts. Reprod Biomed Online 2005; 11: 601–7.
90. Stokes PJ, Hawkhead JA, Fawthrop RK, et al. Metabolism of human embryos following cryopreservation: implications for the safety and selection of embryos for transfer in clinical IVF. Hum Reprod 2007; 22: 829–35.
91. Leese HJ. Metabolism of the preimplantation mammalian embryo. Oxf Rev Reprod Biol 1991; 13: 35–72.
92. Tachataki M, Winston RM, Taylor DM. Quantitative RT-PCR reveals tuberous sclerosis gene, TSC2, mRNA degradation following cryopreservation in the human preimplantation embryo. Mol Hum Reprod 2003; 9: 593–601.
93. Boonkusol D, Gal AB, Bodo S, et al. Gene expression profiles and in vitro development following vitrification of pronuclear and 8-cell stage mouse embryos. Mol Reprod Dev 2006; 73: 700–8.
94. Dhali A, Anchamparuthy VM, Butler SP, et al. Gene expression and development of mouse zygotes following droplet vitrification. Theriogenology 2007; 68: 1292–8.
95. Katz-Jaffe MG, Larman MG, Sheehan CB, Gardner DK. Exposure of mouse oocytes to 1,2-propanediol during slow freezing alters the proteome. Fertil Steril 2008; 89(Suppl 1): 1441–7.
96. Merchant M, Weinberger SR. Recent advancements in surface-enhanced laser desorption/ionization-time of flight-mass spectrometry. Electrophoresis 2000; 21: 1164–77.
97. Gardner DK, Sheehan CB, Rienzi L, Katz-Jaffe M, Larman MG. Analysis of oocyte physiology to improve cryopreservation procedures. Theriogenology 2007; 67: 64–72.
98. Sheehan CB, Lane M, Gardner DK. The Cryloop facilitates re-vitrification of embryos at four successive stages of development without impairing embryo growth. Hum Reprod 2006; 21: 2978–84.
99. Kuleshova LL, Lopata A. Vitrification can be more favorable than slow cooling. Fertil Steril 2002; 78: 449–54.
100. Vajta G, Nagy ZP. Are programmable freezers still needed in the embryo laboratory? Review on vitrification. Reprod Biomed Online 2006; 12: 779–96.
101. Morley P, Whitfield JF. The differentiation inducer, dimethyl sulfoxide, transiently increases the intracellular calcium ion concentration in various cell types. J Cell Physiol 1993; 156: 219–25.
102. Hanslick JL, Lau K, Noguchi KK, et al. Dimethyl sulfoxide (DMSO) produces widespread apoptosis in the developing central nervous system. Neurobiol Dis 2009; 34: 1–10.
103. Stroncek DF, Fautsch SK, Lasky LC, et al. Adverse reactions in patients transfused with cryopreserved marrow. Transfusion 1991; 31: 521–6.
104. Gardner DK, Lane M. Ex vivo early embryo development and effects on gene expression and imprinting. Reprod Fertil Dev 2005; 17: 361–70.
105. Lane M, Gardner DK. Understanding cellular disruptions during early embryo development that perturb viability and fetal development. Reprod Fertil Dev 2005; 17: 371–8.

16 Quality Management of a Low Temperature Storage Program in the Human IVF Laboratory

Lucia De Santis
IVF Unit, Department of Obstetrics and Gynaecology, University Vita-Salute, H.S. Raffaele, Milan, Italy

THE PLACE OF CRYOPRESERVATION IN ASSISTED REPRODUCTION TECHNOLOGY

In the mid-1980s and during the successive decade, the cryopreservation of human mature oocytes was believed to be unfeasible in consideration of the existing knowledge and technology. In effect, the first reports on the use of frozen oocytes in the clinical setting were in many respects rather disappointing and overall unable to prove the applicability of this form of fertility preservation (1,2). In the last few years, however, studies of fundamental cryobiology (3), empirical observations (4) and more systematic clinical experiences (5–10) have generated a renewed interest in this subject. More rational and theoretical approaches could provide the key to the development of methods able to guarantee greater success (11,12). Currently, several clinics offer oocyte cryopreservation as a form of treatment worldwide. Italy provides an extreme example where, as an effect of the introduction of a very restrictive legislation, oocyte cryopreservation has been adopted rather diffusely as an alternative to embryo freezing, and treatment cycles can be counted in their thousands. Oocyte cryopreservation is attracting additional interest as a tool for better management of oocyte donation cycles (see chap. 23). Future studies will need to address the issue of the actual efficiency of oocyte cryopreservation, in order to provide IVF specialists with criteria to enable them to measure performance and, in the final analysis, guide patients in their choices among different treatment options.

GENERAL AND SPECIFIC REQUIREMENTS OF CRYOPRESERVATION PROCEDURES

The implementation of an appropriate program of internal quality control laboratory and the verification of external quality schemes are essential for maintaining a good level of service offered to couples who seek treatment from a centre for assisted reproduction. This is particularly relevant to situations in which cryopreservation is considered, in view of the particular ethical and legal implications originating from the long-term storage of embryos and gametes. The analysis and management of the activity of the IVF laboratory has been a subject of increasing interest since the early eighties (13,14–16). Despite several clinical and technological advances that have significantly increased the results in terms of pregnancies, the application of basic management principles and the standardization of procedures remains a major avenue through which the clinical outcome may be further improved (17,18). In the context of these general considerations, this chapter illustrates some basic elements concerning the cryopreservation of oocytes.

The principle of keeping cells in a "frozen" state is conceptually simple but in reality requires an understanding of a wide variety of biophysical events and careful control of the various practical elements required for cryopreservation. One of the major problems posed by cryopreservation is the risk of inadvertently generating a decrease in the viability of the biological material as an effect of the formation of intracellular ice. Cell damage may also derive from chemical toxicity or osmotic stress subsequent to the addition or removal of cryoprotectant and from "chilling injury" (even in the absence of ice formation), i.e., from direct detrimental effects of low supra-zero temperatures on bio-structures.

The recent interest generated by the approach of vitrification in reproductive biology derives from the possibility of preserving cells in liquid nitrogen in a vitreous state in the absence of the formation of extra- or intracellular ice crystals, thereby reducing the risk of mechanical damage. However, irrespective of whether the slow freezing or the vitrification approach is applied, a rigorous management system should be adopted, taking into account the sensitivity of the oocyte to adverse physical conditions and the influence of different procedural parameters (time, manual handling and experience of the operator) on the survival and viability of the cell after cryostorage.

LABORATORY SPACE, EQUIPMENT, LOGBOOKS, AND DATABASES

In terms of mobilization of resources and time, cryopreservation can represent an important fraction of the IVF laboratory activity. As a consequence, a specific working area should be identified for the manipulation and storage of cryopreserved material. This area should be adjacent to the main lab space, accessible only by authorized members of staff, subjected to a sufficient number of air changes in order to prevent accumulation of nitrogen in its atmosphere, and equipped with a system for O_2 level monitoring. In laboratories in which slow cooling is applied routinely two cryofreezers should be available, to ensure a backup in case of failure of one of the two instruments and to run separate cycles of freezing, if required by the laboratory schedule. For each freezing cycle, cryofreezers generate temperature charts that should be enclosed in the patient's laboratory data. Liquid nitrogen levels in cryotanks for long-term storage should be regularly checked and documented. Cryotanks could be also equipped with automatic alarm devices able to send information to remote users. Safety measures in the cryopreservation area should be always adopted. Members of staff should be instructed how to handle liquid nitrogen properly and have access to dedicate protective equipment.

Quality control and quality assurance programs imply monitoring and recording of the modalities by which gametes and embryos are cryopreserved. The resulting information can afterwards be analyzed for various purposes (e.g., for assessments of efficacy and consistency). Each laboratory should have specific flow charts describing the route that the material (this also applies to semen and embryos) takes from the time of collection to the end of the cryopreservation process and operating instructions illustrating the cryopreservation protocol. All relevant aspects of a cryopreservation procedure (patient identification details, cryopreservation protocol, type and lot of media, containment device, localization in the long-term storage container, etc.) should be stored in an electronic database with a backup in a safe memory area (e.g., the server of the clinic's IT system).

Given below is an example of documentation that can enable the verification of the cryopreservation method, traceability of sample and safety of storage.

Certification standards require that there must be a specific operational instruction for each laboratory procedure including cryopreservation reporting:

1. Purpose
2. Scope
3. Acronyms and definitions
4. Liability
5. Description of activities
6. Registration and storage
7. Informed consent

Detailed description of the procedure is essential to unambiguously identify the cryopreserved material of each patient/couple at any time.

Each operator (embryologist) is responsible for correct sample identification and performance of the procedure. To this end, the operator should mark in a relevant laboratory form each step of the procedure in which he/she has been involved. The operator is also responsible for the conformity of his/her actions with existing legislation (e.g., in Italy Law 40/2004) and related guidelines.

In the event of an error in the procedure, the operator is required to report to his/her manager, fill in a form in which errors and adverse events may be accurately described and recorded for future analysis.

STAFF TRAINING AND CONSISTENCY OF OPERATOR PERFORMANCE

All methods of oocyte cryopreservation, regardless of the approach used (vitrification or slow freezing), are subject to the same strict rules governing embryo freezing. Staff training and experience can largely affect clinical results. Manual skills for oocyte handling, together with adherence to operational times, significantly influence the outcome in terms of survival. Both develop progressively and may be described by a learning curve. Greater the background experience of the laboratory in embryo storage, the shorter is the profile of the learning curve, irrespective of the type of cryopreservation protocol used. Following the introduction of an oocyte cryopreservation method in a laboratory without previous experience, survival rates are initially low, rising to optimal levels only after practice. Once a critical mass of experience in cryopreservation techniques is gained, the same laboratory can immediately achieve ideal survival rates with other newly introduced protocols (19). This is substantiated by the fact that laboratories with proven experience in embryo cryopreservation usually develop a high competence in oocyte cryopreservation (20).

TIME INVESTMENT REQUIRED BY CRYOPRESERVATION PROCEDURES IN THE DAY-TO-DAY ROUTINE

The management of a cryopreservation program involves accurate assessment of the impact that the time required by a specific methodology (19) may have on other laboratory procedures. In slow freezing the actual phase of cooling from physiologic to the liquid nitrogen temperature involves a relatively long period of time (2–3 hours). However, this is not necessarily a drawback, because while the sample temperature is lowered automatically by a cryofreezer, the operator can be involved in other laboratory procedures. Automated cooling also offers the advantage of simultaneously processing a high number of oocytes, often of different patients, once they have been exposed in parallel to the dehydration/loading solutions. This is particularly important for laboratories with large workloads. However, despite this practical advantage, the simultaneous freezing of oocytes of different patients may involve the potential risk of protracting excessively the time of culture between egg pick up and the beginning of the freezing procedure, with possible consequences of ageing in vitro for some of these oocytes (21). Therefore, the precise time at which the freezing procedure is started during the daily schedule should ideally represent a compromise between the different times of retrieval (or human chorionic gonadotropin administration) of each cohort of oocytes and the necessity to cryopreserve at least some of these cohorts simultaneously. A recent study seems to confirm that the time that oocytes spend in culture before freezing has an effect on the clinical outcome (22). Therefore, during a busy day with multiple egg retrievals, it may be appropriate to run more than one freezing cycle in an attempt to standardize the pre-freeze culture time.

Vitrification offers the possibility of repeating sequentially the cryopreservation procedure, which lasts approximately 15 to 20 minutes depending on the amount of material, for each cohort of oocytes. This has the obvious advantage of setting precisely the time intervening between egg retrieval and cryopreservation, but has significant implications in situations in which several cohorts of oocytes need to be cryopreserved. In such cases, the overall operative time could in fact exceed the one required by a slow cooling protocol for the storage of an equivalent number of oocytes.

In light of this, the choice of slow cooling or vitrification as an oocyte cryopreservation methodology should be carefully considered on the basis of the amount of material to be stored, number of embryologists and particular pieces of equipment (flow hoods) available at a specific time, and coordination with other laboratories duties. Slow cooling and vitrification are not necessarily mutually exclusive, instead (once established their ability to give rise to adequate rates of efficiency) one could be preferred over the other case by case, depending on a series of factors.

CHOICE OF CRYOPRESERVATION APPROACH AND METHOD

Established vs. Newly Introduced Methodology

The choice of a specific approach and protocol of oocyte cryopreservation should be obviously inspired by principles of efficacy and safety. As discussed elsewhere in this volume, the lack of current controlled prospective studies does not facilitate decision making based on objective considerations. The adoption of a given method should also be guided by other more general criteria. One important aspect concerns the previous cryopreservation history of the

IVF lab. Normally, all members of an IVF team can easily develop the ability to perform a slow cooling procedure, as the technicalities involved (exposure of embryos to cryopreservation media and loading into 250-µm straws) are not unlike those of other IVF techniques (e.g., embryo transfer). As such, low cooling cryopreservation has been widely implemented since the report of the first human pregnancy from a frozen embryo (23), becoming the norm for over 25 years. Many IVF laboratories have gained extensive experience in this approach, making it an essential part of their IVF program. Thus the introduction of an oocyte slow cooling method, as opposed to the introduction of a vitrification methodology, can occur without particular requirements in terms of training, equipment, and choice and management of materials. Vitrification, described for the first time in 1985 (24), on the other hand has found application in human IVF only in recent times, initially for the preservation of blastocysts. Its reputation and use is increasing, but even now only a few laboratories have developed sufficient expertise. Therefore, in most IVF centers the adoption of vitrification does not take place as a simple extension of a pre-existing ability but would instead generate specific questions. Although vitrification demands specific manual skills that belong to the range of the technical potential of an embryologist, it certainly requires a specific training program. This could present a problem because very few centers have documented oocyte vitrification experience, in qualitative and quantitative terms. In most cases, vitrification also implies the use of specific storage devices (cryotop, cryoloop, etc.) as opposed to the classical straws used in slow cooling cryopreservation. This may appear to be a small difference, but the choice of a containment device can have a significant impact on storage in liquid nitrogen. For example, the first generation of cryotop (25) has been reported to float when placed in a visotube filled with liquid nitrogen, causing an obvious problem in storage. This has necessitated the development of a new version of cryotop that ensures a safer containment. As an approach of recent application, vitrification carries specific problems that can be definitely solved but that demand a particular commitment and can be expected to generate a learning curve whose profile and extension in time may depend on a variety of factors.

Sustainable Margins of Errors

One of the most important aspects of a medical or biological procedure is its reproducibility. In the case of human IVF, the reproducibility of an oocyte cryopreservation method is essential, especially to give patients a realistic estimate of success (or failure) per cycle of stimulation. Reproducibility depends on a number of factors, some of which are applicable to all kinds of protocols, while others are specific to a given approach. An example of a general factor is the fidelity with which a protocol is applied by the different members of an IVF team. To assure reproducibility and leave no room for personal interpretations, protocol instructions should be formulated or approved by a lab director, produced in a written form, always accessible, and periodically updated. Other factors are more relevant to a specific cryopreservation approach. For instance, in slow cooling cryopreservation certain relatively minor deviations from a standardized set of conditions may not necessarily have an effect on reproducibility and therefore on overall success. Particularly, in slow cooling cryopreservation the time for incubation in freezing (dehydration) and thawing (rehydration) solutions is within a range of 5 to 10 minutes. During each of these intervals, the amplitude of the oocyte response to anisosmotic conditions imposed by a given dehydration/rehydration step may be relatively large at the beginning, but tends to be smaller or minimal towards the end. This is the case of the phase of dehydration generated by a freezing mixture including PrOH 1.5 M and 0.2 M sucrose, a treatment representing a foundation of successful slow cooling protocols (5,8). When an oocyte is exposed to such a mixture, it responds with a major decrease in volume (20%) relatively rapidly, within about two minutes (3). This is followed by a slower rate of shrinkage. Under these conditions, it is required that the oocytes are incubated for five minutes. At the end of this interval, dehydration occurs slowly. This implies that if the prescribed incubation time of five minutes is shortened or extended by, for example, 30 seconds, accidentally by an operator, an unimportant difference in the rate of dehydration could be expected. Therefore, although the degree of dehydration is critical for controlling the amount of intracellular ice formation and the associated risk of cell damage, the above described deviation from protocol is likely to cause little or no effect in relation to the survival and viability of the stored material. Similar deviations to vitrification protocols could cause more significant effects. One of the most popular vitrification protocols requires incubation in an equilibration solution containing 7.5% DMSO and 7.5% EG for 10 minutes, followed by incubation for 1 minute in a vitrification solution including 15% DMSO and 15% EG and 0.5 M sucrose. Immediately afterward, oocytes are loaded onto cryotop and directly exposed to liquid nitrogen. Considering the massive cryoprotectant concentration in the vitrification solution, it appears plausible that even small variations (20–30 seconds) in the incubation time could importantly affect the rate of dehydration and/or osmotic stress, with possible effects on the reproducibility of the method. Considerations on the reproducibility of slow cooling and vitrification approaches could be extended to other phases of the cryopreservation process. For examples, vitrification requires an extremely rapid warming (a fraction of a second), while the "rapid thawing" of slow cooling protocols can take place within 30 to 60 seconds (in air and a 30°C water bath). Because the rates of warming/thawing are of utmost importance to achieve maximal survival and viability, it is obvious how an inappropriate timing of recovery from liquid nitrogen could differentially affect the two approaches . In general, it seems that the less extreme and more controllable cryopreservation conditions of slow cooling can tolerate larger margins of errors, assuring a higher reproducibility, irrespective of considerations of efficiency. This appears to be confirmed by the fact that the same slow cooling protocol (4) has given rise to virtually the same rate of implantation rate per thawed oocyte by different IVF laboratories (6,7,26).

Storage Devices: Efficiency and Safety

The design and characteristics of storage devices can have a significant impact on cryopreservation. In fact, straws, open pulled straws (OPS), cryoloops, cryotops, etc., generate differences in volume of the freezing/vitrification solution that can influence cooling/warming rates. They also provide different degrees of containment and, therefore, protection from the risk of contamination.

In slow cooling cryopreservation, the 250-μm straw has been the most widely used storage device. It ensures easy loading, with an advantage of reproducibility. The "high security" version can be loaded and sealed hermetically with no risk of contamination. Straws are rarely used in vitrification protocols where extremely small volumes (<2 μl) are required in order to achieve very high cooling and warming rates. Almost all vitrification devices involve direct exposure of samples to liquid nitrogen during the actual vitrification process and are not compatible with a safe system of sealing. Cryoleaf (27) and cryotop (25) are rather similar in design and use. They are constituted from a holder and a narrow strip at the end of which a very small volume of vitrification solution containing one or a few oocytes are placed. After vitrification in direct contact with liquid nitrogen, the strip is inserted into a shaft while still submerged in liquid nitrogen and the assembled device is moved to a dewar for long term storage. Cryoloop (28) is another vessel that has been used in vitrification, especially for blastocyst cryopreservation. Its design includes a screw cap to which a very small nylon loop is connected. Samples carried in an extremely small volume of vitrification solutions are loaded onto the thin film, which may be created by briefly plunging the loop into a drop of the same vitrification solution. Once the samples are loaded and vitrified by direct exposure to liquid nitrogen, the loop is inserted into a vial and the whole device is transferred into a dewar. Electron microscopy grids (29) may be employed to load a few microliters of solution containing one to three oocytes, but then, again, vitrification requires direct contact with liquid nitrogen, while long term storage occurs in cryovials that do not ensure safe containment. Fine capillaries have been proposed to offer a safer solution for vitrification, considering that they can be loaded with very few microliters of vitrification solution sealed at both ends and protected by a metal shaft or a high security straw (30) for long term storage. However, no clinical data involving the use of capillaries has been published so far. The loading of 2 μl or less of vitrification solution onto one these devices is an action that may readily learned, but it is difficult to estimate whether different vessels are associated with a carry over of comparable amounts of medium together with the oocyte(s). In addition, loading may not always be performed exactly within the same time. Because the volume of the loading solution and exposure time to the same solution have implications in terms of cooling and toxic effects, the choice of the storage device may affect the efficiency of the overall cryopreservation procedure. In this connection, in a pilot study involving human oocytes it has been reported that, while using the same vitrification protocol (25), the cryoleaf system consistently generates much higher survival rates than the cryoloop (>80% and less than 50%, respectively) (31). However, no clinical study has been conducted so far.

Contamination from cryopreservation is a concern that involves various areas of medicine and biotechnology and has prompted the evolution and implementation of specific practices. Yet, this issue does not seem to have been properly addressed in human IVF. In particular, vitrification raises particular questions, as described above. Microbial contamination of cryopreserved samples is not only a theoretical risk (32,33). Sources of contamination may be linked to liquid nitrogen itself or to other biological samples stored in the same dewar. Storage in nitrogen vapors reduces, but does not eliminate risks. Paradoxically, when released from production plants, liquid nitrogen has a very low microbial count. Contamination occurs, in fact, during storage and distribution, where sterilization may be not performed on a laboratory scale. On the other hand, microorganisms are well preserved by simple contact with liquid nitrogen. A particular source of contamination is represented by airborne microorganisms that are captured by minute highly electrostatic ice crystals that form above the liquid nitrogen level and fall again into the same liquid. This phenomenon creates an effect of accumulation of microbial sediment that may account for 1–5% of the content of a dewar. Viable microorganisms of different species, some of clinical relevance, have been isolated from liquid nitrogen and thawed samples (32), drawing particular attention to the issue of safety of cryopreserved oocytes. Concerning this, the current minimum standard requires the separate storage of samples from HBV, HCV, and (where applicable) HIV positive patients.

CONCLUSIONS

Cryopreservation is a well established IVF capability that has found large application for the storage of embryos and spermatozoa. In recent years, it has been progressively extended to the storage of mature oocytes in many laboratories, especially those subjected to laws that prohibit embryo cryopreservation. A 25-year long experience has been gained in slow cooling cryopreservation, leading to the generation of approximately 500,000 children (i.e., 15–20% of all IVF births). This should be taken into account when choosing a novel method for oocyte cryopreservation. However, the apparently high efficacy of vitrification imposes on the embryologist the need to develop new skills and embrace novel technical options. Whatever the method of choice, assurance of quality should be a mandatory goal to achieve consistency, safety and ultimately efficacy. This requires meticulous management of work areas, members of staff, equipment, materials and documentation.

REFERENCES

1. Chen C. Pregnancy after human oocyte cryopreservation. Lancet 1986; 1: 884–6.
2. Al-Hasani S, Diedrich K, van der Ven H, et al. Cryopreservation of human oocytes. Hum Reprod 1987; 2: 695–700.
3. Paynter SJ, Borini A, Bianchi V, et al. Volume changes of mature human oocytes on exposure to cryoprotectant solutions used in slow cooling procedures. Hum Reprod 2005; 20: 1194–9.
4. Fabbri R, Porcu E, Marsella T, et al. Human oocyte cryopreservation: new perspectives regarding oocyte survival. Hum Reprod 2001; 16: 411–16.
5. Porcu E, Fabbri R, Damiano G, et al. Clinical experience and applications of oocyte cryopreservation. Mol Cell Endocrinol 2000; 169: 33–7.
6. Borini A, Sciajno R, Bianchi V, et al. Clinical outcome of oocyte cryopreservation after slow cooling with a protocol utilizing a high sucrose concentration. Hum Reprod 2006; 21: 512–17.
7. De Santis L, Cino I, Rabellotti E, et al. Oocyte cryopreservation: clinical outcome of slow-cooling protocols differing in sucrose concentration. Reprod Biomed Online 2007; 14: 57–63.
8. Bianchi V, Coticchio G, Distratis V, et al. Differential sucrose concentration during dehydration (0.2 mol/L) and rehydration (0.3 mol/L) increases the implantation rate of frozen human oocytes. Reprod Biomed Online 2007; 14: 64–71.
9. Boldt J, Tidswell N, Sayers A, et al. Human oocyte cryopreservation: 5-year experience with a sodium-depleted slow freezing method. Reprod Biomed Online 2006; 13: 96–100.
10. Cobo A, Kuwayama M, Perez S, et al. Comparison of concomitant outcome achieved with fresh and cryopreserved donor oocytes vitrified by the Cryotop method. Fertil Steril 2008; 89: 1657–64.

11. Paynter SJ. A rational approach to oocyte cryopreservation. Reprod Biomed Online 2005; 10: 578–86.
12. Fahy GM. Theoretical considerations for oocyte cryopreservation by freezing. Reprod Biomed Online 2007; 14: 709–14.
13. Dandekar PV, Quigley MM. Laboratory setup for human in vitro fertilization. Fertil Steril 1984; 42: 1–12.
14. Dawson KJ. Quality control and quality assurance in IVF laboratories in the UK. Hum Reprod 1997; 12: 2590–1.
15. Byrd W. Quality assurance in the reproductive biology laboratory. Arch Pathol Lab Med 1992; 116: 418–22.
16. Kiechle FL, Quattrociocchi-Longe TM, Brinton DA. Quality improvement in the laboratory assessment of in vitro fertilization. Arch Pathol Lab Med 1992; 116: 410–17.
17. Matson PL. Internal quality control and external quality assurance in the IVF laboratory. Hum Reprod 1998; 13: 156–65.
18. Magli MC, Van den Abbee IE, Lundin K, et al. Revised guidelines for good practice in IVF laboratories. Hum Reprod 2008; 23: 1253–62. Epub Mar 28, 2008.
19. De Santis L, Cino I, Coticchio G, et al. Objective evaluation of the viability of cryopreserved oocytes. Reprod Biomed Online 2007; 15: 338–45.
20. Borini A, Cattoli M, Bulletti C, et al. Clinical efficiency of oocyte and embryo cryopreservation. Ann N Y Acad Sci 2008; 1127: 49–58.
21. Ducibella T. Biochemical and cellular insights into the temporal window of normal fertilization. Theriogenology 1998; 49: 53–65.
22. Parmegiani L, Cognigni GE, Bernardi S, et al. Freezing within 2 h from oocyte retrieval increases the efficiency of human oocyte cryopreservation when using a slow freezing/rapid thawing protocol with high sucrose concentration. Hum Reprod 2008; 23: 1771–7.
23. Trounson A, Mohr L. Human pregnancy following cryopreservation, thawing and transfer of an eight-cell embryo. Nature 1983; 305: 707–9.
24. Rall WF, Fahy GM. Ice-free cryopreservation of mouse embryos at –196°C by vitrification. Nature 1985; 313: 573–5.
25. Kuwayama M, Vajta G, Kato O, et al. Highly efficient vitrification method for cryopreservation of human oocytes. Reprod Biomed Online 2005; 11: 300–8.
26. Levi Setti PE, Albani E, Novara P, et al. Results of in vitro fertilization in Italy after the introduction of a new law. Fertil Steril 2008; 90: 1081–6.
27. Chian RC, Gilbert L, Huang JY, et al. Live birth after vitrification of in vitro matured human oocytes. Fertil Steril 2009; 91: 372–6.
28. Liebermann J, Tucker MJ, Sills ES. Cryoloop vitrification in assisted reproduction: analysis of survival rates in >1000 human oocytes after ultra-rapid cooling with polymer augmented cryoprotectants. Clin Exp Obstet Gynecol 2003; 30: 125–9.
29. Yoon TK, Kim TJ, Park SE, et al. Live births after vitrification of oocytes in a stimulated in vitro fertilization-embryo transfer program. Fertil Steril 2003; 79: 1323–6.
30. Ciotti PM, Porcu E, Notarangelo L, et al. Meiotic spindle recovery is faster in vitrification of human oocytes compared to slow freezing. Fertil Steril 2008; DOI: 10.1016/j.fertnstert.2008.03.013.
31. Gambardella A, Sciajno R, Coticchio G, et al. Survival and meiotic spindle analysis through Polscope in human oocytes vitrified with either cryoleaf or cryoloop. Hum Reprod 2008; 23(Suppl 1): i151–2.
32. Fountain D, Ralston M, Higgins N, et al. Liquid nitrogen freezers: a potential source of microbial contamination of hematopoietic stem cell components. Transfusion 1997; 37: 585–91.
33. Bielanski A, Nadin-Davis S, Sapp T, Lutze-Wallace C. Viral contamination of embryos cryopreserved in liquid nitrogen. Cryobiology 2000; 40: 110–16.

17 The Contribution of Embryo Cryopreservation to the Clinical Efficacy of ART and to a Reduction of ART Multiple Pregnancies

Karl G. Nygren
Queen Sophia Hospital, Stockholm, Sweden

INTRODUCTION

The development of cryopreservation techniques eventually resulted in the first baby being born in 1983 following freezing-and-thawing of human embryos (1). Today this technique accounts for an increasing proportion of all in vitro fertilization (IVF) treatment cycles. ICMART, the International Committee Monitoring ART, in its latest World Report (2) noticed a worldwide increase of the proportion of frozen-and-thawed embryo replacements (FER) or transfers (FET), from 8% in 1991 to 20% in 2004 (Fig. 1). In Europe in 2004, 25% of all embryo transfers were FER (3). ICMART estimates that, worldwide in 2008, altogether about 3.5 million babies have been born after IVF. Possibly, some 15% to 20% of these come from frozen-and-thawed embryos, corresponding to about 500,000 babies. In recent years, cryopreservation of blastocysts and of oocytes has also been introduced in clinics, although cryopreservation of cleavage stage embryos continues to dominate.

CLINICAL EFFICACY AND MULTIPLE PREGNANCY RATES INTER-RELATE

When IVF was first introduced in 1978, natural cycle treatments were used, with no ovarian stimulation, only one mature egg was harvested and only one single embryo was transferred back to the uterus. The efficacy was low. To enhance efficacy, controlled ovarian hyperstimulation was introduced to enable clinicians to harvest more than one egg and to transfer more than one embryo. It soon became accepted, as good clinical practice, to often transfer all the embryos that resulted from IVF, and transferring up to five to six embryos was not uncommon.

And, indeed, efficacy increased, at least when reported as "pregnancy rates per transfer." But at the same time multiple pregnancy and delivery rates increased dramatically, from the natural rate of about 2% and up 30% and more in most countries, and high order of multiples, triplets or more, were not uncommon.

At first, these developments were not regarded as questionable or harmful, but soon reports were published on serious health problems of these children, primarily arising from pre-maturity and birth defects. The possibility of freezing some of the embryos, to be used for later transfers, opened up, reducing the number of fresh embryos transferred.

Consequently, over the last 10 years or more, the proportion of cryopreservation of embryos increased and the number of fresh embryos transferred actually diminished worldwide, albeit more in some countries, or much more, than in others. Tables 1–3 demonstrate data for Europe on the decreasing number of embryos transferred over time, 1997–2005 (Table 1), the reduction of multiple deliveries over the same time period (Table 2) and increasing pregnancy rates per transfer after IVF, ICSI, FER, and ovum donation (OD) (Table 3). So, as the proportion of FER increased, the number of embryos transferred fresh decreased as did the proportion of multiple deliveries, while pregnancy rates increased following both fresh transfer and FER.

REPORTING EFFICACY—SEVERAL OPTIONS

In the early days of IVF, the attitude of professionals and patients alike was to focus on efficacy rather than safety. This attitude continues to some extent. Often driven by economic realities,

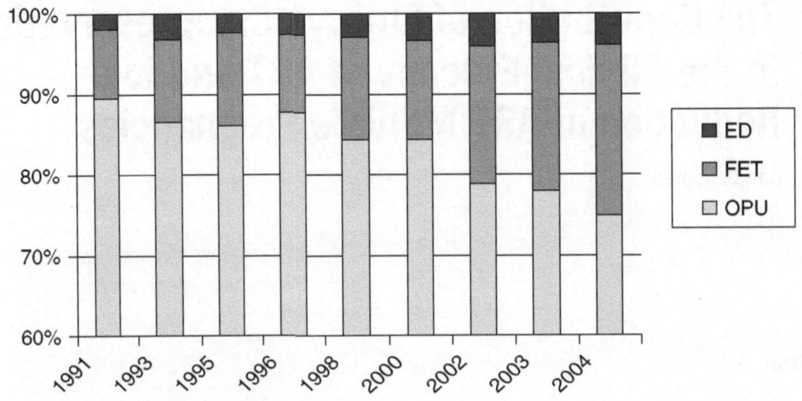

Figure 1 Proportions, over time and world-wide, of frozen-and-thawed embryo transfers (FET) to egg donation (ED) procedures and ovum pick-ups (OPU). *Source*: From ICMART 2008, presentation at ESHRE.

Table 1 Number of Embryos Transferred per Procedure, IVF and ICSI Combined, in Europe (1997–2005)

	2005	2003	2001	1999	1997
1	**20.0**	15.7	12.0	11.9	11.5
2	**56.2**	55.9	51.7	39.2	35.7
3	**21.5**	24.9	30.8	39.6	38.4
4	**2.3**	3.5	5.5	9.3	14.2

Abbreviations: ICSI, intracytoplasmic sperm injection; IVF, in vitro fertilization.
Source: From EIM Consortium 2008, presentation at ESHRE.

Table 2 Proportions of Singleton and Multiple Deliveries, IVF and ICSI Combined in Europe (1997–2005)

	2005	2003	2001	1999	1997
Singleton	**78.2**	76.7	74.5	73.7	70.4
Twin	**19.7**	22.0	24.0	24.0	25.8
Triplet	**0.8**	1.1	1.5	2.2	3.6
Quadruplet	**0.01**	0.08	0.02	0.1	0.15

Abbreviations: ICSI, intracytoplasmic sperm injection; IVF, in vitro fertilization.
Source: From EIM Consortium 2008, presentation at ESHRE.

Table 3 Pregnancy Rates per Transfer in Europe, by Method (1997–2005)

	2005	2001	1997
IVF	**30.4**	29.0	26.1
ICSI	**30.3**	28.3	26.4
FER	**19.3**	16.4	15.2
OD	**42.0**	33.4	27.1

Abbreviations: FER, frozen-and-thawed embryo replacements; ICSI, intracytoplasmic sperm injection; IVF, in vitro fertilization; OD, ovum donation.
Source: From EIM Consortium 2008, presentation at ESHRE.

where IVF treatments remain expensive, patients and doctors tend to set the multiple pregnancy issue aside, to reach maximal efficacy in the one/only treatment cycle affordable.

A consequence of this focus is that efficacy is often reported as "pregnancy per embryo transfer," which gives a higher figure and thereby an impression of a higher efficacy compared to reporting "delivery per started cycle" which actually is a much better and more comprehensive way of describing efficacy. Reporting "delivery" rather than "pregnancy," is more accurate as it represents the true end point of the treatment and of actually having a baby and excludes pregnancy losses (e.g., miscarriage). "Started cycle" is also more relevant than "embryo transfer" because it includes cancellation of treatment cycles, which occurs for a variety of reasons, some to enhance what is perceived as efficacy. Also, "pregnancy rate" is often referred to as "success rate," suggesting success to be equal to efficacy, when in fact, success is rather the balance between several benefit indicators (i.e., efficacy, safety, quality, and cost). The correct reporting of the efficacy per treatment cycle should therefore be delivery per started cycle.

CURRENT TRENDS

In an effort to reduce or eliminate the increased occurrence of ART multiple pregnancy and delivery, there is a current clinical policy transition to use more of elective single embryo transfer (SET) (4), often followed by one or more FER from the same egg harvest procedure (3,5). In recent reports from the European IVF Monitoring (EIM) Consortium (3) and from ICMART (ref) this has been accounted for by introducing "cumulative rates" (i.e., the sum of efficacy rates from fresh embryo transfers and FER from the same year). This way of summarizing results from different women is of course not entirely statistically correct, but is a compromise to get an approximation of a rate that should really summarize the result from several consecutive transfers from the same woman after one egg harvest procedure. Table 4, from the latest EIM report exemplifies this way of reporting efficacy from three selected countries, showing that additional FER from the same initiated cycle and egg harvest does of course enhance the overall figure reported for efficacy, very prominently so in Finland, with a fresh delivery rate per initiated cycle of 18.7% and a cumulative rate of no less than 29.2%.

IN SUMMARY

International reporting shows an increase in the proportion of FER over time, and simultaneously a gradual increase in efficacy, reported as pregnancy per transfer, but even more so when reported as cumulative delivery rates per started cycle. Thus, the technique of cryopreservation of embryos has contributed to, or even been a prerequisite for, better efficacy, as indicated by the recently introduced relevant measure of clinical efficacy, which is the cumulative delivery rate per initiated cycle. During the same time period, the number of fresh and frozen transferred embryos has declined resulting in a substantial reduction of ART multiple pregnancies.

Table 4 Cumulative Rates of Deliveries per Initiated Cycle, Fresh and FER, Europe 2005

	Initiated cycles IVF and ICSI	Deliveries, "fresh" cycles IVF and ICSI	Multiple deliveries fresh	FER cycles (thawings)	Deliveries FER	Multiple deliveries FER	Deliveries fresh	Cumulative deliveries, fresh and FER	All multiples
Finland	4,731	887	104	–	491	52	**18.7%**	**29.15%**	**11.3%**
Sweden	9,415	2,056	127	3,205	211	53	**21.8%**	**24.1%**	**7.94%**
United Kingdom	32,001	7,328	1,903	7,349	1,210	204	**22.9%**	**26.7%**	**24.7%**

Abbreviations: FER, frozen-and-thawed embryo replacements; ICSI, intracytoplasmic sperm injection; IVF, in vitro fertilization.
Source: From EIM Consortium 2008, presentation at ESHRE.

FUTURE SCENARIOS

For the foreseeable future, cryopreservation of cleavage stage embryos will continue to be of utmost importance for the promotion of clinical efficacy combined with a reduction or even elimination of iatrogenic multiple pregnancy and delivery after IVF treatments. Later, clinical policy transitions, currently barely visible, like Mild Stimulation strategies and Low-cost IVF strategies (6), may challenge cryopreservation of embryos, especially so in low resources countries or settings. Sophisticated and expensive laboratory equipment for freezing and for cryopreservation may be replaced by less expensive but still high quality techniques (vitrification may be an example). Cryopreservation of oocytes for legal reasons in some countries and for cancer survivors has the potential to be increasingly important.

REFERENCES

1. Trounson A. Preservation of human eggs and embryos. Fertil Steril 1986; 46: 1–12.
2. Adamson D, de Mouzon J, Lancaster P, et al. World collaborative report on in vitro fertilisation 2000. Fertil Steril 2006; 85: 1586–622.
3. Nyboe Andersen A, Goossens V, Ferraretti AP, et al. Assisted reproductive technology in Europe, 2004: results generated from European registers by ESHRE. Hum Reprod 2008; 23: 756–71.
4. Karlström PO, Bergh C. Reducing the number of embroys transferred in Sweden—impact on delivery and multiple birth rates. Hum Reprod 2007; 22: 2202–7.
5. Tiitinen A, Halttunen M, Härkki P, et al. Elective single embryo transfer: the value of cryopreservation. Hum Reprod 2001; 16: 1140–4.
6. Pennings G, Ombelet W. Coming soon to your clinic: patient friendly ART. Hum Reprod 2007; 22: 2075–9.

18 Factors Influencing the Clinical Efficiency of Oocyte Cryopreservation

Debra A. Gook and David H. Edgar

Reproductive Services, Royal Women's Hospital/Melbourne IVF, and Department of Obstetrics and Gynaecology, University of Melbourne, Melbourne, Victoria, Australia

WHAT DO WE MEAN BY CLINICAL EFFICIENCY IN ASSISTED REPRODUCTION?

The question most frequently asked in relation to assisted reproductive technology (ART) is "What is the success rate?" Although this may seem to be a simple question requiring an equally simple answer, careful consideration brings a realization that any simple response will inevitably ignore a range of important qualifications. While the clear aim of ART is the birth of a healthy child, the efficiency of the process can be assessed in a number of ways.

This is due, in large part, to the widespread use of ovarian stimulation in ART. While the first clinical pregnancies from in vitro fertilization and embryo transfer (IVF/ET) were achieved in the natural menstrual cycle (1), subsequent successes with controlled ovarian stimulation (2) emphasized the advantages of having multiple oocytes available for treatment. Insemination of multiple oocytes in a treatment cycle allows identification and elimination of oocytes that fail to undergo normal fertilization in vitro and also allows selection of embryos with optimal growth and morphological characteristics. The impact of selection of oocytes and embryos at sequential stages of fertilization and development prior to transfer is an important factor in the ultimate efficiency of IVF/ET and this is aptly illustrated when we consider that perhaps less than 5% of the oocytes collected in the course of ART are destined to result in a live baby (3). This phenomenon is particularly important when we come to assess the clinical efficiency of oocyte cryopreservation as little or no selection is applied to oocytes that are cryopreserved and stored prior to fertilization. However, a number of additional general considerations relating to clinical efficiency must be taken into account before we consider those that are specific to oocyte cryopreservation.

A direct consequence of the availability of multiple oocytes following ovarian stimulation for ART has been the practice of replacing multiple embryos in utero in order to increase the probability of an implantation. The problem associated with this approach is the increased risk of multiple pregnancy, which has long been recognized as a major complication of ART (4,5) and one that should be minimized if the definition of clinical efficiency includes a singleton birth as the optimal outcome. Reducing the number of embryos replaced in IVF/ET is facilitated by the cryopreservation of the surplus embryos which can be transferred in subsequent menstrual cycles and this has led to the concept of the cumulative pregnancy rate (including outcomes from the transfer of both fresh and cryopreserved embryos from a single cycle of oocyte collection) as a more accurate indicator of clinical efficiency (6,7). The current trend towards single embryo transfer (SET) has dramatically reduced the incidence of multiple pregnancies arising from ART in many countries (8) and has further emphasized the importance of including outcomes from the transfer of cryopreserved embryos in cumulative pregnancy rates when assessing the clinical efficiency of treatment (9–12).

It is obvious then that the number of embryos transferred in ART and minimizing the incidence of multiple pregnancies should be taken into account when assessing clinical efficiency. In addition to cryopreservation, the importance of selecting the "best" embryo for fresh transfer is critical for efficient ART and this has led to the validation of many markers of implantation potential in embryos, including the use of extended culture to identify embryos that can develop into "top-quality" blastocysts in vitro (13). Blastocyst transfer may increase the implantation rate associated with fresh embryo transfer when compared to transfer of early cleavage stages (14) but will also reduce the number of cryopreserved embryos. The relative clinical efficiency of different strategies for embryo utilization will, again, only be evident if cumulative pregnancy rates are assessed (15).

Perhaps the most important factor affecting the clinical efficiency of ART is female age. There is a clear inverse relationship between the implantation potential of a transferred embryo and female age at the time of oocyte collection. This can be illustrated by analysis of the outcomes obtained from the transfer of a single embryo on day 2 of development in relation to female age at oocyte collection (Table 1). The implantation potential of a single fresh embryo from a woman under 36 years of age is approximately three times that of an embryo from a woman over 39 years. A similar relationship can also be seen for embryos that have been cryopreserved prior to transfer (Table 1). It is, therefore, very important to take this variable into account when assessing the clinical efficiency of ART.

Intraclinic differences in general outcomes may also lead to erroneous conclusions being drawn from comparisons of techniques or approaches based on data from different centers. Comparisons of techniques or approaches should ideally be made by comparing relevant outcomes in women of equivalent age within the same clinic. The paucity of such studies in the literature has hampered the analysis of the clinical efficiency of oocyte cryopreservation relative to the use of fresh oocytes or cryopreserved embryos in ART.

WHAT IS THE CLINICAL CONTEXT OF OOCYTE CRYOPRESERVATION?

The clinical efficiency of oocyte cryopreservation will, inevitably, be dependent on the context in which it is applied. Oocyte cryopreservation may be indicated in a variety of circumstances. In cases where oocyte cryopreservation is used to preserve fertility in women undergoing cytotoxic treatments for malignant disease (16,17), the time available and the number of oocytes that can be stored may be limited by the need to commence chemo- and/or radiotherapy. While many of this group will be relatively young, there is also the possibility that their disease state may adversely affect gamete quality (18). A further impediment to clinical efficiency in some cases, notably estrogen sensitive breast cancers, may also be the contraindication for ovarian stimulation, which results in high levels of circulating estrogen (19). As such, the results likely to be obtained in this group will be heavily dependent on a combination of the above considerations.

Much of the clinical application of oocyte cryopreservation to date has been a direct result of national legislation, which excludes the possibility of embryo cryopreservation in the routine practice of IVF/ET (20,21). The clinical efficiency in this context is affected by factors relating to selection. In circumstances where all embryos must be transferred and embryos cannot be cryopreserved (22), the options for selection of fresh embryos for transfer are likely to be severely limited reducing the probability of a successful implantation. Additionally, there will be no option for selective cryopreservation of embryos that satisfy minimum quality criteria as only unfertilized oocytes are cryopreserved. When thawing cryopreserved oocytes, the number thawed will be restricted by the necessity to transfer all resultant embryos and, again, selection of embryos on the basis of developmental markers will not be possible. The lack of selection associated with application of oocyte cryopreservation in this clinical setting will inevitably reduce the efficiency of treatment in terms of the average number of transfers (and, therefore, time) required to achieve a pregnancy. In this situation, the importance of assessing cumulative pregnancy rates from all the oocytes collected in a cycle is self-evident (23). The extent to which selection is routinely applied in IVF/ET with embryo cryopreservation is in contrast to the lack of selection inherent when oocyte cryopreservation is applied under the above restrictions and is summarized in Table 2.

Table 1 Implantation Rate (IR; FHs/Embryo Transferred) Following Single Embryo Transfer (SET) In Relation To Female Age at Time of Oocyte Collection

	Fresh embryos		Thawed embryos	
Age (yr)	SETs	IR	SETs	IR
<36	833	31.7%	1006	24.6%
36–39	391	19.7%	502	18.3%
>39	507	10.7%	572	10.3%

Abbreviation: FH, fetal heart.

Table 2 Summary of Selection Criteria Used in Routine IVF/ET with Embryo Cryopreservation Compared to Lack of Selection Associated with IVF/ET and Legally Imposed Oocyte Cryopreservation

	Embryo cryopreservation		Oocyte cryopreservation	
	Fresh embryos	Frozen embryos	Fresh embryos	Frozen oocytes
Developmental rate	+	+	–	–
Morphology	+	+	–	–
Early syngamy/cleavage	+	+	–	–
Post-thaw survival	n/a	+	n/a	+
Post-thaw development	n/a	+	n/a	–

Abbreviations: ET, embryo transfer; IVF, in vitro fertilization.

Oocyte cryopreservation offers a number of potential advantages when applied in conjunction with oocyte donation programs, particularly in relation to simplification of synchrony between donor and recipient and the option of quarantining donated gametes. The efficiency of treatment will be dependent on the number of oocytes available and the degree of selection applied to the resultant embryos. For example, high pregnancy and implantation rates have been reported when multiple donated oocytes are warmed after vitrification and cultured to the blastocyst stage prior to transfer (24). The age profile of oocyte donors is also likely to be younger than in the ART population and may be expected to result in apparently improved outcomes.

Banking of cryopreserved oocytes for young women, usually without partners, who wish to defer their fertility options has often been referred to as "social egg freezing." While this use of the technology remains controversial, its likely efficiency will, again, be dependent on the number and quality of oocytes, the age of the woman at time of storage and the extent of selection applied when the gametes are eventually used to form embryos for transfer.

The above indications for oocyte cryopreservation will, therefore, impact on the clinical efficiency of treatment in addition to the biological effects of cryopreservation on the oocyte and resultant embryos. It is, however, the efficiency of the technology that will determine the eventual success and we shall now consider the effects of oocyte quality and current methods of cryopreservation on the human oocyte.

OOCYTE QUALITY

One inherent problem associated with oocyte cryopreservation is the lack of non-invasive markers of oocyte quality, which limits the ability to apply selection criteria prior to cryopreservation. As discussed above, a number of objective and subjective criteria based on morphology and timing of development are routinely applied to embryos used in ART but in many of the clinical situations where oocyte cryopreservation is used these developmental criteria are not applicable. Assuming appropriate culture conditions, it is highly likely that poor quality embryos, which do not meet selection criteria, arise from poor quality oocytes. Therefore, if poor quality oocytes can be identified and eliminated prior to cryopreservation, clinical efficiency will be improved. The question is whether there are clear morphological selection criteria that can be applied to oocytes without the risk of "throwing the baby out with the bath water."

Exclusion of oocytes from cryopreservation on the basis of maturational status is justifiable in the case of germinal vesicle (GV) stage oocytes retrieved following ovarian stimulation but less so in the case of metaphase I (MI) oocytes. Although three births have been reported following in vitro maturation of GV oocytes (25–27), the low developmental potential of these oocytes (<1% reaching blastocyst stage) (28), the high incidence of spindle and chromosome anomalies (29–31) and fragmentation of nuclei/multinucleation in resultant blastomeres (32) suggest that these oocytes should be excluded from all ART treatments.

Improved clinical efficiency in IVM programs where natural cycles are manipulated solely by hCG administration early in the follicular phase (33), is achieved by in vitro maturation prior to cryopreservation. Notwithstanding the additional impact of cryopreservation (34), however, the clinical efficiency of these IVM programs is reduced relative to the use of mature oocytes from stimulated cycles.

Evidence to support exclusion of metaphase I oocytes from the oocyte pool for cryopreservation, especially in patients such as young women with cancer is less clear. Some births have been reported from the use of these oocytes (28,35–38) including one following oocyte cryopreservation (39). These oocytes are used routinely in clinical ICSI when they mature in vitro in the few hours following cumulus removal although their developmental potential is extremely low [4% to blastocyst (28) and 0.5% fetal heart (37)]. The overall contribution of immature oocytes to clinical efficiency is likely to be minimal unless they occur in significant numbers and they should only be considered for clinical use as a last resort. Caution is still warranted when using these oocytes, as a report, based on assessment of five chromosomes, showed that only 26% of these oocytes were euploid (40).

Selection of oocytes for cryopreservation based on intra and extracytoplasmic features may improve clinical efficiency but the impact of different anomalies is difficult to quantify and assessment can be highly subjective. Although some oocytes may appear intuitively unattractive to an embryologist, there is no clear evidence that irregular shape, a large perivitelline space or a dark zona pellucida have an impact on subsequent development to the blastocyst stage (41). Time lapse video visualization of fertilized oocytes and embryos has demonstrated that some anomalies such as vacuolation may be expelled during cleavage. However, in a large study of over 5000 individually assessed good quality day 3 embryos, only 2% of these had arisen from oocytes with vacuolation (41). Similarly, less than 2% were derived from oocytes with central granulation, which is a likely indication of asynchronous nuclear and cytoplasmic development and, as observed with GV oocytes, is associated with an elevated incidence of aneuploidy (42). The presence of other organelles within the oocyte is often mistakenly referred to as vacuolation. Clusters of smooth endoplasmic reticulum (sERC) in the oocyte are associated with cleavage arrest in resultant embryos but a small proportion will reach the morula stage (43). Although oocytes with cytoplasmic anomalies may be excluded from the cohort of cryopreserved oocytes, concerns remain that the presence of such oocytes is indicative of a poor cycle or a poor prognosis patient. Otsuki et al. (43) showed that cycles yielding oocytes with sERC have compromised implantation potential (2%) despite the transfer of high quality embryos derived from sibling oocytes without sERC. An extremely low implantation rate (1.7%) was also observed in patients showing the same dysmorphic phenotype in more than two cycles (44). Selection of only morphologically "perfect" oocytes (both cytoplasmic and extra cytoplasmic) for cryopreservation would not, therefore, necessarily overcome this phenomenon. In fact, applying very strict selection criteria for oocyte cryopreservation did not appear to significantly enhance the implantation rate (45). On the contrary, exclusion from cryopreservation based on irregular zona, debris in the perivitelline space or a fragmented polar body may have resulted in the rejection of some implantation potential.

Although high implantation rates can result from transfer of good quality blastocysts derived from cryopreserved oocytes (24), others have shown that a third of blastocysts from fresh oocytes are chromosomally abnormal (46). Selection for oocyte cryopreservation based on normal genetic compliment would theoretically improve the clinical efficiency and this is now possible by applying comparative genomic hybridization to the first polar body (47). Using this approach, Sher et al. (48) determined that only a third of oocytes collected from young donors were euploid. Selection for cryopreservation of only euploid oocytes resulted in a high blastocyst development rate (65%) and a high implantation rate (61%). It should be noted, however, that in another study, 78% of oocytes from IVF patients aged below 37 years were found to be euploid (49).

It may be possible to improve clinical efficiency without discarding oocytes by adopting a mild stimulation regimen (50). Although fewer oocytes were retrieved following mild stimulation in this randomized control trial, the rate of embryonic aneuploidy was reduced, resulting in the same number of euploid embryos. This milder approach may increase clinical efficiency and reduce the time taken to achieve pregnancy and may be particularly attractive in countries where legal restrictions necessitate the transfer of all embryos which are created.

It is clear from the discussion above that the quality of oocytes cryopreserved will impact clinical efficiency. This serves to emphasize the importance of controlling oocyte quality when determining the effects of cryopreservation, with parallel fresh controls being an ideal component of any study. Unfortunately, data that conform to this rigorous requirement are all too rare in oocyte cryopreservation studies.

SURVIVAL OF CRYOPRESERVED OOCYTES

It is self-evident that an important determinant of the clinical efficiency of any form of cryopreservation in ART will be the proportion of cells that survive the cryopreservation process. Lysis of the cell membrane as a consequence of ice formation and solution effects during slow cooling or only the latter during vitrification (51) is the most obvious form of cryodamage. Unlike embryo cryopreservation where a proportion of the embryonic cells may survive cryopreservation, the all or nothing situation associated with oocyte cryopreservation has posed a difficult challenge. This appears to relate to the wide variation in permeability properties of the human mature oocyte membrane (52).

Over time the evolution of oocyte cryopreservation methodology has resulted in significant improvements in the survival rate for the human mature oocyte. Initial reports of human oocyte cryosurvival with dimethyl sulfoxide (DMSO) and slow cooling ranged from 20% to 30% (53–58). Almost double this survival rate was achieved by changing the permeating cryoprotectant to propanediol and the addition of 0.1 M sucrose (20,21,59–65). Increasing the concentration of the non-permeating cryoprotectant, sucrose, from 0.1 to 0.2 M or 0.3 M during dehydration increased survival to between 60% and 80%, with the most commonly used method (0.3 M) generally achieving survival rates over 70% (66–77). In house comparisons of 0.1 M and elevated sucrose concentration have consistently shown an improvement in survival with higher sucrose concentrations (76,78–80).

Controlled comparisons between other methodological approaches have not been reported in the literature. Similar survival rates were achieved when the basal medium was modified to a sodium depleted medium (81–86), and following vitrification using various methods (34,87,88). However, recent developments in vitrification using the cryoprotectants DSMO, ethylene glycol (EG) and sucrose have achieved encouraging results with survival rates over 90% (24,89). With potential survival rates now equivalent to or better than those reported for embryo cryopreservation, the relative clinical efficiency therefore depends on the subsequent developmental potential of the cryopreserved oocytes.

FERTILIZATION OF CRYOPRESERVED OOCYTES

Improved fertilization rates have also been observed using modified methodologies with increases from approximately 50% with 0.1 M sucrose to 60–70% with 0.3 M and 80% with 0.2 M sucrose. In the few studies that compared fertilization of fresh oocytes and oocytes cryopreserved using the elevated sucrose procedures, equivalent rates were observed (74,77,79,85,90,91). Similar conclusions can be drawn from studies involving vitrification using the DSMO plus EG method (24,89). These similar fertilization rates observed with fresh and cryopreserved oocytes mean that the cryopreservation procedures currently in use are not likely to have an impact on clinical efficiency by reducing the proportion of oocytes that undergo fertilization. It should be noted that, almost without exception, fertilization of cryopreserved oocytes has been achieved using intracytoplasmic sperm injection (ICSI).

DEVELOPMENT AND QUALITY OF EMBRYOS FROM CRYOPRESERVED OOCYTES

Based solely on the proportion of normally fertilized oocytes which undergo subsequent cleavage relative to fresh controls, oocyte cryopreservation does not appear to reduce clinical efficiency by limiting the number of cleaving embryos available (24,74,77,85,89,91). However, the rate of development, an important correlate of implantation potential, does appear to be compromised following slow cooling in the presence of 0.3 M sucrose.

In one study that reported developmental rate in more detail (71), the majority of the embryos generated on day 2 were at the two-cell stage (68%) and only 14% were at the four-cell stage. However, no comparative fresh data were presented for this study and, therefore, the poor developmental rate cannot be conclusively attributed to the cryopreservation procedure. Although apparently retarded in their development, the embryos in this study did not exhibit high levels of fragmentation with only 20% of the embryos classified as grade III or IV.

A recent study by Konc et al. (77) has clearly shown a significant delay in development on day 2 in embryos derived from oocytes cryopreserved using the 0.3 M sucrose procedure when compared to embryos from fresh oocytes (17% at the four-cell stage vs. 66% in fresh controls). There is only one clinical report of a small number of embryos from cryopreserved (slow cooling in 0.3 M sucrose) oocytes which were cultured for an extended period and transferred at the blastocyst stage (70), making further analysis of the impact of cryopreservation on developmental rate difficult.

Further evidence which suggests that slow cooling in 0.3 M sucrose may retard subsequent development is the observation that very few resultant embryos (7%) had undergone early cleavage at 25 hours post insemination compared to the majority (59%) of embryos from fresh oocytes (92). Early cleavage, when assessed at 25 to 27 hours post insemination, is an independent indicator of developmental potential (93,94).

Embryos from oocytes cryopreserved in sodium depleted media also appear to exhibit retarded development with a higher proportion of two cells (55%) on day 2 of development relative to embryos from fresh oocytes (28%) (85) and a lower than expected mean cell number of only 5.4 on day 3 (84).

In contrast, 50% of embryos generated from oocytes cryopreserved in 0.2 M sucrose were at least at the four-cell stage on day 2 (91) suggesting no impact on embryo development with this procedure. Similarly, embryo development to the blastocyst stage appeared to unaffected following oocyte vitrification (24) using the most frequently used method initially reported by Katayama et al. (95).

IMPLANTATION OF EMBRYOS FROM CRYOPRESERVED OOCYTES

As in vitro developmental rate is a strong predictor of outcome following embryo transfer (96,97), any retardation of development following oocyte cryopreservation would be expected to have an impact on implantation potential and therefore clinical efficiency. In the Borini et al. study (71) using 0.3 M sucrose, in which the majority of embryos (68%) were at the two-cell stage on day 2, a poor implantation rate of only 5.2% was reported. Similar low implantation rates have been observed in a number of studies using this method although information regarding cell numbers at transfer is not provided in the other publications (72,73,76). In a large study of over 1000 oocytes cryopreserved using the 0.3 M sucrose procedure, a significantly lower implantation rate (5.7 %) was observed compared to 17% with control fresh oocytes (74). Although it had been, in general, abandoned due to poor clinical efficiency, the initial propanediol method using 0.1 M sucrose, in fact resulted in higher implantation rates (65,76) than the method employing 0.3 M sucrose (71). In contrast, following cryopreservation of oocytes using the 0.2 M sucrose procedure, in which a higher proportion of 4 cell embryos (43%) was generated, an implantation rate of 13% was reported (91). An implantation rate of 15.7% was also reported following cryopreservation in 0.2 M sucrose in association with developmental rates similar to those observed in parallel fresh oocytes (98).

The clinical efficiency of oocyte cryopreservation, therefore, will be dependent on a method which not only results in high survival rates but which also has a negligible effect on embryo development in vitro and, consequently, on implantation potential following transfer.

ATTRITION RATES AFTER OOCYTE CRYOPRESERVATION

The above data suggest that, although survival has improved with the higher sucrose methods, this has been offset by the poor implantation rate observed with the 0.3 M sucrose procedure resulting in no real improvement in the clinical efficiency relative to the initial 0.1 M sucrose method. Therefore, to accurately estimate the clinical efficiency of a procedure, a calculation of the attrition resulting from each step of the process is necessary. To overcome bias and the small sample numbers in many studies, the total published data for each method has been pooled to compare the clinical efficiency which is ultimately assessed as the number of implantations resulting from 100 oocytes cryopreserved and thawed/warmed using each method. The relative clinical efficiency of each method is shown in Table 3 which is an update of data previously reported by Gook and Edgar (99).

Over double the number of embryos are generated using the 0.3 M sucrose relative to the 0.1 M sucrose method but similar clinical efficiency is achieved (2.9 and 2.3 implantations per 100 thawed oocytes, respectively). In comparison, the number of embryos generated with some of the newer methodological changes is similar to that calculated for the 0.3 M sucrose method. However, due to improved implantation rates, these result in a clinical efficiency approximately double that of 0.3 M sucrose (Table 3).

A recent publication by Cobo et al. (24) has reported very high survival and implantation rates (97 % and 41%, respectively) following vitrification using the EG + DMSO + sucrose procedure. It has been suggested by some (100,101), that slow freezing should be abandoned on the basis of results such as these, but not all vitrification procedures have shown an improvement in the clinical efficiency relative to controlled rate freezing. In fact, although the implantation rate in the Cobo et al. study (24) is very high, partly due to transfer after selection by extended culture to the blastocyst stage, the clinical efficiency (8.7 implantations per 100 warmed oocytes) is similar to that with controlled rate freezing using the 0.2 M sucrose procedure (9.0 implantations per 100 thawed oocytes). It is also relevant to note that the vitrified oocytes were from an oocyte donor population with a mean age of 26.7 years. At present, however, the EG + DMSO + sucrose vitrification procedure appears to be associated with the highest clinical efficiency.

The potential importance of oocyte selection prior to cryopreservation has been emphasized in the case of vitrification of only euploid oocytes (48) in which the clinical efficiency was reported as 26 implantations per 100 warmed oocytes. However, in this study only 51 embryos per 100 warmed oocytes were generated suggesting that the improved clinical efficiency is attributable to the selection and not vitrification with EG + DMSO + Ficoll + sucrose per se. In fact, if we assume that the discarded aneuploid oocytes would not have implanted, the overall clinical efficiency of the vitrification procedure (19 implantations per 317 warmed oocytes; 6%) would be less than the controlled rate freezing with 0.2 M sucrose and vitrification with EG + DMSO + sucrose.

RELATIVE CLINICAL EFFICIENCY OF OOCYTE CRYOPRESERVATION

There has been a general perception that oocyte cryopreservation is an experimental and relatively inefficient approach that has limited application in ART and this has resulted in a reluctance to adopt the technology other than as a last resort. The legislative framework that was introduced in Italy and resulted in widespread adoption of oocyte cryopreservation as an alternative to embryo storage led to comparisons being made between the two approaches. Given the pregnancy/implantation rates which can be achieved following transfer of embryos cryopreserved at early cleavage stages (102,103) or at the blastocyst stage (104), the clinical efficiency of oocyte cryopreservation may appear to be significantly reduced. However, as discussed earlier in this chapter, much of the attrition which occurs after oocyte cryopreservation has already occurred prior to embryo cryopreservation.

For example, the proportion of oocytes that do not fertilize or fertilize abnormally will have been excluded, as will embryos that do not develop normally or whose morphology renders them unsuitable for cryopreservation. It can be realistically estimated that, from 100 fresh oocytes, 65 to 70 embryos may be derived, 40 to 45 of which may be suitable for cryopreservation. This is an ambitious target which may not be the case in many laboratories where the proportions of embryos deemed suitable for cryopreservation are significantly lower. This figure will, of course, also be lower if embryos are cryopreserved after extended culture to the blastocyst stage. Assuming cryosurvival rates consistent with those observed in many centers, we may predict that 30 to 35 embryos will be available for transfer and that 5 to 10 implantations may result. This range, derived from realistic estimates, reflects the clinical efficiency of the overall process (i.e., a yield of between 5 and 10 implantations per 100 oocytes). This yield from 100 oocytes is similar to that observed using currently optimal methodology for oocyte

Table 3 Summary of Clinical Outcomes from Oocyte Cryopreservation Using Various Protocols

	Controlled rate freezing			Controlled rate freezing			Vitrification			
Permeating cryoprotectant	1.5 M PrOH			1.5 M PrOH (Na depleted)			5 M EG	2.7 M EG + 2.1 M DMSO	2.7 M EG + 2.0 M PrOH	3.6 M EG + 2.7 M DMSO + 0.1 M ficoll[b]
Sucrose	0.1 M	0.2 M	0.3 M	0.1 M	0.2 M	0.3 M	1.0 M	0.5 M	0.5 M	1.0 M
Survival (no. of thawed oocytes)	51% (4027)	71% (1451)	73% (7595)	52% (127)	60% (815)	70% (890)	75% (838)	83%[a] (1454)	80% (395)	96% (111)
Fertilization	54%	80%	73%	56%	66%	72%	74%	87%	70%	–
Cleavage	85%	93%	90%	100%	84%	88%	94%	93%	53%	76%
Embryos per 100 thawed oocytes	23	53	48	29	33	44	52	76	30	51
Implantation rate	10%	17%	6%	21%	11%	13%	10%	16%	13%	61%
Implantations per 100 thawed oocytes	2.3	9.0	2.9	6.1	3.7	5.8	5.2	10.7	3.8	26

Pooled data from all published reports to December 2008.

[a]Including reported frequency of failed recovery.

[b]Only euploid oocytes cryopreserved.

Abbreviations: DMSO, dimethyl sulfoxide; EG, ethylene glycol; PrOH, 1,2-propanediol.

cryopreservation. The importance of attrition and selection in determining the relative clinical efficiency of approaches/technology used in ART cannot be underestimated.

Returning to the observation, made at the beginning of this chapter, that only a relatively small proportion of all oocytes collected for ART go on to result in a live birth, and the varying extent to which attrition and selection may manifest at different stages of a process depending on the approach adopted (or imposed), the value of assessing cumulative pregnancy rates (23) should be self-evident. Estimations of clinical efficiency must take into account a number of diverse factors if valid comparisons and/or conclusions are to be made.

REFERENCES

1. Edwards RG, Steptoe PC, Purdy JM. Establishing full-term human pregnancies using cleaving embryos grown in vitro. Br J Obstet Gynaecol 1980; 87: 737–56.
2. Trounson AO, Leeton JF, Wood C, Webb J, Wood J. Pregnancies in humans by fertilization in vitro and embryo transfer in the controlled ovulatory cycle. Science 1981; 212: 681–2.
3. Edgar DH, Gook DA. How should the clinical efficiency of oocyte cryopreservation be measured? Reprod Biomed Online 2007; 14: 430–5.
4. Cohen J. How to avoid multiple pregnancies in assisted reproduction. Hum Reprod 1998; 13(Suppl 3): 197–214; discussion 215–18.
5. Cohen J, Jones HW Jr. How to avoid multiple pregnancies in assistive reproductive technologies. Semin Reprod Med 2001; 19: 269–78.
6. Jones HW Jr, Out HJ, Hoomans EH, et al. Cryopreservation: the practicalities of evaluation. Hum Reprod 1997; 12: 1522–4.
7. Schnorr JA, Doviak MJ, Muasher SJ, Jones HW Jr. Impact of a cryopreservation program on the multiple pregnancy rate associated with assisted reproductive technologies. Fertil Steril 2001; 75: 147–51.
8. Gerris J. The near elimination of triplets in IVF. Reprod Biomed Online 2007; 15(Suppl 3): 40–4.
9. Tiitinen A, Halttunen M, Harkki P, Vuoristo P, Hyden-Granskog C. Elective single embryo transfer: the value of cryopreservation. Hum Reprod 2001; 16: 1140–4.
10. Tiitinen A, Hyden-Granskog C, Gissler M. What is the most relevant standard of success in assisted reproduction? The value of cryopreservation on cumulative pregnancy rates per single oocyte retrieval should not be forgotten. Hum Reprod 2004; 19: 2439–41.
11. Thurin A, Hausken J, Hillensjo T, et al. Elective single-embryo transfer versus double-embryo transfer in in vitro fertilization. N Engl J Med 2004; 351: 2392–402.
12. Lundin K, Bergh C. Cumulative impact of adding frozen-thawed cycles to single versus double fresh embryo transfers. Reprod Biomed Online 2007; 15: 76–82.
13. Gardner DK, Lane M, Stevens J, Schlenker T, Schoolcraft WB. Blastocyst score affects implantation and pregnancy outcome: towards a single blastocyst transfer. Fertil Steril 2000; 73: 1155–8.
14. Gardner DK, Schoolcraft WB, Wagley L, et al. A prospective randomized trial of blastocyst culture and transfer in in-vitro fertilization. Hum Reprod 1998; 13: 3434–40.
15. Edgar DH, Speirs AL, McBain JC. Strategies for embryo utilization in assisted reproduction—how do we assess their relative effectiveness? J Assist Reprod Genet 1999; 16: 460–1.
16. Yang D, Brown SE, Nguyen K, et al. Live birth after the transfer of human embryos developed from cryopreserved oocytes harvested before cancer treatment. Fertil Steril 2007; 87: 1469.e1–4.
17. Porcu E, Venturoli S, Damiano G, et al. Healthy twins delivered after oocyte cryopreservation and bilateral ovariectomy for ovarian cancer. Reprod Biomed Online 2008; 17: 265–7.
18. Dolmans MM, Demylle D, Martinez-Madrid B, Donnez J. Efficacy of in vitro fertilization after chemotherapy. Fertil Steril 2005; 83: 897–901.
19. Oktay K, Hourvitz A, Sahin G, et al. Letrozole reduces estrogen and gonadotropin exposure in women with breast cancer undergoing ovarian stimulation before chemotherapy. J Clin Endocrinol Metab 2006; 91: 3885–90.
20. Porcu E, Fabbri R, Marsella T, et al. Clinical experience and applications of oocyte cryopreservation. Mol Cell Endocrinol 2000; 169: 33–7.
21. Borini A, Bonu MA, Coticchio G, et al. Pregnancies and births after oocyte cryopreservation. Fertil Steril 2004; 82: 601–5.
22. Boggio A. Italy enacts new law on medically assisted reproduction. Hum Reprod 2005; 20: 1153–7.
23. Borini A, Bianchi V, Bonu MA, et al. Evidence-based clinical outcome of oocyte slow cooling. Reprod Biomed Online 2007; 15: 175–81.
24. Cobo A, Kuwayama M, Perez S, et al. Comparison of concomitant outcome achieved with fresh and cryopreserved donor oocytes vitrified by the Cryotop method. Fertil Steril 2008; 69: 1657–64.

25. Nagy ZP, Cecile J, Liu J, et al. Pregnancy and birth after intracytoplasmic sperm injection of in vitro matured germinal-vesicle stage oocytes: case report. Fertil Steril 1996; 65: 1047–50.
26. Tucker MJ, Wright G, Morton PC, Massey JB. Birth after cryopreservation of immature oocytes with subsequent in vitro maturation. Fertil Steril 1998; 70: 578–9.
27. Menezo YJ, Nicollet B, Rollet J, Hazout A. Pregnancy and delivery after in vitro maturation of naked ICSI-GV oocytes with GH and transfer of a frozen thawed blastocyst: case report. J Assist Reprod Genet 2006; 23: 47–9.
28. Chen SU, Chen HF, Lien YR, et al. Schedule to inject in vitro matured oocytes may increase pregnancy after intracytoplasmic sperm injection. Arch Androl 2000; 44: 197–205.
29. Racowsky C, Kaufman ML. Nuclear degeneration and meiotic aberrations observed in human oocytes matured in vitro: analysis by light microscopy. Fertil Steril 1992; 58: 750–5.
30. Li Y, Feng HL, Cao YJ, et al. Confocal microscopic analysis of the spindle and chromosome configurations of human oocytes matured in vitro. Fertil Steril 2006; 85: 827–32.
31. Lundin K, Ziebe S, Bergh C, et al. Effect of rescuing donated immature human oocytes derived after FSH/hCG stimulation following in vitro culture with or without follicular fluid meiosis activating sterol (FF-MAS)—an embryo chromosomal and morphological analysis. J Assist Reprod Genet 2007; 24: 87–90.
32. DeScisciolo C, Wright DL, Mayer JF, et al. Human embryos derived from in vitro and in vivo matured oocytes: analysis for chromosomal abnormalities and nuclear morphology. J Assist Reprod Genet 2000; 17: 284–92.
33. Chian RC, Buckett WM, Tulandi T, Tan SL. Prospective randomized study of human chorionic gonadotrophin priming before immature oocyte retrieval from unstimulated women with polycystic ovarian syndrome. Hum Reprod 2000; 15: 165–70.
34. Chian RC, Son WY, Huang JY, et al. High survival rates and pregnancies of human oocytes following vitrification: preliminary report. Fertil Steril 2005; 84(Suppl 1): S36.
35. Veeck LL, Wortham JW Jr, Witmyer, et al. Maturation and fertilization of morphologically immature human oocytes in a program of in vitro fertilization. Fertil Steril 1983; 39: 594–602.
36. Edirisinghe WR, Junk SM, Matson PL, Yovich JL. Birth from cryopreserved embryos following in-vitro maturation of oocytes and intracytoplasmic sperm injection. Hum Reprod 1997; 12: 1056–8.
37. De Vos A, Van de Velde H, Joris H, Van Steirteghem A. In-vitro matured metaphase-I oocytes have a lower fertilization rate but similar embryo quality as mature metaphase-II oocytes after intracytoplasmic sperm injection. Hum Reprod 1999; 14: 1859–63.
38. Vanhoutte L, De Sutter P, Van der Elst J, Dhont M. Clinical benefit of metaphase I oocytes. Reprod Biol Endocrinol 2005; 3: 71.
39. Kan A, Kilani S, Tilia L, et al. Pregnancy from intracytoplasmic injection of a frozen-thawed oocyte. Aust N Z J Obstet Gynaecol 2004; 44: 262–3.
40. Magli MC, Ferraretti AP, Crippa A, et al. First meiosis errors in immature oocytes generated by stimulated cycles. Fertil Steril 2006; 86: 629–35.
41. Balaban B, Ata B, Isiklar A, Yakin K, Urman B. Severe cytoplasmic abnormalities of the oocyte decrease cryosurvival and subsequent embryonic development of cryopreserved embryos. Hum Reprod 2008; 23: 1778–85.
42. Van Blerkom J, Henry G. Oocyte dysmorphism and aneuploidy in meiotically mature human oocytes after ovarian stimulation. Hum Reprod 1992; 7: 379–90.
43. Otsuki J, Okada A, Morimoto K, Nagai Y, Kubo H. The relationship between pregnancy outcome and smooth endoplasmic reticulum clusters in MII human oocytes. Hum Reprod 2004; 19: 1591–7.
44. Meriano JS, Alexis J, Visram-Zaver S, Cruz M, Casper RF. Tracking of oocyte dysmorphisms for ICSI patients may prove relevant to the outcome in subsequent patient cycles. Hum Reprod 2001; 16: 2118–23.
45. Parmegiani L, Cognigni GE, Bernardi S, et al. Freezing within 2 h from oocyte retrieval increases the efficiency of human oocyte cryopreservation when using a slow freezing/rapid thawing protocol with high sucrose concentration. Hum Reprod 2008; 23: 1771–7.
46. Fragouli E, Lenzi M, Ross R, et al. Comprehensive molecular cytogenetic analysis of the human blastocyst stage. Hum Reprod 2008; 23: 2596–608.
47. Gutierrez-Mateo C, Benet J, Wells D, et al. Aneuploidy study of human oocytes first polar body comparative genomic hybridization and metaphase II fluorescence in situ hybridization analysis. Hum Reprod 2004; 19: 2859–68.
48. Sher G, Keskintepe L, Mukaida T, et al. Selective vitrification of euploid oocytes markedly improves survival, fertilization and pregnancy-generating potential. Reprod Biomed Online 2008; 17: 524–9.
49. Fragouli E, Wells D, Thornhill A, et al. Comparative genomic hybridization analysis of human oocytes and polar bodies. Hum Reprod 2006; 21: 2319–28.

50. Baart EB, Martini E, Eijkemans MJ, et al. Milder ovarian stimulation for in-vitro fertilization reduces aneuploidy in the human preimplantation embryo: a randomized controlled trial. Hum Reprod 2007; 22: 980–8.
51. Mazur P, Leibo SP, Chu EHY. A two-factor hypothesis of freezing injury evidence from Chinese hamster tissue culture cells. Exp Cell Res 1972; 71: 345–55.
52. Hunter J, Bernard A, Fuller B, McGrath J, Shaw RW. Plasma membrane water permeabilities of human oocytes: the temperature dependence of water movement in individual cells. J Cell Physiol 1992; 150: 175–9.
53. Al-Hasani S, Diedrich K, van der Ven H, et al. Cryopreservation of human oocytes. Hum Reprod 1987; 2: 695–700.
54. Van Uem JFHM, Siebzehnrubl ER, Schuh B, et al. Birth after cryopreservation of unfertilised oocytes. Lancet 1987; i: 752–3.
55. Mandelbaum J, Junca AM, Tibi C, et al. Cryopreservation of immature and mature hamster and human oocytes. Ann N Y Acad Sci 1988; 541: 550–61.
56. Mandelbaum J, Junca AM, Plachot M, et al. Cryopreservation of human embryos and oocytes. Hum Reprod 1988; 3: 117–19.
57. Siebzehnruebl ER, Todorow S, van Uem J, et al. Cryopreservation of human and rabbit oocytes and one-cell embryos: a comparison of DMSO and propanediol. Hum Reprod 1989; 4: 312–17.
58. Todorow SJ, Siebzehnrubl ER, Spitzer M, et al. Comparative results on survival of human and animal eggs using different cryoprotectants and freeze-thawing regimens. II. Human. Hum Reprod 1989; 4: 812–16.
59. Gook DA, Osborn SM, Johnston WI. Cryopreservation of mouse and human oocytes using 1,2-propanediol and the configuration of the meiotic spindle. Hum Reprod 1993; 8: 1101–9.
60. Antinori S, Dani G, Selman HA, et al. Pregnancies after sperm injection into cryopreserved human oocytes. Hum Reprod 1998; 13: 157–8. Abst P-55, 14th Annual ESHRE Meeting.
61. Borini A, Bafaro MG, Bonu MA, et al. Pregnancies after oocyte freezing and thawing, preliminary data. Hum Reprod 1998; 13: 124–5. Abst P-241, 14th Annual ESHRE Meeting.
62. Porcu E, Fabbri R, Seracchioli R, et al. Birth of six healthy children after intracytoplasmic sperm injection of cryopreserved human oocytes. Hum Reprod 1998; 13: 124. Abst O-240, 14th Annual ESHRE Meeting.
63. Porcu E, Fabbri R, Ciotti P, et al. Four healthy children from frozen human oocytes and frozen human sperms. Fertil Steril 2001; 76(Suppl 1): S76. Abst O-203, ASRM Meeting.
64. Porcu E, Fabbri R, Ciotti P, et al. Oocytes or embryo storage. Fertil Steril 2002; 78(3 Suppl 1): S15, Abst O-38.
65. Borini A, Lagalla C, Bonu MA, et al. Cumulative pregnancy rates resulting from the use of fresh and frozen oocytes: 7 years' experience. Reprod Biomed Online 2006; 12: 481–6.
66. Chen SU, Lien YR, Tsai YY, et al. Successful pregnancy occurred from slowly freezing human oocytes using the regime of 1.5 mol/L 1,2 propanediol with 0.3 mol/L sucrose. Hum Reprod 2002; 17: 1412–13.
67. Fosas N, Marina F, Torres PJ, et al. The births of five Spanish babies from cryopreserved donated oocytes. Hum Reprod 2003; 18: 1417–21.
68. Chen SU, Lien YR, Chen HF, et al. Observational clinical follow-up of oocyte cryopreservation using a slow-freezing method with 1,2-propanediol plus sucrose followed by ICSI. Hum Reprod 2005; 20: 1975–80.
69. Li XH, Chen SU, Zhang X, et al. Cryopreserved oocytes of infertile couples undergoing assisted reproductive technology could be an important source of oocyte donation: a clinical report of successful pregnancies. Hum Reprod 2005; 20: 3390–4.
70. Tjer GC, Chiu TT, Cheung LP, Lok IH, Haines CJ. Birth of a healthy baby after transfer of blastocysts derived from cryopreserved human oocytes fertilized with frozen spermatozoa. Fertil Steril 2005; 83: 1547–9.
71. Borini A, Sciajno R, Bianchi V, et al. Clinical outcome of oocyte cryopreservation after slow cooling with a protocol utilizing a high sucrose concentration. Hum Reprod 2006; 21: 512–17.
72. Chamayou S, Alecci C, Ragolia C, et al. Comparison of in-vitro outcomes from cryopreserved oocytes and sibling fresh oocytes. Reprod Biomed Online 2006; 12: 730–6.
73. La Sala GB, Nicoli A, Villani MT, et al. Outcome of 518 salvage oocyte-cryopreservation cycles performed as a routine procedure in an in vitro fertilization program. Fertil Steril 2006; 86: 1423–7.
74. Levi Setti PE, Albani E, Novara PV, Cesana A, Morreale G. Cryopreservation of supernumerary oocytes in IVF/ICSI cycles. Hum Reprod 2006; 21: 370–5.
75. Barritt J, Luna M, Duke M, et al. Report of four donor-recipient oocyte cryopreservation cycles resulting in high pregnancy and implantation rates. Fertil Steril 2007; 87: 189.e13–17.
76. De Santis L, Cino I, Rabellotti E, et al. Oocyte cryopreservation: clinical outcome of slow-cooling protocols differing in sucrose concentration. Reprod Biomed Online 2007; 14: 57–63.

77. Konc J, Kanyo K, Varga E, Kriston R, Cseh S. Births resulting from oocyte cryopreservation using a slow freezing protocol with propanediol and sucrose. Syst Biol Reprod Med 2008; 54: 205–10.
78. Fabbri R, Porcu E, Marsella T, et al. Human oocyte cryopreservation: new perspectives regarding oocyte survival. Hum Reprod 2001; 16: 411–16.
79. Chen ZJ, Li M, Li Y, et al. Effects of sucrose concentration on the developmental potential of human frozen-thawed oocytes at different stages of maturity. Hum Reprod 2004; 19: 2345–9.
80. De Santis L, Cino I, Coticchio G, et al. Objective evaluation of the viability of cryopreserved oocytes. Reprod Biomed Online 2007; 15: 338–45.
81. Quintans CJ, Donaldson MJ, Bertolino MV, Pasqualini RS. Birth of two babies using oocytes that were cryopreserved in a choline-based freezing medium. Hum Reprod 2002; 17: 3149–52.
82. Boldt J, Cline D, McLaughlin D. Human oocyte cryopreservation as an adjunct to IVF-embryo transfer cycles. Hum Reprod 2003; 18: 1250–5.
83. Azambuja R, Badalotti M, Teloken C, Michelon J, Petracco A. Successful birth after injection of frozen human oocytes with frozen epididymal spermatozoa. Reprod Biomed Online 2005; 11: 449–51.
84. Boldt J, Tidswell N, Sayers A, Kilani R, Cline D. Human oocyte cryopreservation: 5-year experience with a sodium-depleted slow freezing method. Reprod Biomed Online 2006; 13: 96–100.
85. Petracco A, Azambuja R, Okada L, et al. Comparison of embryo quality between sibling embryos originating from frozen or fresh oocytes. Reprod Biomed Online 2006; 13: 497–503.
86. Stachecki JJ, Cohen J, Garrisi J, et al. Cryopreservation of unfertilized human oocytes. Reprod Biomed Online 2006; 13: 222–7.
87. Yoon T, Chung H, Lim J, et al. Pregnancy and delivery of healthy infants developed from vitrified oocytes in a stimulated in vitro fertilization-embryo transfer program. Fertil Steril 2000; 74: 180–1.
88. Kuwayama M, Vajta G, Kato O, Leibo SP. Highly efficient vitrification method for cryopreservation of human oocytes. Reprod Biomed Online 2005; 11: 300–8.
89. Antinori M, Licata E, Dani G, et al. Cryotop vitrification of human oocytes results in high survival rate and healthy deliveries. Reprod Biomed Online 2007; 14: 72–9.
90. Yang D, Winslow K, Blohm P, et al. Oocyte donation using cryopreserved donor oocytes. Fertil Steril 2002; 78(3, Suppl 1): S14, Abst O-37.
91. Bianchi V, Coticchio G, Distratis V, et al. Differential sucrose concentration during dehydration (0.2 mol/L) and rehydration (0.3 mol/L) increases the implantation rate of frozen human oocytes. Reprod Biomed Online 2007; 14: 64–71.
92. Bianchi V, Coticchio G, Distratis V, Di Giusto N, Borini A. Early cleavage delay in cryopreserved human oocytes. Hum Reprod 2005; 20(Suppl 1): i54.
93. Lundin K, Bergh C, Hardarson T. Early embryo cleavage is a strong indicator of embryo quality in human IVF. Hum Reprod 2001; 16: 2652–7.
94. Sakkas D, Percival G, D'arcy Y, Sharif K, Afnan M. Assessment of early cleaving in vitro fertilized human embryos at the 2-cell stage before transfer improves embryo selection. Fertil Steril 2001; 76: 1150–6.
95. Katayama KP, Stehlik J, Kuwayama M, Kato O, Stehlik E. High survival rate of vitrified human oocytes results in clinical pregnancy. Fertil Steril 2003; 80: 223–4.
96. Ziebe S, Petersen K, Lindenberg S, et al. Embryo morphology or cleavage stage: how to select the best embryos for transfer after in-vitro fertilization. Hum Reprod 1997; 12: 1545–9.
97. Ziebe S, Bech B, Petersen K, et al. Resumption of mitosis during post-thaw culture: a key parameter in selecting the right embryos for transfer. Hum Reprod 1998; 13: 178–81.
98. Coticchio G, Distratis V, Bianchi V, Bonu A, Borini A. Fertilization and early developmental ability of cryopreserved human oocytes is not affected compared to sibling fresh oocytes. Fertil Steril 2007; 88(Suppl 1): P-700.
99. Gook DA, Edgar DH. Human oocyte cryopreservation. Hum Reprod Update 2007; 13: 591–605.
100. Vajta G, Nagy ZP. Are programmable freezers still needed in the embryo laboratory? Review on vitrification. Reprod Biomed Online 2006; 12: 779–96.
101. Al-hasani S, Ozmen B, Koutlaki N, et al. Three years of routine vitrification of human zygotes: is it still fair to advocate slow-rate freezing? Reprod Biomed Online 2007; 14: 288–93.
102. Hyden-Granskog C, Unkila-Kallio L, Halttunen M, Tiitinen A. Single embryo transfer is an option in frozen embryo transfer. Hum Reprod 2005; 20: 2935–8.
103. Edgar DH, Archer J, McBain J, Bourne H. Embryonic factors affecting outcome from single cryopreserved embryo transfer. Reprod Biomed Online 2007; 14: 718–23.
104. Youssry M, Ozmen B, Zohni K, Diedrich K, Al-Hasani S. Current aspects of blastocyst cryopreservation. Reprod Biomed Online 2008; 16: 311–20.

19 Early and Recent History of Oocyte Cryopreservation in Human IVF

Eleonora Porcu, Leonardo Notarangelo, Antonia Bazzocchi, Chiara Landolfo, and Stefano Venturoli
Reproductive Medicine Unit, Infertility and IVF Center, University of Bologna, Bologna, Italy

INTRODUCTION

The history of human oocyte cryopreservation is quite peculiar as compared to the development of other assisted reproduction technologies.

Since the first successful case of human embryo cryopreservation 25 years ago, it has often been pointed out that freezing of excess embryos is only a temporary solution with many disadvantages and that it would often be more acceptable to store human oocytes to avoid ethical and legal problems. Indeed, shortly after the first birth with frozen human embryos, a similar clinical success with frozen human oocytes was announced. However, the enthusiasm for the announcement of the first pregnancies with frozen eggs was not followed by a rapid integration of the technique into the IVF clinical routine and, unlike the early and fast spread of embryo freezing, oocyte cryopreservation was and still is considered an experimental technique. Lessons learned from the general history of IVF show us that techniques indispensable to solving a problem have been rapidly adopted in the clinical routine before displaying evidence of safety. Embryo cryopreservation was developed inside the unavoidable dilemma of "freezing or wasting." On the contrary, oocyte cryopreservation was an optional choice. Ethical concerns were not powerful enough to push clinical research and application in oocyte cryopreservation. This is the main reason why those early fully successful clinical applications remain anecdotal. The postulated intrinsic cryo-vulnerability of the mature oocyte raised safety concerns and became a major constraint. It was thanks to the basic biological studies of excellent researchers in reproductive cryobiology that many of those concerns were alleviated in the 1990s allowing the clinical resumption of oocyte cryopreservation by the end of the decade. A relatively intensive and, in some cases, routine application has been performed only in Italy, due to the recent restrictive legal regulation of IVF in that country. Variations in the results reported from different countries and IVF teams makes it still difficult to measure the true success of the procedure even though, in some reports, it seems to equal the embryo freezing efficiency. In the past few years, the clinical application of egg storage has partially shifted to vitrification, despite the absence of sufficient basic biological studies addressing safety issues. Recently, the proposal of novel applications, such as the debated egg storage for non-medical indications to postpone maternity in young women, as well as the development of commercial interests, have prompted an acceleration in the spread of oocyte cryopreservation.

EARLY ANIMAL STUDIES

Studies concerning the cryopreservation of oocytes and the effect of low temperatures on cell biology have been undertaken since the end of 1940s, yielding knowledge that would have practical as well as theoretical significance.

Chang (1,2) documented the ability of rabbit oocytes stored either at 10°C and 0°C to generate embryos and healthy offspring. Afterwards, however, the author observed that pregnancy occurred more frequently with embryos kept at low temperatures (11–37% after storage for 24 hours) than with eggs (3–11% after storage for 6–31 hours). Subsequently, Chang (3) published further investigations on egg storage at low temperature. Rabbit eggs were taken from the tubes of albino rabbits and placed in a suspension containing an equivalent percentage of "Ringer" solution and rabbit serum. Part of the oocytes (130) were kept in test tubes for 30 to 31 hours at a temperature of 10°C, while the remaining 137 oocytes were kept at 0°C, after an initial acclimatization time of 1 hour at 10°C. Oocytes stored at 10°C were transferred to the left tube and oocytes stored at 0°C to the right tube. No statistically significant difference was seen between the effects of conservation at 0°C or 10°C. Both groups registered a low percentage of normal fertilization, 7% of transferred oocytes in the first group and 4% in the second group, after 31 hours of preservation. This percentage was statistically lower than that of a control group (27%). These observations raised the possibility of damage resulting from cell storage at low temperatures.

Sherman (4) evaluated the effect of glycerol supplementation on low temperature preservation of guinea pig eggs. He observed that some germ cells can survive exposure to temperatures of –10°C, if 5% glycerol is added to the storage solution consisting of 500 ml of distilled water, 3.38 g of sodium chloride, 0.09 g of calcium chloride, 0.16 g of potassium chloride, 0.05 g sodium bicarbonate, and 0.36 g dextrose (modified Locke solution). The glycerol lowers the freezing point preventing the medium from freezing. One hundred eighty oocytes were transferred into 30 guinea pig recipients, with 17% embryo development after one hour of storage and 5% after two hours. The same study analyzed the effects of glycerol presence inside the cell on the ability of oocytes to generate offspring after transfer to recipients. The penetration of glycerol in the cells was achieved through exposure in 5% glycerol solution for 15 minutes. Exposure to glycerol solution first caused oocyte shrinkage followed by their recovery to their initial volume after 8 to 10 minutes. Sixty oocytes were transferred into 10 guinea pig recipients and 6 pregnancies were achieved. Therefore, the presence of glycerol inside the cells did not jeopardize fertilization and embryo development.

In another study, Sherman (5) carried out an assessment of the survival and function of mouse egg cells after rapid reduction in temperature and storage at low temperatures. One thousand eight hundred and fifty-eight oocytes, harvested from the tubes of brown female guinea pigs, after hyperstimulation, were transferred into 310 white guinea pig recipients. The egg cells of the study group were released into the medium of storage; some were cooled to 0°C at a rate of 0.2°C/sec and the remainder to a temperature of –21°C at a speed of 1°C/sec. The cells were then restored to a temperature of 22°C and transferred to recipients. In the control group, cells were preserved at a temperature of 22°C. The rate of pregnancy and survival of egg cells, along with their reproductive capacity, were unaffected by the rapid lowering of temperature.

However, egg cells proved extremely sensitive to prolonged exposure to low temperatures: with increasing exposure time, the rate of survival dramatically decreases.

Shortly later, Burks (6) reported on the morphology of cryopreserved human and rabbit egg cells subject to freezing. Four hundred and fifty-five rabbit eggs were divided into groups and subjected to increasing concentrations of glycerol (0%, 10%, and 35%) in Locke's solution at different temperatures (5°C, 22°C, and 37°C for 15 minutes). The cells were then stored in sealed glass containers and immersed in liquid nitrogen. Morphological evaluation was performed after 1 to 15 days of storage. The survival cytological criteria were: (*i*) integrity of the vitelline membrane and the zona pellucida and (*ii*) absence of granulation and/or deformation of the cytoplasm.

The highest survival rate (96%) was recorded in egg cells exposed to 35% glycerol at a temperature of 5°C. In the early 1960s, Mazur (7) conducted various studies on the kinetics of cellular dehydration in relation to changes in temperature. He worked out a differential equation relating cooling rate, surface and cell volume, membrane permeability to water and the temperature coefficient of the permeability constant. Through this equation it is possible to derive the intracellular water content and predict the probability of intracellular ice growing. The equation allows a number of conclusions.

- Larger cells with a lower surface/volume ratio have a slower dehydration speed than smaller cells because they retain more water and are more likely to encounter ice formation;
- A high rate of cooling does not allow enough water leakage from the cell, and makes intracellular crystal formation more likely;
- A cell characterized by high permeability to water achieves faster dehydration.

The fate of intracellular water as temperature changes in egg cells was evaluated in a study by Asahina (8) which observed how the intracellular content of the sea urchin egg cell did not freeze until it reached a temperature of –5°C; if cooling continued at a rate of 1°C/min, cells did not freeze but suffered a gradual decline in volume, while increasing the cooling rate to 10°C/min caused freezing and blocked the escape of water. In keeping with the Mazur equation, the trend to form intracellular crystals seems much more related to rapid changes in temperature.

In later studies, however, Mazur (9–11) pointed out that inhibition of intracellular ice formation is necessary but not sufficient for the survival of the cell. Slow-freezing can be injurious in itself. As ice develops outside the cells, the residual unfrozen medium forms channels of decreasing size and increasing solute concentration. The cells lie in the channels and shrink in osmotic response to the rising solute concentration. Previous theories ascribed slow freezing injury to the concentration of solutes and/or cell shrinkage. Recent experiments, however, indicate that the damage is due more to a decrease in the size of the unfrozen channels. In addition, the rate of warming can have as much effect on survival as the cooling rate.

The first birth from animal oocytes cryopreserved in liquid nitrogen was announced by Parkening (12) who obtained three normal mouse offspring. Whittingham (13), too, achieved successful cryopreservation of unfertilized mouse oocytes resulting in offspring. Hamster egg cryopreservation was also investigated. Choung (14) evaluated the viability and fertilizability of unfertilized hamster oocytes in different freezing conditions: in the first group oocytes were placed in medium at room temperature, while in the second group they stayed at 0°C. They were cooled to –6°C at a rate of –2°C/min (method 1) or at –0.5°C/min (method 2). After thawing, 52% (method 1) and 64% (method 2) of the oocytes showed normal morphologic features. The fertilization rate of frozen-thawed oocytes was 26%, while the penetration rate proved to be one third of fresh oocytes. Critser (15) conducted three experiments to evaluate the effects of vitrification (16) or slow freezing on human sperm penetration of zona-free hamster oocytes. The survival of hamster oocytes, as defined by observation of morphologic features, did not differ between the vitrification and the freezing group. Vitrified oocytes showed a lower frequency of sperm penetration than frozen oocytes (15).

EARLY HUMAN STUDIES

Burks (6), encouraged by his good results with rabbit oocyte cryopreservation, performed a subsequent experiment with human oocytes. Ten human eggs were exposed to 35% glycerol at a temperature of 5°C and then frozen in liquid nitrogen. After thawing, 9 out of 10 eggs survived (90%).

Trounson (17) examined different methods of human oocyte cryopreservation. Part of the experiment was done with aged oocytes that failed to fertilize in the routine IVF program. Slow cooling with 1.5 M dimethyl sulfoxide (DMSO) gave 18% of survival. Rapid cooling with 1.5 M DMSO gave 25% of survival. Ultrarapid cooling with DMSO 1.5 M and 3.0 M resulted in 52% of egg survival. Vitrification produced 75% survival rate. The second part of the experiment was performed with mature human oocytes. No survival was registered after slow cooling with DMSO 1.5 M. Slow cooling with 1.5 M PrOH gave good results with 67% survival and 100% fertilization. Vitrification after removal of cumulus cells resulted in 67% survival rate and 75% fertilization rate without subsequent cleavage. Vitrification with an intact cumulus was followed by fertilization and embryo development to at least the eight-cell stage. Therefore, even though small numbers of mature oocytes were involved in these experiments, it has been shown that human oocytes may survive freezing and thawing and may be fertilized after insemination. Furthermore, those early experiments gave indications about the most efficient methods of human egg cryopreservation. Unfortunately, despite the development of these methods for egg cryopreservation in mid-1985, Trounson's group has been unable to transfer developing embryos to patients because of legal restrictions in the Victorian State.

Chen (18) reported the first human pregnancy with cryopreserved oocytes. In his oocyte cryopreservation program, Chen adopted a slow freezing-rapid thawing protocol with DMSO as a cryoprotectant. After partial removal of surrounding cumulus cells, DMSO 1.5 M was added as a one-step procedure. Seeding was performed at –7°C followed by slow cooling between –7°C and –36°C, rapid freezing to –196°C and storage in liquid nitrogen. Rapid thawing in a 37°C water bath was followed by dilution of DMSO in a single step. This was achieved by adding four times the sample volume of phosphate-buffered saline. The oocyte was examined for morphological evidence of survival. Further development of the oocyte required the transfer to the regular culture medium, and at the appropriate time insemination was carried out.

Thirty-eight out of 50 (80%) cryopreserved oocytes survived thawing (76%), 71% fertilized, and 85% cleaved even to six to eight cells (19). Embryo transfers of between two and three embryos derived from frozen-thawed oocytes were performed in seven patients and two pregnancies (one twin and one single) resulted.

In the first patient who became pregnant, a 29-year-old woman, six eggs were frozen, three were thawed, survived and fertilized. Three embryos were transferred at the three- to four-cell stages, about 46 hours after insemination. A twin pregnancy was achieved with the birth of two healthy children. The second patient, aged 37 years, conceived with the same technique using her spare eggs, which were kept frozen for four months. She delivered a healthy female.

Therefore, Chen's clinical results led to a pregnancy rate of 29% embryo transfers and a birth rate of 6% per thawed oocytes.

Al-Hasani et al. (20) cryopreserved 133 oocytes of varying quality from 22 patients. The survival rate achieved after thawing was 34%, while fertilization rate was 75%. Out of a total of seven embryos transfers there was one pregnancy following the replacement of a six-cell embryo. Van Uem (21) published the second birth reported in the literature after oocyte cryopreservation with an alternative freezing technique. In an attempt to overcome cell damage due to supercooling, he developed a computer-controlled "open-vessel" freezing device (CTE 8100). This device, using tailed plastic straws, permits seeding to take place automatically in the ideal temperature range around the freezing point of the medium ("self-seeding"). Further, van Uem adopted the technique of slow-freezing and the slow-thawing. As Chen did, van Uem also removed the cumulus by needle dissection. A freezing medium of phosphate-buffered saline containing 10% heat-inactivated fetal cord serum and 1.5 mol/L DMSO not chilled before the addition to the oocyte was used. Seven out of 28 (25%) frozen oocytes survived after thawing.

A further study by Al-Hasani et al. (22) describes the results of three freezing methods used for cryopreserving 320 human oocytes. The first method was a slow freezing protocol and adopted DMSO 1.5 M and 20% inactivated fetal calf serum. In the second method of slow freezing the cryoprotectant was PrOH 1.5 M and sucrose 0.1 M. The third method with ultrarapid freezing used DMSO 3.0 M and sucrose 0.25 M. The survival and fertilization rates of the first method were 28% and 50%, respectively, with 20% of polyploidy. The survival and fertilization rates of the second method were 32% and 75%, respectively, with 40% of polyploidy. The ultracooling method gave poor results with a survival rate of 4%.

Diedrich (23) published a study where two hundred eighty–three excess oocytes were frozen from 48 patients. One hundred and thirty-six, out of 157, oocytes survived after thawing (87% survival rate). The viability rate of oocytes after the procedure, evaluated through morphologic criteria, was 32% (43 oocytes out of 136) and fertilization rate was 58%. Embryo transfer was performed in 11 patients, giving rise to two pregnancies, though both ended in abortions.

As a matter of fact, within a few years during the 1980s, the knowledge about human oocyte cryopreservation progressed greatly. It clearly demonstrated the feasibility of human egg freezing, provided the oocytes were mature and of good quality. Oocytes could survive freezing and thawing by both slow freezing and vitrification with high survival and fertilization rates. Embryos generated from thawed oocytes could develop and implant normally and healthy children could be delivered. It is surprising and disconcerting that, despite the proof of principle obtained in both basic research and clinical application, no pregnancies and births were achieved in the eight years following these early successes.

SUBSEQUENT RESEARCH

At the time the possibility of increased aneuploidy resulting from exposure of mature eggs to cryoprotectants and freezing and thawing caused concern.

Depolymerization of the spindle microtubules by cryoprotectants or by cooling may prevent the normal separation of sister chromatids at fertilization and thus lead to chromatid nondisjunction and the state of aneuploidy after the extrusion of the second polar body.

Magistrini and Szöllösi (24) and Pickering and Johnson (25) described depolymerization of the spindle microtubules induced by cooling in mouse eggs and even by transient cooling to room temperature in human oocytes (26). However, the same authors observed that to some extent the depolymerization was reversible. Vincent et al. (27) found that the addition of cryoprotectant was responsible for the depolymerization of the microtubules but these changes were reversible after the removal of cryoprotectant and a short period of culture. Glenister et al. (28) found that oocytes frozen by slow cooling in DMSO did not significantly increase the incidence of aneuploid zygotes in the mouse. These data contradict the findings of Kola et al. (29) who found a two- to three-fold increase of aneuploidy after slow cooling in DMSO and after vitrification. Sathananthan et al. (30) suggested a cautious approach to the integration of egg freezing in clinical IVF because of the sensitivity of meiotic spindle to simple cooling, even at 0°C, leading to extensive depolymerization of the microtubules. However, widespread displacement of chromosomes was not apparent in that study. These authors affirm that the results of the literature taken together clearly indicate that not all embryos or fetuses derived from frozen oocytes are abnormal. It is possible that depolymerization and repolymerization may occur in some oocytes in a way to least disturb the normal distribution and behavior of chromosomes that are suspended on the spindle during freezing and thawing.

The consensus arising from all those studies appears to be that inappropriate exposure to cryoprotectants and cooling may induce anomalies in the spindle microtubules (31–34).

Because of the concern that had been raised over mature oocyte cryopreservation, some teams addressed their research towards freezing immature oocytes at the germinal vesicle (GV) stage as it was thought to be less sensitive to cryoinjury (35–39).

Actually, the literature did not show any true advantage of immature oocyte freezing as regards survival and fertilization rates as well as embryo developmental capacity.

In the 1990s, in parallel, the effects of cryopreservation on mature human eggs were reappraised. Several authors agreed on the fact that appropriate protocols allow the cryopreservation of mature oocytes without any significant change in the meiotic spindle and without any increase in the rate of aneuploid embryos (40–44).

These reassuring fundamental studies should have opened the way to new clinical trials. In reality, additional doubts about possible injury to the zona pellucida (45,46) and cortical granules (22,47,48)leading to alterations in the mechanism of fertilization and increased incidence of polyspermy continued to cause concern.

A turning point in the efficiency of human oocytes cryopreservation was reached with the introduction of intracytoplasmic sperm injection. Gook et al. (49) compared traditional IVF and ICSI for thawed oocyte insemination. Although no differences were seen in the percentage of normal fertilization, ICSI was found to be associated with a significantly higher cleavage rate (50). A similar study was done by Kazem et al. (46) who reached a percentage of fertilization of 45.9% in the ICSI group and 13.5% in the IVF group; in the IVF group, only one fertilized oocyte cleaved while in the ICSI group all the fertilized eggs cleaved (46).

RECENT CLINICAL APPLICATION

In spite of the great increase in basic knowledge, the clinical application of oocyte cryopreservation has been stagnating for a long time. Unlike other techniques such as embryo cryopreservation or intracytoplasmic sperm injection, which were introduced fairly quickly in the clinical routine before displaying evidence of safety, oocyte cryopreservation was still regarded as an experimental procedure, potentially dangerous and unsafe.

As a matter of fact, there was no urgent need for implementation of oocyte freezing in current assisted reproductive technology, as embryo freezing was a reliable alternative. On the other hand, ethical indications were not powerful enough to urge clinical research in oocyte cryopreservation.

It was only in the second half of the 1990s that further clinical studies were undertaken. Tucker et al. (51) cryopreserved 285 aged unfertilized oocytes with slow freezing using 1,2-propanediol (PrOH) 1.5 M and sucrose 0.1 M. Fifty-five percent of oocytes survived thawing and were inseminated with ICSI achieving 41% fertilization rate. In the second part of the study, seven couples from their donor oocyte program consented to receive cryopreserved donated oocytes in eight thaw cycles. Cryosurvival of the 81 fresh donated oocytes was relatively poor

(24.7%) but normal fertilization after ICSI was 65% and embryo development was 100%. Uterine transfer of 13 embryos in five cycles gave rise to three early pregnancies (60%), but none went successfully to term.

In the following years, a number of first-ever achievements changed the clinical background of oocyte cryopreservation. Among the giants in reproductive cryobiology, Gook's studies (40,41,49,50) were especially useful in alleviating concerns about safety of oocyte cryopreservation.

The first live birth of a healthy female conceived from oocytes cryopreserved with PrOH 1.5 M + sucrose 0.2 M and inseminated with ICSI was published by Porcu et al. (52) at the University of Bologna. The announcement of this live birth after nearly a decade since the first reports of births following cryopreservation of human oocytes (18,21) reawakened interest in this field. In the following year, several reports announced pregnancies and births from oocytes cryopreserved with PrOH and inseminated with ICSI (53–55). In the same year, Tucker et al. (56) published the first birth from immature cryopreserved oocytes. In this study Tucker collected and cryopreserved twenty nine oocytes after ovarian stimulation in a 28-year-old woman; none of the16 mature oocytes survived thawing, while three of 13 GV-stage oocytes survived. Two out of three oocytes reached full maturity in culture and fertilized normally after ICSI. A single pregnancy was achieved after transferring both embryos. An uneventful full term gestation ended with the birth of a healthy female weighting 3300 g. Shortly after, Kuleshova et al. (57) published the first birth of a healthy female from oocytes cryopreserved with vitrification.

From then on, two main lines of clinical research were followed: slow freezing and vitrification. Initially, slow freezing was the most popular.

Since 1997, ICSI had been almost invariably used to inseminate cryopreserved oocytes and proved to be a suitable means of optimizing fertilization after thawing and subsequent embryo development. This was probably the key to success that opened a new era in the clinical application of oocyte cryopreservation.

Porcu's team at the University of Bologna thought it was the right time for undertaking an intensive investigation and documenting the actual degree of egg freezing clinical efficiency in humans. In the first trial 1769 supernumerary oocytes were frozen using slow freezing with PrOH and sucrose 0.2 M. One thousand five hundred two oocytes were thawed. The overall survival rate was 54%, the fertilization rate after ICSI was 58% and the cleavage rate 91%. In a part of thawed oocytes, the effects of duration of cryopreservation were evaluated: no difference was seen in the survival rate between 24 hours and 3 months of storage (59% and 60%, respectively). Sixteen pregnancies were achieved in 112 thawing cycles. Seven singleton and two twin pregnancies ended with the birth of eleven healthy children (58). In that early trial it was also documented that elective cryopreservation of all the oocytes retrieved might be a valid strategy to avoid severe ovarian hyperstimulation syndrome (OHSS). In 20 patients at risk of OHSS, a total of 384 oocytes were frozen. In the subsequent postponed ICSI cycles with thawed oocytes the pregnancy rate was 17.8%.

With revived interest in the field from the 1990s, several case reports and small series were published. Common procedures adopted in those studies were slow freezing using PrOH 1.5 M and sucrose and ICSI to inseminate thawed eggs. Within this basic protocol, some lines of research investigated the role of different concentration of sucrose and the role of sodium depleted medium in an attempt to improve oocyte survival and competence. Porcu's team discovered that sucrose 0.3 M seemed to improve egg survival when compared to sucrose 0.1 M and 0.2 M (59). With PrOH + 0.1 M sucrose pregnancies and births were reported (54,55,60–67). Freezing protocols with 0.2 M sucrose were used by several authors (52,58,68–77) as well as 0.3 M sucrose (73,78—88). In general, 0.3 M sucrose seems to improve egg survival. However, differential sucrose concentrations during dehydration (0.2 mol/L) and rehydration (0.3 mol/L) seem to increase the implantation rate of frozen oocytes (76). Another modification of the cryopreservation technique includes the replacement of sodium with choline in the freezing medium. Quintans et al. (89) used this freezing protocol and obtained 63% survival rates, 59% fertilization rate, and 25% implantation rate. They achieved six pregnancies, with the birth of two babies. Boldt et al. (90,91) reported oocyte cryopreservation outcomes after use of a sodium deleted protocol: survival rate was 60.4%, fertilization rate 62%, and pregnancy rate 32.4% per embryo transfer.

Until recently, it has been impossible to calculate the actual efficiency of oocyte freezing because only a low number of cases were done and sporadic pregnancies obtained. Variables to be taken into consideration are egg survival, fertilization, cleavage and implantation. Survival rate reported in the early literature varies widely from 25% to more than 80% probably because a generally low number of frozen eggs were evaluated in most studies and several different procedures were used (21,22,40,51,48,45,50,92). Recently, percentages of post-thaw egg viability are relatively more consistent. Chen et al. (93) demonstrated a significantly higher MII oocyte survival when using sucrose 0.2 Mol compared to 0.1 M (78.3 vs. 48.9%). One of the most active groups working in oocyte cryopreservation is that lead by Borini (66,72,83,94,95). This team showed 37% egg survival with 0.1 M (66) and 74.1% with 0.3 M sucrose concentration (72). The good results with 0.3 M sucrose solution are confirmed by Fosas et al. (80) with 89.8% survival, by Chen et al. (79) who achieved 75% post-thaw egg viability, and by Levi Setti et al. (73) reporting 69.9% post-thaw intact oocytes. However, it must be kept in mind the concern expressed by Paynter et al. (96) about the possible injury derived by extreme cell shrinkage occurring in 0.3 M sucrose solution.

In general, the rate of success of oocyte cryopreservation continues to be difficult to be estimate considering the variability among various studies depending more on the number of oocytes and thawing cycles and type of patient population than on the freezing protocols.

The number of births from frozen eggs increased progressively after the first case reports. Among others, the largest series should be cited. Porcu et al. (58) reported the birth of 13 healthy children. Winslow (71) announced the birth of 16 children in 2001, after thawing 324 egg cells, resulting in a survival rate of 68.5%, a fertilization rate of 81% and a pregnancy rate of 26.2% per transfer. Yang et al. (74) obtained the birth of 14 children after thawing 158 oocytes from donor cycles, with a survival percentage of 71% and a pregnancy rate per transfer of 45.8%. Fosas (80) announced the birth of five children. Borini et al. (66) documented the birth of 13 children. Porcu (97) reported the birth of 70 children from 2750 frozen eggs, with a survival rate after thawing of 69.6%, a fertilization rate of 74% and a pregnancy rate per transfer of 17%. Levi Setti et al. (73) reported the birth of 13 children. La Sala et al. (85) achieved the birth of seven children. Borini et al. published the birth of 4 additional children (72) and then of 105 babies (94). Konc et al. who described the birth of three children in Hungary, with a survival rate and a fertilization rate of, respectively, 80% and 81% (88).

Oocyte cryopreservation proved to increase the flexibility of ART procedure allowing the use of epidydimal and testicular sperm (68,69), the use of both the male and female gametes frozen (58) and the possibility of refreezing zygotes and embryos derived from cryopreserved eggs and sperm (75,77,82,98,99).

Oocyte cryopreservation proved also to be a suitable choice for fertility preservation in oncological patients. Porcu et al. (100) documented the collection and storage of an average of fifteen oocytes in young oncological patients undergoing chemotherapy.

The first births recently reported demonstrate that oocyte cryopreservation in oncology is a reliable option. Yang et al. (101) documented the first birth in a gestational carrier who conceived with frozen oocytes belonging to a patient with Hodgkin's lymphoma. This patient underwent egg retrieval and cryopreservation before radiotherapy. Ten of 12 cryopreserved oocytes survived after thawing and 9 were fertilized and developed into good quality embryos, which were transferred in three different cycles to a gestational carrier. A singleton pregnancy was achieved after the third transfer, resulting in the delivery of a healthy male. Surrogate motherhood is however not allowed in a majority of countries because of ethical concerns. Porcu et al. (102) reported the first delivery of healthy twins by a patient with bilateral ovariectomy for ovarian cancer who conceived using her own cryopreserved oocytes. Seven oocytes were retrieved before ovariectomy. After four years three oocytes were thawed. The survival rate of 100% as well as the fertilization and cleavage rate of 100% demonstrated that eggs may be safely stored for several years. The twin pregnancy progressed uneventfully to term. Oocyte storage may be then a concrete, pragmatic alternative also for oncological patients. The duration of oocytes storage does not seem to interfere with the oocytes survival as pregnancies occurred even after several years of gametes cryopreservation in liquid nitrogen.

Despite the difficulties in evaluating the true efficiency of cryopreservation with slow freezing, taking together all the clinical data available so far, a rough estimate of 5 implantations per 100 thawed oocytes may be done.

In the general awakening of interest about oocyte cryopreservation, the late 1990s also brought the revival of attention towards vitrification as an alternative technique for storing human eggs. Vitrification was originally proposed by Rall and Fahy (16) in mouse embryos.

Unlike slow freezing, the application of vitrification techniques to human oocytes occurred in the absence of basic biological studies addressing safety issues. Kuleshova et al. (57) obtained the first birth of a healthy girl from vitrified human oocytes and reported an egg survival rate of 65%. This survival rate was not higher than that achieved in slow freezing. However, vitrification appeared attractive as potentially it would avoid the development of ice crystals and would be a simpler and cheaper technique. The most important variables that influence the results of vitrification are the choice of cryoprotectants, the exposure time, the prefreezing temperature and the devices for storing samples in liquid nitrogen. Several modifications have been introduced to improve the efficiency of the technique by increasing the cooling rates, the cryoprotectant concentration and reducing cryoprotectants toxicity. The cooling rate can be increased by reducing the volume of the cryoprotectant solution and directly exposing the biological sample to liquid nitrogen in either an open straw or an electron microscopy grid. Very high concentration of ethylene glycol (EG) (5.5 M) and sucrose (1.0 M) apparently displayed an apparently tolerable toxicity (57). The same cryoprotectant solution in combination with grids, permitted to achieve a pregnancy from immature human oocytes (103) and seven births from mature oocytes (104) with a survival rate, respectively, of 59% and 69%. Kim et al. (105,106) achieved, respectively, one pregnancy and four births by using Kuleshova cryoprotectant solution. Ruvalcaba et al. (107) obtained the birth of five children. Chian et al. (108) used an especially designed device (Cryoleaf) and obtained a 93.9% survival rate, 74.6% fertilization rate, and seven pregnancies. Another type of vitrification device is the Cryotop, which was designed by Kuwayama et al. (109): with this device and a slightly lower concentration of EG (5.0 M) this author obtained 91% egg survival rate and 90% fertilization rate. Cobo et al. (110) published the birth of a healthy boy in a donation cycle. The birth of additional 28 children in 57 ovum donation cycles were also reported (111). In this study, survival rate was 96.1% and fertilization rate 73.1%. In a further study (112), a comparison between fresh and vitrified oocytes was done in term of survival, fertilization, cleavage, pregnancy and implantation rates. When evaluated simultaneously, the potential of vitrified oocytes to fertilize and further develop resulted similar to fresh counterparts, with a survival rate of 96.9%, a fertilization rate around 76.3%, and 11 ongoing pregnancies.

A different vitrification protocol with 2.7 M EG + 2.1 M DMSO + 0.5 M sucrose resulted in pregnancies and births (113–118). An alternative vitrification protocol, the super rapid cooling using slush nitrogen (SN2), was recently introduced. By applying negative pressure with a vacuum, liquid nitrogen will freeze and will be transformed into a slush state (SN2) avoiding the development of gas bubbles when plunging a biological sample. With this vitrification strategy, Yoon et al. (119) achieved 85.1% survival rate (302 out of 364 frozen eggs), 77.4% fertilization 94.3% cleavage rate and thirteen pregnancies (43.3%) from 30 uterine transfers of 120 embryos.

Studies on oocyte vitrification after in vitro maturation of immature oocytes have also been carried out. Huang and Buckett (120) reported the successful vitrification of three in vitro matured oocytes, as a strategy of fertility preservation in women with borderline ovarian malignancy. The same author (121) published the results of four cryopreservation cycle after in vitro maturation. After achieving a mean maturation rate of 79%, a total of eight oocytes were vitrified. In the author's opinion, this technique seems to be associated with satisfying fertilization rate (121).

The first live birth of a healthy baby after vitrification of in vitro matured oocytes was recently reported (122).

Until recently, slightly more than 60 births from oocyte vitrification were published. Then, suddenly, the birth of 200 children was reported (122). Rather unusually, the paper does not report details on the number of warming cycles, the number of vitrified-warmed oocytes, the egg survival, fertilization and cleavage rates and the pregnancy rate. Vitrification medium composed of 15% EG and 1,2-PrOH or 15% EG and DMSO, plus 0.5 M sucrose were used. The oocytes were then loaded onto the McGill Cryoleaf or Cryotop and directly immersed into liquid nitrogen. Obstetric and perinatal outcomes of 165 pregnancies and 200 infants conceived with vitrified oocytes were evaluated. The multiple pregnancy rate was 17%. The mean gestational age was 37 weeks + 1 day, the percentage of preterm deliveries (34–37 weeks) was 30%, and the high preterm delivery rate (<34 weeks) was 6%. The mean birth weights were 2920 ± 37 g for singletons and 2231 ± 55 g for multiples. The incidence of congenital anomalies was 2.5%. These results indicate that the mean birth weight and the incidence of congenital anomalies are comparable to that of spontaneous conceptions.

A few studies addressed the issue of meiotic spindle integrity in human oocytes after vitrification (see also chap. 13). Noyes et al. (123) established an association between the presence of the meiotic spindle after thawing and warming and the subsequent development of competent embryos. Larman et al. (124) documented the maintenance of the meiotic spindle during vitrification in human and mouse oocytes. Ciotti et al. (125) investigated the behavior of meiotic spindle during vitrification and demonstrated that meiotic spindle recovery is faster in vitrification of human oocytes as compared to slow freezing. Cobo et al. (126) reported no influence of vitrification by the cryotip methodology on the proportion of oocytes with a normal spindle apparatus. These data support a possible protective effect of vitrification on the meiotic spindle structure and, therefore, on the subsequent clinical outcome of the procedure even though additional basic biological studies as well as comparative clinical studies are needed.

CONCLUSIONS

To date, it is still difficult to estimate the true efficacy of egg storage in the ART technologies background as sufficient consistency and reproducibility of the results are still to be achieved.

Implantation rates per thawed or warmed oocyte vary considerably depending on the patient population, clinical conditions, oocyte quality and number of embryos transferred than to the cryopreservation technique itself. Slow freezing seems to produce apparently less favorable results than those obtained with vitrification. Nevertheless, it must be taken into consideration that the largest series of egg slow freezing have been performed in Italy were the insemination of no more than three eggs is allowed. In these studies, oocyte freezing is used in the general ART population and therefore also in cases with less favorable prognosis. On the other hand, most papers on vitrification included selected populations, frequently with donor programs, very young patients and a very high number of embryos transferred.

As regards safety, a study by Noyes et al. (127), pointed out that in more than 900 births from oocyte cryopreservation, congenital anomalies do not seem to be higher than with natural conception.

REFERENCES

1. Chang MC. Transplantation of fertilized rabbit ova: the effect on viability of age, in vitro storage period, and storage temperature. Nature 1948; 159: 602.
2. Chang MC. Probability of normal development after transplantation of fertilized rabbit ova stored at different temperatures. Proc Soc Exp Biol Med 1948; 161: 978.
3. Chang MC. Storage of unfertilized rabbit ova: subsequent fertilization and the probability of normal development. Nature 1953; 172: 353–4.
4. Sherman JK. Effect of glycerol and low temperature on survival of unfertilized mouse eggs. Nature 1958; 181: 785.
5. Sherman JK. Temperature shock and cold-storage of unfertilized mouse Eggs. Fertil Steril 1959; 10: 384–96.
6. Burks JL. Morphologic evaluation of frozen rabbit and human ova. Fertil Steril 1965; 16: 638–41.
7. Mazur P. Kinetics of water loss from cells at subzero temperatures and the likelihood of intracellular freezing. J Gen Physiol 1963; 47: 347–51.
8. Asahina E. Intracellular freezing and frost resistance in egg-cells of the sea urchin. Nature 1961; 191: 1263–65.
9. Mazur P. Cryobiology: the freezing of biological systems. Science 1970; 168: 939–49.
10. Mazur P. Freezing of living cells: mechanisms and implications. Am J Physiol 1984; 16: c125–42.
11. Mazur P. The freezing of living cells. Ann N Y Acad Sci 1988; 9: 514–31.
12. Parkening TA. Effects of various low temperatures, cryoprotective agents and cooling rates on the survival, fertilizability and development of frozen—thawed mouse eggs. J Exp Zool 1976; 197: 369–74.
13. Whittingham DG. Fertilization in vitro and development to term of unfertilized mouse oocytes previously stored at −196°C. J Reprod Fertil 1977; 49: 89–94.
14. Choung CJ. Effects of cryopreservation on the viability and fertilizability of unfertilized hamster oocytes. Am J Obstet Gynecol 1986; 155: 1240–5.
15. Critser JK. Cryopreservation of hamster oocytes: effects of vitrification or freezing on human sperm penetration of zona-free hamster oocytes. Fertil Steril 1986; 46: 277–84.

16. Rall WF, Fahy GM. Ice-free cryopreservation of mouse embryos at –196°C by vitrification. Nature 1985; 313: 573–5.
17. Trounson A. Preservation of human eggs and embryos. Fertil Steril 1986; 46: 1–12.
18. Chen C. Pregnancy after human oocyte cryopreservation. Lancet 1986; 1: 884–6.
19. Chen C. Pregnancies after human oocyte cryopreservation. Ann N Y Acad Sci 1988; 541: 541–9.
20. Al-Hasani S, Diedrich K, van der Ven H, Krebs D. Initial results of the cryopreservation of human oocytes. Geburthshilfe Franuehilkild 1986; 46: 643–4.
21. Van Uem JF, Siebzehnrubl ER, Schuh B, et al. Birth after cryopreservation of unfertilised oocytes. Lancet 1987; i: 752–3.
22. Al-Hasani S, Diedrich K, van der Ven H, et al. Cryopreservation of human oocytes. Hum Reprod 1987; 2: 695–700.
23. Diedrich K, Al-Hasani S, van der Ven H, Krebs D. Successful *in vitro* fertilization of frozen-thawed rabbit and human oocytes. Ann N Y Acad Sci 1988; 541: 562–71.
24. Magistrini M, Szöllösi D. Effects of cold and of isopropyl-N-phenylcarbamate on the second meiotic spindle of mouse oocytes. Eur J Cell Biol 1980; 22: 699–707.
25. Pickering SJ, Johnson MH. The influence of cooling on the organization of the meiotic spindle of the mouse oocyte. Hum Reprod 1987; 2: 207–16.
26. Pickering SJ, Braude PR, Johnson MH, Cant A, Currie J. Transient cooling to room temperature can cause irreversible disruption to the meiotic spindle in human oocytes. Fertil Steril 1990; 54: 102–8.
27. Vincent C, Garnier V, Heyman Y, Renard JP. Solvent effects on cytoskeletal organization and in-vivo survival after freezing of rabbit oocytes. J Reprod Fertil 1989; 87: 809–20.
28. Glenister PH, Wood MJ, Kirby C, Whittingham DG. Incidence of chromosome anomalies in first-cleavage mouse embryos obtained from frozen-thawed oocytes fertilized in vitro. Gamete Res 1987; 16: 205–16.
29. Kola I, Kirby C, Shaw J, Davey A, Trounson A. Vitrification of mouse oocytes results in aneuploid zygotes and malformed fetuses. Teratology 1988; 38: 467–74.
30. Sathananthan AH, Trounson A, Freemann L, Brady T. The effects of cooling human oocytes. Hum Reprod 1988; 3: 968–77.
31. Aigner S, Van der Elst J, Siebzehnrubl E, et al. The influence of slow and ultra-rapid freezing on the organization of the meiotic spindle in the mouse oocyte. Hum Reprod 1992; 7: 857–64.
32. Joly C, Bchini O, Boulekbache H, Testart J, Maro B. Effects of 1,2-propanediol on the cytoskeletal organisation of the mouse oocyte. Hum Reprod 1992; 3: 374–8.
33. Bernard A, Fuller BJ. Cryopreservation of human oocytes: a review of current problems and perspectives. Hum Reprod Update 1996; 2: 193–207.
34. Boiso I, Mati M, Santalo J, et al. A confocal microscopy analysis of the spindle and chromosome configurations of human oocytes cryopreserved at the germinal vesicle and metaphase II stage. Hum Reprod 2002; 17: 1885–91.
35. Mandelbaum J, Belasch-Allart J, Junca AM, et al. Cryopreservation in human assisted reproduction is now routine for embryos but remains a research procedure for oocytes. Hum Reprod 1998; 13(Suppl 3): 161–77.
36. Toth TL, Baka SG, Veeck LL, et al. Fertilization and in vitro development of cryopreserved human prophase I oocytes. Fertil Steril 1994; 61: 891–4.
37. Baka SG, Toth TL. Evaluation of the spindle apparatus of in-vitro matured human oocytes following cryopreservation. Hum Reprod 1995; 10: 1816–20.
38. Son WY, Park SE, Lee KA, et al. Effects of 1,2-propanediol and freezing on the in vitro developmental capacity of human immature oocytes. Fertil Steril 1996; 66: 996–9.
39. Park SE, Son WI. Chromosome and spindle configurations of human oocytes matured in vitro after cryopreservation at the germinal vesicle stage. Fertil Steril 1997; 68: 920–6.
40. Gook DA, Osborn SM, Johnston WI. Cryopreservation of mouse and human oocytes using 1,2-propanediol and the configuration of the meiotic spindle. Hum Reprod 1993; 8: 1101–19.
41. Gook DA, Osborn SM, Bourne H, Johnston WIH. Fertilization of human oocytes following cryopreservation normal karyotypes and absence of stray chromosomes. Hum Reprod 1994; 9: 684–91.
42. Van Blerkom J, Davis PW. Cytogenetic, cellular, and developmental consequences of cryopreservation of immature and mature mouse and human oocytes. Microsc Res Tech 1994; 27: 165–3.
43. George MA, Pickering SJ, Braude PR, Johnson MH. The distribution of α- and γ-tubulin in fresh and aged mouse and human oocytes exposed to cryoprotectant. Mol Hum Reprod 1996; 2: 445–56.
44. Cobo A, Rubio C, Gerli S, et al. Use of fluorescence in situ hybridization to assess the chromosomal status of embryos obtained from cryopreserved oocytes. Fertil Steril 2001; 75: 354–60.
45. Dumoulin JC, Bergers-Janssen JM, Pieters MH, et al. The protective effect of polymers in the cryopreservation of human and mouse zonae pellucidae and embryos. Fertil Steril 1994; 62: 793–8.

46. Kazem R, Thompson LA, Srikantharajah A, et al. Cryopreservation of human oocytes and fertilization by two techniques: in-vitro fertilization and intracytoplasmic sperm injection. Hum Reprod 1995; 10: 2650–4.
47. Vincent C, Pickering SJ, Johnson MH. The hardening effect of dimethylsulphoxide on the mouse zona pellucida requires the presence of an oocyte and is associated with a reduction in the number of cortical granules present. J Reprod Fertil 1990; 89: 253–9.
48. Al-Hasani S, Diedrich K. Oocyte storage. In: Gruzzinskas JG, Yovich JL, eds. Gametes—The Oocyte. Cambridge: Cambridge University Press, 1995.
49. Gook DA, Osborn SM, Johnston WI. Parthenogenetic activation of human oocytes following cryopreservation using 1,2-propanediol. Hum Reprod 1995; 10: 654–8.
50. Gook D, Schiewe MC, Osborn S, et al. Intracytoplasmic sperm injection and embryo development of human oocytes cryopreserved using 1,2-propanediol. Hum Reprod 1995; 10: 2637–41.
51. Tucker MJ, Wright G, Morton P, et al. Preliminary experience with human oocyte cryopreservation using 1,2-propanediol and sucrose. Hum Reprod 1996; 11: 1513–15.
52. Porcu E, Fabbri R, Seracchioli R, et al. Birth of a healthy female after intracytoplasmic sperm injection of cryopreserved human oocytes. Fertil Steril 1997; 68: 724–6.
53. Polak de Fried E, Notrica J, Rubinsten M, et al. Pregnancy after human donor oocyte cryopreservation and thawing in association with intracytoplasmic sperm injection in a patient with ovarian failure. Fertil Steril 1998; 69: 555–7.
54. Nawroth F, Kissing K. Pregnancy after intracytoplasmatic sperm injection (ICSI) of cryopreserved human oocytes. Acta Obstet Gynecol Scand 1998; 77: 462–3.
55. Young E, Kenny A, Puigdomenech E, et al. Triplet pregnancy after intracytoplasmic sperm injection of cryopreserved oocytes: case report. Fertil Steril 1998; 70: 360–1.
56. Tucker MJ, Wright G, Morton PC, Massey JB. Birth after cryopreservation of immature oocytes with subsequent in vitro maturation. Fertil Steril 1998; 70: 578–9.
57. Kuleshova L, Gianaroli L, Magli C, et al. Birth following vitrification of a small number of human oocytes: case report. Hum Reprod 1999; 14: 3077–9.
58. Porcu E, Fabbri R, Marsella T, et al. Clinical experience and application of oocyte cryopreservation. Mol Cell Endocrinol 2000; 169: 33–7.
59. Fabbri R, Porcu E, Marsella T, et al. Human oocyte cryopreservation: new perspectives regarding oocyte survival. Hum Reprod 2001; 16: 411–16.
60. Wurfel W, Schleyer M, Krusmann G, Hertwig IV, Fiedler K. Fertilization of cryopreserved and thawed human oocytes (Cryo-Oo) by injection of spermatozoa (ICSI)—medical management of sterility and case report of a twin pregnancy. Zentralbl Gynakol 1999; 121: 444–8.
61. Chia C, Chan W, Quah E, Cheng L. Triploid pregnancy after ICSI of frozen testicular spermatozoa into cryopreserved human oocytes. Case report. Hum Reprod 2000; 15: 1962–4.
62. Huttelova R, Becvarova V, Brachtlova T. More successful oocyte freezing. J Assist Reprod Genet 2003; 20: 293.
63. Notrica J, Kanzepolsky L, Divita A, et al. A healthy female born after ICSI of a cryopreserved oocyte and cryopreserved spermatozoa banked prior to radiotherapy in a patient with a seminoma: a case report. Fertil Steril 2003; 80(Suppl 3): S149.
64. Allan J. Re: Case report: Pregnancy from intracytoplasmic injection of a frozen-thawed oocyte. Aust N Z J Obstet Gynaecol 2004; 44: 588.
65. Miller KA, Elkind-Hirsch K, Levy B, et al. Pregnancy after cryopreservation of donor oocytes and preimplantation genetic diagnosis of embryos in a patient with ovarian failure. Fertil Steril 2004; 82: 211–14.
66. Borini A, Bonu MA, Coticchio G, et al. Pregnancies and births after oocyte cryopreservation. Fertil Steril 2004; 82: 3.
67. De Geyter M, Steimann S, Holzgreve W, De Geuter C. First delivery of healthy offspring after freezing and thawing of oocytes in Switzerland. Swiss Med Wkli 2007; 137: 443–7.
68. Porcu E, Fabbri R, Ciotti PM, et al. Ongoing pregnancy after intracytoplasmic sperm injection of epididymal spermatozoa into cryopreserved human oocytes. J Assist Reprod Genet 1999; 16: 283–5.
69. Porcu E, Fabbri R, Petracchi S, et al. Ongoing pregnancy after intracytoplasmic injection of testicular spermatozoa into cryopreserved human oocytes. Am J Obstet Gynecol 1999; 4: 1044–5.
70. Yang D, Blohm P, Winslow K, Cramer L. A twin pregnancy after microinjection of human cryopreserved oocyte with a specially developed oocyte cryopreservation regime. Fertil Steril 1998; 70(Suppl 1): S239.
71. Winslow K, Yang D, Blohm P, et al. Oocyte cryopreservation—a three year follow up of sixteen births. Fertil Steril 2001; 76(Suppl 1): S120–1.
72. Borini A, Lagalla C, Bonu MA, et al. Cumulative pregnancy rates resulting from the use of fresh and frozen oocytes: 7 years' experience. Reprod Biomed Online 2006; 12: 481–6.

73. Levi Setti PE, Albani E, Novara PV, Cesana A, Morreale G. Cryopreservation of supernumerary oocytes in IVF/ICSI cycles. Hum Reprod 2006; 21: 370–5.
74. Yang D, Winslow K, Blohm P, et al. Oocyte donation using cryopreserved donor oocytes. Fertil Steril 2002; 78(Suppl 1): S14.
75. Montag M, van der Ven K, Dorn C, et al. Birth after double cryopreservation of human oocytes at metaphase II and pronuclear stages. Fertil Steril 2006; 85: 751.e5–7.
76. Bianchi V, Coticchio G, Distratis V, et al. Differential sucrose concentration during dehydration (0.2 mol/L) and rehydration (0.3 mol/L) increases the implantation rate of frozen human oocytes. Reprod Biomed Online 2007; 14: 64–71.
77. Gook DA, Hale L, Edgar DH. Live birth following transfer of a cryopreserved embryo generated from a cryopreserved oocyte and a cryopreserved sperm: case report. J Assist Reprod Genet 2007; 24: 43–5.
78. Chen SU, Lien YR, Tsai YY, et al. Successful pregnancy occurred from slowly freezing human oocytes using the regime of 1.5 mol/L 1,2-propanediol with 0.3 mol/L sucrose. Hum Reprod 2002; 17: 1412–13.
79. Chen SU, Lien YR, Chen HF, et al. Observational clinical follow-up of oocyte cryopreservation using a slow freezing method with 1,2-propanediol plus sucrose followed by ICSI. Hum Reprod 2005; 20: 1975–80.
80. Fosas N, Marina F, Torres PJ, et al. The births of five Spanish babies from cryopreserved donated oocytes. Hum Reprod 2003; 18: 1417–21.
81. Li XH, Chen SU, Zhang X, et al. Cryopreserved oocytes of infertile couples undergoing assisted reproductive technology could be an important source of oocyte donation: a clinical report of successful pregnancies. Hum Reprod 2005; 20: 3390–4.
82. Tjer GC, Chiu TT, Cheung LP, Lok IH, Haines CJ. Birth of a healthy baby after transfer of blastocysts derived from cryopreserved human oocytes fertilized with frozen spermatozoa. Fertil Steril 2005; 83: 1547–9.
83. Borini A, Sciajno R, Bianchi V, et al. Clinical outcome of oocyte cryopreservation after slow cooling with a protocol utilizing a high sucrose concentration. Hum Reprod 2006; 21: 512–17; 2005; 11: 300–8.
84. Chamayou S, Alecci C, Ragolia C, et al. Comparison of in-vitro outcomes from cryopreserved oocytes and sibling fresh oocytes. Reprod Biomed Online 2006; 12: 730–6.
85. La Sala GB, Vicoli A, Villani MT, et al. Outcome of 518 salvage oocyte-cryopreservation cycles performed as a routine procedure in an in vitro fertilization program. Fertil Steril 2006; 86: 1423–7.
86. Barritt J, Luna M, Duke M, et al. Report of four donor-recipient oocyte cryopreservation cycles resulting in high pregnancy and implantation. Fertil Steril 2007; 87: 189.e13–17.
87. De Santis L, Cino I, Rabellotti E, et al. Oocyte cryopreservation: clinical outcome of slow-cooling protocols differing in sucrose concentration. Reprod Biomed Online 2006; 14: 57–63.
88. Konc J, Kanyo K, Cseh S. Does oocyte cryopreservation have a future in Hungary? Reprod Biomed Online 2007; 14: 11–13.
89. Quintans CJ, Donaldson MJ, Bertolino MV, Pasqualini RS. Birth of two babies using oocytes that were cryopreserved in a choline-based freezing medium. Hum Reprod 2002; 17: 3149–52.
90. Boldt J, Cline D, McLaughlin D. Human oocyte cryopreservation as an adjunct to IVF-embryo transfer cycles. Hum Reprod 2003; 18: 1250–5.
91. Boldt J, Tidswell N, Sayers A, Kilani R, Cline D. Human oocyte cryopreservation: 5-year experience with a sodium-depleted slow freezing method. Reprod Biomed Online 2006; 13: 96–100.
92. Chen. 1987.
93. Chen ZJ, Li M. Effects of sucrose concentration on the developmental potential of human frozen-thawed oocytes at different stages of maturity. Hum Reprod 2004; 19: 2345–9. Epub Aug 6, 2004.
94. Borini A, Bianchi V, Bonu MA, et al. Evidence-based clinical outcome of oocyte slow cooling. Reprod Biomed Online 2007; 15: 175–81.
95. Borini A, Cattoli M, Mazzone S, et al. Survey of 105 babies born after slow cooling oocyte cryopreservation. Fertil Steril 2007; 88: S13.
96. Paynter SJ, Borini A, Bianchi V, et al. Volume changes of mature human oocytes on exposure to cryoprotectant solution used in slow cooling procedures. Hum Reprod 2005; 20: 1194–9.
97. Porcu E. Cryopreservation of oocytes: indications, risks and outcomes. Hum Reprod 2005; 20: 1: S50.
98. Azambuja R, Badalotti M, Teloken C, Michelon J, Petracco A. Successful birth after injection of frozen human oocytes with frozen epididymal spermatozoa. Reprod Biomed Online 2005; 11: 449–51.
99. Levi Setti PE, Albani E, Novara PV, et al. Normal birth after transfer of cryopreserved human embryos generated by microinjection of cryopreserved testicular spermatozoa into cryopreserved human oocytes. Fertil Steril 2005; 83: 1041.

100. Porcu E, Fabbri R, Damiano G, et al. Oocyte cryopreservation in oncological patients. Eur J Obstet Gynecol Reprod Biol 2004; 113: S14–16.
101. Yang D, Brown SE, Nguyen K, et al. Live birth after the transfer of human embryos developed from cryopreserved oocytes harvested before cancer treatment. Fertil Steril 2007; 87: 1469e4.
102. Porcu E, Venturoli S, Damiano G, et al. Healthy twins delivered after oocyte cryopreservation and bilateral ovariectomy for ovarian cancer. Reprod Biomed Online 2008; 17: 267–9.
103. Wu J, Zhang L, Wang X. In vitro maturation, fertilization and embryo development after ultrarapid freezing of immature human oocytes. Reproduction 2001; 121: 389–93.
104. Yoon TK, Kim TJ, Park SE, et al. Live births after vitrification of oocytes in a stimulated in vitro fertilization-embryo transfer program. Fertil Steril 2003; 79: 1323–6.
105. Kim TJ, Hong SW, Park SE, Cha KY. Pregnancy after vitrification of human oocytes and blastocysts using same cryoprotectant solution, ethylene glycol and sucrose. Fertil Steril 2003; 80: S143.
106. Kim T, Hong S, Cha K. Pregnancies from cryopreserved oocytes using vitrification protocol. Fertil Steril 2005; 84(Suppl 1): S179.
107. Ruvalcaba L, Marínez R, Cuneo S, et al. Improving donor programs with an oocyte bank using vitrification. Fertil Steril 2005; 84(Suppl 1): S70.
108. Chian RC, Son WY, Huang JY, et al. High survival rates and pregnancies of human oocytes following vitrification: preliminary report. Fertil Steril 2005; 84: S36.
109. Kuwayama M, Vajta G, Kato O, Leibo SP. Highly efficient vitrification method for cryopreservation of human oocytes. Reprod Biomed Online 2005; 11: 300–8.
110. Cobo A, Kuwayama M, Perez S, et al. Comparison of concomitant outcome achieved with fresh and cryopreserved donor oocytes vitrified by the Cryotop method. Fertil Steril 2008; 89: 1657–64.
111. Cobo A, Bellver J, Domingo J, et al. New options in assisted reproduction technology: the Cryotop method of oocyte vitrification. Reprod Biomed Online 2008; 17: 68–72.
112. Cobo A, Domingo J, Perez S, et al. Vitrification: an effective new approach to oocyte banking and preserving fertility in cancer patients. Clin Transl Oncol 2008; 10: 268–73.
113. Katayama KP, Stehlik J, Kuwayama M, et al. High survival rate of vitrified human oocytes results in clinical pregnancy. Fertil Steril 2003; 80: 223–4.
114. Kyono K, Fuchinoue K, Yagi A, et al. Successful pregnancy and delivery after transfer of a single blastocyst derived from a vitrified mature human oocyte. Fertil Steril 2005; 84: 1017.
115. Okimura T, Kato K, Zhan Q, et al. Update on clinical efficiency of the vitrification method for human oocytes in an in vitro fertilization program. Fertil Steril 2005; 84(Suppl 1): S174.
116. Lucena E, Bernal DP, Lucena C, et al. Successful ongoing pregnancies after vitrification of oocytes. Fertil Steril 2006; 85: 108–11.
117. Selman H, Angelini A, Barnocchi N, et al. Ongoing pregnancies after vitrification of human oocytes using a combined solution of ethylene glycol and dimethyl sulfoxide. Fertil Steril 2006; 86: 997–1000.
118. Antinori M, Licata E, Dani G, et al. Cryotop vitrification of human oocytes results in high survival rate and healthy deliveries. Reprod Biomed Online 2007; 14: 72–9.
119. Yoon TK, Lee DR, Cha SK, et al. Survival rate of human oocytes and pregnancy outcome after vitrification using slush nitrogen in assisted reproductive technologies. Fertil Steril 2007; 88: 925–6.
120. Huang JY, Buckett WM. Retrieval of immature oocytes followed by in vitro maturation and vitrification: a case report on a new strategy of fertility preservation in women with borderline ovarian malignancy. Gynecol Oncol 2007; 105: 542–4. Epub Mar 26, 2007.
121. Huang JY, Tulandi T. Combining ovarian tissue cryobanking with retrieval of immature oocytes followed by in vitro maturation and vitrification: an additional strategy of fertility preservation. Fertil Steril 2008; 89: 567–72. Epub Jun 4, 2007.
122. Chian RC, Huang JY, Lucena E, Saa A, et al. Obstetric and perinatal outcome in 200 infants conceived from vitrified oocytes. Reprod Biomed Online 2008; 16: 608–10.
123. Noyes N, Chang C, Liu H, et al. Presence of meiotic spindle predicts embryo competence following oocyte cryopreservation. Fertil Steril 2006; 86: S2.
124. Larman MG, Minasi MG, Rienzi L, Gardner DK. Maintenance of the meiotic spindle during vitrification in human and mouse oocytes. Reprod Biomed Online 2007; 15: 692–700.
125. Ciotti PM, Porcu E, Notarangelo L, et al. Meiotic spindle recovery is faster in vitrification of human oocyte compared to slow freezing. Fertil Steril 2008; (in press).
126. Cobo A, Perez S, De los Santos MJ. Effect of different cryopreservation protocols on the metaphase II spindle in human oocytes. Reprod Biomed Online 2008; 17: 350–9.
127. Noyes N, Porcu E, Borini A. With more than 900 oocyte cryopreservation babies born, congenital defects do not appear more frequent than in natural conceptions. Reprod Biomed Online 2009; (in press).

128. Chian RC, Gilbert L, Huang JYJ, et al. Live birth after vitrification of in vitro matured human oocytes. Fertil Steril 2009; 91: 372–6. Epub Jun 2, 2008.
129. Fukinoue K, Fukanaga N. Freezing of human immature oocytes using cryoloops with Taxol in the vitrification solution. J Assisst Reprod Genet 2004; 21: 307–9.
130. Porcu E, Fabbri R, Ciotti P, et al. Four healthy children from frozen human oocytes and frozen human sperms. Fertil Steril 2001; 76: S76.
131. Porcu E, Fabbri R, Seracchioli R, et al. Obstetric, perinatal outcome and follow up of children conceived from cryoprserved oocytes. Fertil Steril 2000; 74: S48.

20 Success Rates from Oocyte Cryopreservation

Andrea Borini and Maria Antonietta Bonu
Tecnobios Procreazione, Centre for Reproductive Health, Bologna, Italy

INTRODUCTION

Although still considered to be an experimental procedure by many, oocyte cryopreservation is rapidly expanding, boosted by advances in cryopreservation methodologies. Especially in Italy, probably several thousands patients have been treated in the last few years by different slow cooling and vitrification protocols. This imposes an urgent need to assess the clinical efficiency of oocyte cryopreservation in comparison to other treatment strategies. The definition of clinical efficiency in IVF has been a heatedly debated matter. In the United States, where costs of treatment are very high and not covered by national health agencies, the pregnancy rate per embryo transfer has been adopted as the most significant parameter of efficiency. Because it is clear that only a very small minority of oocytes (5–10%) has the potential to develop into a baby (1), in that country many clinics implement treatment strategies involving strict criteria of post-fertilization selection (e.g., blastocyst culture) and/or the simultaneous use of multiple embryos, to maximize the overall chances of at least one event of implantation per transfer. Such an approach can certainly give rise to comparatively high pregnancy rates, but also inevitably generates a high incidence of multiple pregnancies whose implications, in terms of prematurity, birth defects and healthcare costs, are increasingly considered unacceptable (see chap. 17). In Europe, especially in northern and Scandinavian countries, a different perception is emerging, according to which the rate of single healthy deliveries per cycle of stimulation is considered as the ultimate parameter to measure the clinical efficiency of IVF treatments. This is currently leading to a decrease in the number of transferred embryos per transfer and to a more extensive use of the cryopreservation of "surplus" embryos. Fresh and frozen-thawed deliveries can be therefore taken into account together to express a comprehensive (cumulative) definition of clinical efficiency. From these and other considerations discussed in this chapter, it shows the cumulative pregnancy rate is probably the most correct approach to assess also the efficiency of oocyte cryopreservation. Unfortunately, no large prospective controlled studies have been conducted so far with the aim of comparing the cumulative delivery rates of treatment cycles where embryo or oocyte cryopreservation (the latter in its various methodological versions) has been used. However, no matter how imperfect, current evidence suggests that the clinical efficiency of oocyte cryopreservation is rapidly approaching the rates normally achieved by cryopreserved embryos. This is also indicated by a comparison of the rates of implantations elaborated on the number of oocytes used for a treatment cycle.

A SYNOPSIS OF OOCYTE CRYOPRESERVATION BY THE SLOW COOLING APPROACH

The initial attempts to freeze oocytes were described by Whittingham about 30 years ago using mouse oocytes (2). After 1986, when the first pregnancy from frozen human oocyte was obtained (3) other live births were reported, but remained sporadic (4,5).

After about a decade, pregnancies, after oocyte cryopreservation, were reported again (6,7) using the slow-cooling technique developed for early cleavage embryo freezing (8). These successes were obtained after that, in 1995. Gook et al. suggested the use of intracytoplasmic sperm injection (ICSI), instead of standard IVF, for thawed oocytes with the aim to avoid possible fertilization failure caused by premature cortical granules release and zona pellucida hardening (9). During the following years, numerous small reports of pregnancies and births have been published (10–15). These initial results showed that conventional cryopreservation techniques were not able to guarantee the successful storage of oocytes, especially as a consequence of low post-thaw survival rates.

Compared to embryos, the mature oocyte is less suitable to freezing as a consequence of its size and cellular characteristics. Poor post-thaw survival rate has prevented oocyte freezing from being adopted as an established form of treatment. New developed cryopreservation protocols can now guarantee higher survival rates, almost comparable to the ones obtained after embryo freezing. Fertilization rates are also high. In this chapter, we will try to summarize the success rates obtained with oocyte cryopreservation, in particular focusing on: (a) freezing protocols; (b) cumulative pregnancy rates; (c) comparison with 2PN and early cleavage stage embryos results; (d) number of implantations compared with fresh oocytes.

CLINICAL SUCCESS RATE BASED ON FREEZING PROTOCOL

Usually, during slow cooling the cryoprotectants used are propane-1,2-diol (PROH) and sucrose. During last decade, different freezing protocols have been developed on the basis of different sucrose concentrations used at freezing and thawing.

0.1 mol/L Sucrose at Freezing and 0.2 mol/L Sucrose at Thawing

As already mentioned, the first protocol used was based on 1.5 mol/L PROH and 0.1 mol/L sucrose. Gook et al. have the merit to have tested this protocol on human oocytes describing survival, fertilization and cleavage rates, although they did not assess its clinical performance (9,16,17). Using this protocol, Porcu obtained the first pregnancy utilizing ICSI to achieve fertilization of thawed oocytes (6). Before that experience, Tucker et al. reported a low survival rate (25%) and a fertilization rate of 65%. After the transfer of 13 embryos, two implantations with fetal heartbeat were obtained. Unfortunately, both pregnancies miscarried during the first trimester (18).

However larger studies reported a survival rate from 25% (19) to 50% (16), presenting this procedure as relatively inefficient.

In 2004, our group reported results, using 0.1 mol/L sucrose, on 68 patients, whose treatment required oocyte freezing in response to a variety of conditions, during the period from January 1997 to December 2000 (20).

Survival, fertilization and cleavage rate were 37%, 45.4%, and 86.3%, respectively. The pregnancy rate per patients, per thawing cycle and per transfer were 22%, 17.4%, and 25.4%, respectively. The implantation rate was 16.4% and the implantation rate per thawed oocytes was 2.3% (Table 1).

These data confirmed that inadequate survival (37%) was the outcome when mature oocytes were frozen with a slow freezing protocol that on the contrary ensured excellent recovery when used for the storage of early cleavage embryos. Despite reduced survival and fertilization rates, we observed a cleavage rate of 86%, similar to the value achieved with embryos from fresh oocytes. Other reports have confirmed these findings, describing similar cleavage rates (21,22).

Table 1 Clinical Outcome of Oocyte Cryopreservation after Slow Cooling with a Protocol Utilizing a Low Sucrose Concentration (0.1 mol/L) at Freezing (20)

Age (yr)	33 ± 4.3
Thawing cycles	86
Oocytes thawed	737
Oocytes survived (%)	273 (37%)
2PN (%)	124 (45.4%)
Cleaved (%)	107 (86.3%)
Pregnancy rate/patient	22%
Pregnancy rate/transfer	25.4%
Pregnancy rate/thawing	17.4%
Implantation rate	16.4%
Abortion (%)	3 (20%)

0.3 mol/L Sucrose at Freezing and 0.3 mol/L Sucrose at Thawing

An increase in the extent of dehydration of human oocytes prior to cooling was considered as a possible approach for improving the outcome of cryopreservation. Fabbri et al. used higher sucrose concentration (0.2 and 0.3 mol/L) in the freezing solution, achieving proportionally improved survival rates (58% and 83%, respectively) (23).

Chen confirmed the beneficial effect of this protocol in term of survival (24). In his study, 75% of 159 oocytes were found to be intact after thawing. Fertilization and cleavage rates were within the normal ranges; pregnancy and implantation rates were 33% and 11%, respectively, and the implantation per thawed oocyte was 5% (Table 2). The author concluded that in his experience oocyte cryopreservation with high sucrose protocol was comparable to cryopreservation at the 2PN stage.

Our group, using essentially the same protocol, presented a much larger number of thawing cycles (201 and 927 thawed oocytes) (25). Survival, fertilization and cleavage rates were 74.1%, 76.1%, and 90.2%, respectively. Besides, pregnancy rate per patient, per thawing cycle, and per transfer was 12.3%, 8.9%, and 9.7%, respectively. Finally, implantation rate was 5.2% and implantation rate per thawed oocyte 2.6% (Table 3).

So, we obtained high survival and fertilization rates, as well as apparently undisturbed cleavage, but in our experience this high sucrose concentration protocol did not generate high pregnancy rates.

In 2007, De Santis et al. (26) presented a retrospective study evaluating the clinical outcome of two alternative slow freezing protocols involving 0.1 and 0.3 mol/L sucrose. Survival rates were, respectively, 24.3% and 71.2%. Whilst fertilization rate was higher in 0.3 mol/L group, enhanced results for 0.1 mol/L group were achieved overall. Implantation rate per transferred embryos was 12.2% versus 5.7%, pregnancy rate per transfer 16.7% versus 9.5%. This suggests that while the 0.1 mol/L protocol severely compromised the survival rate, at the same time it is able to better preserve the developmental potential of survived oocytes.

Table 2 Clinical Outcome of Oocyte Cryopreservation after Slow Cooling with a Protocol Utilizing a High Sucrose Concentration (0.3 mol/L) at Freezing Reported by Chen et al. (24)

Age (yr)	32 ± 5
Oocytes thawed	159
Oocytes survived (%)	119 (75%)
2PN (%)	80 (67%)
Pregnancy rate/transfer	33%
Implantation rate	11%

Table 3 Clinical Outcome of Oocyte Cryopreservation after Slow Cooling with a Protocol Utilizing a High Sucrose Concentration (0.3 mol/L) at Freezing (25)

Age (yr)	34.9 ± 4.2
Thawing cycles	201
Oocytes thawed	927
Oocytes survived (%)	687 (74.1%)
2PN (%)	448 (76.1%)
Pregnancy rate/patient	12.3%
Pregnancy rate/transfer	9.7%
Pregnancy rate/thawing	8.9%
Implantation rate	5.2%
Abortion (%)	3 (14.2%)

Other experiences, using this freezing protocol, were reported in the literature (27–29). Substantially, all the authors reported the same survival, fertilization and cleavage rates confirming the high survival and fertilization rates, but the low development potential of the embryos derived from frozen oocytes. The substantially comparable results obtained by different groups suggest, at least, that the procedure is repeatable and easily accessible by different hands.

Confocal microscopy observations suggested that the different developmental ability of oocytes stored with the two protocols might not be explained by meiotic spindle damage. In fact, oocytes cryopreserved with 0.3 mol/L exhibit a frequency of normal spindle morphology comparable to fresh controls (30).

On the other hand, higher degree of alterations of the cytoplasm were found (31). In fact, Nottola et al., comparing 0.1 and 0.3 mol/L frozen oocytes to control fresh ones, described a higher rate of loss of cortical granules in oocytes frozen with both protocols and a higher presence of cytoplasmic vacuoles when the 0.3 mol/L protocol was used. Such alterations of the cytoplasm may explain the lower development potential of the embryos derived from oocyte cryopreserved using the 0.3 mol/L protocol.

0.2 mol/L at Freezing and 0.3 mol/L at Thawing

Trying to improve results, our group introduced a novel protocol involving 1.5 mol/L PROH and 0.2 M sucrose concentration in the freezing solution that allows a minor dehydration of the oocyte and a 0.3 M sucrose concentration in the thawing solutions (32).

This idea was developed on the rationale of the first original protocol that used a difference in the sucrose gradient between the freezing and the thawing procedures, so as to reduce the osmotic stress that the egg has to undergo during this delicate procedure. In this way, we succeeded in maintaining a higher survival rate and, at the same time achieving a more than satisfying pregnancy rate as well as a high number of cycles that resulted in embryo transfer.

Four hundred and three oocytes were thawed, with a survival rate of 75.9%. Among the 306 surviving oocytes, 252 were microinjected and 192 (76.2%) showed two pronuclei. Cleavage rate was 93.8%. Pregnancy rates per transfer, per patient and per thawing cycle were 21.3%, 21.8%, and 18.9%, respectively. The implantation rate was 13.5% and the implantation rate per thawed oocyte was 5.9% (Table 4).

It appeared that the protocol applied in this study gave rise to survival, fertilization, and cleavage rates significantly higher than ones obtained with the 0.1 mol/L protocol. On the other hand, no major differences were observed in comparison to the use of 0.3 mol/L protocol. It seems that 0.1 mol/L protocol is not sufficient to ensure an adequate dehydration before cooling; on the other hand, the 0.2 mol/L protocol could represent an appropriate compromise, especially if associated with a higher concentration (0.3 mol/L) of the same CPA in the thawing solution with the aim of preventing excessive osmotic stress.

A different approach has been used on mouse oocytes by Stachecki, who replaced sodium with choline, a less toxic organic ion, in the freezing solutions (90% survival rate and 40% pregnancy rate) (33–35).

Table 4 Oocyte Cryopreservation Outcome Derived from a Method Based on Differential Sucrose Concentrations at Freezing (0.2 mol/L) and Thawing (0.3 mol/L) (32)

No. of patients	141
Thawing cycles	90
Oocytes thawed	403
Oocytes survived (%)	306 (75.9%)
2PN (%)	192 (76.2%)
Pregnancy rate/patient	21.8%
Pregnancy rate/transfer	21.3%
Pregnancy rate/thawing	18.9%
Implantation rate	13.5%
Abortion rate	11.8%

Table 5 Oocyte Cryopreservation Outcome Derived from the Application of a Sodium-Depleted Method (37)

Mean age (yr)	31.3
No. of patients	46
Thawing cycles	53
Oocytes thawed	361
Oocytes survived (%)	218 (60.4%)
2PN (%)	134 (62%)
Pregnancy rate/transfer	33%
Pregnancy rate/thawing	30%
Implantation rate	13.3%

This protocol was proposed also by Quintans (36) and Boldt (37) with good clinical results: Quintans obtained survival, fertilization and implantation rates of 63%, 59%, and 25%, respectively; Boldt on the other hand presented pregnancy per transfer, per thawed oocyte and implantation rates of 33%, 30%, and 13.3%, respectively. The two studies, however, involved only small groups of patients (12 and 46, respectively) (Table 5).

CUMULATIVE PREGNANCY RATES

Assessment of the safety and clinical efficiency of oocyte cryopreservation represents a difficult task. The overall clinical outcome is largely determined by well-recognized factors, such as oocyte quality, post-thaw survival and subcellular effects of cryopreservation conditions. However, other elements, for example oocyte and embryo selection criteria, reproducibility and adherence to cryopreservation protocols, and strategy of use of the stored material, can make particularly difficult the interpretation of the clinical significance of oocyte cryopreservation.

In the final analysis, evaluation of the efficiency of the different protocols will require the comparison of implantation rate derived from the treatment of homogeneous groups of patients and estimated on the basis of the number of frozen-thawed oocytes.

While there is little doubt that the ultimate therapeutic goal of IVF consists in the delivery of a healthy baby, the criteria that should be adopted to measure this standard of success remain a matter of debate (38,39).

The current practice in many clinics that employ oocyte cryopreservation is determined by legal restriction and results in loss of selection when transferring fresh embryos and those derived from cryopreserved oocytes.

Besides, the relative efficiency of IVF treatment cycles involving oocyte cryopreservation can only be fully determined when all oocytes have been thawed.

In 2006 our group published a paper reporting cumulative results after fresh and, in case of no pregnancy, after thawing cycles during the period between December 1996 and December 2003. Eighty patients started a fresh cycle and expressed their wish to use only a few oocytes for the fresh cycle and to store the remaining ones, so as to prevent or at least limit the formation of supernumerary embryos. The oocytes were cryopreserved using the 0.1 mol/L sucrose protocol. Twenty-four pregnancies were obtained in the fresh cycle, eight of which miscarried. After thawing all the stored oocytes, other 14 clinical pregnancies were obtained, 4 of which miscarried. The cumulative pregnancy rate was 47.5% (40).

After February 2004, as an effect of the Italian IVF law, our program contemplated oocyte cryopreservation as an alternative to embryo freezing, to maximize the potential of all oocytes retrieved from individual stimulation cycles. We have adopted oocyte cryopreservation as a measure which could surrogate embryo cryopreservation in the endeavor to offer our patients the best chances of pregnancy per started cycle. Only three fresh oocytes were inseminated and the others were frozen for later use. Frozen cycles were conducted in order to obtain three viable oocytes after thawing, although this was not possible in all cases because some of the oocytes had been frozen in groups rather than individually.

In 2007, we reported results obtained in our programme between March 2004 and December 2006 (41). Seven hundred and forty-nine patients underwent an ART cycle, 702 had embryos transferred after the use of three fresh oocytes and 267 pregnancies were achieved. Clinical pregnancy and implantation rates were 38% and 21.8%, respectively. After that, 510 patients underwent 660 thawing cycles obtaining 88 pregnancies. The clinical pregnancy and implantation rates after thawing was 14.9% and 8.1%, respectively. A total of 355 pregnancies were cumulatively achieved resulting in a cumulative pregnancy rate of 47.4%. Considering different age groups, we calculated the cumulative pregnancy rate of all patients treated from March 2004 until March 2008. After 1477 pick up, a maximum of three oocytes were inseminated in fresh cycles, whereas the remaining mature oocytes were frozen with two alternative freezing protocols (0.3 or 0.2M sucrose) and subsequently, all or some were used in successive thawing cycles (unpublished data). Overall, we obtained a pregnancy rate per embryo transfer of 40.4% in fresh cycles. After 1206 thawing cycles the clinical pregnancy rate per cycle and per embryo transfer was 14.4% and 16.3%, respectively and the implantation rate per transferred embryo and per injected oocyte was 9.7% and 6.8%, respectively. The overall cumulative pregnancy rate after fresh and thawed cycles was 48.8%.

Table 6 shows results obtained in fresh cycles in the different age groups. Table 7 details the number of thawing cycles performed and the mean of oocytes thawed per each cycle in the attempt to obtain 3 good oocytes to inseminate. In a total of 1206 cycles the overall survival rate was 69.7%. Table 8 reports per each group the number of pregnancies achieved, the clinical pregnancy per cycle and per embryo transfer, and the implantation rate per transferred embryos and injected oocytes. The cumulative pregnancies were 346, 253, 88, 28, and 7 in women aged ≤34, 35–38, 39–40, 41–42, and ≥43 years, respectively, with cumulative pregnancy rates of 53.2%, 52.4%, 47%, 25.4%, and 14.9% (Table 9).

These data confirmed the outcome of our previous studies (40,41) and moreover suggest that oocyte cryopreservation increases significantly the chances of clinical success per started cycle.

Table 6 Pregnancy Rates from Fresh Cycles Analyzed by Female Age

	Female age (yr)					
	≤34	35–38	39–40	41–42	≥43	Total
Ovum pick-up	650	483	187	110	47	1477
Clinical pregnancies	261	199	61	24	4	549
% Per pick-up	40.2	41.2	32.6	21.8	8.5	37.2
% Per embryo transfer	44.0	44.5	34.7	24.5	8.7	40.4
% Singleton	62.1	65.3	78.7	87.5	100	66.5
% Twins	28.4	28.6	18.0	12.5	0.0	26.4
% High order	9.6	6.0	3.3	0.0	0.0	7.1

Source: Tecnobios procreazione data.

Table 7 Mean Number of Frozen-Thawed Oocytes and Survival Rates Analyzed by Female Age

	Female age (yr)					
	≤34	35–38	39–40	41–42	≥43	Total
Cycles	474	376	163	126	67	1206
Mean	5.2	4.9	5.2	4.7	4.8	5.0
% Survival	69.3	69.5	68.8	71.2	73.4	69.7

Source: Tecnobios procreazione data.

Table 8 Pregnancy and Implantation Rates from Frozen Oocytes Cycles Analyzed by Female Age

	Female age (yr)					
	≤34	35–38	39–40	41–42	≥43	Total
Clinical pregnancies	80	55	29	6	4	174
% Per cycles	17	14.6	17.8	4.8	6	14.4
% Per embryo transfer	19	17	20	5.3	7	16.3
% Implantation per transferred embryos	11.1	10.3	11.8	2.6	4.3	9.7
% Implantation per injected oocytes	7.8	6.9	9.0	1.8	2.7	6.8

Source: Tecnobios procreazione data.

Table 9 Cumulative Pregnancy Rates Analyzed by Female Age from Cycles in Which Embryos or Oocytes Were Cryopreserved

	Female age (yr)					
	≤34	35–38	39–40	41–42	≥43	Total
Cumulative pregnancy rates from frozen embryos 1992–2004 (54)	54	50	34	26		50
Cumulative pregnancy rates from frozen oocytes 2004–2008	53	52	47	25	15	49

Results are similar between periods before and after the introduction of the Italian IVF law.
Source: Tecnobios procreazione data.

COMPARISON WITH 2PN AND EARLY CLEAVAGE STAGE EMBRYO FREEZING

Pronuclear-stage oocytes may be frozen and thawed with high survival rates. Damario et al. reported a recovery rate as high as 90% (42), but more commonly this rate ranges from 75% to 80% (43–45).

The survival rate of cleavage-stage day 2 embryos is not considered a limiting factor. An embryo is considered survived when at least 50% of the original cell mass is recovered intact after thawing. This event occurs in 20–25% of frozen-thaw four-cell embryos (46). Edgar et al. reported an implantation rate of 26% when a frozen-thawed fully intact embryo is transferred (47). They also reported that a post-thaw loss of one out of four blastomeres does not affect the implantation ability (27.5%). Instead, loss of a further blastomere (two out of four) causes a major decline in the implantation rate (9.4%).

It is almost impossible to compare results of a thawing cycle with oocytes, versus 2PN oocytes versus day 2 embryos. In fact, the latter two are results of a biological selection. I will try to make this comparison considering the cumulative pregnancy rate starting from a single stimulation cycle.

Germond et al. reported results of about 17% delivery rate on the fresh and 32% of cumulative delivery rate after the thawing of pronucleate stage oocytes (48).

In other countries where it is allowed to generate as many embryos as possible, usually embryo freezing is used to preserve spare embryos not selected for the "fresh transfer." In this case, the best quality one or two embryos are transferred and the others, with good morphology, are frozen. Using this strategy, Tiitinen et al. (49) were able to obtain pregnancy and delivery rates of 38.6% and 26.8%, respectively, after the transfer of the "best fresh embryo." Following the transfer of one or two frozen-thawed embryos the cumulative pregnancy and delivery rates were 62.4% and 52.8%, respectively. However, these very high results depend on the high selected study population included. In fact, the possibility to perform a transfer with a "high quality embryo" is limited to a good prognosis group of patients in which those embryos are generated.

Before the introduction of the IVF law, we used to freeze embryos in the attempt to preserve spare good quality embryos (grade I or II in a scale I to IV) and in the mean time, reducing the number of transferred embryos, to limit the incidence of multiple pregnancies. Considering all the 3570 fresh cycles (with spare embryos frozen) performed between 1992 and February 2004 we could obtain 1014 pregnancies (28.4%). Other 758 pregnancies were achieved after thawing. The total pregnancies were 1772 resulting in a cumulative pregnancy rate of 50%.

Table 7 details results obtained in the different age groups. Briefly, in the group of patients below the age of 35 the fresh and cumulative pregnancy rates were 29.7 and 54%, respectively.

If we compare cumulative pregnancy rates, in young women, after embryo (54%) or oocyte (53.2%) freezing, we can speculate that there are not differences between the two methods, at least in our hands (Table 9).

IMPLANTATION RATES FROM CRYOPRESERVED OOCYTES AND EMBRYOS

It remains difficult to appraise the relative efficiency of oocyte cryopreservation. However, this does not rule out critical assessment of oocyte cryopreservation results that are coming to light from several studies reported in the last few years. The modalities of use of cryopreserved material can vary considerably, with effects on clinical outcome and the possibility of comparing the efficiency of different approaches. Sometimes, religious, ethical or legal restrictions exclude the possibility of generating more embryos (generally one to three) than those strictly used for each embryo transfer. In frozen oocyte cycles, this requirement may be met by thawing only a few oocytes at each attempt in order to have two to three oocytes viable and suitable for insemination (20). Following the introduction of an IVF law in 2004, in Italy the approach of thawing a small number of oocytes has become a prescribed standard (25–27,32). In other countries, frozen oocyte cycles have been performed without limitations to the number of embryos available for selection and transfer and therefore the number of oocytes that may be thawed at each cycle (37,50,51). The two options have rather different implications. The thawing of only a few oocytes per cycle rules out the option of embryo selection. Besides, in some cases the number of transferred embryos may be inadequately low as an effect of attrition at the fertilization or cleavage stage. These conditions can significantly affect the clinical outcome in terms of pregnancy rate per thawing cycle and per transfer, as well as the implantation rate classically defined as the ratio between implantations and embryos transferred. A rather different scenario derives from situations in which an excess of oocytes may be thawed in single attempts and the transferred embryos represent a more abundant and selected material. In diverse oocyte cryopreservation studies, this can lead to mean numbers of embryos transferred that can be as different as 1.1 (20) and 4.6 (52), with predictable consequences in terms of pregnancy rates. Gook and Edgar (53) suggested a more objective approach for the assessment of the relative efficiency of embryo and oocyte cryopreservation cycles (see also chap. 18 by the same authors in this book), based on the number of frozen-thawed oocytes destined to a treatment and on the events of attrition at pre- and post-storage stages, rather than the mere implantation potential of the transferred embryos. Considering, therefore, the loss of material occurring at fertilization, cleavage, selection for cryopreservation, and thawing, they calculated that about 5 implantations from frozen-thawed embryos may be obtained per 100 oocytes collected. Quite correctly, they also argued that the appraisal of the efficiency of a given oocyte cryopreservation method should be founded on a similar exercise. Under such conditions, certain differences between alternative methods are only apparent. For example it is known that a major improvement (from 35–40% to 70–75%) in the survival rate of oocyte frozen via slow cooling may be obtained by raising the sucrose concentration in the freezing solution from 0.1 to 0.3 mol/L. This change also improves the rate of fertilization (20,25–27). However, the relatively low attrition at the survival and fertilization steps in the 0.3 mol/L sucrose protocol is counterbalanced by a higher implantation rate of embryo generated by the protocol involving the lower sucrose concentration (20,25–27). The overall outcome, pondered as the proportion of implantations per thawed oocytes ultimately makes the two protocols very similar (2.4–2.6%) and in any case insufficient for competing with embryo freezing. A more recent version of the slow cooling approach generated a much higher implantation rate per thawed oocyte (6–8%, depending on age) (32) (Table 10). Recently, it has been reported that vitrification also can produce similar or higher rates (51). Implantation rates per thawed oocyte of 5% or

Table 10 Implantation Rates Per Oocyte Thawed Reported in Different Oocyte Cryopreservation Studies

Study	No. of oocytes thawed	Implantations	Implantation rate per embryo transfer (%)	Implantation rate per oocyte used (%)
Borini et al. (20)	737	17	16.4	2.3
Borini et al. (25)	927	21	5.2	2.6
Levi Setti et al. (27)	1087	19	5.7	1.9
Boldt et al. (37)	190	10	15.9	5.3
De Santis et al. (26)	396	7	5.7	1.8
Bianchi et al. (32)	403	24	13.5	5.9

higher would be competitive in efficiency with embryo freezing. However, these initial findings require scrupulous verification through large, prospective controlled studies.

CONCLUSIONS

The storage of unfertilized oocytes offers an alternative to other forms of cryopreservation and can contribute considerably to the overall clinical success, avoiding the ethical and legal complications that may occur when frozen embryos remain unused after the achievement of a fresh pregnancy or in the countries where embryo cryopreservation is forbidden.

Multicentre studies should also be conducted to facilitate the collection of clinical data and test the reproducibility of alternative methods. Despite some uncertainties, it appears oocyte cryopreservation will become a valid option for IVF patients.

REFERENCES

1. Patrizio P, Sakkas D. From oocyte to baby: a clinical evaluation of the biological efficiency of in vitro fertilization. Fertil Steril 2008; 91: 1061–6.
2. Whittingham DG. Fertilization in vitro and development to term of unfertilized mouse oocytes previously stored at –196°C. J Reprod Fertil 1977; 49: 89–94.
3. Chen C. Pregnancy after human oocyte cryopreservation. Lancet 1986; 1: 884–6.
4. Ai-Hasani S, Diedrich K, van der Ven H, et al. [Initial results of the cryopreservation of human oocytes]. Geburtshilfe Frauenheilkd 1986; 46: 643–4.
5. Diedrich K, al-Hasani S, van der Ven H, et al. Successful in vitro fertilization of frozen-thawed rabbit and human oocytes. Ann N Y Acad Sci 1988; 541: 562–70.
6. Porcu E, Fabbri R, Seracchioli R, et al. Birth of a healthy female after intracytoplasmic sperm injection of cryopreserved human oocytes. Fertil Steril 1997; 68: 724–6.
7. Borini A, Bafaro G, Bonu MA, et al. Pregnancies after oocyte freezing and thawing: preliminary data. Hum Reprod 1998; 13(Suppl 1): 124–5.
8. Lassalle B, Testart J, Renard JP. Human embryo features that influence the success of cryopreservation with the use of 1,2-propanediol. Fertil Steril 1985; 44: 645–51.
9. Gook DA, Schiewe MC, Osborn SM, et al. Intracytoplasmic sperm injection and embryo development of human oocytes cryopreserved using 1,2-propanediol. Hum Reprod 1995; 10: 2637–41.
10. Nawroth F, Kissing K. Pregnancy after intracytoplasmatic sperm injection (ICSI) of cryopreserved human oocytes. Acta Obstet Gynecol Scand 1998; 77: 462–3.
11. Young E, Kenny A, Puigdomenech E, et al. Triplet pregnancy after intracytoplasmic sperm injection of cryopreserved oocytes: case report. Fertil Steril 1998; 70: 360–1.
12. Chia CM, Chan WB, Quah E, et al. Triploid pregnancy after ICSI of frozen testicular spermatozoa into cryopreserved human oocytes: case report. Hum Reprod 2000; 15: 1962–4.
13. Huttelova R, Becvarova V, Brachtlova T. More successful oocyte freezing. J Assist Reprod Genet 2003; 20: 293.
14. Notrica J, Divita A, Neuspiller F, et al. Healthy girl born after cryopreservation of gametes and ICSI in a patient with seminoma. Reprod Biomed Online 2004; 9: 620–2.
15. Miller KA, Elkind-Hirsch K, Levy B, et al. Pregnancy after cryopreservation of donor oocytes and preimplantation genetic diagnosis of embryos in a patient with ovarian failure. Fertil Steril 2004; 82: 211–14.

16. Gook DA, Osborn SM, Johnston WI. Cryopreservation of mouse and human oocytes using 1,2-propanediol and the configuration of the meiotic spindle. Hum Reprod 1993; 8: 1101–9.
17. Gook DA, Osborn SM, Bourne H, et al. Fertilization of human oocytes following cryopreservation: normal karyotypes and absence of stray chromosomes. Hum Reprod 1994; 9: 684–91.
18. Tucker M, Wright G, Morton P, et al. Preliminary experience with human oocyte cryopreservation using 1,2-propanediol and sucrose. Hum Reprod 1996; 11: 1513–15. abs.html
19. Kazem R, Thompson LA, Srikantharajah A, et al. Cryopreservation of human oocytes and fertilization by two techniques: in-vitro fertilization and intracytoplasmic sperm injection. Hum Reprod 1995; 10: 2650–4.
20. Borini A, Bonu MA, Coticchio G, et al. Pregnancies and births after oocyte cryopreservation. Fertil Steril 2004; 82: 601–5.
21. Porcu E, Fabbri R, Damiano G, et al. Clinical experience and applications of oocyte cryopreservation. Mol Cell Endocrinol 2000; 169: 33–7.
22. Boldt J, Cline D, McLaughlin D. Human oocyte cryopreservation as an adjunct to IVF-embryo transfer cycles. Hum Reprod 2003; 18: 1250–5.
23. Fabbri R, Porcu E, Marsella T, et al. Human oocyte cryopreservation: new perspectives regarding oocyte survival. Hum Reprod 2001; 16: 411–16.
24. Chen SU, Lien YR, Chen HF, et al. Observational clinical follow-up of oocyte cryopreservation using a slow-freezing method with 1,2-propanediol plus sucrose followed by ICSI. Hum Reprod 2005; 20: 1975–80.
25. Borini A, Sciajno R, Bianchi V, et al. Clinical outcome of oocyte cryopreservation after slow cooling with a protocol utilizing a high sucrose concentration. Hum Reprod 2006; 21: 512–17.
26. De Santis L, Cino I, Rabellotti E, et al. Oocyte cryopreservation: clinical outcome of slow-cooling protocols differing in sucrose concentration. Reprod Biomed Online 2007; 14: 57–63.
27. Levi Setti PE, Albani E, Novara PV, et al. Cryopreservation of supernumerary oocytes in IVF/ICSI cycles. Hum Reprod 2006; 21: 370–5.
28. La Sala GB, Nicoli A, Villani MT, et al. Outcome of 518 salvage oocyte-cryopreservation cycles performed as a routine procedure in an in vitro fertilization program. Fertil Steril 2006; 86: 1423–7.
29. Chamayou S, Alecci C, Ragolia C, et al. Comparison of in-vitro outcomes from cryopreserved oocytes and sibling fresh oocytes. Reprod Biomed Online 2006; 12: 730–6.
30. Coticchio G, De Santis L, Rossi G, et al. Sucrose concentration influences the rate of human oocytes with normal spindle and chromosome configurations after slow-cooling cryopreservation. Hum Reprod 2006; 21: 1771–6.
31. Nottola SA, Macchiarelli G, Coticchio G, et al. Ultrastructure of human mature oocytes after slow cooling cryopreservation using different sucrose concentrations. Hum Reprod 2007; 22: 1123–33.
32. Bianchi V, Coticchio G, Distratis V, et al. Differential sucrose concentration during dehydration (0.2 mol/L) and rehydration (0.3 mol/L) increases the implantation rate of frozen human oocytes. Reprod Biomed Online 2007; 14: 64–71.
33. Stachecki JJ, Cohen J, Willadsen S. Detrimental effects of sodium during mouse oocyte cryopreservation. Biol Reprod 1998; 59: 395–400.
34. Stachecki JJ, Cohen J, Willadsen SM. Cryopreservation of unfertilized mouse oocytes: the effect of replacing sodium with choline in the freezing medium. Cryobiology 1998; 37: 346–54.
35. Stachecki JJ, Willadsen SM. Cryopreservation of mouse oocytes using a medium with low sodium content: effect of plunge temperature. Cryobiology 2000; 40: 4–12.
36. Quintans CJ, Donaldson MJ, Bertolino MV, et al. Birth of two babies using oocytes that were cryopreserved in a choline-based freezing medium. Hum Reprod 2002; 17: 3149–52.
37. Boldt J, Tidswell N, Sayers A, et al. Human oocyte cryopreservation: 5-year experience with a sodium-depleted slow freezing method. Reprod Biomed Online 2006; 13: 96–100.
38. Min JK, Breheny SA, Maclachlan V, et al. What is the most relevant standard of success in assisted reproduction? The singleton, term gestation, live birth rate per cycle initiated: the BESST endpoint for assisted reproduction. Hum Reprod 2004; 19: 3–7.
39. Messinis IE, Domali E. What is the most relevant standard of success in assisted reproduction? Should BESST really be the primary endpoint for assisted reproduction? Hum Reprod 2004; 19: 1933–5.
40. Borini A, Lagalla C, Bonu MA, et al. Cumulative pregnancy rates resulting from the use of fresh and frozen oocytes: 7 years' experience. Reprod Biomed Online 2006; 12: 481–6.
41. Borini A, Bianchi V, Bonu MA, et al. Evidence-based clinical outcome of oocyte slow cooling. Reprod Biomed Online 2007; 15: 175–81.
42. Damario MA, Hammitt DG, Galanits TM, et al. Pronuclear stage cryopreservation after intracytoplasmic sperm injection and conventional IVF: implications for timing of the freeze. Fertil Steril 1999; 72: 1049–54.

43. Queenan JT Jr, Veeck LL, Toner JP, et al. Cryopreservation of all prezygotes in patients at risk of severe hyperstimulation does not eliminate the syndrome, but the chances of pregnancy are excellent with subsequent frozen-thaw transfers. Hum Reprod 1997; 12: 1573–6.
44. Horne G, Critchlow JD, Newman MC, et al. A prospective evaluation of cryopreservation strategies in a two-embryo transfer programme. Hum Reprod 1997; 12: 542–7.
45. Senn A, Vozzi C, Chanson A, et al. Prospective randomized study of two cryopreservation policies avoiding embryo selection: the pronucleate stage leads to a higher cumulative delivery rate than the early cleavage stage. Fertil Steril 2000; 74: 946–52.
46. Edgar DH, Bourne H, Speirs AL, et al. A quantitative analysis of the impact of cryopreservation on the implantation potential of human early cleavage stage embryos. Hum Reprod 2000; 15: 175–9.
47. Edgar DH, Archer J, McBain J, et al. Embryonic factors affecting outcome from single cryopreserved embryo transfer. Reprod Biomed Online 2007; 14: 718–23.
48. Germond M, Urner F, Chanson A, et al. What is the most relevant standard of success in assisted reproduction? The cumulated singleton/twin delivery rates per oocyte pick-up: the CUSIDERA and CUTWIDERA. Hum Reprod 2004; 19: 2442–4.
49. Tiitinen A, Halttunen M, Harkki P, et al. Elective single embryo transfer: the value of cryopreservation. Hum Reprod 2001; 16: 1140–4.
50. Kuwayama M, Vajta G, Kato O, et al. Highly efficient vitrification method for cryopreservation of human oocytes. Reprod Biomed Online 2005; 11: 300–8.
51. Cobo A, Kuwayama M, Perez S, et al. Comparison of concomitant outcome achieved with fresh and cryopreserved donor oocytes vitrified by the Cryotop method. Fertil Steril 2008; 89: 1657–64.
52. Lucena E, Bernal DP, Lucena C, et al. Successful ongoing pregnancies after vitrification of oocytes. Fertil Steril 2006; 85: 108–11.
53. Gook DA, Edgar DH. Cryopreservation of the human female gamete: current and future issues. Hum Reprod 1999; 14: 2938–40.
54. Borini A, Cattoli M, Bulletti C, et al. Clinical efficiency of oocyte and embryo cryopreservation. Ann N Y Acad Sci 2008; 1127: 49–58.

21 Clinical Evidence of Oocyte Vitrification

Ri-Cheng Chian[1] and Yun-Xia Cao[2]

[1] *Department of Obstetrics and Gynecology, McGill University, Montreal, Canada and*

[2] *Center for Reproductive Medicine, Department of Obstetrics and Gynecology, The First Affiliated Hospital of Anhui Medical University, Hefei, P.R. China*

INTRODUCTION

The development of an effective oocyte-freezing program will have a major impact on clinical practice in reproductive medicine and will serve as a powerful tool to preserve fertility for teenage girls and young women without male partners, or for those individuals who are affected by malignancies. It will also be beneficial to infertile couples who are considering moral or religious objections about embryo cryopreservation. In addition, a successful oocyte cryopreservation program will eliminate the need for donor–recipient menstrual cycle synchronization and will enable the establishment of oocyte banks, which would facilitate the logistics of coordinating egg donors with recipients.

The clinical options for preserving female fertility can be divided into two broad categories: cryopreservation of cells/tissues and surgical interventions. Cryopreservation of embryos, oocytes, and ovarian tissues are the three main options. For women who wish to retain their choice of a male partner and are not good candidate for embryos and ovarian tissue cryopreservation, cryopreservation of oocytes is the only clinically effective option.

Recent advances in vitrification techniques have markedly improved the efficacy of oocyte cryopreservation (1–4), demonstrating that vitrification may be more effective than the conventional slow-cooling method. Several groups have reported high survival rates (5–9). The purpose of this chapter is to review the recent advances in clinical evidence of oocyte vitrification. An emphasis will be made on the recent development of oocyte cryopreservation by vitrification method. It is important to provide clinical evidence of oocytes vitrification for its efficiency and safety issue.

VITRIFICATION METHODS

Vitrification involves the transformation of oocytes into a vitreous- or glass-like state by rapid freezing and thawing in order to prevent ice-crystal formation. Although this approach may improve viability of the oocytes, a very high concentration (up to 8.0 M) of cryoprotectant is used to prevent ice-crystal formation (10). The definition of vitrification is glass-like solidification of a solution at low temperature as the result of an extreme increase in viscosity during the cooling process. In simpler terms, vitrification means ice-free solidification of an aqueous solution. Luyet et al. (1937) proposed the advantages of achieving an ice-free, glass-like state for cryopreservation of cells and tissues (11). The concept is that living systems can be successfully cryopreserved if they are cooled so rapidly that ice crystallization does not occur.

The proposed mechanism of vitrification is the use of high concentration of cryoprotectant and extremely rapid cooling and warming rates in order to avoid intra- and extracellular ice formation. Strictly speaking, vitrification also occurs in cells cryopreserved by slow-freezing procedures, since the cell cytoplasm is transformed into a glass-like state or vitrified at the glass transition temperature (–130°C). Nonetheless, in the field of cryopreservation of cells and tissues, the distinction between vitrification and slow-freezing procedures is that vitrification involves ultra-rapid cooling rates.

Vitrification represents an attractive cryopreservation strategy for oocytes compared with conventional slow-cooling methods. In contrast to slow-freezing techniques, vitrification procedure is precise and every step of the procedure can be visualized. More importantly, vitrification procedure minimizes the duration of exposure to sub-physiological environment. Typically,

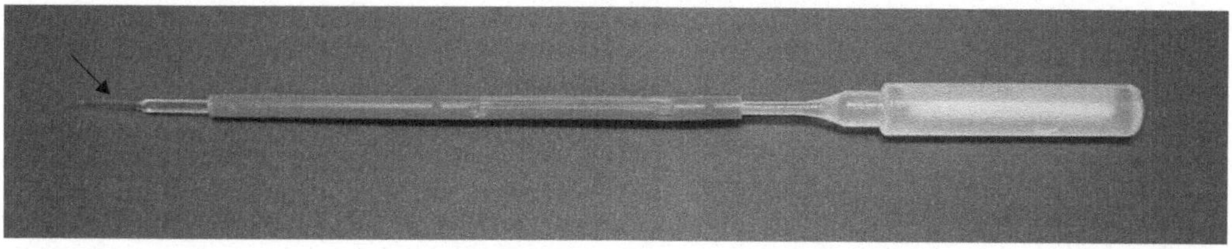

(**A**)

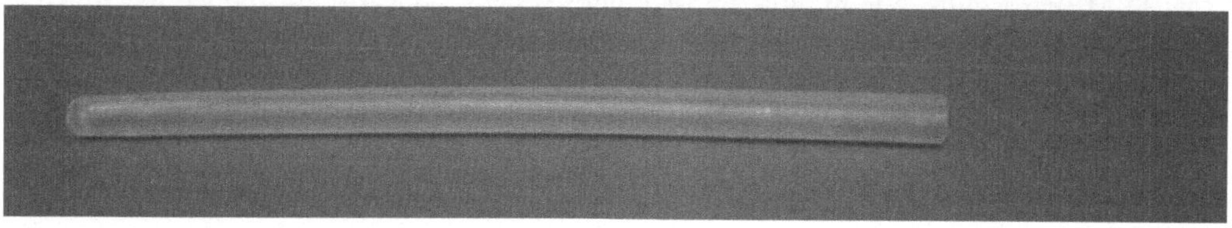

(**B**)

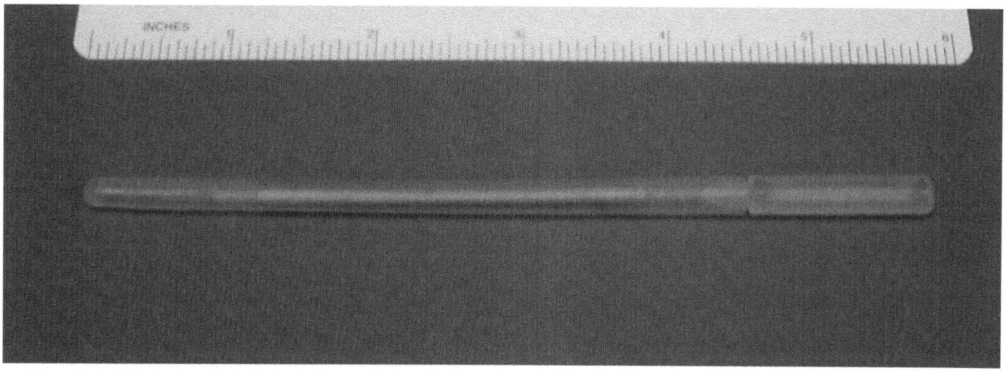

(**A**) and (**B**)

Figure 1 Actual view of the McGill Cryoleaf: (**A**) Core part of the McGill Cryoleaf; the arrow indicates the loading portion that is made of a thin polyethylene stick; (**B**) protection straw for storage; (**A+B**) the completed parts of the McGill Cryoleaf.

vitrification procedures require few minutes to perform compared with more than two hours for the slow-freezing methods.

Numerous groups attempted to identify a less toxic vitrification solution involving various combinations of permeable cryoprotectants and macromolecules (12–16). One possibility of minimizing cryoprotectant toxicity is replacing permeable cryoprotectant with polymers such as sugar molecules and polyvinylpyrrolidone (17). Another approach to reduce cryoprotectant toxicity is the use a mixture of cryoprotectants and pre-equilibrating the oocytes or embryos in a mixture of cryoprotectants at low concentration before being transferred to a high-concentration mixture (18,19).

Different carriers have been designed to minimize the volume of cryoprotectant solution and to allow the sample to submerge quickly in liquid nitrogen (LN_2). These devices include the Cryotop (19), McGill Cryoleaf (20) (Fig. 1), Cryotip (21), nylon loop or Cryoloop (22,23), thin capillary or open-pulled straws (OPS) (24), hemi-straws (25), and electron microscope copper/gold grids (26). Using the McGill Cryoleaf, it has been estimated that the freezing speed can be increased up to 20,000°C/min when plunging the device into LN_2 (19).

A common protocol for vitrification is to first equilibrate the oocytes in a solution containing a lower concentration of permeable cryoprotectant before transferring them to a solution containing the full strength permeable cryoprotectant and a non-permeable cryoprotectant such as a disaccharide or macromolecules.

Using McGill Cryoleaf, the vitrification procedure is as follows (Fig. 2): Two or three oocytes are first suspended in equilibration medium containing 7.5% ethylene glycol (EG) and 7.5% 1,2-propanediol (PrOH) for five minutes and transferred to vitrification medium containing 15% EG, 15% PrOH, and 0.5M sucrose for one minute at room temperature. The oocytes are quickly loaded onto the vitrification device, the McGill Cryoleaf, and immediately plunged into LN_2 for storage.

For thawing (Fig. 3), the McGill Cryoleaf is directly inserted into a 37°C thawing medium (TM; 1.0M sucrose) for one minute. Subsequently, thawed oocytes are transferred to 0.5M and 0.25M sucrose solutions for three minutes, respectively, and then washed twice with culture medium before being transferred to an incubator containing 5% CO_2 at 37°C for in vitro fertilization (IVF).

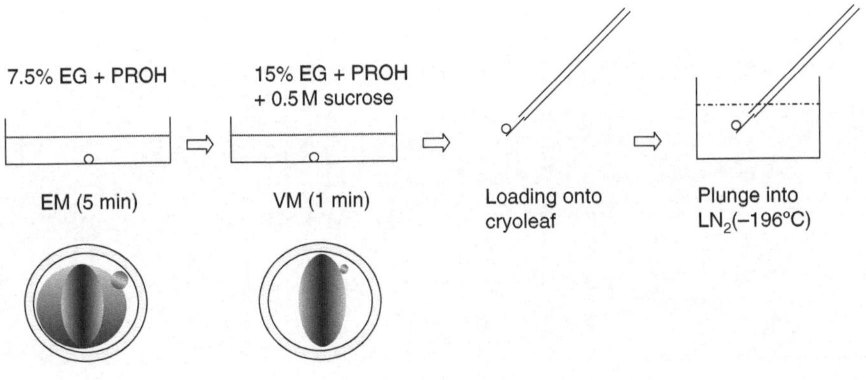

Figure 2 Vitrification procedure: the egg shrinks initially and then returns to its shape within 5 min in equilibration medium (EM) at room temperature. The egg shrinks again in vitrification medium (VM) and is then loaded onto the McGill Cryoleaf for cryopreservation by directly plunging into liquid nitrogen (LN_2).

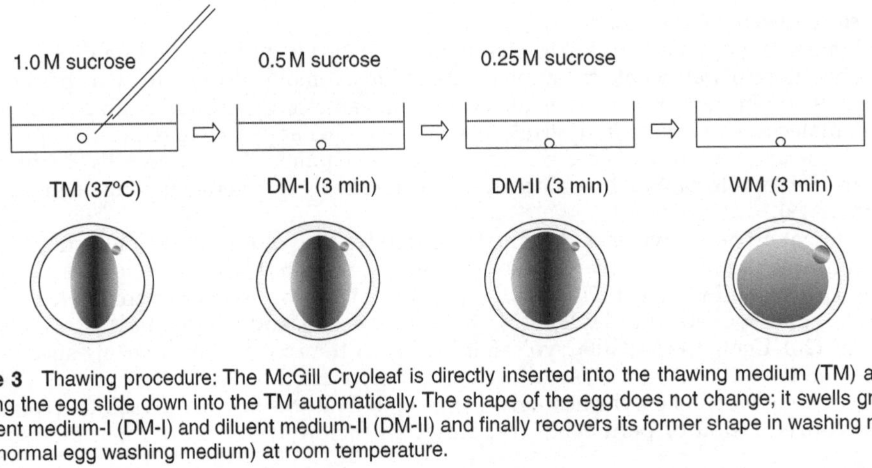

Figure 3 Thawing procedure: The McGill Cryoleaf is directly inserted into the thawing medium (TM) at 37°C, allowing the egg slide down into the TM automatically. The shape of the egg does not change; it swells gradually in diluent medium-I (DM-I) and diluent medium-II (DM-II) and finally recovers its former shape in washing medium (WM; normal egg washing medium) at room temperature.

Table 1 Clinical Outcomes Following Vitrification of Oocytes Obtained from Ovarian Stimulation and In Vitro Maturation Cycles

	IVF group	IVM group	P-value
Patients who underwent thawing and embryo transfer	38	20	
Mature (MII) oocytes retrieved	399	6	
Immature (GV) oocytes retrieved	91	290	
Oocyte maturation rate following IVM (±SD)	70.3 ± 20.0	67.3 ± 19.3	NS
Oocytes matured in vitro	64	209	
MII oocytes vitrified and thawed (mean per patient ± SD)	463 (12.2 ± 5.7)	215 (10.8 ± 5.9)	NS
Oocytes survived (mean % ± SD)	383 (81.4 ± 22.6)	148 (67.5 ± 26.1)	Mann–Whitney, $P < 0.001$
Oocytes fertilized (mean % ± SD)	287 (75.6 ± 22.5)	96 (64.2 ± 19.9)	Mann–Whitney, $P < 0.05$
Embryos transferred (mean ± SD)	133 (3.5 ± 1.1)	64 (3.2 ± 1.5)	NS
Cumulative embryo scores (mean ± SD)	38.4 ± 23.3	20.0 ± 13.8	Mann–Whitney, $P < 0.05$
Implantation per embryo (mean % ± SD)	25/133 (19.1 ± 25.8)	4/64 (9.6 ± 24.1)	Mann–Whitney, $P = 0.07$
Pregnancy rate per cycle started (%)	19 (50)	4 (20.0)	χ^2 test, NS
Biochemical pregnancy	2	0	
Clinical pregnancy rate per cycle started (%)	17 (44.7)	4 (20.0)	χ^2 test, NS
Miscarriages	2	0	
Singleton pregnancies	9	4	
Twin pregnancies	5	0	
Triplet pregnancies	1	0	
Live birth/pregnancy rate per cycle started (%)	15 (39.5)	4 (20.0)	χ^2 test, NS
Newborns	22	4	

Abbreviations: GV, germinal vesicle; IVF, in vitro fertilization; IVM, in vitro maturation; MII, metaphase II; NS, not significant.

VITRIFICATION OF MATURE OOCYTES

It has been believed that vitrification is a promising novel technique and may be more effective than slow-freezing procedure for oocyte cryopreservation (1). Vitrification of human oocytes has resulted in relatively high survival rates (13–16,25,27–33). In fact, promising results of pregnancies and live births have been reported after cryopreservation of mature human oocytes using vitrification procedures (4–9,34–36).

In 1999, Kuleshova et al. (1999) reported the first human live birth following vitrification of 17 mature eggs using OPS and high concentration of EG and sucrose (14). Eleven eggs survived (65%) after vitrification and five pronuclear zygotes (46%) were obtained after intracytoplasmic sperm injection (ICSI). Recently, tremendous progress is achieved for mature oocyte cryopreservation using vitrification procedure. As part of a clinical trial at McGill University, a total of 463 mature oocytes were vitrified from 38 women. Of those, 383 oocytes survived post-thawing (4) (Table 1). Following insemination by ICSI, 287 oocytes were fertilized normally (74.9%). Seventeen patients became pregnant (44.7%) after transferring the resulted embryos. The implantation rate was 18.8% (25/133). This trial has resulted in 22 healthy live births. Results from our trial suggest that vitrification of mature oocytes allow pregnancy rates comparable to fresh oocytes with IVF/ICSI in many IVF programs.

VITRIFICATION OF IMMATURE OOCYTES

Oocyte cryopreservation has been performed by using the conventional slow-freezing method that has been associated with relatively low survival rate, and only a limited number of live

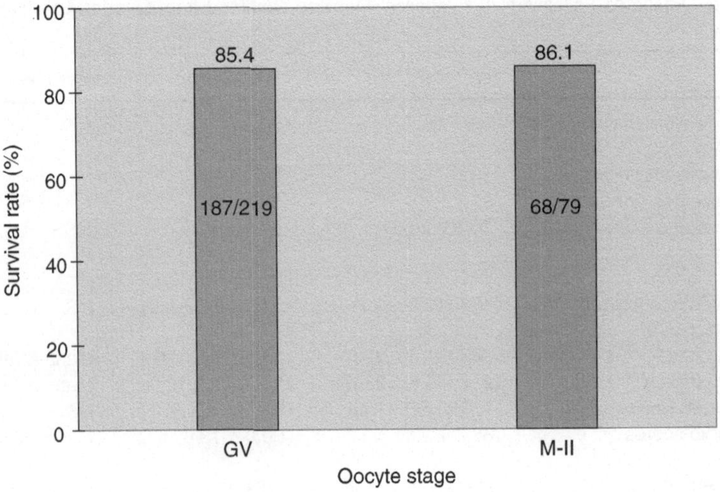

Figure 4 Survival rates of human oocytes vitrified at the immature germinal vesicle (GV) stage and at the mature metaphase-II (M-II) stage. A total of 298 oocytes (219 immature oocytes and 79 mature oocytes) were vitrified and thawed. There were no differences in the survival rates between the two groups.

births were achieved (37–39). Recently, the modified slow-freezing method has been introduced (40–52). Although the survival and fertilization rates seem to have improved slightly, the clinical outcome still needs to be confirmed.

With the conventional slow-freezing method, an alternative strategy to avoid spindle depolymerization is to cryopreserve immature germinal vesicle (GV) stage oocytes. These immature oocytes are arrested in diplotene state of prophase I. Theoretically, the use of immature GV stage oocytes circumvents the risk of polyploidy and aneuploidies because the chromosomes are diffuse and surrounded by a nuclear membrane (53,54). Although the oocyte survival rate seems improved, the poor maturation, fertilization, and embryonic development were the main problems associated with immature oocyte freezing (55–57) by the slow-freezing method. So far, only one live birth has been reported following the slow-freezing of immature oocytes at GV stage (58). Therefore, cryopreservation of the mature oocytes [metaphase-II (M-II) stage] using the slow-freezing method seems more efficient than immature oocytes.

There has been limited information on the oocyte survival, fertilization, and early embryonic development from the oocytes vitrified at immature GV stage. Recently, with experiments, we found that the survival rates were not significantly different between the oocytes vitrified at GV stage and mature M-II stage (Fig. 4). However, oocyte maturation rates were significantly reduced when the oocytes were vitrified at immature GV stage followed by in vitro maturation (IVM) in comparison with the control group (Fig. 5). Following insemination by ICSI, there were no differences in fertilization, cleavage, and blastocyst development rates between these two groups. These results suggest that better results can be obtained by vitrifying the mature oocytes instead of the immature oocytes.

VITRIFICATION OF IN VITRO MATURED OOCYTES

Immature oocyte retrieval without preceding gonadotropin stimulation to the ovaries, IVM culture, and fertilization of the oocytes is a safe and effective treatment option for a large population of infertile patients and has yielded clinical pregnancy rates of 35% (59–62), which are comparable to those achieved from standard IVF treatment cycles reported by several national registries (63–65). It has also been recently demonstrated that babies born after IVM do not have any increased fetal abnormality rates compared with regular IVF or spontaneous conceptions in

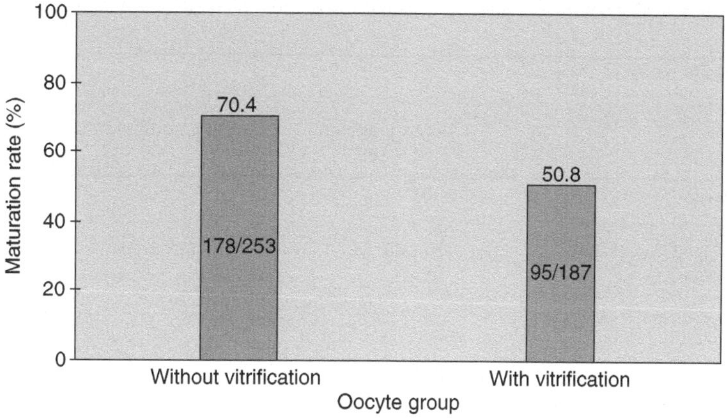

Figure 5 In vitro maturation rates of immature human oocytes with or without vitrification for IVM. A total of 440 immature oocytes (253 immature oocytes without vitrification and 187 immature oocytes with vitrification) were cultured in vitro for maturation. There were significant differences ($P < 0.05$) between the two groups.

fertile women (66). The advantages of immature oocyte retrieval without ovarian stimulation include the avoidance of the expensive medications and ovarian hyperstimulation syndrome (OHSS). The incidence of severe OHSS is 0.6% to 1.9% but may be as high as 6% in high-risk groups such as young women with polycystic ovaries (67). Although a number of strategies have been proposed to predict and prevent this potentially life-threatening complication, the only preventive strategy that is certain to eliminate OHSS is avoidance of ovarian stimulation altogether.

At present, there are limited options for fertility preservation that are applicable to young cancer patients, many of whom do not have a male partner and present with time constraints and concerns regarding the long-term effects of gonadotropin stimulation. A standard ovarian stimulation treatment protocol entails a four- to six-week delay in order to downregulate the pituitary, followed by gonadotropin stimulation and oocyte retrieval (68). Even if a short protocol was used, there would be a delay of at least two to three weeks. On the other hand, the IVM oocyte cryopreservation protocol takes only 2 to 10 days. Other reasons for wishing to avoid repeated gonadotropin stimulation include concerns about the substantial cost of the drugs and the long-term effects of repeated doses of gonadotropins on the risk of developing ovarian, endometrial, and breast cancers (69,70).

Although it has been demonstrated for the first time that a series of healthy live births can be achieved from the combination of IVM and oocyte vitrification (Table 1), the reverse was noted; that is, vitrification of in vitro matured oocytes is less effective than vitrification of in vivo matured oocytes. However, the combination of IVM and oocyte vitrification has several advantages, eliminating the cost of drugs and the need for close monitoring. Most importantly, this novel treatment protocol has the greatest potential to help cancer patients. In estrogen-receptor-positive breast cancer patients, the IVM protocol eliminates the risk of stimulating hormone-sensitive tumors. Moreover, IVM involves minimal delay in treatment and may be completed in as few as two days (71).

Chemotherapy has an adverse effect on ovarian reserve, which may lead to premature ovarian failure and infertility (72). Nevertheless, cancer treatment is often commenced without pursuing fertility preservation options as neither cancer patients nor their oncologists wish to delay chemotherapy. However, there is evidence to suggest that cancer patients cope better emotionally with their treatment if they feel that the option of having a biological child is available to them in the future (73). Therefore, based on the results of a series of healthy live births achieved from the combination of IVM and oocyte vitrification, the fertility preservation with oocyte IVM followed by vitrification should be considered for some cancer patients before chemotherapic treatment.

Table 2 Obstetric and Perinatal Outcomes and Incidence of Congenital Malformations in Children Born from Vitrified Oocytes

Characteristics	All pregnancies (n = 165)	Singleton pregnancies (n = 137)	Multiple gestation pregnancies (n = 28)
Obstetric Outcomes			
Mean gestational age (wk + days)	37 + 1	37 + 3	35 + 5
No. of deliveries 34–37 wk (%)	46 (30)	30 (22)	16 (57)
No. of deliveries <34 wk (%)	10 (6)	6 (4)	4 (14)
	All newborns (n = 200)	Singleton newborns (n = 141)	Multiple gestation newborns (n = 59)
Perinatal Outcomes			
Mean birth weight (g ± SEM)	2784 ± 37	2920 ± 37	2231 ± 55
No. of LBW (1500–2500 g) (%)	68 (34)	24 (17)	44 (74)
No. of VLBW (<1500 g) (%)	4 (2)	1 (0.7)	3 (5)
Median Apgar score at 1 min	8	9	8
Median Apgar score at 5 min	10	10	10
Incidence of Congenital Anomalies			
Biliary atresia	1	0	1
Club foot	1	1	0
Skin hemangioma	1	1	0
Ventricular septal defect	2	0	2
Total (%)	5 (2.5)	2 (1.4)	3 (5.1)

Abbreviations: LBW, low birth weight; VLBW, very low birth weight.

OBSTETRIC AND PERINATAL OUTCOMES FROM VITRIFIED OOCYTES

The application of vitrification techniques to human oocytes has been questioned because of the absence of basic biological studies addressing safety issues (74). Indeed, reviewing current vitrification protocols found that relatively higher concentrations of cryoprotectant have been used to vitrify the oocytes, and the cytotoxicity of cryoprotectant and osmotic changes of vitrification solutions are the main concerns (75). Therefore, it has been indicated that the clinical efficiency and safety issue of vitrified oocytes cannot be precisely assessed because of the lack of well-controlled clinical trials (74). Although it is important to assess viability and abnormality of the vitrified oocytes via embryology parameters (76), the most important is to evaluate the health of live births produced from vitrified oocytes.

Recently, we have analyzed the obstetric and perinatal outcomes in 165 pregnancies and 200 infants conceived following oocyte vitrification cycles in three IVF centers (Table 2) (77). The results indicate that the mean birth weight and the incidence of congenital anomalies are comparable to that of spontaneous conceptions in fertile women or infertile women undergoing IVF treatment. These preliminary findings may provide reassuring evidence that pregnancies and infants conceived following oocyte vitrification are not associated with increased risk of adverse obstetric and perinatal outcomes.

CONCLUSIONS

1. Vitrification of in vivo matured oocytes results in a satisfactory clinical pregnancy rate.
2. Healthy live births can be achieved from the combination of IVM and oocyte vitrification.
3. There are no differences in the survival rates between the oocytes vitrified at immature GV stage and at the mature M-II stage. However, the potential of the oocyte maturation is reduced by vitrification of immature GV stage oocytes.

4. Vitrification of in vitro matured oocytes represents a novel option for fertility preservation.
5. The preliminary findings provide reassuring clinical evidence that pregnancies and infants conceived following oocyte vitrification are not associated with increased risk of adverse obstetric and perinatal outcomes.

ACKNOWLEDGMENTS

The authors are very thankful to the colleagues who participated in the related experiments that generated the data presented in this chapter.

REFERENCES

1. Kuleshova LL, Lopata A. Vitrification can be more favorable than slow cooling. Fertil Steril 2002; 78: 449–54.
2. Oktay K, Cil AP, Bang H. Efficiency of oocyte cryopreservation: a meta-analysis. Fertil Steril 2006; 86: 70–80.
3. Huang JY, Tan SL, Chian RC. Fertility preservation for female. J Reprod Contracept 2006; 17: 109–28.
4. Chian RC, Huang JYJ, Gilbert L, et al. Obstetric outcomes following vitrification of in-vitro and in-vivo matured oocytes. Fertil Steril 2007; DOI: 10.10.16/j.fertinstert.2007.11.088.
5. Katayama KP, Stehlik J, Kuwayama M, et al. High survival rate of vitrified human oocytes results in clinical pregnancy. Fertil Steril 2003; 80: 223–4.
6. Kuwayama M, Vajta G, Kato O, et al. Highly efficient vitrification method for cryopreservation of human oocytes. Reprod Biomed Online 2005; 11: 300–8.
7. Kuwayama M. Highly efficient vitrification for cryopreservation of human oocytes and embryos: the cryotop method. Theriogenology 2007; 67: 73–80.
8. Lucena E, Bernal DP, Lucena C, et al. Successful ongoing pregnancies after vitrification of oocytes. Fertil Steril 2006; 85: 108–11.
9. Cobo A, Kuwayama M, Perez S, et al. Comparison of concomitant outcome achieved with fresh and cryopreserved donor oocytes vitrified by the cryotop method. Fertil Steril 2008; 89: 1657–64.
10. Kasai M, Zhu SE, Pedro PB, et al. Fracture damage of embryos and its prevention during vitrification and warming. Cryobiology 1996; 33: 459–64.
11. Luyet B. The vitrification of organic colloids and protoplasm. Biodynamica 1937; 1: 1–14.
12. Ali J, Shelton JN. Vitrification of preimplantation stages of mouse embryos. J Reprod Fertil 1993; 98: 459–65.
13. Mukaida T, Wada S, Takahashi K, et al. Vitrification of human embryos based on the assessment of suitable conditions for 8-cell mouse embryos. Hum Reprod 1998; 13: 2874–9.
14. Kuleshova L, Gianaroli L, Magli C, et al. Birth following vitrification of a small number of human oocytes. Hum Reprod 1999; 14: 3077–9.
15. Kuleshova LL, Shaw JM. A strategy for rapid cooling of embryos within a double straw to eliminate the risk of contamination during cryopreservation and storage. Hum Reprod 2000; 15: 2604–9.
16. Fahy GM, Wowk B, Wu J, Paynter S. Improved vitrification solutions based on the predictability of vitrification solution toxicity. Cryobiology 2004; 48: 22–35.
17. Kasai M, Komi JH, Takakamo A, et al. A simple method for mouse embryo cryopreservation in a low toxicity vitrification solution, without appreciable loss of viability. J Reprod Fertil 1990; 89: 91–7.
18. Vajta G, Holm P, Kuwayama M, et al. Open pulled straw (OPS) vitrification: a new way to reduce cryoinjuries of bovine ova and embryos. Mol Reprod Dev 1998; 51: 53–8.
19. Chian RC, Kuwayama M, Tan L, et al. High survival rate of bovine oocytes matured in vitro following vitrification. J Reprod Dev 2004; 50: 685–96.
20. Chian RC, Son WY, Huang J, et al. High survival rates and pregnancies of human oocytes following vitrification: preliminary report. Fertil Steril 2005; 84: S36.
21. Kuwayama M, Vajta G, Ieda S, et al. Comparison of open and closed methods for vitrification of human embryos and the elimination of potential contamination. Reprod Biomed Online 2005; 11: 608–14.
22. Lane M, Bavister BD, Lyons EA, et al. Container-less vitrification of mammalian oocytes and embryos. Nat Biotechnol 1999; 17: 1234–6.
23. Lane M, Schoolcraft WB, Gardner DK. Vitrification of mouse and human blastocysts using a novel cryoloop container-less technique. Fertil Steril 1999; 72: 1073–8.

24. Vajta G, Holm P, Kuwayama M, et al. Open pulled straw (OPS) vitrification: a new way to reduce cryoinjuries of bovine ova and embryos. Mol Reprod Dev 1998; 51: 53–8.
25. Vanderzwalmen P, Bertin G, Debauche CH, et al. Vitrification of human blastocysts with the Hemi-Straw carrier: application of assisted hatching after thawing. Hum Reprod 2003; 18: 1504–11.
26. Martino A, Songsasen N, Leibo SP. Development into blastocysts of bovine oocytes cryopreserved by ultra-rapid cooling. Biol Reprod 1996; 54: 1059–69.
27. Mukaida T, Nakamura S, Tomiyama T, et al. Successful birth after transfer of vitrified human blastocysts with use of a cryoloop containerless technique. Fertil Steril 2001; 76: 618–20.
28. Hong SW, Chung HM, Lim JM, et al. Improved human oocyte development after vitrification: a comparison of thawing methods. Fertil Steril 1999; 72: 142–6.
29. Yokota Y, Yokota H, Yokota M, et al. Birth of healthy twins from in vitro development of human refrozen embryos. Fertil Steril 2001; 76: 1063–5.
30. Chung HM, Hong SW, Lim JM, et al. In vitro blastocyst formation of human oocytes obtained from unstimulated and stimulated cycles after vitrification at various maturational stages. Fertil Steril 2000; 73: 545–51.
31. Chi HJ, Koo JJ, Kim MY, et al. Cryopreservation of human embryos using ethylene glycol in controlled slow freezing. Hum Reprod 2002; 17: 2146–51.
32. Cho HJ, Son WY, Yoon SH, et al. An improved protocol for dilution of cryoprotectants from vitrified human blastocysts. Hum Reprod 2002; 17: 2419–22.
33. Son WY, Yoon SH, Yoon HJ, et al. Pregnancy outcome following transfer of human blastocysts vitrified on electron microscopy grids after induced collapse of the blastocoele. Hum Reprod 2003; 18: 137–9.
34. Yoon TK, Kim TJ, Park SE, et al. Live births after vitrification of oocytes in a stimulated in vitro fertilization-embryo transfer program. Fertil Steril 2003; 19: 1323–36.
35. Katayama KP, Stehlik J, Kuwayama M, et al. High survival rate of vitrified human oocytes results in clinical pregnancy. Fertil Steril 2003; 80: 223–4.
36. Kyono K, Fuchinoue K, Yagi A, et al. Successful pregnancy and delivery after transfer of a single blastocyst derived from a vitrified mature human oocyte. Fertil Steril 2004; 84: 1017.
37. Chen C. Pregnancy after human oocyte cryopreservation. Lancet 1986; 2: 884–6.
38. Tucker M, Wright G, Morton P, et al. Preliminary experience with human oocyte cryopreservation using 1,2-propanediol and sucrose. Hum Reprod 1996; 11: 1513–5.
39. Porcu E, Fabbri R, Seracchioli R, et al. Birth of a healthy female after intracytoplasmic sperm injection of cryopreserved human oocytes. Fertil Steril 1997; 68: 724–6.
40. Yang DS, Blohm PL, Winslow KL, et al. A twin pregnancy after microinjection of human cryopreserved oocyte with a specially developed oocyte cryopreservation regime. Fertil Steril 1998; 70: S86.
41. Stachecki JJ, Cohen J, Willadsen SM. Cryopreservation of unfertilized oocytes: the effect of replacing sodium with choline in the freezing medium. Cryobiology 1998; 37: 346–55.
42. Fabbri R, Porcu E, Marsella T, et al. Human oocyte cryopreservation: new perspectives regarding oocyte survival. Hum Reprod 2001; 16: 411–6.
43. Quintans CJ, Donaldson MJ, Bertolino MV, et al. Birth of two babies using oocytes that were cryopreserved in a choline-based freezing medium. Hum Reprod 2002; 17: 3149–52.
44. Boldt J, Cline D, McLaughlin D. Human oocyte cryopreservation as an adjunct to IVF-embryo transfer cycles. Hum Reprod 2003; 18: 1250–5.
45. Boldt J, Tidswell N, Sayers A, et al. Human oocyte cryopreservation: 5-year experience with a sodium-depleted slow freezing method. Reprod Biomed Online 2006; 13: 96–100.
46. Borini A, Bonu MA, Coticchio G, et al. Pregnancies and births after oocyte cryopreservation. Fertil Steril 2004; 82: 601–5.
47. Borini A, Sciajno R, Bianchi V, et al. Clinical outcome of oocyte cryopreservation after slow cooling with a protocol utilizing a high sucrose concentration. Hum Reprod 2006; 21: 512–7.
48. Stachecki JJ, Cohen J. An overview of oocyte cryopreservation. Reprod Biomed Online 2004; 9: 152–63.
49. Azambuja R, Badalotti M, Teloken C, et al. Successful birth after injection of frozen human oocytes with frozen epididymal spermatozoa. Reprod Biomed Online 2005; 11: 449–51.
50. Bianchi V, Coticchio G, Fava L, et al. Meiotic spindle imaging in human oocytes frozen with a slow freezing procedure involving high sucrose concentration. Hum Reprod 2005; 20: 1078–83.
51. Coticchio G, Bonu MA, Bianchi V, et al. Criteria to assess human oocyte quality after cryopreservation. Reprod Biomed Online 2005; 11: 421–7.
52. Levi-Setti PE, Albani E, Novara PV, et al. Cryopreservation of supernumerary oocytes in IVF/ICSI cycles. Hum Reprod 2006; 21: 370–5.
53. Cooper A, Paynter SJ, Fuller BJ, et al. Differential effects of cryopreservation on nuclear or cytoplasmic maturation in vitro in immature mouse oocytes from stimulated ovaries. Hum Reprod 1998; 13: 971–8.

54. Isachenko EF, Nayudu PL. Vitrification of mouse germinal vesicle oocytes: effect of treatment temperature and egg yolk on chromosomal normality and cumulus integrity. Hum Reprod 1999; 14: 400–8.
55. Toth TL, Lanzendorf SE, Sandow BA, et al. Cryopreservation of human prophase I oocytes collected from unstimulated follicles. Fertil Steril 1994; 61: 1077–82.
56. Toth TL, Baka SG, Veeck LL, et al. Fertilization and in vitro development of cryopreserved human prophase I oocytes. Fertil Steril 1994; 61: 891–4.
57. Son WY, Park SE, Lee KA, et al. Effects of 1,2-propanediol and freezing-thawing on the in vitro developmental capacity of human immature oocytes. Fertil Steril 1996: 66: 995–9.
58. Tucker MJ, Wright G, Morton PC, et al. Birth after cryopreservation of immature oocytes with subsequent in vitro maturation. Fertil Steril 1998; 70: 578–9.
59. Chian RC. In-vitro maturation of immature oocytes for infertile women with PCOS. Reprod BioM Online 2004; 8: 547–52.
60. Chian RC, Lim JH, Tan SL. State of the art in in-vitro oocyte maturation. Curr Opin Obstet Gynecol 2004; 16: 211–19.
61. Le Du A, Kadoch IJ, Bourcigaux N, et al. In vitro oocyte maturation for the treatment of infertility associated with polycystic ovarian syndrome: the French experience. Hum Reprod 2005; 20: 420–4.
62. Jurema MW, Nogueira D. In vitro maturation of human oocytes for assisted reproduction. Fertil Steril 2006; 86: 1277–91.
63. Gunby J, Daya S. IVF Directors Group of the Canadian Fertility and Andrology Society. Assisted reproductive technologies (ART) in Canada: 2003 results from the Canadian ART Register. Fertil Steril 2007; 88: 550–9.
64. Centers for Disease Control and Prevention. Assisted Reproductive Technology (ART) Report: Section 2—ART Cycles Using Fresh, Nondonor Eggs or Embryos (Updated on March 27, 2006; cited on October 19, 2006). Available from: http://www.cdc.gov/ART/ART2003/sect2_fig3-13.htm#Figure%2010. Accessed on March 29, 2007.
65. The European IVF-Monitoring Programme (EIM) for the European Society of Human Reproduction and Embryology (ESHRE); Andersen AN, Gianaroli L, Felberbaum R, de Mouzon J, Nygren KG. Assisted reproductive technology in Europe, 2002. Results generated from European registers by ESHRE. Hum Reprod 2006; 21: 1680–97.
66. Buckett WM, Chian RC, et al. A comparison of congenital abnormalities and perinatal outcome in pregnancies conceived following IVM, IVF, and ICSI with spontaneously conceived pregnancies delivered in a single centre: a case-controlled study. Obstet Gynecol 2007; 110: 885–91.
67. Brinsden PR, Wada I, Tan SL, et al. Diagnosis, prevention and management of ovarian hyperstimulation syndrome. Br J Obstet Gynaecol 1995; 102: 767–72.
68. Tan SL, Maconochie N, Doyle P, et al. Cumulative conception and live-birth rates after in vitro fertilization with and without the use of long, short, and ultrashort regimens of the gonadotropin-releasing hormone agonist Buserelin. Am J Obstet. Gynecol 1994; 171: 513–20.
69. Venn A, Jones P, Quinn M, et al. Characteristics of ovarian and uterine cancers in a cohort of in vitro fertilization patients. Gynecol Oncol 2001; 82: 64–8.
70. Venn A, Watson L, Bruinsma F, et al. Risk of cancer after use of fertility drugs with in-vitro fertilization. Lancet 1999; 354: 1586–90.
71. Rao GD, Chian RC, Son WS, et al. Fertility preservation in women undergoing cancer treatment. Lancet 2004; 363: 1829–30.
72. Bines J, Oleske DM, Cobleigh MA. Ovarian function in premenopausal women treated with adjuvant chemotherapy for breast cancer. J Clin Oncol 1996; 14: 1718–29.
73. Partridge AH, Gelber S, Peppercorn J, et al. Web-based survey of fertility issues in young women with breast cancer. J Clin Oncol 2004; 22: 4174–83.
74. Gook DA, Edgar DH. Human oocyte cryopreservation. Hum Reprod Update 2007; 13: 591–605.
75. Fahy G. Theoretical considerations for oocyte cryopreservation by freezing. Reprod BioM Online 2007; 14: 709–14.
76. De Santis L, Cino I, Coticchio G, et al. Objective evaluation of the viability of cryopreserved oocytes. Reprod Biomed Online 2007; 15: 338–45.
77. Chian RC, Huang JY, Tan SL, et al. Obstetric and perinatal outcome in 200 infants conceived from vitrified oocytes. Reprod Biomed Online 2008; 16: 608–10.

22 Oocyte Storage for Fertility Preservation Before Radio- and Chemotherapy

Michael von Wolff

Department of Gynecological Endocrinology and Reproductive Medicine, Women's Hospital, University of Bern, Bern, Switzerland

INTRODUCTION

Every year about 100 out of 1 million women under the age of 40 years develop cancer in Western countries. In Europe, this number amounts to 50,000 women. Girls up to the age of 14 years most frequently develop leukemia (34%), CNS diseases (20%), and lymphoma (13%), whereas in higher age groups (up to 45 years) the incidence of breast cancer (26%) and cervical cancer (15%) increases (1).

The survival rate after malignant disease has greatly improved in recent years. In Germany, for example, the five-year survival rate for the total number of 1800 children below 15 years who develop cancer each year has now reached 74% (1). As a result, the number of women who have survived cancer during childhood or in their reproductive years is on the rise. Consequently, the importance of quality of life and fertility after oncological treatment is also increasing. According to a study by Schover et al. (2), 76% of women and men who have survived cancer want to have children at a later stage. However, many of these women are unable to conceive because the chemo- and/or radiotherapy treatments they have undergone have damaged their gonads. Cytotoxic therapy frequently leads to the immediate loss of ovarian function or to subsequent premature ovarian failure.

In this situation, the rapid advances in reproductive medicine in recent years and the current development of new techniques to preserve fertility are opening up new opportunities for fertility protection in patients who are to undergo cytotoxic therapy.

In this article, we shall introduce some general aspects and then describe the risk of ovarian damage caused by different cancer therapies and diseases. Finally, we outline the therapeutic options for fertility protection focusing on the retrieval and cryopreservation of fertilized and unfertilized oocytes.

PREGNANCY—A NEGATIVE PROGNOSTIC FACTOR FOR TUMOR DISEASE?

When discussing the possibilities for fertility protection after cytotoxic therapy, the risks of a tumor relapse caused by pregnancy should also be critically evaluated. However, such an evaluation has severe limitations as experience in this field is based on a few studies with a small number of cases.

In a major Japanese study of 50 patients who became pregnant in the first remission phase after acute leukemia (3), no increased risk of pregnancy complications or leukemia relapse was found.

In another study, out of 44 patients with a borderline ovarian tumor, 14 became pregnant on an average of 37 months after the first diagnosis (4). In 76% of the patients, the tumor was limited to the ovaries. None of the patients had a relapse after the pregnancy.

The data on the frequency of relapse after breast cancer therapy and subsequent pregnancy is slightly better. According to several studies, around 5% of the women with previous breast cancer carry a pregnancy to term. Theoretically, because of the hormone dependency of breast cancer, the prognosis was that the patient should get worse after a pregnancy. However, in several studies covering more than 700 post-delivery patients with previous breast cancer, no worsening of the prognosis was found (5). Nevertheless, when interpreting these encouraging figures we should bear in mind that no detailed analysis of women with hormone-receptor-positive tumor tissue is available and that a relapse has a different ethical dimension for young mothers even if it does not seem to affect pregnant women more frequently.

RISK OF OVARIAN FAILURE

Radiation

The radiation damage sustained by the ovaries seems to be strictly dose-dependent; however, there is no threshold dose. Even a radiation dose of only 2 Gy is reported to destroy 50% of the oocytes (6). A dose of 10 Gy is highly likely to result in amenorrhea.

The uterus and the endometrium are much less radiosensitive than the germ cells. Although radiation of the pelvis reduces the uterine volume and the thickness of the endometrium, the consequences for a subsequent pregnancy are limited.

A radiation dose of below 10 Gy did not produce a miscarriage in five patients after chemotherapy and whole body irradiation (7). Irradiation with 10 to 14 Gy led to an abortion in 6 out of 16 pregnancies (38%), whereas 20 to 35 Gy applied to treat Wilms' tumor resulted in abortion in 9 out of 40 pregnancies (22%) (8). The risk of premature birth is also increased after radiotherapy. Five out of eight pregnancies ended prematurely after chemotherapy and whole body irradiation (7). After radiotherapy for Wilms' tumor, the weight of the babies was 300 g lower on average than that of non-irradiated controls (8). The cause of these miscarriages and premature births are presumed to be reduced uterine elasticity due to irradiation-induced fibrosis and damage to the uterine vessels.

Chemotherapy

There is a proven link between the ovariotoxic effect of chemotherapeutic agents and age. From an age of 35 years onwards, the pool of primordial follicles and the quality of the oocytes is reduced to such an extent that even small ovariotoxic effects can result in a premature end of the reproductive period.

Even if the patient does not suffer from amenorrhea immediately after the cytotoxic therapy, the pool of primordial follicles may still be so reduced that premature ovarian failure will result a few years after the end of therapy. This risk is very acute particularly in young women as many years will normally go by before they want to get pregnant. However, it is very hard to estimate the individual risk. There are estimates that the risk of premature ovarian failure, which amounts to around 2% of all women below the age of 40 years, increases by a factor of 9 if young women are treated with alkylating agents and by a factor of 27 if the alkylating agents are combined with radiation (9). However, these figures are rough estimates and cannot be used as a basis for an individual risk estimate.

Hodgkin's Lymphoma

According to the register of the FertiPROTEKT network, 33% of patients seeking advice to perform fertility preservation in 2007 were diagnosed with Hodgkin's lymphoma (HL) (10) (Fig. 1).

However, due to limited data it is difficult to estimate therapy-induced ovarian damage. Some published articles are based on small case numbers or date back to the 1980s and cannot therefore be interpreted because of the change in chemotherapeutic regimens.

- In a study published in 1981, 11 out of 24 patients (46%) developed secondary amenorrhea after therapy for HL based on the MOPP regimen [mechlorethamine, vincristine (Oncovin), procarbacine, and prednisone] (11).
- According to a study published in 1983, 17 out of 44 patients (39%) with HL developed a secondary amenorrhea after being treated with the MVPP regimen (mechlorethamine, vincristine, procarbacine, and prednisone) (12).
- In 2003, a study was published in which 34 out of 84 patients (40%) developed amenorrhea after receiving treatment for HL or non-Hodgkin's lymphoma (13).

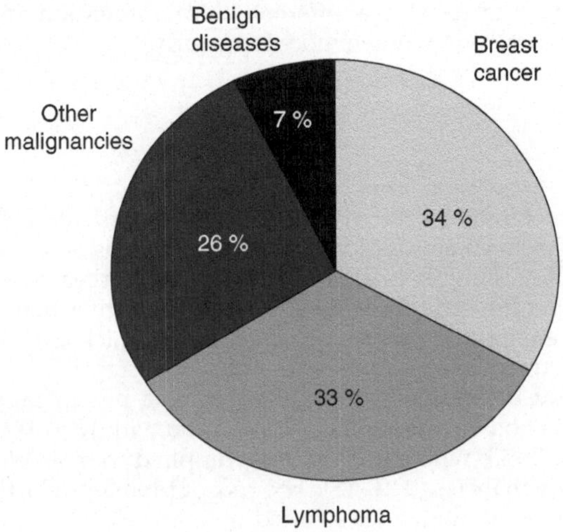

Figure 1 Spectrum of patients most frequently counseled for fertility preserving techniques [according to the FertiPROTEKT network (10)]. *See Color Plates on Page xxii.*

- In 2005, a major study was carried out with 405 patients in an attempt to determine the amenorrhea rate as a function of age and the type of chemotherapy applied (14). In women <30 years, the rate of amenorrhea was between 5% and 40% and in women ≥30 years, it was between 5% and 70%. Low amenorrhea rates were found after chemotherapies using the ABVD regimen [doxorubicin hydrochloride (Adriamycin), bleomycin, vinblastine, and dacarbazine]. The rate was much higher after treatment with the COPP and BEACOOP regimens [bleomycin, etoposide, doxorubicin hydrochloride (Adriamycin), cyclophosphamide, vincristine sulfate (Oncovin), procarbazine, and prednisone] and especially after the intensified-dose BEACOOP regimen. This study supports other data (15,16) showing that, after treatment with the ABVD regimen applied in the early stages of the lymphoma, the rate of amenorrhea is low. Therefore, it may be possible to avoid fertility protection measures when the ABVD regimen is followed.

Breast Cancer

In 2007, the FertiPROTEKT network documented that 34% of patients seeking advice to perform fertility preservation were diagnosed with breast cancer (10).

Data concerning ovarian damage following chemotherapy are limited. In the studies published, the women have been allocated into a group <40 years and another ≥40 years. There is no further subdivision into an age group <35 years, where less ovarian damage can be expected.

A good summary of the risks of therapy-induced amenorrhea in women <40 years as a function of the type of chemotherapy applied can be found in a review by Stearns et al. (17).

- In the case of chemotherapy based on the CMF regimen (×6), the rate of amenorrhea was estimated at 30% to 80% after analyzing seven studies.
- If the AC regimen was followed (×4), the rate of amenorrhea was estimated at 13% to 30% (six studies).
- If the FEC/FAC regimen was followed (×6–8), the rate of amenorrhea was estimated at 13% to 35% (three studies).
- If the AC regimen was used (×4) and followed by P × 4, the rate of amenorrhea was estimated at 35% (one study).
- If the AC regimen was used (×4) and followed by T × 4, the rate of amenorrhea was estimated at 29% to 42% (one study).

(A, doxorubicin; C, cyclophosphamide; E, epirubicin; F, 5-fluorouracil; M, methotrexate; P, paclitaxel; T, docetaxel).

Autoimmune Diseases

Various autoimmune diseases such as systemic lupus erythematosus (SLE), Behcet's disease, chronic inflammatory demyelinizing polyradiculoneuropathy, Goodpasture's syndrome, Wegener's granulomatosis, and various forms of systemic vasculitis are treated by chemotherapy with the alkylating agent cyclophosphamide, which may cause irreversible damage to the gonads. However, only few of these patients are currently counseled for fertility preservation. The FertiPROTEKT network documented only 7% of patients seeking advice with benign diseases.

- In a study published in 2002, 84 female patients (29 ± 10 years) with autoimmune diseases (56 with SLE) received a mean intravenous cyclophosphamide dose of 0.9 ± 0.14 g every 4 weeks over 13 ± 6.5 cycles. In a follow-up of 5.1 ± 3.7 years, 27% developed an amenorrhea. On average, premature ovarian failure occurred at the age of 40 ± 7.6 years. The risk correlated highly significantly with age at the onset of therapy (18).
- In a study published in 2004, 67 female patients with SLE (28.9 ± 8.3 years) were treated intravenously with 0.5 to 0.75 mg/m^2 cyclophosphamide every 4 weeks over 6 cycles followed by the same dose every 3 months over a further 18 months. In a follow-up of 74.4 ± 20.6 months, 37.3% developed an amenorrhea over a variable period. However, in 14.9% of the patients, this amenorrhea also persisted afterwards (19).
- In a 2004 review, the influence of cyclophosphamide on the ovarian function in SLE patients was analyzed. The studies were designed extremely heterogeneously (n between 8 and 92, different dosages and cycle numbers) and showed an incidence of premature ovarian insufficiency of between 11% and 55%. The route of administration (oral vs. intravenous) seems to have been of no influence. However, the higher cumulative doses applied orally seem to increase ovarian toxicity. The likelihood of persisting amenorrhea rose with patient age, and the authors specified a cut-off of >32 years (20).

The age-dependent ovarian toxicity was confirmed by Manger et al. (21), describing premature ovarian failure after therapy with cyclophosphamide for 50% of SLE patients aged <30 years and 60% of patients aged between 30 and 40 years.

FERTILITY PRESERVATION TECHNIQUES

As cryopreservation of oocytes can be combined with other fertility preserving techniques such as cryopreservation of ovarian tissue and gonadotropin-releasing hormone analogues (GnRH-a), these techniques are briefly described and discussed first (Fig. 2; Table 1).

Cryopreservation of Ovarian Tissue

The extraction and cryopreservation of ovarian tissue can be performed shortly before beginning cytotoxic therapy. It can also be combined with the cryopreservation of oocytes and with GnRH-a. It is usually an outpatient procedure, so it does not cause a lot of stress to the patient. If complete functional loss of the ovaries is expected (e.g., when the pelvis has been irradiated or the bone marrow has been transplanted), it may make sense to retrieve the whole ovary. Otherwise about half of the ovarian cortex is usually retrieved from one ovary. Immediately after the intervention, the tissue is transferred to a tissue bank, which carries out the preparation and cryopreservation using cryoprotective agents.

If, after a sufficiently long relapse-free interval, the patient wishes to get pregnant but suffers from ovarian insufficiency, the tissue is usually transplanted orthotopically to the remaining ovary (22,23) or into a peritoneal pocket (24).

The birth of five children and several more pregnancies after the orthotopic transplantation of only about 25 patients indicate the very high potential of this technique (25).

The cryopreservation of ovarian tissue is especially suitable for younger patients as their ovaries still contain a large number of oocytes; so the chances of successful transplantation are

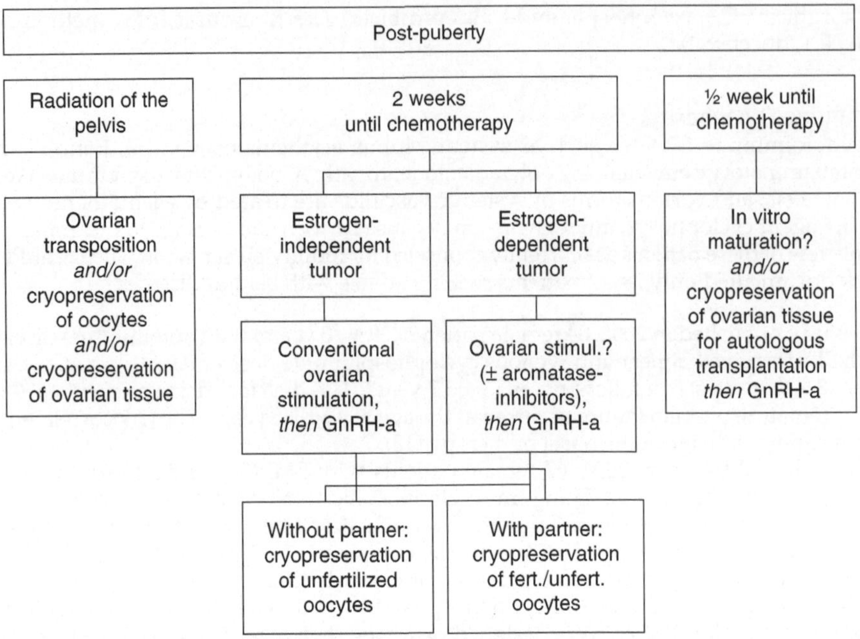

Figure 2 Flow chart of fertility preservation techniques to be offered to women before chemo- and radiotherapy. *Abbreviation*: GnRH-a, gonadotropin-releasing hormone analogues.

Table 1 Indications and Criteria for Different Fertility Preservation Techniques

Cryopreservation of fertilized or unfertilized oocytes
- Minimal patient's age: around 14–18 yr
- Maximum patient's age: 40 yr
- Required time frame: 2 wk
- Not suitable for hormone-dependent breast cancer with residual tumor or metastasis
- Technique well established (cryopreservation of fertilized oocytes)

Cryopreservation of ovarian tissue
- Minimal patient's age: no limit
- Maximum patient's age: 35–37 yr
- Required time frame: 1/2 wk
- Full anesthesia required
- Not suitable for patients with high risk of full anesthesia such as patients with large mediastinal tumors, etc.
- Retransplantation of tissue still experimental

GnRH analogues (GnRH-a)
- Minimal patient's age: around 12–14 yr (menarche)
- Maximum patient's age: no limit
- Required time frame: 1 wk
- Efficacy not yet definitely proven

Abbreviation: GnRH-a, gonadotropin-releasing hormone analogues.

higher. The upper age limit is around 35 years. If the ovarian reserve is still very high, the age limit may be raised on a case-by-case basis. This method is also suitable if there is only about half a week left until the beginning of cytotoxic therapy; it is not possible to stimulate the ovaries in such a short span of time and extract oocytes, which would take around two weeks.

Theoretically, there is a risk that malignant cells are transplanted along with the tissue, but this transplantation risk has not been confirmed in practice. However, the possibility of cryopreserving the ovarian tissue of patients with leukemia or a borderline tumor of the ovary should be treated with great caution.

Before the tissue is retransplanted, various tests to exclude micrometastases are carried out. Attempts have also been made to perform xenologous transplantation (i.e., the transplantation of tissue into immunodeficient mice). In this method, it seems to be easier to evaluate the risk of re-metastasis in cases of high-risk transplantations (e.g., after leukemia) (25).

GnRH Analogues

Despite encouraging progress in the cryopreservation of oocytes and ovarian tissue, it would certainly be ideal to preserve fertility by effective medicinal ovarian protection under chemotherapy. The first tests with high-dose progestins have been replaced by GnRH-a, showing encouraging effects first in animal models and then also in humans. The basis for such a therapy is the fact that in pre-adolescent girls the ovaries sustain much less damage by chemotherapy. As quiescent ovaries [i.e., ovaries without follicle-stimulating hormone (FSH) stimulation] are less receptive to toxic chemotherapies, the secretion of FSH in post-adolescent women is inhibited using GnRH-a.

However, the effectiveness of this therapy has not been conclusively proven. Until now, there are nine published controlled studies all showing a protective effect of GnRH agonists on ovarian function (26), but there is still no scientific proof.

Nevertheless, in view of the above data on the effectiveness of GnRH-a as ovarian protection agents, this medication can be prescribed to women undergoing chemotherapy, especially if there is no alternative fertility protection method available. GnRH-a can also be combined with other techniques such as the cryopreservation of unfertilized and fertilized oocytes and the cryopreservation of ovarian tissue.

Cryopreservation of Fertilized Oocytes

In vitro fertilization (IVF) and intracytoplasmic sperm injection (ICSI) are fully established methods of assisted reproduction, which can be used in patients awaiting cytotoxic therapy. After stimulating the ovaries, oocytes are transvaginally retrieved and fertilized. The oocytes are cryopreserved at the pronuclear stage, or as blastocysts.

Based on data from IVF registers such as the German IVF register, it is predicted that an average of 10 oocytes can be retrieved per stimulation cycle with a fertilization rate of around 55% (27). As the average pregnancy rate per transfer of around two cryopreserved pronuclear cells is about 20% (27), an average cumulative pregnancy rate of up to 40% can be expected. However, as cancer patients cannot be compared with average infertility patients, these numbers can only be used as a rough estimation for the success of this technique for the following reasons.

First, ovarian stimulation usually requires a time window of two to five weeks during which no cytotoxic therapy must be applied. As five weeks is far too long for cancer patients to wait, new stimulation protocols have been developed to allow ovarian stimulation during the luteal phase of the menstrual cycle (28). Even though these protocols have been proven to work, the number of oocytes retrieved is slightly lower than in conventional stimulation regimes, used for average infertility patients. Luteal phase stimulation resulted in the collection of 10.0 with a fertilization rate of 75.6% (28).

Second, the data in the IVF registers are based on many women whose chance of pregnancy is limited due to their average age of around 35 years and due to other factors limiting their fertility. Many cancer patients are younger and the fertility of the couple is not impaired, thus theoretically allowing the generation of more fertilized oocytes. By using conventional flare-up or long protocols in cancer patients, 13.1 oocytes were retrieved (28).

Third, patients diagnosed with hormone-dependent breast cancer are stimulated using a combination of gonadotropins and aromatase inhibitors (29). As these stimulation protocols have been developed recently, data about their efficacy are still limited. Oktay et al. reported the collection 12.4 oocytes using this treatment regimen.

Fourth, cryopreservation of oocytes can be combined with other techniques such as cryopreservation of ovarian tissue. The combination of both techniques might, however, reduce the number and quality of aspirated oocytes. In 10 patients ovarian stimulation was started a few days after cryopreservation of around 50% of ovarian cortex in one ovary. An average of 10.5 oocytes was retreived (37).

Cryopreservation Unfertilized Oocytes

If a woman does not live in a stable partnership, the only way to fertilize the oocytes at the time of retrieval would be to use donated sperm. However, such a procedure, which is called heterologous IVF, is unacceptable to most patients and illegal for unmarried patients in some countries. Progress in the preservation of unfertilized oocytes has opened up new opportunities in this field.

The different techniques to cryopreserve oocytes are described in detail in various chapters in this book.

The success rate for the cryopreservation of unfertilized oocytes seems to depend on the freezing protocols used. According to recent findings based on a meta-analysis, the survival and implantation rates seem to be lower than that of fertilized oocytes when slow freezing protocols are used (29).

According to this meta-analysis, the so-called vitrification process promises higher success rates, a finding confirmed by recent studies by other groups. In vitrification, the cells are immersed directly in liquid nitrogen, so they are frozen ultrafast and without crystallization. The resulting crystal-free solutions are crystal-clear (vitreous). By vitrifying unfertilized oocytes, the survival and pregnancy rates achieved seem to be similar to that of fertilized oocytes.

Irrespective of the freezing technique used, the success rate depends heavily on the expertise of the center. Therefore, unfertilized oocytes should be cryopreserved only if the center can prove it is experienced in performing this technique and can show own data to support this experience.

OVARIAN STIMULATION IN PATIENTS WITH BREAST CANCER

In hormone-dependent carcinoma such as breast cancer, which is receptor-positive (estrogen receptor, progesterone receptor) in around 50% of young women, the hormonal stimulation must be discussed at length. It is theoretically possible that tumor cell growth is accelerated under ovarian stimulation, which could increase the likelihood of a relapse.

What mitigates against this assumption is that the young patient will maintain her menstrual cycle right up to chemotherapy even without fertility protection measures, so she will have a high endogenous estrogen level. It is unlikely that the short-term increase of these estrogen levels caused by ovarian stimulation will significantly accelerate tumor growth. However, there are no scientific data to support this assumption. As data supporting the assumption that patients with hormone-dependent breast cancer can safely be stimulated are still unavailable, stimulation protocols that limit estrogen production should be favored.

Oktay et al. (29) have introduced a technique that combines gonadotropins with aromatase inhibitors or tamoxifen. Both techniques strongly reduce the estrogen levels under stimulation treatment. Ovarian stimulation with on average 1317 units of FSH in combination with 5 mg letrozole resulted in estrogen levels of 483.4 pg/ml in comparison to 1464 pg/ml in conventional long protocols. The number of oocytes collected was 12.4, which was comparable to conventional protocols (n = 11.1). Following initial criticism concerning the potential risks of using aromatase inhibitors, Azim et al. have published a study (30) that demonstrates the safety of this medication.

Alternatively, it is possible to retrieve immature oocytes with no or very mild gonadotropin stimulation and to mature them before the cryopreservation [in vitro maturation (IVM)]. However, this technique, which is described in detail below, is only performed in a small number of centers and the number of oocytes retrieved is low. Therefore, this technique should not be used as a sole measure but combined with another method such as the cryopreservation of ovarian tissue.

A third, rather theoretical option is to perform the so-called natural cycle IVF. Follicles are aspirated just before ovulation without any prior ovarian stimulation (31). As the pregnancy rate per treatment cycle is only 10% in fresh cycles (after cryopreservation, the success rate will be even lower), this technique is too inefficient to be offered to cancer patients.

OVULATION INDUCTION AFTER STIMULATION

If oocytes are to be retrieved for cryopreservation, high-dosage ovarian stimulation is necessary in order to collect the maximum number of oocytes. On one hand, the patient must not develop an ovarian hyperstimulation syndrome (OHSS), which could mean that the cytotoxic therapy would have to be postponed. It is true that the complication register of the FertiPROTEKT network (10) only documented one severe hyperstimulation syndrome in 54 patients after conventional ovarian stimulation. However, many patients were only stimulated with relatively low gonadotropin doses for fear of causing hyperstimulation, so only ≤6 oocytes were retrieved in 25% of the patients.

One way to resolve this apparent conflict between high-dose stimulation to retrieve as many oocytes as possible and a minimal risk of OHSS seems to be to develop special stimulation regimens. Bodri et al. (32) carried out high-dose gonadotropin stimulation in 2077 stimulation cycles as part of an oocyte donation program in order to collect as many oocytes as possible. The patients received recombinant FSH, and the pituitary glands were suppressed with GnRH-a (antagonist protocol). About 36 hours before oocyte retrieval, the patients received a short-acting GnRH agonist (triptorelin 0.2 mg) instead of chorionic gonadotropin. The immediate flare-up effect led to the secretion of luteinizing hormone (LH) and subsequent ovulation induction. As the flare-up effect is very short-lived with this medication and is followed by suppression of the LH release with luteolysis, the risk of OHSS is much lower. The effectiveness of this technique and the avoidance of OHSS have been confirmed by Griesinger et al. (33).

LUTEAL PHASE STIMULATION

Oocyte collection usually requires a time frame of up to six weeks to collect oocytes as conventional regimes using GnRH-a are time-consuming, requiring pituitary downregulation before gonadotropin administration. In many cancer diseases such as breast cancer and HL, chemotherapy needs to be started within two weeks. Therefore, new stimulation protocols were developed, which allow ovarian stimulation and oocyte collection in all patients within two weeks irrespective of the menstrual phase at which ovarian stimulation was started (28). In the follicular phase ovarian stimulation was performed according to the classical short flare-up protocols, applying both GnRH-a and gonadotropins at the beginning of menstrual bleeding or antagonist protocols applying gonadotropins just after menstrual bleeding, followed by GnRH antagonists a few days later to suppress endogenous LH increase.

In the luteal phase, ovarian stimulation combined the application of GnRH antagonists to induce immediate luteolysis and at the same time initiation of ovarian stimulation with recombinant FSH. Pure recombinant FSH needs to be applied as LH or human chorionic gonadotropin (HCG) delays luteolysis (34).

von Wolff et al. (28) demonstrated that in patients stimulated during the follicular phase compared with stimulation during the luteal phase, the duration of gonadotropin administration (10.6 days vs. 11.4 days), the FSH dosage (2255 IE vs. 2720 IE, the number of collected oocytes (13.1 vs. 10.0), the percentage of mature oocytes (84% vs. 80%), and the fertilization rate by ICSI (61% vs. 76%) were not significantly different.

IN VITRO MATURATION

Some authors have suggested (35) IVM in order to retrieve oocytes. In IVM, the follicles are collected as soon as the lead follicle has grown to a maximum size of 12 to 14 mm. If necessary, the patients are given a three-day FSH stimulation and/or ovulation induction with HCG prior to retrieval. The number of oocytes retrieved depends strongly on the number of secondary and tertiary follicles visible on ultrasound. Accordingly, in patients with polycystic ovaries, about 13.6 oocytes can be collected (36). As the maturation process results in the loss of around 50% of the oocytes, the number of fertilized oocytes was only 5.1 in women with polycystic ovaries. If you consider the low implantation rates of 6.9% (36), the oocyte losses on cryopreservation and thawing, and the higher abortion rate of around 25% after IVM, it becomes apparent that the initial euphoria about the effectiveness of this technique must be put into perspective. Furthermore, these data are derived from patients with polycystic ovaries. In cancer patients without polycystic ovaries, the number of oocytes is far lower. For this reason, this procedure—especially prior to cytotoxic therapy—should only be offered by highly specialized centers and in combination with other techniques (e.g., the cryopreservation of ovarian tissue).

COMBINATION WITH OTHER TECHNIQUES

The chances of achieving pregnancy with cryopreserved fertilized or unfertilized oocytes are limited. On one hand, only one IVF cycle can usually be carried out before cytotoxic therapy; on the other hand, only a small number of oocytes can be retrieved in many patients. For this reason, this technique should be combined with other fertility protection methods to increase the effectiveness of fertility protection (Fig. 3).

Combination with GnRH-a

When stimulation is combined with GnRH-a, the question is at what time treatment with GnRH-a-depot preparations should be started. The GnRH-a-initiated flare-up effect with increased LH and FSH concentrations for five to seven days may increase the risk of OHSS due to LH. Furthermore, the fertility-protecting effect of GnRH-a could be reduced during the increased FSH secretion, while the first cycle of chemotherapy is administered.

If the patient receives gonadotropin stimulation as part of a conventional long or flare-up protocol, GnRH-a are applied during the gonadotropin stimulation as part of the regular stimulation protocol. In this way, GnRH-a can be continuously applied throughout the stimulation, and no additional flare-up effect occurs.

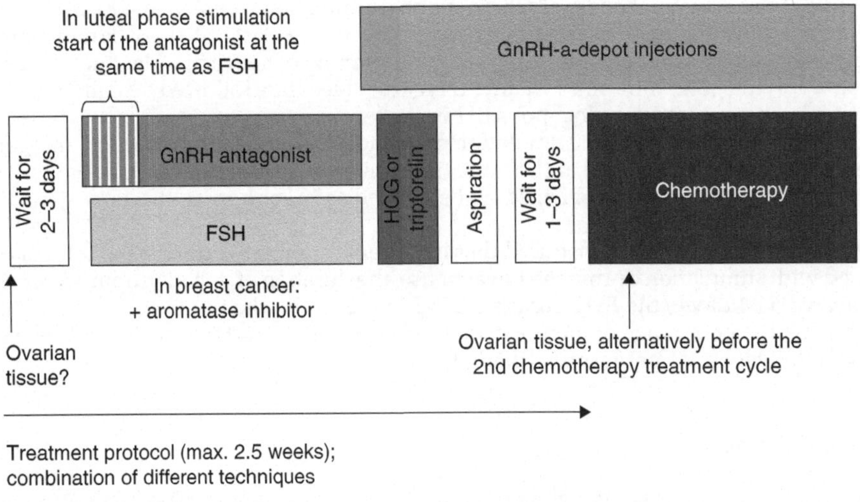

Figure 3 Protocol for the combination of the three most important techniques for fertility preservation such as ovarian stimulation and oocyte collection, laparoscopic removal of ovarian tissue, and depot injections of GnRH analogues. The stimulation protocol can be applied to all patients before cytotoxic therapy irrespective of the phase of the menstrual cycle. *Abbreviations*: FSH, follicle-stimulating hormone; GnRH-a, gonadotropin-releasing hormone analogues; HCG, human chorionic gonadotropin.

When antagonist protocols are used (e.g., as part of the above-mentioned luteal phase stimulation), the patient receives either HCG or short-acting GnRH-a as shown above. Therefore, in this case, the administration of the first GnRH-a-depot preparation must be scheduled with the associated risks and mechanisms of action in mind.

Combination with the Cryopreservation of Ovarian Tissue

There are two ways to combine oocyte retrieval with the cryopreservation of ovarian tissue. However, their effectiveness still has to be confirmed in larger studies.

One way is to retrieve ovarian tissue three to four weeks after the collection of follicles (i.e., 2–3 weeks after the first cycle of chemotherapy and just before the second cycle). By this time, the follicles have receded and the patient's coagulation parameters have normalized so that tissue retrieval is possible at a low risk.

The other possibility is to first retrieve ovarian tissue laparoscopically and to start gonadotropin stimulation a few days later. This technique has been evaluated recently. In six patients so far, an average of nine oocytes were retrieved. No complications such as ovarian bleeding occurred. However, no data on the pregnancy rate after the transfer of these embryos are available as yet.

Some centers have suggested a third option, that is, tissue retrieval at the time of follicle retrieval. However, at this time, the ovarian tissue is very thin and therefore difficult to use. Furthermore, there is a substantial risk of bleeding from the stimulated ovaries.

OUTLOOK

As cancer survival rates are increasing, fertility preservation is becoming increasingly important. However, intense, interdisciplinary scientific and clinical initiatives are necessary in order to use fertility protection techniques effectively.

On the one hand, the available techniques must be optimized and their effectiveness proven by documenting the cases in registers. On the other hand, highly specialized centers are required to provide universal, decentralized care.

Successful national and international initiatives already show that these goals can be achieved. The efficiency of the various fertility protection techniques is increasing, so a pregnancy rate of 50% seems realistic after ovarian function loss if the techniques are used individually or in combination in a customized way. Moreover, national structures such as the FertiPROTEKT network show that universal and efficient care provided by highly specialized centers is possible.

REFERENCES

1. Krebs in Deutschland. Arbeitsgemeinschaft Bevölkerungsbezogener Krebsregister in Deutschland, 3 erweiterte, aktualisierte Ausgabe. Saarbrücken, 2002.
2. Schover LR, Rybicki LA, Martin BA, et al. Having children after cancer. A pilot survey of survivors' attitudes and experiences. Cancer 1999; 86: 697–709.
3. Kawamura S, Ueda R, Ohno R, et al. Pregnancy among long-term survivors of acute leukemia. A second nationwide survey. Int J Hematol 1998; 67: 37–43.
4. Morice P, Camatte S, El Hassan J, Pautier P, et al. Clinical outcomes and fertility after conservative treatment of ovarian borderline tumors. Fertil Steril 2001; 75: 92–6.
5. Kroman N, Jensen MB, Melbye M, et al. Should women be advised against pregnancy after breast-cancer treatment? Lancet 1997; 350: 319–22.
6. Wallace WH, Thomson AB, Kelsey TW. The radiosensitivity of the human oocyte. Hum Reprod 2003; 18: 117–21.
7. Sanders JE, Hawley J, Levy W, et al. Pregnancies following high-dose cyclophosphamide with or without high-dose busulfan or total-body irradiation and bone marrow transplantation. Blood 1996; 87: 3045–52.
8. Hawkins MM, Smith RA. Pregnancy outcomes in childhood cancer survivors: probable effects of abdominal irradiation. Int J Cancer 1989; 43: 399–402.
9. Larsen EC, Müller J, Schmiegelow K, et al. Reduced ovarian function in long-term survivors of radiation- and chemotherapy-treated childhood cancer. J Clin Endocrinol Metab 2003; 88: 5307–14.

10 FertiPROTEKT. Website of the "Network for fertility preservation in radio- and chemotherapy," www.fertiprotekt.eu. Responsible webmaster: von Wolff M. Installed January 2007, latest update September 2008.
11. Schilsky RL, Sherins RJ, Hubbard SM, et al. Long-term follow up of ovarian function in women treated with MOPP chemotherapy for Hodgkin's disease. Am J Med 1981; 71: 552–6.
12. Whitehead E, Shalet SM, Blackledge G, et al. The effect of combination chemotherapy on ovarian function in women treated for Hodgkin's disease. Cancer 1983; 52: 988–93.
13. Franchi-Rezgui P, Rousselot P, Espié M, et al. Fertility in young women after chemotherapy with alkylating agents for Hodgkin and non-Hodgkin lymphomas. Hematol J 2003; 4: 116–20.
14. Behringer K, Breuer K, Reineke T, et al. German Hodgkin's Lymphoma Study Group. Secondary amenorrhea after Hodgkin's lymphoma is influenced by age at treatment, stage of disease, chemotherapy regimen, and the use of oral contraceptives during therapy: a report from the German Hodgkin's Lymphoma Study Group. J Clin Oncol 2005; 23: 7555–64.
15 André M, Brice P, Cazals D, et al. Results of three courses of adriamycin, bleomycin, vindesine, and dacarbazine with subtotal nodal irradiation in 189 patients with nodal Hodgkin's disease (stage I, II and IIIA). Hematol Cell Ther 1997; 39: 59–65.
16. Brusamolino E, Lunghi F, Orlandi E, et al. Treatment of early-stage Hodgkin's disease with four cycles of ABVD followed by adjuvant radio-therapy: analysis of efficacy and long-term toxicity. Haematologica 2000; 85: 1032–9.
17. Stearns V, Schneider B, Henry NL, et al. Breast cancer treatment and ovarian failure: risk factors and emerging genetic determinants. Nat Rev Cancer. 2006; 6: 886–93. Epub Oct 12, 2006.
18. Huong DL, Amoura Z, Duhaut P, et al. Risk of ovarian failure and fertility after intravenous cyclophosphamide. A study in 84 patients. J Rheumatol 2002; 29: 2571–6.
19. Park MC, Park YB, Jung SY, et al. Risk of ovarian failure and pregnancy outcome in patients with lupus nephritis treated with intravenous cyclophosphamide pulse therapy. Lupus 2004; 13: 569–74.
20. Katsifis GE, Tzioufas AG. Ovarian failure in systemic lupus erythematosus patients treated with pulsed intravenous cyclophosphamide. Lupus 2004; 13: 673–8.
21. Manger K, Wildt L, Kalden JR, Manger B. Prevention of gonadal toxicity and preservation of gonadal function and fertility in young women with systemic lupus erythematosus treated by cyclophosphamide: the PREGO-Study. Autoimmun Rev 2006; 5: 269–72.
22. Donnez J, Dolmans MM, Demylle D, et al. Livebirth after orthotopic transplantation of cryopreserved ovarian tissue. Lancet 2004; 364: 1405–10.
23. Meirow D, Levron J, Eldar-Geva T, et al. Pregnancy after transplantation of cryopreserved ovarian tissue in a patient with ovarian failure after chemotherapy. N Engl J Med 2005; 353: 318–21.
24. Andersen CY, Rosendahl M, Byskov AG, et al. Two successful pregnancies following autotransplantation of frozen/thawed ovarian tissue. Hum Reprod 2008; 23: 2266–72. [Epub ahead of print].
25. von Wolff M, Donnez J, Hovatta O, et al. Cryopreservation and autotransplantation of human ovarian tissue prior to cytotoxic therapy – a technique in its infancy but already successful in fertility preservation. Eur J Cancer 2009; Mar 3. [Epub ahead of print].
26. Blumenfeld Z, von Wolff M. GnRH-analogues and oral contraceptives for fertility preservation in women during chemotherapy. Hum Reprod Update 2008; 14: 543–52.
27. German IVF-Register. [Available from: http://www.deutsches-ivf-register.de]
28. von Wolff M, Thaler CJ, Frambach T, et al. Ovarian stimulation to cryopreserve fertilized oocytes in cancer patients can be started in the luteal phase. Fertil Steril 2008; Oct 16. [Epub ahead of print]
29. Oktay K, Hourvitz A, Sahin G, et al. Letrozole reduces estrogen and gonadotropin exposure in women with breast cancer undergoing ovarian stimulation before chemotherapy. J Clin Endocrinol Metab 2006; 91: 3885–90.
30. Azim AA, Costantini-Ferrando M, Oktay K. Safety of fertility preservation by ovarian stimulation with letrozole and gonadotropins in patients with breast cancer: a prospective controlled study. J Clin Oncol 2008; 26: 2630–5.
31. Schimberni M, Morgia F, Colabianchi J, et al. Natural-cycle in vitro fertilization in poor responder patients: a survey of 500 consecutive cycles. Fertil Steril. 2008; Sep 13. [Epub ahead of print]
32. Bodri D, Guillén JJ, Galindo A, et al. Triggering with human chorionic gonadotropin or a gonadotropin-releasing hormone agonist in gonadotropin-releasing hormone antagonist-treated oocyte donor cycles: findings of a large retrospective cohort study. Fertil Steril. 2008; Mar 24. [Epub ahead of print]
33. Griesinger G, von Otte S, Schroer A, et al. Elective cryopreservation of all pronuclear oocytes after GnRH agonist triggering of final oocyte maturation in patients at risk of developing OHSS: a prospective, observational proof-of-concept study. Hum Reprod 2007; 22: 1348–52.

34. Niswender GD, Juengel JL, Silva PJ, et al. Mechanisms controlling the function and life span of the corpus luteum. Physiol Rev 2000; 80: 1–29.
35. von Wolff M, Strowitzki T, van der Ven H, et al. In vitro maturation is an efficient technique to generate oocytes and should be considered in combination with cryopreservation of ovarian tissue for preservation of fertility in women. J Clin Oncol 2006; 24: 5336–7; author reply 5337–8.
36. Cha KY, Han SY, Chung HM, et al. Pregnancies and deliveries after in vitro maturation culture followed by in vitro fertilization and embryo transfer without stimulation in women with polycystic ovary syndrome. Fertil Steril 2000; 73: 978–83.
37. von Wolff M, Zeeb C, Lawrenz B , et al. Cryopreservation of ovarian tissue and cryopreservation of oocytes can be efficiently combined and performed within 2 weeks before chemotherapy. Hum Reprod. Abstract Book, 2009, in press).

23 Impact of Oocyte Storage in Oocyte Donation Treatments

Ana Cobo[1] **and José Remohí**[2]
[1] *Cryobiology Unit, IVI-Valencia, Valencia, Spain*
[2] *Obstetrics and Gynaecology, Faculty of Medicine, University of Valencia, Valencia, Spain*

INTRODUCTION

Oocyte donation represents an exceptional situation in assisted reproduction technologies (ART), since the oocytes are derived from young fertile women, and the resulting embryos are transferred to the uterus of a different woman who has been previously prepared with a controlled regimen of exogenous estrogen and progesterone optimizing the endometrial receptivity. This model in turn makes possible the optimization of oocyte quality and therefore embryo quality, since oocytes are recruited from young healthy women. This fact has contributed to increase the number of indications for this therapeutic option in infertile patients, initially intended for young women with premature ovarian failure (POF) (1). As a result, this strategy is nowadays a well-established practice, currently applied in cases of heritable maternal genetic disorders (2), early perimenopausal and/or menopausal women (3), poor responder patients (4), or in cases of multiple unsuccessful in vitro fertilization (IVF) attempts (5).

Ovum donation has consistently produced the highest pregnancy rates reported for any assisted reproduction methods, basically due to the reduction in poor oocyte quality because of the selection of healthy young donors (6). Accordingly, as reported by the Assisted Reproduction Technology Registry in 2004 (7), the delivery rate per transfer was 29.9% versus 43.7% in fresh nondonor (N = 73,406) and donor (N = 7581) cycles respectively, performed in 2000 by 383 IVF centers in United States.

We published our first five-year experience with ovum donation conducted in 767 cycles between 1991 and 1995 (8). This report indicated that the technique was successful regardless of the cause of infertility and patient's age. Accordingly, ovum donation provides similar success rates when applied to women suffering from different reproductive disorders, including previously failed IVF cycles (four attempts) with a cumulative birth rate of 88%. Life tables for low responders and endometriosis have shown an effect on the oocyte and the resulting embryo rather than on the uterine environment at implantation.

These observations have been confirmed in the same ovum donation program, which is one of the largest in Europe (9). Moreover, a considerable improvement in the overall outcome has been achieved (9). This study shows the outcome of our ovum donation program over a 10-year period. Major indications were advanced maternal age, defined as older than 40 years (27%), low responder patients (24%), POF (15%), recurrent IVF failure (6%), endometriosis (6%), poor oocyte quality (4%), menopause (3%), genetic disorders (3%), and recurrent miscarriage (2%). Cycle outcome parameters including pregnancy rate (PR), implantation rate (IR), clinical and ongoing PR, miscarriage rate (MR), and ectopic PR, as well as twin and high-order PRs, were calculated for every two- to three-year consecutive period between September 1995 and September 2005 to detect any trend in decrease or increase in these parameters.

From the 10,537 cycles initiated, a total of 8430 donations and 7186 embryo transfer procedures were performed and included in the study. After donation, embryo transfer could not be performed in 1244 (14.7%) cycles because of fertilization failure, poor embryo quality, or unavailability of euploid embryos for transfer. Overall PR, IR, clinical PR, and MR per embryo transfer performed were 54.9%, 27%, 50.3%, and 19%, respectively. Ongoing PR per transfer was 40.2%, and twin and high-order multiple PRs were 39% and 6%, respectively. Mean number of embryos transferred was reduced from 3.6 ± 0.8 to 1.9 ± 0.3, IR improved from 16.7% to 38.3% and ongoing PR improved from 31% to 44.3%. Cumulative PRs did not differ significantly among different indications for oocyte donation, age groups, or origin of sperm used for oocyte insemination. Overall cumulative PRs after three and five cycles were calculated as 87% and 96.8%, respectively. Similar results have been published by others (6,10,11).

Ovum donation represents a perfect model to assess the impact of advancing age on uterine receptiveness, since embryos develop from oocytes of a young fertile woman. In our first large report on oocyte donation overall outcome (8), we were not able to observe statistically significant differences regarding patient's age. However, a trend of higher rate of success in women younger than 30 years, and an increase in the rate of spontaneous miscarriage in older patients were observed. These observations have been confirmed later (12,13). This study reveals that PR and IR are significantly reduced and MR is significantly increased from 45 years of age onward. Comparisons of women aged 44 or less (N = 2686 cycles) with those aged 45 or more (N = 406 cycles) demonstrated that PR (49.8% vs. 44.4%), IR (20.7% vs. 16.8%), and MR (16.8% vs. 23.3%) were significantly lower among the older women. Notwithstanding, these patients showed a highly acceptable PR (44.4%). On the other hand, in the same study, the univariate and linear regression analyses of the association between age and PR, IR, and MR showed that age was not significantly relevant to any of these outcomes. These evidences indicate that age is not really a predicting factor of the outcome of the oocyte donation cycles.

These results absolutely confirm that ovum donation programs are currently a well-established and successful strategy, very useful in the management of infertility. It is interesting to note that the concept of ovum donation is very similar to the donation of organs. Such programs are defined as the removal of healthy organs and tissues from one person for transplantation into another. Both organ and ovum donation are carried out in a very similar way, taking in to account that both consist of patients waiting for a proper donor. The patients are allocated within waiting lists and receive the donation once a compatible donor is available. Unfortunately, sometimes the donation is not possible for a variety of reasons, mainly because of the unavailability of adequate donors or large waiting lists. This situation could end with fatal consequences for patients waiting for a vital organ. Of course, this extreme circumstance does not occur in cases of infertile patients waiting for an oocyte donation. However, undesirable difficulties can arise as a consequence of the long waiting lists due to the difficulty in finding qualified donors and the synchronization with the recipient, which are common inconveniences responsible for the reduction of the efficiency of ovum donation programs. This situation could cause greatly stress and discomfort to the patients.

Our protocol of endometrial preparation for recipients has been described previously as "prolonged follicular phase protocol" (8,9,12,14). Patients with ovarian function are desensitized with a single IM administration of triptorelin 3.75 mg depot (Decapeptyl 3.75 mg; Ipsen-Pharma, Barcelona, Spain) administered between days 18 and 21 of the previous cycle. Estrogen replacement is started on day 3–5 of the next menstruation after confirming ovarian quiescence and a thin endometrium with transvaginal ultrasound examination. Oral estradiol (E2) valerate administration of 2 mg/day (Progynova; Schering, Madrid, Spain) is initially given for eight days, increased to 4 mg/day for the following three days, and then maintained at 6 mg/day until the pregnancy test. After two weeks of E2 valerate administration, recipients are ready to receive embryos, and then they wait until a suitable donation became available.

This scheme allows successful implantation after 100 days of unopposed E2 valerate administration (14). However, a later study has demonstrated that estrogen therapy lasting more than seven weeks is associated with reduced PR and IR (12). Currently, our average time of waiting is around 35 to 40 days. Nonetheless, vaginal spotting or bleeding in the recipient during E2 valerate administration is the main cause of cancellation (25%) in our oocyte donation program. In these cases, recipients usually loose the opportunity to receive the oocytes from a suitable donor and they have to start a new cycle for endometrial preparation.

Additionally, according to current regulations, donors must be tested to avoid possible transmission of infectious diseases. So far, ovum donation cycles have been conducted with fresh oocytes. With this modality, it has not been possible to keep a proper quarantine period as in the case of semen banks. Obviously, a more accurate screening for transmission of infectious diseases would be quite welcome. All these limitations could be solved with an efficient egg-banking program, which in turn obviously requires an efficient and reproducible oocyte cryopreservation methodology.

MAIN CRYOPRESERVATION STRATEGIES

The essential role of cryopreservation in ART has become obvious since the commencement of the infertility treatment, becoming a more flexible and efficient practice. Semen and embryo cryopreservation has been a successful strategy, routinely applied in IVF procedures for a long time. However, in spite of numerous studies conducted over the last 20 years, the reliability of oocyte cryopreservation is just being confirmed currently. All efforts that have been made are clearly justified, mostly because an efficient oocyte cryopreservation program would be quite welcome and widely applied in ART, as there are many other indications for oocyte cryopreservation that are different from the above-mentioned egg banking for ovum donation. Potential beneficiaries of this technology would be cancer patients who need an option for fertility preservation before undergoing the potential sterilizing treatment (15), or women who wish to delay their motherhood due to a variety of reasons (16): government restrictions on IVF (17,18), ethical reasons against embryo cryopreservation, and practical reasons such as unavailability of the male gamete on the day of pickup (19,20).

Despite this wide diversity of potential applications, egg-banking has not been a routine procedure until very recently; in fact, there are still very few centers worldwide that have incorporated this practice in their clinical routine. This can be explained by the fact that the methodology to cryopreserve human oocytes has been disappointing, with results that have not always been reproducible. The purpose of this chapter in not to deeply assess the fundamental principles of cryobiology; however, a brief description of this topic could be useful.

There are some reasons that could explain the low rate of successes that has been traditionally observed. Some of them include the size, shape, and number of oocytes. These gametes are the largest cells of the human body, and this could explain, at least in part, the great differences in cryotolerance between oocytes and, for example, spermatozoa. In addition, the spherical shape of the oocyte could disturb the uniform distribution of cryoprotectants (CPAs). Finally, oocytes can be considered unique cells because they have only one chance of success and there are few possibilities to restore them after serious damage; this situation is completely different from that observed in tissues composed of million of cells, in which the damage could be compensated in different proportions.

Other major factors responsible for the high oocyte sensitivity to cryopreservation include chilling injury and intracellular ice-crystal formation. These factors are related to the cryopreservation method. There are two main strategies in cryobiology: slow cooling and vitrification. During the former, the cells are gradually dehydrated in the presence of CPAs, and the temperature is lowered at a very slow cooling rate (–0.3°C) (21). In this way, the cells are exposed to low temperatures during an extended period of time, which can lead to what is known as chilling injury (22). At the final stage, water solidifies into ice crystals. Chilling injury can be defined as the irreversible damage following exposure of cells to low temperatures, from +15°C to –5°C before the nucleation of ice (23). This detrimental event affects mainly the cytoskeleton (24) and cell membranes (25). The ice-crystal formation within the cytoplasm must be avoided at all cost in order to guarantee the survival and integrity of the cells when they are later thawed. Chilling injury can be minimized during vitrification by use of high cooling rates. Moreover, as the sample vitrifies, ice-crystal formation is avoided (26).

Vitrification, in particular by means of methodologies that use a minimum volume, is altering this situation by producing results that have not been achieved with slow cooling protocols (20). The physical phenomenon of vitrification takes place when the solidification of the solution occurs not by ice crystallization but by extreme elevation in viscosity, which is achieved by using high cooling rates from –15,000°C to –30,000°C per minute, therefore avoiding the risk of chilling injury (26). This ice-free cryopreservation method has undergone modifications in order to optimize results. One such modification has been to reduce the volume of the vitrification solution containing oocytes, which allows the CPA concentration and, consequently, the cytotoxicity to be decreased (27,28). As mentioned above, this procedure circumvents the two major limiting factors for achieving optimal cryopreservation: chilling injury (27) and ice formation (29). Chilling injury can be minimized during vitrification by use of high cooling rates (26). The velocity of the process is dependent on the volume of the vitrification solution. Thus, the smaller the volume of the sample, the higher the cooling rate. On the other hand, direct contact with liquid nitrogen also contributes to increase the cooling rate. To avoid ice formation, the vitrification technique makes use of high CPA concentrations (26), despite the fact that such high concentrations are considered toxic to cells (30). Nonetheless, an appropriate, phased composition of CPA could mitigate the toxic and osmotic consequences of highly concentrated CPAs (20). In this way, a combination of two or three of these agents can decrease the individual specific toxicity. The most common mixture employed for this purpose consists of ethylene glycol, dimethylsulfoxide, and sucrose (20). To optimize the results, in addition to an appropriate selection of CPAs, it is also helpful to use these agents at lower concentration, while maintaining the necessary concentration to achieve vitrification. By dramatically increasing the cooling rate, the CPA concentration could be reduced. As a result, a high cooling rate avoids chilling injury and allows the reduction of the concentration of CPA, thereby preserving the cells at nontoxic concentrations of CPA. Several approaches fit these conditions. The "minimum drop vitrification" method proposed by Arav (31) has been tested successfully for bovine and porcine oocytes; it uses a very small volume of vitrification solution, and samples are placed on a special device that is cooled very quickly (31). A wide variety of approaches have been reported in the literature (32–37). The Cryotop device is one of them, consisting of a fine transparent polypropylene film attached to a plastic handle and equipped with a cover straw, onto which very small volumes of medium containing oocytes can be loaded (up to ~0.1 µl), ten times lower than the capacity of other minimum volume devices (38). This method has proven its efficacy by leading to the birth of more healthy babies than other methodologies (39).

CLINICAL OUTCOME

Since the very first frozen-oocyte pregnancy was achieved in humans, the slow cooling method has been the most widely applied technique with varying success (40–50). A recent meta-analysis showed that IVF success rates with slow-frozen oocytes in terms of clinical PR and IR per transfer are 20.6% and 10.1%, respectively, whereas, these values are 68.5% and 39.8%, respectively for IVF cycles using unfrozen oocytes (51).

To improve the results achieved by slow cooling methods, several modifications of the original method have been introduced (52–55). With the combined use slow freezing and intracytoplasmic sperm injection (ICSI), some investigators have obtained pregnancy rates around 22% to 25% per cycle (52,56). One of the most significant improvements has been achieved by the increase in the sucrose concentration within the freezing media (52). With this strategy, survival rates have been improved to values near 75% (50,57–59). However, PR (10–12%) and IR (~5%) were lower than the expected. Another study showed a significant improvement using a differential sucrose concentration during dehydration (0.2 M) and rehydration (0.3 M), in terms of PR and IR (21.3% and 13.5%, respectively) (55). High sucrose concentrations have also been used within sodium-depleted choline-based media with different grades of success (60–62).

Another interesting study shows the cumulative clinical outcome derived from 80 treatments in which few fresh oocytes were inseminated to prevent the generation of "surplus" embryos, while all the remaining oocytes were cryopreserved. PR and IR for fresh cycles were 30.0% and 22.6%, respectively. Survival after the slow-cooling method using 1.5 M 1,2-propanediol and 0.1 M sucrose was 43.4%. PR and IR for thawed cycles were 19.2% and 12.3%, respectively. The cumulative PR was 47.5% per patient. Authors concluded that despite a low survival rate achieved, oocyte storage significantly improves the number of pregnancies per treatment cycle in cases in which only a minority of oocytes are destined for the fresh treatment (63).

More recent study reports and our own experience have indicated that efficiency of oocyte cryopreservation is improving especially when employing vitrification. Other methodologies using low volumes for loading the samples have been developed. Yoon et al. (2000) reported the birth of two infants from vitrified oocytes using electron microscopy grids (64). Three years later, the same group reported that 325 out of 474 oocytes survived (68.6%) after warming, and 142 of those (43%) were fertilized normally (65). PR and IR were 21.4% and 6.4%, respectively. The open-pulled straw (OPS) method has been designed and tested in bovine by Vajta (37). Kuleshova et al. reported a live birth after the transfer of one chromosomally normal embryo as assessed by FISH analysis, generated from a vitrified human oocyte using the OPS system (66). A very recent study also reports two ongoing pregnancies after vitrification of human oocytes applying this method (67), although this study was performed with a low number of oocytes.

The oocyte survival rate was 75% (18 out of 24 oocytes), the fertilization rate was 77.7%, PR per transfer was 33.3%, and the IR was 21.4% (3 out of 14 replaced embryos).

The most recently developed vitrification system, the Cryotop, designed by Kuwayama (68), allows the samples to be loaded in a very small volume of vitrification solution. Once oocytes are loaded onto the Cryotop, almost the entire loading solution is removed by aspiration before direct immersion into liquid nitrogen; thus, the final volume is approximately 0.1 µl. Therefore, the best advantage of this device is that an extremely fast cooling rate is achieved (23,000°C/min) so that chilling injury is avoided (38). This extremely low volume is also useful to achieve higher warming rates (42,000°C/min), thereby avoiding ice-crystal formation during warming. Another advantage of this method is that the permeable CPA concentration is reduced to 30%, thus reducing potentially toxic effects (38).

Applying the Cryotop method to human oocytes, Kuwayama et al. have reported a 91% survival rate, 81% cleavage rate, and a 50% blastocyst rate with a 41% PR per embryo transfer, resulting in 11 live births (38). Other groups have also published their experiences with the Cryotop method in humans. Katayama et al. (69) have reported excellent survival, fertilization, and cleavage rates of 94%, 91%, and 90%, respectively, and the first baby from vitrified oocytes in the United States. Another publication reports a 90.1% post-thaw survival and 34.1% pregnancy rates using the Cryotop method (70). These authors have obtained 34 healthy babies until now (unpublished data). Other authors have published survival, fertilization, and pregnancy rates of 89%, 87%, and 57%, respectively (71), with up to 117 babies born (unpublished data). More recently, another publication has supported these results (72). In this study, 99.4% of oocytes survived and the fertilization, pregnancy, and implantation rates were 92.9%, 32.5%, and 13.2%, respectively. To date, more than 400 healthy babies have been born as a result of the Cryotop vitrification method (28).

We confirmed the potential of vitrification of oocytes using this methodology, through the simultaneous comparison of the outcome of both vitrified and fresh oocytes from the same ovarian stimulation cycle (73). This model is extremely valuable to assess the possible effects of vitrification procedure on the oocytes' potential, since we were able to simultaneously generate embryos from vitrified and fresh oocytes from the same cohort and using the same semen sample.

All oocytes retrieved from a single donor were donated to a single compatible recipient. After ovum pickup, oocytes were maintained in human tubal fluid media for two hours. They were then enzymatically denuded. Metaphase II (MII) oocytes assigned to a single recipient were randomly allocated to either one of two groups: "vitrified oocytes" and "fresh oocytes." After being vitrified and stored for at least one hour, cryopreserved oocytes were warmed and rehydrated, while the fresh oocytes remained in culture for two hours. This meant that ICSI could be carried out simultaneously on both vitrified and fresh oocytes.

In these series, embryo transfer was performed on day 3. Good quality embryos derived from vitrified oocytes were transferred in preference to those derived from fresh oocytes. All embryos not transferred on day 3 were cultured till day 5–6. We obtained a 96.7% survival rate and detected no significant difference in fertilization rates (76.3% and 82.2%), day 2 embryo cleavage rates (94.2% and 97.8%), day 3 embryo cleavage rates (77.6% and 84.6%), or blastocyst formation rates (48.7% and 47.5%) for vitrified and fresh oocytes, respectively. The ratios of good quality embryos on day 3 and at blastocyst stage were similar in vitrification and fresh groups. Pregnancy, implantation, miscarriage, and ongoing pregnancy rates/transfer (65.2%, 40.8%, 20%, and 47.8%, respectively) are comparable with results obtained in our oocyte donation program conducted with fresh oocytes (12).

After this encouraging experience and taking into consideration that the most essential prerequisite for a successful egg cryobanking program is to have an efficient oocyte freezing/warming technology, we decided to establish our donor's oocyte bank, intended to provide a more efficient approach in an egg donor–recipient program. With this in mind, we have evaluated the outcome of oocyte donation cycles (N = 150 donors and recipients) conducted with fresh and vitrified oocytes (N = 170 donors and recipients).

Donor ovarian stimulation was carried out as described above. Endometrial preparation for recipients receiving fresh (Fig. 1) and vitrified (Fig. 2) oocytes has been performed according to our current protocol. From the first day of spontaneous menses patients were given an anti-contraceptive pill, for 16/18 days. Three days before interrupting the pill, an intramuscular injection of a single dose of gonadotropin releasing hormone (GnRH) agonist depot was administered.

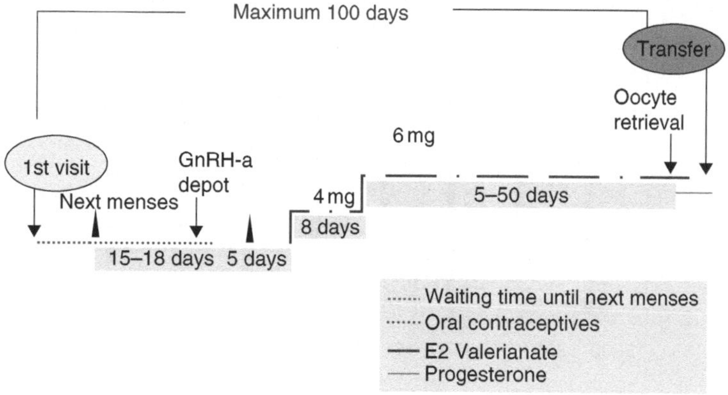

Figure 1 Recipient managing to receive fresh oocytes.

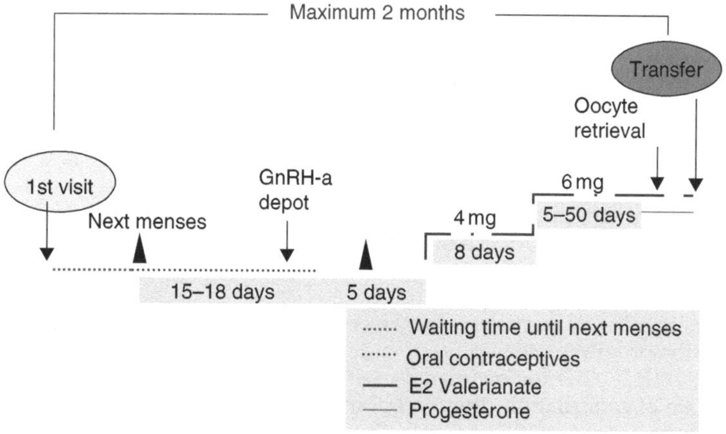

Figure 2 Egg-banking program. Recipient managing to receive vitrified oocytes.

Endometrial preparation was started five days after the last contraceptive pill, with a starting dose of 4 mg of estradiol valerate daily for eight days and after that 6 mg daily. A transvaginal ultrasound was performed between day 8 and 10; if endometrial thickness is >6 mm, the program schedule was maintained. At this point, patients receiving fresh donor oocytes were included within a waiting list until a proper donor is available. Patients were cancelled if presented with endometrial bleeding or if estrogen replacement was longer than 50 days.

The same schedule for endometrial preparation was followed for those patients receiving vitrified oocytes. However, in these cases patients do not have to wait for an appropriate donor, since matching with donor's vitrified oocytes is performed when they start oral contraceptive administration (Fig. 2). With this scheme, the warming procedure was performed between day 12 and 14 of estrogen therapy replacement. Matching was carried out by taking into consideration physical characteristics and blood groups of donors and recipients for patients receiving either fresh or egg-bank oocytes. Cryostored oocytes were considered for donation after at least six months of quarantine period, once donor's serology was confirmed.

A total of 1729 and 1814 vitrified and fresh MII oocytes, respectively, were donated to each group, respectively. A total of 1665 (96.3%) oocytes survived after vitrification. The mean number of donated oocytes was 11.3 ± 0.7 and 10.3 ± 0.9 to egg-banking recipients and fresh oocytes recipients, respectively. No statistical differences were observed regarding fertilization (73.1% vs. 79.8%, $P < 0.05$), cleavage rate on day 3 (91.5% vs. 89.7%, $P < 0.05$), and blastocyst formation (42% vs. 42.2%, $p < 0.05$). Clinical outcome was also similar for both types of

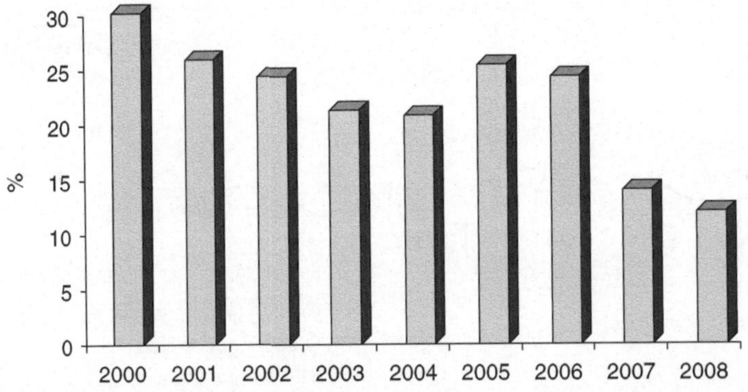

Figure 3 Recipient cancellation (from 2000 to 2008) due to endometrial bleeding or long E2 therapy.

recipients: PR per embryo transfer (59.9% vs. 60.3%, $P < 0.05$), IR (39.3% vs. 39.7%, $P < 0.05$), and MR (16.1% vs. 15.3%).

These results indicate that Cryotop vitrification method is currently a reliable option to successfully cryopreserve oocytes as a routine practice in an IVF program, such as ovum donation. In our previous study we were able to confirm the potential of oocytes vitrified by Cryotop method (73). However, in this study oocytes remained vitrified for just one hour, thus, we were not able to confirm the outcome after long-term storage, which obviously is an essential requisite for egg banking. The current results confirm the potential of vitrified oocytes after at least six months of cryostorage, providing evidences of the usefulness of this technology for egg-banking establishment. On the other hand, these results also confirm the reproducibility of this strategy, providing the confidence to apply it for other indications different from egg banking with high chances of success.

This study aimed to evaluate how oocyte cryopreservation would change the dynamics of our ovum donation program. Mean days of endometrial preparation were 12.3 ± 2.7 (7–13) and 32.7 ± 12.4 (11–54) ($P > 0.05$) for recipients receiving vitrified and fresh oocytes, respectively. Moreover, the rate of cancellation due to endometrial bleeding or long estrogen replacement therapy for fresh oocytes was 19%, whereas it was 0% for recipients of the egg-banking program. Figure 3 shows the rates of cancellation from 2000 to 2008 due to the same reasons. With the establishment of our egg-banking program there has been a significant drop in the rate of recipients' cancellation in our oocyte donation program (Fig. 3). These data clearly show the impact of egg banking on the management of an oocyte donation program, which becomes easier and much more efficient, achieving high clinical results similar to those obtained with fresh donor oocytes. Furthermore, egg banking for oocyte donation will be a safer procedure as it would allow a more accurate screening for possible donor infectious diseases, as is already the case for cryopreserved semen.

In our experience the egg-banking practice has a positive impact on our job schedule, allowing a more efficient management of both donors and recipients with a high clinical outcome. Currently, the donation is performed as soon as recipients are prepared to receive the embryos and, moreover, we are able to schedule the date of embryo transfer on the day the recipients start with the GnRH-a.

These evidences demonstrate that oocyte cryopreservation can be considered a useful tool to provide highly successful outcomes in an egg donor program. Our results validate the use of oocyte cryobanking for egg donation purposes.

REFERENCES

1. Lutjen P, Trounson A, Leeton J, et al. The establishment and maintenance of pregnancy using in vitro fertilization and embryo donation in a patient with primary ovarian failure. Nature 1984; 307: 174–5.
2. Rosenwaks Z. Donor eggs: their application in modern reproductive technologies. Fertil Steril 1987; 47: 895–909.

3. Sauer MV, Paulson RJ, Lobo RA. Pregnancy after age 50: application of oocyte donation to women after natural menopause. Lancet 1993; 341: 321–3.
4. Remohi J, Vidal A, Pellicer A. Oocyte donation in low responders to conventional ovarian stimulation for in vitro fertilization. Fertil Steril 1993; 59: 1208–15.
5. Burton G, Abdalla HI, Kirkland A, et al. The role of oocyte donation in women who are unsuccessful with in-vitro fertilization treatment. Hum Reprod 1992; 7: 1103–5.
6. Sauer MV, Kavic SM. Oocyte and embryo donation 2006: reviewing two decades of innovation and controversy. Reprod Biomed Online 2006; 12: 153–62.
7. Assisted reproductive technology in The United States: 2000 results generated from the American Society for Reproductive Medicine/Society for Assisted Reproductive Technology Registry. Fertil Steril 2004; 81: 1207–20.
8. Remohi J, Gartner B, Gallardo E, et al. Pregnancy and birth rates after oocyte donation. Fertil Steril 1997; 67: 717–23.
9. Budak E, Garrido N, Soares SR, et al. Improvements achieved in an oocyte donation program over a 10-year period: sequential increase in implantation and pregnancy rates and decrease in high-order multiple pregnancies. Fertil Steril 2007; 88: 342–9.
10. Paulson RJ, Hatch IE, Lobo RA, et al. Cumulative conception and live birth rates after oocyte donation: implications regarding endometrial receptivity. Hum Reprod 1997; 12: 835–9.
11. Noyes N, Hampton BS, Berkeley A, et al. Factors useful in predicting the success of oocyte donation: a 3-year retrospective analysis. Fertil Steril 2001; 76: 92–7.
12. Soares SR, Troncoso C, Bosch E, et al. Age and uterine receptiveness: predicting the outcome of oocyte donation cycles. J Clin Endocrinol Metab 2005; 90: 4399–404.
13. Borini A, Bianchi L, Violini F, et al. Oocyte donation program: pregnancy and implantation rates in women of different ages sharing oocytes from single donor. Fertil Steril 1996; 65: 94–7.
14. Remohi J, Gutierrez A, Cano F, et al. Long oestradiol replacement in an oocyte donation programme. Hum Reprod 1995; 10: 1387–91.
15. Cobo A, Domingo J, Pérez S, et al. Vitrification, an effective new approach to oocyte banking in healthy women, could be applied in cancer patients to preserve their fertility. Clin Trans Oncol 2008; 10: 268–73.
16. Homburg R, van der Veen F, Silber SJ. Oocyte vitrification—women's emancipation set in stone. Fertil Steril 2009; 91: 1319–20.
17. Boggio A. Italy enacts new law on medically assisted reproduction. Hum Reprod 2005; 20: 1153–7.
18. Borini A, Lagalla C, Bonu MA, et al. Cumulative pregnancy rates resulting from the use of fresh and frozen oocytes: 7 years' experience. Reprod Biomed Online 2006; 12: 481–6.
19. Cobo A, Bellver J, Domingo J, et al. New options in assisted reproduction technology: the Cryotop method of oocyte vitrification. Case reports. Reprod Biomed Online 2008; 17: 68–72.
20. Vajta G, Nagy ZP. Are programmable freezers still needed in the embryo laboratory? Review on vitrification. Reprod Biomed Online 2006; 12: 779–96.
21. Woods EJ, Benson JD, Agca Y, et al. Fundamental cryobiology of reproductive cells and tissues. Cryobiology 2004; 48: 146–56.
22. Mazur P. Principles of cryobiology. In: Fuller B, Lane M, Benson E, eds. Life in the Frozen State. New York: CRC Press, 2004; 4–55.
23. Watson PF, Morris GJ. Cold shock injury in animal cells. Symp Soc Exp Biol 1987; 41: 311–40.
24. Pickering SJ, Braude PR, Johnson MH, et al. Transient cooling to room temperature can cause irreversible disruption of the meiotic spindle in the human oocyte. Fertil Steril 1990; 54: 102–8.
25. Ghetler Y, Yavin S, Shalgi R, et al. The effect of chilling on membrane lipid phase transition in human oocytes and zygotes. Hum Reprod 2005; 20: 3385–9.
26. Liebermann J, Dietl J, Vanderzwalmen P, et al. Recent developments in human oocyte, embryo and blastocyst vitrification: where are we now? Reprod Biomed Online 2003; 7: 623–33.
27. Vajta G, Kuwayama M. Improving cryopreservation systems. Theriogenology 2006; 65: 236–44.
28. Kuwayama M, Vajta G, Cobo A. Vitrification of oocytes: general considerations and the use of the Cryotop method. In: Tucker MJ, Liebermann J, eds. Vitrification in Assisted Reproduction: A User's Manual and Trouble-Shooting Guide. London: Informa Healthcare, 2007: 119–28.
29. Mazur P. Freezing of living cells: mechanisms and implications. Am J Physiol 1984; 247: C125–42.
30. Fuller B, Paynter S. Fundamentals of cryobiology in reproductive medicine. Reprod Biomed Online 2004; 9: 680–91.
31. Arav A. Vitrification of oocyte and embryos. In: Lauria A, Gandolfi F, eds. New Trends in Embryo Transfer. Cambridge, England: Portland Press, 1992: 255–64.
32. Martino A, Songsasen N, Leibo SP. Development into blastocysts of bovine oocytes cryopreserved by ultra-rapid cooling. Biol Reprod 1996; 54: 1059–69.

33. Liebermann J, Nawroth F, Isachenko V, et al. Potential importance of vitrification in reproductive medicine. Biol Reprod 2002; 67: 1671–80.
34. Lane M, Gardner DK. Vitrification of mouse oocytes using a nylon loop. Mol Reprod Dev 2001; 58: 342–7.
35. Matsumoto H, Jiang JY, Tanaka T, et al. Vitrification of large quantities of immature bovine oocytes using nylon mesh. Cryobiology 2001; 42: 139–44.
36. Papis K, Shimizu M, Izaike Y. Factors affecting the survivability of bovine oocytes vitrified in droplets. Theriogenology 2000; 54: 651–8.
37. Vajta G, Holm P, Kuwayama M, et al. Open Pulled Straw (OPS) vitrification: a new way to reduce cryoinjuries of bovine ova and embryos. Mol Reprod Dev 1998; 51: 53–8.
38. Kuwayama M, Vajta G, Kato O, et al. Highly efficient vitrification method for cryopreservation of human oocytes. Reprod Biomed Online 2005; 11: 300–8.
39. Kuwayama M. Highly efficient vitrification for cryopreservation of human oocytes and embryos: the Cryotop method. Theriogenology 2007; 67: 73–80.
40. van Uem JF, Siebzehnrubl ER, Schuh B, et al. Birth after cryopreservation of unfertilized oocytes. Lancet 1987; 1: 752–3.
41. Gook DA, Osborn SM, Bourne H, et al. Fertilization of human oocytes following cryopreservation: normal karyotypes and absence of stray chromosomes. Hum Reprod 1994; 9: 684–91.
42. Kazem R, Thompson LA, Srikantharajah A, et al. Cryopreservation of human oocytes and fertilization by two techniques: in-vitro fertilization and intracytoplasmic sperm injection. Hum Reprod 1995; 10: 2650–4.
43. Porcu E, Fabbri R, Seracchioli R, et al. Birth of a healthy female after intracytoplasmic sperm injection of cryopreserved human oocytes. Fertil Steril 1997; 68: 724–6.
44. Nawroth F, Kissing K. Pregnancy after intracytoplasmatic sperm injection (ICSI) of cryopreserved human oocytes. Acta Obstet Gynecol Scand 1998; 77: 462–3.
45. Young E, Kenny A, Puigdomenech E, et al. Triplet pregnancy after intracytoplasmic sperm injection of cryopreserved oocytes: case report. Fertil Steril 1998; 70: 360–1.
46. Yang DS, Blohm PL, Winslow KL, et al. Oocyte donation using cryopreserved donor oocytes. Fertil Steril 2002; 78(Suppl): S14 (Abstract).
47. Allan J. Re: Case report: pregnancy from intracytoplasmic injection of a frozen-thawed oocyte. Aust N Z J Obstet Gynaecol 2004; 44: 588.
48. Notrica J, Divita A, Neuspiller F, et al. Healthy girl born after cryopreservation of gametes and ICSI in a patient with seminoma. Reprod Biomed Online 2004; 9: 620–2.
49. Borini A, Sciajno R, Bianchi V, et al. Clinical outcome of oocyte cryopreservation after slow cooling with a protocol utilizing a high sucrose concentration. Hum Reprod 2006; 21: 512–17.
50. Levi Setti PE, Albani E, Novara PV, et al. Cryopreservation of supernumerary oocytes in IVF/ICSI cycles. Hum Reprod 2006; 21: 370–5.
51. Oktay K, Cil AP, Bang H. Efficiency of oocyte cryopreservation: a meta-analysis. Fertil Steril 2006; 86: 70–80.
52. Fabbri R, Porcu E, Marsella T, et al. Human oocyte cryopreservation: new perspectives regarding oocyte survival. Hum Reprod 2001; 16: 411–16.
53. Stachecki JJ, Cohen J, Willadsen SM. Cryopreservation of unfertilized mouse oocytes: the effect of replacing sodium with choline in the freezing medium. Cryobiology 1998; 37: 346–54.
54. Eroglu A, Toner M, Toth TL. Beneficial effect of microinjected trehalose on the cryosurvival of human oocytes. Fertil Steril 2002; 77: 152–8.
55. Bianchi V, Coticchio G, Distratis V, et al. Differential sucrose concentration during dehydration (0.2 mol/L) and rehydration (0.3 mol/L) increases the implantation rate of frozen human oocytes. Reprod Biomed Online 2007; 14: 64–71.
56. Borini A, Bonu MA, Coticchio G, et al. Pregnancies and births after oocyte cryopreservation. Fertil Steril 2004; 82: 601–5.
57. Chen CK, Wang CW, Tsai WJ, et al. Evaluation of meiotic spindles in thawed oocytes after vitrification using polarized light microscopy. Fertil Steril 2004; 82: 666–72.
58. De Santis L, Cino I, Rabellotti E, et al. Oocyte cryopreservation: clinical outcome of slow-cooling protocols differing in sucrose concentration. Reprod Biomed Online 2007; 14: 57–63.
59. La Sala GB, Nicoli A, Villani MT, et al. Outcome of 518 salvage oocyte-cryopreservation cycles performed as a routine procedure in an in vitro fertilization program. Fertil Steril 2006; 86: 1423–7.
60. Boldt J, Cline D, McLaughlin D. Human oocyte cryopreservation as an adjunct to IVF-embryo transfer cycles. Hum Reprod 2003; 18: 1250–5.
61. Stachecki JJ, Cohen J, Schimmel T, et al. Fetal development of mouse oocytes and zygotes cryopreserved in a nonconventional freezing medium. Cryobiology 2002; 44: 5–13.

62. Petracco A, Azambuja R, Okada L, et al. Comparison of embryo quality between sibling embryos originating from frozen or fresh oocytes. Reprod Biomed Online 2006; 13: 497–503.
63. Borini A, Bianchi V, Bonu MA, et al. Evidence-based clinical outcome of oocyte slow cooling. Reprod Biomed Online 2007; 15: 175–81.
64. Yoon TK, Chung HM, Lim JM, et al. Pregnancy and delivery of healthy infants developed from vitrified oocytes in a stimulated in vitro fertilization-embryo transfer program. Fertil Steril 2000; 74: 180–1.
65. Yoon TK, Kim TJ, Park SE, et al. Live births after vitrification of oocytes in a stimulated in vitro fertilization-embryo transfer program. Fertil Steril 2003; 79: 1323–6.
66. Kuleshova L, Gianaroli L, Magli C, et al. Birth following vitrification of a small number of human oocytes: case report. Hum Reprod 1999; 14: 3077–9.
67. Selman H, Angelini A, Barnocchi N, et al. Ongoing pregnancies after vitrification of human oocytes using a combined solution of ethylene glycol and dimethyl sulfoxide. Fertil Steril 2006; 86: 997–1000.
68. Kuwayama M, Vajta G, Ieda S, et al. Comparison of open and closed methods for vitrification of human embryos and the elimination of potential contamination. Reprod Biomed Online 2005; 11: 608–14.
69. Katayama KP, Stehlik J, Kuwayama M, et al. High survival rate of vitrified human oocytes results in clinical pregnancy. Fertil Steril 2003; 80: 223–4.
70. Ruvalcaba L, Marínez R, Cuneo S, et al. Improving donor programs with an oocyte bank using vitrification. Fertil Steril 2005; 84(Suppl 1): S70.
71. Lucena E, Bernal DP, Lucena C, et al. Successful ongoing pregnancies after vitrification of oocytes. Fertil Steril 2006; 85: 108–11.
72. Antinori M, Licata E, Dani G, et al. Cryotop vitrification of human oocytes results in high survival rate and healthy deliveries. Reprod Biomed Online 2007; 14: 72–9.
73. Cobo A, Kuwayama M, Perez S, et al. Comparison of concomitant outcome achieved with fresh and cryopreserved donor oocytes vitrified by the Cryotop method. Fertil Steril 2008; 89: 1657–64.

24 Health of Children Born from Cryopreserved Oocytes and Embryos

Audrey Soo and Alastair Sutcliffe

General and Adolescent Unit, Institute of Child Health, University College London, London, U.K.

INTRODUCTION

Assisted reproductive technologies (ART) can be regarded as a means to an end, where the goal is to produce healthy offspring. In order to fully measure the success of this process, it is not only important to ensure that the pregnancy progresses and results in live birth, but also that the health and well-being of the child are not adversely affected by the techniques themselves. Specific measurements of this latter aspect have been discussed extensively in the literature in two main ways. First, the proportion of children who have congenital abnormalities as a possible consequence of a particular type of ART technique. Second, the postnatal development of the child and whether it is comparable with their peers. This includes measurements of growth, neurodevelopment, and psychosocial well-being. Put in this way, it may seem as if the safety and efficacy of each type of ART technique can be easily determined. However, there are several issues that confound this seemingly straightforward process.

This chapter is divided into two main sections. The first will deal generally with what is known about the health of children born through ART. This will include a brief history of ART, a discussion on the difficulties faced by studies of ART children as well as possible confounding factors that need to be taken into consideration. In the second part of this chapter, we will be focusing on a particular type of ART technique, specifically cryopreservation, where embryos or oocytes are frozen and subsequently thawed for later use. Studies designed to examine the health of children conceived from cryopreserved oocytes and embryos will be reviewed. The chapter ends with some conclusions and suggestions for future work in this field.

ASSISTED REPRODUCTIVE TECHNOLOGIES: AN OVERVIEW

In this section, a brief history of ART will be outlined. The report of the first human child born after ART, specifically in vitro fertilization (IVF), was published in 1978 (1). Thirty years on, it is estimated that there are now eight million IVF babies that have been born worldwide (2). The success of that first pregnancy was built on the foundation of knowledge laid down by previous work, including the discovery of the possibility of oocyte and embryo culture in animals at the end of the 19th century as well as an understanding of the developmental stages of mammalian oocytes (3,4).

In 1934, Pincus and Enzmann (5) reported the first live birth in mammals following IVF. These scientists mixed oocytes and sperms in vitro and transferred them into the fallopian tubes of rabbits. It is, however, possible (and in fact probable) that the oocytes were fertilized in the fallopian tubes as opposed to being fertilized in vitro (6). Therefore, the first successful mammalian live birth following IVF can actually be attributed to Chang (7). Nonetheless, Gregory Pincus and his work partly influenced John Rock to decide that the time had come to attempt IVF of human oocytes (6). Menken and Rock (8) retrieved more than 800 human oocytes and exposed 138 of them to spermatozoa. It was thought that cleavages were observed in three oocytes but none was transferred back to the uterus.

Professor Robert Edwards, one of the IVF pioneers, first started working on human IVF in 1966 (9). He was involved in the first preimplantation genetic diagnosis (PGD) conducted in rabbits in 1968 (10). The year 1971 saw the first published report of human blastocysts seen in vitro (11). Edwards later teamed up with Steptoe, whose expertise and work in laparoscopy would prove invaluable. En route to the milestone achievement in 1978, ovarian stimulation was initially tried but was abandoned for natural cycle (2). Urinary tests were used to detect the luteinizing hormone surge. After several failed pregnancies, Louise Brown (the first IVF baby) was eventually born close to midnight on July 25, 1978 (1).

There are several other key developments in ART that are worthy of mention. In 1972, two independent groups were able to achieve live births following the cryopreservation of mice embryos (12,13). Whittingham was also the first to demonstrate the ability for cryopreserved oocytes to undergo IVF and progress to a live birth (14). These achievements would pave the way for human cryopreservation in assisted reproduction in the later years. (For more information on human cryopreservation, please refer to the section "Introduction to Human Cryopresevation.")

Although PGD was first conducted in 1968 (10), it was not until after human IVF was established and the development of polymerase chain reaction that PGD was conducted in humans (15). The first live birth of a child who had PGD done was reported in 1992 (16). In the same year, the first successful human pregnancies following intracytoplasmic sperm injection (ICSI) were reported (17).

Health of Children Born After ART

Before discussing the health of children conceived from cryopreservation, one needs to consider more generally the health of children born after ART compared to spontaneously conceived children. This is particularly relevant as scientific advances are allowing more and more possible combinations of ART techniques. Take for example, the report of the first birth of a healthy child after the transfer of cryopreserved embryos generated by ICSI with cryopreserved testicular spermatozoa into cryopreserved human oocytes (18). This example alone involves six (including IVF) additional procedures experienced early on in life compared to a spontaneously conceived child. Although it is important to evaluate any potential long-term effect of each individual procedure on the child, one needs to take a further step back to examine if merely having any form of assisted reproduction to begin with has associated differences when compared with spontaneously conceived controls. What kind of study design should be used to evaluate possible health differences between ART children and non-ART children? Once that has been addressed, the next question that then begs to be answered is how does one evaluate the safety of each of the procedures on the subsequent development of the child?

The main difficulties faced by most studies examining the health of ART children are the lack of power and the inability to find suitable controls. Ideally, a control group would be virtually identical to the ART group, with the exception of the mode of conception. It would then be possible to safely conclude that any differences found in ART children can be attributed to the assisted reproductive techniques themselves. Although it is extremely difficult to find a perfect control group, many good studies have attempted to match the exposed and control groups as closely as possible, while adjusting for confounding factors. Examples of poor study design include having a large percentage of the study group that was lost to follow-up and analyzing singletons with those coming from higher order birth.

A summary of findings in ART (especially IVF and ICSI) children will now follow. The greatest risk posed to those seeking ART treatment is the increased risk of multiple pregnancies (19). A popular practice in the past has been to transfer a number of embryos in order to increase the chances of a successful implantation. This culminated in a large proportion of high order (and hence, high risk) pregnancies. Multiple pregnancies are associated with increased risk of preterm labor, low birthweight, and intrauterine growth retardation (19). Fortunately, there has been growing recognition that single–embryo transfer (SET) could reduce the number of multiple gestations without significantly affecting pregnancy rates (20). This is particularly true when SET is combined with cryopreservation, demonstrating just one of the many advantages of this particular technique.

However, multiple pregnancies are only part of the problem as, after having adjusted for this possible confounder, meta-analyses (21–23) have shown that singletons born after IVF, ICSI, or gamete intrafallopian transfer have an increased risk of having a low birthweight and are frequently born small for gestational age (19). In addition, it has been found that there is an increased risk of preterm delivery and low birthweight associated with the time-to-pregnancy interval (24,25). This is an indication that these outcome measure differences could possibly be related more to the underlying subfertility than to the specific ART that is being examined.

Another risk that is frequently discussed in literature is the risk of major malformation in children after ART. "Is there an increased risk of malformations after assisted reproductive technologies?" is both the title of a paper (26) and an important question when it comes to evaluating the health of ART children. Studies of this type face important difficulties including differing definitions of a major malformation. In addition, in some studies, the criteria and methods of assessment differ between the study group and the control group, particularly when it comes to retrospective studies comparing with data from a registry. At present, meta-analyses (27–29) show an approximately 30% increase in risk of major malformations after IVF or ICSI compared with spontaneously conceived controls. A large, prospective study found similar values when comparing ICSI children with spontaneously conceived children (30). No significant differences have been found when comparing between IVF and ICSI groups (29,31). This suggests that the invasiveness of the particular procedure chosen for infertility treatment does not significantly impact the risk of major malformation (26). Nonetheless, the question that still needs to be answered is why there is an increased risk of major malformation between the IVF or ICSI group compared with the control group. Theoretical reasons that have been suggested include other steps involved in assisted reproduction and genetic factors that are related to subfertility (26). Nonetheless, a relative increased risk of 1.3 times is unsurprising in the context that couples needing ART are skewed in a genetic sense being less healthy/fertile than naturally conceiving couples.

There is a special form of malformation known as imprinting disorders. Imprinting disorders are a result of changes in gene function that are inherited in a non-random, parent-of-origin specific manner (19). Examples of imprinting disorders that have been implicated among children conceived via IVF or ICSI include Angelman's syndrome (AS) and Beckwith-Wiedemann syndrome (BWS). In order to study the validity of these associations more closely, Sutcliffe and colleagues (32) conducted a case series study in the United Kingdom. The authors contacted families with four types of imprinting disorders: AS, BWS, Prader-Willi syndrome, and transient neonatal diabetes mellitus. These families were contacted to obtain what was called a "conception history." An association was found only between ART and BWS (32). This association was also found in other studies (33,34). However, the absolute risk of an ART child having an imprinting disorder is still relatively small (32). On the whole, this specific type of disorder requires further studies, in particular to confirm the reasons for the association.

When it comes to neurodevelopment, numerous studies have been undertaken to explore the development of children born after different types of ART. Examples of these include studies examining the outcome after IVF (35), ICSI (36,37), and cryopreservation (please refer to the sections below). On the whole, if a singleton ART child is born at full term and is not of low birthweight, the child faces similar risks of having developmental difficulties as their spontaneously conceived peers (19). In terms of neurological disorders, a large multi-center European study examined children born after IVF, ICSI, and spontaneous conception and found no difference between these groups (38). Children who were born very prematurely (below 32 weeks gestation), and hence at high risk of cerebral palsy, were not included in the study (38). In several other large registry-based studies, children born after IVF were found to face an increased risk of having cerebral palsy compared with naturally conceived children (39–41). This might be due to prematurity, low birthweight, or twin pregnancies (41). In singleton IVF children, the increased risk of cerebral palsy might be due to early intrauterine death of a twin counterpart (42).

Although most studies of the health of ART children have focused on physical attributes, it is also important to consider the psychosocial well-being of the child. In the European Study of Assisted Reproduction Families (43), children conceived through IVF and donor insemination, children who were adopted, and those who were spontaneously conceived were compared. Information was collected via interviews and questionnaires completed by parents, teachers, and the children themselves. The results were similar between groups and there is no evidence to suggest that ART has a negative effect on either the parent-child relationship or on the self-esteem of the children, even when comparing between countries. A later follow-up study was completed as the children transited into early adolescence, and again no differences were found between the groups on any measures of psychosocial adjustment (44). Similar results were found in another large European study looking at IVF, ICSI, and naturally conceived children (45).

Points for Consideration When Counselling Couples Seeking Treatment for Subfertility

1. The biggest risk for ART is that of multiple births.
2. The abortion rate is about 20–34% higher compared with couples spontaneously conceiving. This increase seems to be linked to known anamnestic differences such as older maternal age, endocrine disorders (e.g., thyroid disease, polycystic ovary syndrome), or organic abnormalities (e.g., tubal disease, uterine malformations). Additional independent risk factors include subfertility, the invasiveness of the treatment procedure (i.e., ICSI is more invasive than IVF), and the degree of ovarian stimulation.
3. The rate of numerical sex-chromosome abnormalities is enhanced in pregnancies after ICSI, analyzed by invasive parental diagnosis (0.63% vs. 0.2%). This growth might be caused, at least in part, by a high proportion of chromosomally abnormal sperm in these couples.
4. Risk of preeclampsia is increased by 55%, risk of preterm delivery by about twofold, and risk of placenta previa by about threefold. There is a greater risk of stillbirth (2.55-fold). Babies are more usually born with a low or very low birthweight (1.70–1.77-fold and 2.70–3.00-fold, respectively) and have a higher risk of being small for gestational age (1.40–1.60-fold).
5. Risk of major malformations is increased by about 30% (1.3 times normal value), which seems to be technique independent. The specific risk of two rare imprinting disorders—Angelman's syndrome and Beckwith–Wiedemann syndrome—might also be raised.
6. Overall, there is a higher risk of cerebral palsy in children born after ART, partly because of the risk associated with increased premature birth and partly because some twin pregnancies undergo early in-utero loss, and this event is associated with an enhanced risk of cerebral palsy. This risk is reduced if SET takes place.
7. Neurodevelopmentally mature-term babies born after ART progress healthily in relation to naturally conceived children.
8. No concerns exist about the family relationships and psychosocial issues after ART conception.

So far, little evidence exists about other health problems in children born after ART.

Abbreviations: ART, assisted reproductive technologies; ICSI, intracytoplasmic sperm injection; IVF, in vitro fertilization; SET, single–embryo transfer.

Source: From Ref. 19.

In this section, an attempt has been made to discuss and summarize the findings of studies examining the health of children born after ART. After discounting multiple births as the largest risk factor, it remains that there are still health risks for singleton children conceived through ART. This information is invaluable when it comes to counselling couples seeking infertility treatment who have concerns about the health of their prospective children (see Panel). It is also important to recognize that several risks posed to children born after ART might not be due to the techniques themselves, but instead are more related to the underlying causes of subfertility.

INTRODUCTION TO HUMAN CRYOPRESERVATION

Cryopreservation is a technique that allows the cooling of biological entities to subzero temperatures. Cryopreservation has contributed greatly to ART by allowing embryos or gametes to be stored and thawed for later use (46). Advantages of this include the ability to reduce the rate of multiple pregnancies (by reducing the number of fresh embryos transferred per cycle), postponement of embryo transfer for various reasons (particularly to reduce the risk of ovarian hyperstimulation syndrome (47) and to preserve the potential for future childbearing among women who are about to undergo chemotherapy for cancer), and to allow repeat virus (like HIV) testing before donation (48). As mentioned previously, two independent groups were able to achieve live births of mice from cryopreserved embryos in 1972 (12,13). However, it was not until 1983 when the first human pregnancy following the transfer of a cryopreserved embryo was reported (46).

Cryopreservation of human oocytes, on the other hand, has faced more technical challenges in the past compared with cryopreservation of spermatozoa or embryos. There is great interest in developing this technique of cryopreserving oocytes, because it is void of some of the legal and ethical issues faced by the cryopreservation of embryos. The first report of human pregnancy from cryopreserved oocytes was in 1986 (49), but until the relatively recent development of ICSI, cryopreservation of oocytes was considered a low efficacy technique. Since then, it has been found that a similar pregnancy rate can be achieved by either cryopreserved oocytes or embryos (49).

As more and more children are conceived from cryopreserved embryos and/or oocytes, it is important to conduct long-term, follow-up studies among this cohort of children. The obstetric and perinatal outcomes as well as the postnatal development of children need to be compared with an appropriate control group to ensure that the cryopreservation process has no adverse long-term effects. Studies that have been conducted in this field will be discussed below.

Health of Children Born from Cryopreserved Oocytes (Including Vitrified Oocytes)

Although the first pregnancy from cryopreserved oocytes was reported in 1986 (49), there were only a handful of successful pregnancies conceived through this method for almost a decade. However, with the development of ICSI, the clinical efficacy of conceiving from cryopreserved oocytes increased substantially. This resulted in many more successful pregnancies and births, starting with the first birth of a healthy female in 1997 (50). This same group in Italy reported many subsequent pregnancies and births of healthy babies conceived from cryopreserved oocytes (51–56). Other groups have also been able to achieve successful pregnancies and births (49,57–65).

There is also a specific type of rapid freezing technique known as vitrification that is gaining popularity. The highly concentrated solution of cryoprotectants freezes without the formation of ice crystals and is in direct contact with the liquid nitrogen (66). A recent international study has reported on the obstetric and perinatal outcomes of a cohort of children born from oocytes that have been cryopreserved by vitrification (67).Of the 200 infants studied, 17.7% of singletons and 81% of those from multiple pregnancies were found to be of low or very low birthweight. The incidence of congenital anomalies in this cohort was 2.5%. No control group was used, and the findings were compared with previously reported rates in spontaneously conceived controls and IVF controls. The authors conclude from these preliminary findings that children born from vitrified oocytes are not associated with increased risk of adverse obstetric or perinatal outcomes (67).

On the whole, observations indicate that children conceived from cryopreserved oocytes, albeit small in numbers, are usually born healthy. However, more long-term, large follow-up studies with carefully matched controls will need to be conducted. As it stands, there is nothing to suggest that the technique itself has any adverse effects on the children conceived from cryopreserved oocytes.

Health of Children Born from Cryopreserved Embryos

Only few studies specifically designed to examine the health of babies born from cryopreserved embryos have been carried out, and even fewer still are the number of studies following the health of these children as they progress through life. In this section, we will be examining these studies (Table 1) in chronological order from when they were first published. The strengths and weaknesses of each study will be discussed.

In 1994, Wada and colleagues (68) conducted a retrospective study looking at the perinatal outcome of 283 babies conceived through IVF of cryopreserved embryos. The outcome measures used for comparison were multiplicity of birth, presence of major congenital malformations, gestational age, and birthweight at delivery. The control group consisted of babies conceived through IVF of fresh embryos. The use of this control group (as opposed to using spontaneously conceived children) is a strength of this study as it minimizes the risk of the IVF treatment itself being a confounding factor. Also, the authors analyzed the results after adjusting for multiplicity of births. It was found that the cryopreserved group had a similar average

Table 1 Studies Designed to Examine the Health of Children Born from Cryopreserved Embryos

Authors	Year of publication	Study design	Main findings
Wada et al. (68)	1994	Retrospective study. SG: 283 babies born after IVF of cryopreserved embryos (from 232 consecutive births). CG: 961 babies born after IVF of fresh embryos. OM: Multiplicity of birth, presence of major congenital malformations, gestational age, and birthweight at delivery.	Incidence of twins and triplets were similar between the groups. Birth characteristics of babies conceived from cryopreserved embryos were similar to those conceived from fresh embryos. There were fewer major malformations in the cryopreserved group.
Sutcliffe et al. (69)	1995	First prospective study following the postnatal growth and health of children from cryopreserved embryos. SG: 91 children conceived from cryopreserved embryos. CG: 93 spontaneously conceived children (matched for social class and age). OM: Birth characteristics, detailed clinical assessment including a medical examination and a developmental assessment using the Griffith's scales of mental development.	Children born from cryopreserved embryos had a lower mean birthweight and mean gestational age compared to controls. On the whole, the development of the SG was similar to CG.
Sutcliffe et al. (70)	1995	Using a previously described cohort (79), the major and minor malformations found among the children were further analyzed. SG: 91 children conceived from cryopreserved embryos. CG: 93 spontaneously conceived children (matched for social class and age). OM: Incidence of minor and major malformation and development assessed using Griffith's scales of mental development.	The rates of minor and major malformations were similar in both groups (matched for age, sex, and social class). The development of the children was not adversely affected by cryopreservation.
Olivennes et al. (48)	1996	Prospective study following a total cohort of children conceived between 1986 and 1994 from cryopreserved embryos at one hospital. SG: 82 children conceived from cryopreserved embryos aged 1–9 yr(4.3% lost to follow-up). CG: None. OM: Children's development.	The prematurity rates for singletons and twins were 14.7% and 85.7% respectively, with half of these deliveries occurring in the 36th week of gestation. 8% of singletons and 28.6% of twins were small for gestational age. The total malformation rate was 3.4%. The medical and surgical illness, principal acquisitions for children aged below 5 yr as well as scholastic performance for school-aged children did not show pathological features.

(Continued)

Table 1 Studies Designed to Examine the Health of Children Born from Cryopreserved Embryos (*Continued*)

Authors	Year of publication	Study design	Main findings
Wennerholm et al.(71)	1997	Retrospective study evaluating the obstetric and neonatal outcome of children conceived from cryopreserved embryos. SG: 267 infants conceived from cryopreserved embryos. CG: one group of children born after IVF with fresh embryos and one group of children born after spontaneous conception (both groups matched to SG for maternal age, parity, plurality of birth, and date of delivery).	Gestation age at delivery, birthweight, incidence of malformations and perinatal mortality in SG were similar to that of the two control groups. There were a significantly greater number of singletons from the SG who were delivered via cesarean section compared with the spontaneously conceived group.
Wennerholm et al. (72)	1998	Prospective study investigating the postnatal health and growth of children (up to 18 months) conceived from cryopreserved embryos. SG: 255 children from cryopreserved embryos. CG: 255 children born after IVF with fresh embryos and 252 children from spontaneous pregnancies (both groups matched for maternal age, parity, plurality of birth, and date of delivery). OM: Growth, prevalence of chronic illness, major malformations, cumulative incidence of common diseases, and development during the first 18 months.	Growth features, incidence of major malformations, and prevalence of chronic diseases were similar when comparing between groups.

Abbreviations: CG, control group; IVF, in vitro fertilization; OM, outcome measures; SG, study group.

gestational age and birthweight when compared to the control group. In addition, there was a significantly lower percentage of major congenital malformations in the cryopreserved group. However, it was not mentioned as to how this was assessed. The authors concluded that there was nothing to suggest that the cryopreservation process itself adversely affects the development of the embryo postimplantation.

In 1995, a separate group in the United Kingdom, led by Dr. Sutcliffe, conducted the first prospective study following the postnatal growth and health of children conceived from cryopreserved embryos (69). The findings from this study were reported in two publications (69,70). Ninety-one children conceived from cryopreserved embryos were enrolled in the study. The control group used consisted of 83 spontaneously conceived children. Although the control group was matched for age, sex, and social class, it was not possible to match the children by parental age, parity, multiplicity of pregnancy, history of infertility, mode of delivery, and preterm birth. This was admittedly a drawback when trying to come to firm conclusions (69). However, a great strength of this study was that the principal investigator saw all the children involved, minimizing the risk of inter-investigator bias. Also, in addition to a general medical examination, a standardized developmental assessment test (Griffith's scales of mental development) was used. Although it was found that the children from the cryopreserved group had a lower mean birthweight and mean gestational age (69), the incidence of minor and major congenital malformations was similar with that of the control group (70). For the Griffith's assessment, it was found that the children from the cryopreserved group performed on a similar functional level as the control group (69,70). Although one needs to exercise cautious optimism when viewing these results because of the above mentioned control group issue, the study does paint quite a reassuring picture with regard to the health of children conceived from cryopreserved embryos.

The findings from Sutcliffe and colleagues were largely confirmed by a French study designed to follow-up on a whole cohort of children conceived from cryopreserved embryos from one hospital (48). Olivennes et al. studied children aged one to nine years born from cryopreserved embryos. It was found that the children were of normal intelligence, with some even outperforming their peers at school, despite the increased risk of prematurity and multiple births (48). However, this study was limited by the fact that there was no control group and by using parents as the primary source of assessing development.

In 1997, Wennerholm and colleagues reported the obstetric and perinatal outcomes of a total cohort of children conceived from cryopreserved embryos compared to two control groups—a group of children born after IVF with fresh embryos and a group of spontaneously conceived children (71). It was found that the gestational age at delivery, birthweight, incidence of malformations, and the perinatal mortality were similar with that of the two control groups.

Soon after, Wennerholm et al. reported on the postnatal growth and health of children conceived from cryopreserved embryos (72). Similar control groups were used, with 255 children conceived from IVF of fresh embryos and 252 spontaneously conceived children. The main outcome measure used was growth, measured in terms of weight, length, and head circumference. This study was excellent in comparing the growth of children between groups, particularly in length, as standard growth curves were used for comparison. It was found that the growth of children from the cryopreserved group was within normal range. The secondary endpoints of chronic diseases and major malformations were found to be of similar rates to that of the control groups. However, due to the method of data collection, it is likely that only major malformations would have been picked up, and "minor handicaps, behavioral disturbances, learning difficulties, and dysfunction of attention and perception" could not be ruled out based on this study alone (72). Nonetheless, it was reasonably safe to conclude that children born after cryopreservation were not adversely affected in terms of growth and health up to the age of 18 months.

It must be mentioned that there is one caveat in drawing conclusions about congenital anomaly rates exclusively from studies like the ones mentioned above, which is that these studies are very underpowered. For example, it has been calculated that for a condition that affects 1 in 1000 births, the corresponding number of children required in each group would be 20,000 (73). What is required is a birth registry with sufficient power that applies the same specific parameters to each child. One country that has taken on this mammoth task is Sweden. In the latest Swedish registry study of children born after various types of IVF methods, it was found that children from cryopreserved embryos (n = 1055) had better outcome variables, with a lower risk of preterm birth and low Apgar score (74). In addition, with regard to congenital malformation risk, the same group reported no difference when comparing between different IVF techniques (75).

CONCLUSIONS

In conclusion, the limited number of studies that exist (described above) suggest that there is no evidence at the moment to substantiate the claim that the cryopreservation technique itself has an adverse effect on the health of the child conceived subsequently. The greatest risk from ART is still the risk of multiple pregnancies. As SET continues to gain popularity together with the increasing efficacy of cryopreservation, the rate of multiple pregnancies and their accompanying complications should continue to decline. As described previously, there are still some health risks faced by singleton children conceived through ART. The reasons for the risks are still not entirely clear but may be related to the underlying causes of subfertility. Neurodevelopmentally, children born at full term through ART (including those from cryopreserved embryos) seem to progress on a similar level as their spontaneously conceived peers. There are also no recorded adverse effects on the psychosocial well-being of ART children to date. Now, at the 30-year point after the original IVF birth (76–78), there is substantial knowledge about the health of these children but less is known about adults as this first generation of ART children (79,80) grow up.

Much more future work is required in this field, as there are still a great number of unknowns about the health of babies conceived after cryopreservation as well as the health of these children as they progress into adulthood. An interesting topic to explore is the rate of subfertility among children born from cryopreservation (and indeed among ART children in general). At present, only a limited number of studies have been conducted looking at the obstetric and perinatal outcome of children born from cryopreserved embryos and even fewer still are the number of studies following up on the health of these children beyond the neonatal period. A considerable number of studies suffer from the common difficulties in studies of this nature, such as insufficient power and the lack of an appropriate control group. To date, no case-controlled study investigating the health of children conceived from cryopreserved oocytes as they grow up has been conducted. Large, high-quality, prospective studies with carefully matched controls need to be conducted to follow-up on the children born from cryopreservation. Another possible solution is to establish large registries, such as those available in Sweden. These types of studies would allow one to conclude once and for all on any long-term effects on the health of children conceived from cryopreserved oocytes or embryos.

REFERENCES

1. Steptoe PC, Edwards RG. Birth after reimplantation of a human embryo. Lancet 1978; 2: 366.
2. Hartshorne G. Thirty years of IVF. Human Fertility 2008; 11: 77–83.
3. Pincus G, Neumann EV. The growth, maturation and artesian of ovarian eggs in the rabbit. J Morphol 1937; 61: 351–83.
4. Hertig AT, Rock J, Adams EC, et al. Thirty-four fertilized human ova, good, bad and indifferent, recovered from 210 women of known fertility—a study of biologic wastage in early human pregnancy. Pediatrics 1959; 23: 202–11.
5. Pincus G, Enzmann EV. Can mammalian eggs undergo normal development in vitro? Proc Natl Acad Sci U S A 1934; 20: 121–2.
6. Cohen J, Trounson A, Dawson K, et al. The early days of IVF outside the UK. Hum Reprod Update 2005; 11: 439–59.
7. Chang MC. Fertilization of rabbit ova in vitro. Nature 1959; 184: 466–7.
8. Menken I, Rock J. In vitro fertilization and cleavage of human ovarian oocytes. Am J Obstet Gynecol 1948; 55: 440–51.
9. Edwards RG, Donahue RP, Baramki TA, et al. Preliminary attempts to fertilize human oocytes matured in vitro. Am J Obstet Gynecol 1966; 96: 192–200.
10. Gardner RL, Edwards RG. Control of sex ratio at full term in rabbit by transferring sexed blastocysts. Nature 1968; 218: 346–8.
11. Steptoe PC, Edwards RG, Purdy JM. Human blastocysts grown in culture. Nature 1971; 229: 132–3.
12. Whittingham DG, Leibo SP, Mazur P. Survival of mouse embryos frozen to –196° and –269°C. Science 1972; 178: 411–14.
13. Wilmut I. Effect of cooling rate, warming rate, cryoprotective agent and stage of development on survival of mouse embryos during freezing and thawing. Life Sci 1972; 11: 1071–9.
14. Whittingham DG. Fertilization in vitro and development to term of unfertilized mouse oocytes previously stored at –196°C. J Reprod Fertil 1977; 49: 89–94.
15. Coutelle C, Williams C, Handyside A, et al. Genetic analysis of DNA from single human oocytes—a model for preimplantation diagnosis of cystic fibrosis. BMJ 1989; 299: 22–24.
16. Handyside AH, Lesko JG, Tarin JJ, et al. Birth of a normal girl after invitro fertilization and preimplantation diagnostic testing for cystic fibrosis. N Engl J Med 1992; 327: 905–9.
17. Palermo G, Joris H, Devroey P, et al. Pregnancies after intracytoplasmic injection of single spermatozoon into an oocyte. Lancet 1992; 340: 17–18.
18. Levi Setti PE, Albani E, Novara PV, et al. Normal birth after transfer of cryopreserved human embryos generated by microinjection of cryopreserved testicular spermatozoa into cryopreserved human oocytes. Fertil Steril 2005; 83: 1041.
19. Sutcliffe AG, Ludwig M. Outcome of assisted reproduction. Lancet 2007; 370: 351–9.
20. Thurin A, Hausken J, Hillensjo T, et al. Elective single-embryo transfer versus double-embryo transfer in in vitro fertilization. N Engl J Med 2004; 351: 2392–402.

21. Helmerhorst FM, Perquin DAM, Donker D, et al. Perinatal outcome of singletons and twins after assisted conception: a systematic review of controlled studies. BMJ 2004; 328: 261–4B.
22. Jackson RA, Gibson KA, Wu YW, et al. Perinatal outcomes in singletons following in vitro fertilization: a meta-analysis. Obstet Gynecol 2004; 103: 551–63.
23. McGovern PG, Llorens AJ, Skurnick JH, et al. Increased risk of preterm birth in singleton pregnancies resulting from in vitro fertilization-embryo transfer or gamete intrafallopian transfer: a meta-analysis. Fertil Steril 2004; 82: 1514–20.
24. Ghazi HA, Spielberger C, Kallen B. Delivery outcome after infertility—a registry study. Fertil Steril 1991; 55: 726–32.
25. Henriksen TB, Baird DD, Olsen J, et al. Time to pregnancy and preterm delivery. Obstet Gynecol 1997; 89: 594–9.
26. Ludwig M. Is there an increased risk of malformations after assisted reproductive technologies? Reprod Biomed Online 2005; 10: 83–9.
27. Hansen M, Bower C, Milne E, et al. Assisted reproductive technologies and the risk of birth defects—a systematic review. Hum Reprod 2005; 20: 328–38.
28. Lie RT, Lyngstadaas A, Orstavik KH, et al. Birth defects in children conceived by ICSI compared with children conceived by other IVF-methods: a meta-analysis. Int J Epidemiol 2005; 34: 696–701.
29. Rimm AA, Katayama AC, Diaz M, et al. A meta-analysis of controlled studies comparing major malformation rates in IVF and ICSI infants with naturally conceived children. J Assist Reprod Genet 2004; 21: 437–43.
30. Katalinic A, Rosch C, Ludwig M. Pregnancy course and outcome after intracytoplasmic sperm injection: a controlled, prospective cohort study. Fertil Steril 2004; 81: 1604–16.
31. Bonduelle M, Liebaers I, Deketelaere V, et al. Neonatal data on a cohort of 2889 infants born after ICSI (1991–1999) and of 2995 infants born after IVF (1983–1999). Hum Reprod 2002; 17: 671–94.
32. Sutcliffe AG, Peters CJ, Bowdin S, et al. Assisted reproductive therapies and imprinting disorders—a preliminary British survey. Hum Reprod 2006; 21: 1009–11.
33. Gicquel C, Gaston V, Mandelbaum J, et al. In vitro fertilization may increase the risk of Beckwith-Wiedemann syndrome related to the abnormal imprinting of the KCNQ1OT gene. Am J Hum Genet 2003; 72: 1338–41.
34. Maher ER, Brueton LA, Bowdin SC, et al. Beckwith-Wiedemann syndrome and assisted reproduction technology (ART). J Med Genet 2003; 40: 62–4.
35. Saunders K, Spensley J, Munro J, et al. Growth and physical outcome of children conceived by in vitro fertilization. Pediatrics 1996; 97: 688–92.
36. Bonduelle M, Joris H, Hofmans K, et al. Mental development of 201 ICSI children at 2 years of age. Lancet 1998; 351: 1553.
37. Sutcliffe AG, Taylor B, Saunders K, et al. Outcome in the second year of life after in-vitro fertilisation by intracytoplasmic sperm injection: a UK case-control study. Lancet 2001; 357: 2080–4.
38. Bonduelle M, Wennerholm UB, Loft A, et al. A multi-centre cohort study of the physical health of 5-year-old children conceived after intracytoplasmic sperm injection, in vitro fertilization and natural conception. Hum Reprod 2005; 20: 413–19.
39. Ericson A, Nygren KG, Olausson PO, et al. Hospital care utilization of infants born after IVF. Hum Reprod 2002; 17: 929–32.
40. Lidegaard O, Pinborg A, Andersen AN. Imprinting diseases and IVF: Danish National IVF cohort study. Hum Reprod 2005; 20: 950–4.
41. Stromberg B, Dahlquist G, Ericson A, et al. Neurological sequelae in children born after in-vitro fertilisation: a population-based study. Lancet 2002; 359: 461–5.
42. Pinborg A, Lidegaard O, Freiesleben NL, et al. Consequences of vanishing twins in IVF/ICSI pregnancies. Hum Reprod 2005; 20: 2821–9.
43. Golombok S, Brewaeys A, Cook R, et al. The European study of assisted reproduction families: family functioning and child development. Hum Reprod 1996; 11: 2324–31.
44. Golombok S, Brewaeys A, Giavazzi MT, et al. The European study of assisted reproduction families: the transition to adolescence. Hum Reprod 2002; 17: 830–40.
45. Barnes J, Sutcliffe AG, Kristoffersen I, et al. The influence of assisted reproduction on family functioning and children's socio-emotional development: results from a European study. Hum Reprod 2004; 19: 1480–7.
46. Trounson A, Mohr L. Human-pregnancy following cryopreservation, thawing and transfer of an 8-cell embryo. Nature 1983; 305: 707–9.
47. Wada I, Matson PL, Troup SA, et al. Does elective cryopreservation of all embryos from women at risk of ovarian hyperstimulation syndrome reduce the incidence of the condition. Br J Obstet Gynaecol 1993; 100: 265–9.

48. Olivennes F, Schneider Z, Remy V, et al. Perinatal outcome and follow-up of 82 children aged 1-9 years old conceived from cryopreserved embryos. Hum Reprod 1996; 11: 1565–8.
49. Chen C. Pregnancy after human oocyte cryopreservation. Lancet 1986; 1: 884–6.
50. Porcu E, Fabbri R, Seracchioli R, et al. Birth of a healthy female after intracytoplasmic sperm injection of cryopreserved human oocytes. Fertil Steril 1997; 68: 724–6.
51. Porcu E, Fabbri R, Seracchioli R, et al. Birth of six healthy children after intracytoplasmic sperm injection of cryopreserved human oocytes. Hum Reprod 1998; 13: 124.
52. Porcu E, Fabbri R, Ciotti PM, et al. Ongoing pregnancy after intracytoplasmic sperm injection of epididymal spermatozoa into cryopreserved human oocytes. J Assist Reprod Genet 1999; 16: 283–5.
53. Porcu E, Fabbri R, Petracchi S, et al. Ongoing pregnancy after intracytoplasmic injection of testicular spermatozoa into cryopreserved human oocytes. Am J Obstet Gynecol 1999; 180: 1044–5.
54. Porcu E, Fabbri R, Ciotti PM, et al. Cycles of human oocyte cryopreservation and intracytoplasmic sperm injection: results of 112 cycles. Fertil Steril 1999; 72(Suppl 1): S2.
55. Porcu E, Fabbri R, Damiano G, et al. Clinical experience and applications of oocyte cryopreservation. Mol Cell Endocrinol 2000; 169: 33–7.
56. Porcu E, Fabbri R, Ciotti P, et al. Four healthy children from frozen human oocytes and frozen human sperms. Fertil Steril 2001; 76: S76–7.
57. Antinori S, Dani G, Selman HA, et al. Pregnancy after sperm injection into cryopreserved human oocytes. Hum Reprod 1998; 13: 157–8.
58. Boldt J, Cline D, McLaughlin D. Human oocyte cryopreservation as an adjunct to IVF-embryo transfer cycles. Hum Reprod 2003; 18: 1250–5.
59. Borini A, Bonu MA, Coticchio G, et al. Pregnancies and births after oocyte cryopreservation. Fertil Steril 2004; 82: 601–5.
60. Fosas N, Marina F, Torres PJ, et al. The births of five Spanish babies from cryopreserved donated oocytes. Hum Reprod 2003; 18: 1417–21.
61. Nawroth F, Kissing K. Pregnancy after intracytoplasmatic sperm injection (ICSI) of cryopreserved human oocytes. Acta Obstet Gynecol Scand 1998; 77: 462–3.
62. Polak De Fried E, Notrica J, Rubinstein M, et al. Pregnancy after human donor oocyte cryopreservation and thawing in association with intracytoplasmic sperm injection in patients with ovarian failure. Fertil Steril 1998; 69: 555–7.
63. Quintans CJ, Donaldson MJ, Bertolino MV, et al. Birth of two babies using oocytes that were cryopreserved in a choline-based freezing medium. Hum Reprod 2002; 17: 3149–52.
64. Yang DS, Winslow KL, Blohm PM. Improved survival rate after cryopreservation of human fresh and aged unfertilized oocytes using a specially developed oocyte cryopreservation regime. Fertil Steril 1998; 70(Suppl 1): S86.
65. Young E, Kenny A, Puigdomenech E, et al. Triplet pregnancy after intracytoplasmic sperm injection of cryopreserved oocytes: case report. Fertil Steril 1998; 70: 360–1.
66. Porcu E, Venturoli S. Progress with oocyte cryopreservation. Curr Opin Obstet Gynecol 2006; 18: 273–9.
67. Chian RC, Huang JYJ, Tan SL, et al. Obstetric and perinatal outcome in 200 infants conceived from vitrified oocytes. Reprod Biomed Online 2008; 16: 608–10.
68. Wada I, Macnamee MC, Wick K, et al. Birth characteristics and perinatal outcome of babies conceived from cryopreserved embryos. Hum Reprod 1994; 9: 543–6.
69. Sutcliffe AG, Dsouza SW, Cadman J, et al. Outcome in children from cryopreserved embryos. Arch Dis Child 1995; 72: 290–3.
70. Sutcliffe AG, Dsouza SW, Cadman J, et al. Minor congenital anomalies, major congenital malformations and development in children conceived from cryopreserved embryos. Hum Reprod 1995; 10: 3332–7.
71. Wennerholm UB, Hamberger L, Nilsson L, et al. Obstetric and perinatal outcome of children conceived from cryopreserved embryos. Hum Reprod 1997; 12: 1819–25.
72. Wennerholm UB, Albertsson-Wikland K, Bergh C, et al. Postnatal growth and health in children born after cryopreservation as embryos. Lancet 1998; 351: 1085–90.
73. Sutcliffe AG. Follow-up of children conceived from cryopreserved embryos. Mol Cell Endocrinol 2000; 169: 91–3.
74. Kallen B, Finnstrom O, Nygren KG, et al. In vitro fertilization (IVF) in Sweden: infant outcome after different IVF fertilization methods. Fertil Steril 2005; 84: 611–17.
75. Kallen B, Finnstrom O, Nygren KG, et al. In vitro fertilization (IVF) in Sweden: risk for congenital malformations after different IVF methods. Birth Defects Research Part A-Clinical and Molecular Teratology 2005; 73: 162–9.

76. Pearson H. Making babies: the next 30 years. Nature 2008; 454: 260–2.
77. Deech R. 30 years: from IVF to stem cells. Nature 2008; 454: 280–1.
78. Life after SuperBabe [Editorial]. Nature 2008; 454: 253.
79. Sutcliffe AG. IVF Children: The First Generation. London: Taylor & Francis, 2002.
80. Sutcliffe AG. Health and Welfare of ART Children. London: Taylor & Francis, 2006.

25 The Ethics of Fertility Preservation

Carlo Flamigni
National Council of Bioethics, Rome, Italy

On first examination, oocyte freezing should raise no objections of an ethical nature, particularly if its aim is to improve the chances of success of assisted reproduction techniques (ARTs). In practice, however, this is not the case, and the technique is challenged on at least two grounds. The first is that, as it is always part of a medical plan for assisted conception, at least some people consider it as offending the dignity of procreation. The second argument addresses the experimental nature of the technique and the fact that, at least in its early stages, it was applied in spite of unfavorable preliminary data suggesting that the product of conception would be subject to a significant risk of malformation. There is a third point, which might open up a bioethical debate, namely that the technique offers the opportunity of preserving the fertility of individuals for a more or less distant future. Currently, however, this topic attracts little attention, particularly due to the limited efficiency of oocyte cryopreservation techniques. To gain a better understanding of the criticisms addressed against ARTs, mostly formulated by the Roman Catholic Magisterium, we need to refer back to the first debate held on this topic in 1984 by the Pontifical Academy of Science in Rome. On that occasion, half a dozen scientists specializing in this field gathered in the Vatican gardens to discuss the biological and moral problems raised by in vitro fertilization (IVF) before a wide audience of theologians. Howard Jones, who published a lengthy report on the meeting, wrote: "Towards the end of the conference it became clear that there was general agreement on declaring in vitro fertilization ethically acceptable. There was, however, one dissenter, namely Monsignor Carlo Caffarra, the President of the John Paul II Institute for Studies on Marriage and Family, appointed to that office by the Pope. He held staunchly to his position that in vitro fertilization is illegal since the resulting conception falls outside the bonds of conjugal love and does not result from the natural process of sexual intercourse." This was in fact the key issue underpinning all Catholic criticism of the new technique, based on the view that separating sexual and reproductive life is unlawful as it offends the dignity of procreation.

Monsignor Caffarra's position stirred up a hornet's nest. Not all Catholics agreed with it, and many religious hospitals, foremost among them the Obstetrics Clinic of the Catholic University of Rome, continued to use a number of ARTs such as the IVF method, restricted to clearly defined "simple cases" and intrafallopian insemination.

March 22, 1987, saw the publication of what was then called the Ratzinger Instruction, now known as the Donum Vitae (Gift of Life) (1). This lengthy document, drafted by the Congregation for the Doctrine of the Faith, set out the official position of the Catholic Church on the issue. The document reaffirmed all the traditional prohibitions and was a great disappointment to all those who were expecting some sign of openness towards the new techniques (among other things). As far as the embryo is concerned, the Donum Vitae reiterates that it should be respected "as a human person" but does not deal with the problem of when this person actually comes into being. The Instruction, moreover, prohibits any form of assisted fertilization, including insemination with the husband's semen, consistent with the doctrine concerning the bond between the meaning of the conjugal act and the goods of marriage. The doctrine states that "it is never permitted to separate these different aspects to such a degree as positively to exclude either the procreative intention or the conjugal relation." It argues that contraception deliberately deprives the conjugal act of its openness to procreation and "in this way brings about a voluntary dissociation of the ends of marriage. Artificial fertilization within the couple, in seeking a procreation which is not the fruit of a specific act of conjugal union, objectively effects an analogous separation between the goods and the meanings of marriage."

It is therefore clear that, like contraception, ART is seen to violate the principle of the inseparableness of the unitive and procreative meaning of the conjugal act. This principle is regarded as underpinning the originality of transmission of human life in matrimony, deriving from the idea that human life is sacred because it involves the creative action of God from the time of its inception, and remains forever in a special relationship with the Creator—this being its sole end.

This interpretation of the so-called natural laws opens up a conflict between the relationship that ought to exist between the ethical order and the legal system. Once again, the objections on this subject have been raised by Monsignor Carlo Caffarra who, in his address entitled "La procreazione artificiale: riflessione etico-politica" (Artificial Procreation: an Ethical-Political Reflection) delivered at a conference held in Ferrara on September 16, 1999, rejected the lawfulness of a "democratic" system of negotiation, which would consider as being right only that which the majority of people deem to be right (2). Such a legal system, argued Caffarra, would increasingly come to represent a mechanism of prior empirical regulation of different and opposing interests. If we claim that everything is negotiable, the argument goes on, either we accept that negotiation too is negotiable (and we reject the principle—democracy) or we exclude that which makes negotiation itself possible, namely the existence of any human truth and the possibility that human reason is able to grasp such a truth. To believe that no human truth exists, and that communal life should be regulated on precepts bearing no reference to absolute values, concluded Caffarra, is the premise for corrupting democracy and turning it into a dictatorship.

Leaving the naivety of this final consideration aside (it would have been more logical for Monsignor Caffarra to see anarchy as the final result of the degeneration of democracy), this argument simply reintroduces the conflict between religious ethics and secular ethics. The key difference lies in the fact that, in the former, norms are given independently of the human will, while in the latter, they are established according to that will. In the first case, they are a natural—or divine—fact, in the second case, they are a human creation of a social nature. They are actually two different paradigms, that is, two different ways of conceiving reality.

The accusation most commonly levelled at secular morality is that it inevitably leads to ethical relativism, in which everything is subjective. In actual fact, however, the secular notion of ethics to which I am referring denies subjectivism, as it relies on rational justification. Bearing in mind that rationality implies universality, it becomes clear that ethical judgment is based on a perspective that eschews subjectivism and instead presents itself as something that all rational beings can agree to. As David Hume argued, the notion of morality implies some sentiment common to all humanity, which commends the same object to general approval. Ethics, which includes rationality among its characteristics, cannot be relativistic. To say "This is right for me" is a contradiction in terms because "right" implies a universal point of view, while "for me" implies something peculiar to myself, and these two aspects are irreconcilable. To suppose that what is morally right depends on one's own individual taste is to deny the very notion of morality.

In concluding his conference address, Monsignor Caffarra returned to the lawfulness of ARTs. "The state—he argued—might tolerate a situation in which two spouses resort to homologous IVF where only one ovum is fertilized. This kind of legitimising action—he went on-while accepting a serious violation to the dignity of the embryo, since it is produced rather than generated, seems not to undermine the foundations of matrimony and the family, nor to violate the fundamental rights of the embryo." The additional points of criticism levelled against ARTs by Catholicism thus become clear: one is concerned with the waste of embryos and the other challenges the use of gametes originating from outside the couple. There transpires, however, an acceptance "under duress" of the "simple case." While this is certainly a significant opening, it runs counter to the contents of other recent documents released by the Roman Catholic Magisterium. The Pontifical Academy for Life (on the occasion of its Tenth General Assembly in February 2004) wrote as follows: "an even more worrying phenomenon is looming on the horizon (3). We refer here to the progressive emergence of a new mentality, according to which recourse to artificial reproductive techniques constitutes a *preferential* route—compared to the 'natural' route—to bring a child into this world, because it is possible through these techniques to exercise a more effective 'control' over the quality of the conceived child in line with the wishes of those who ask for such a child." This is an accusation of "eugenetics," and—as a moral criticism—it is highly appropriate in the context of the proposal by those who believe that freezing oocytes might be a useful option for individuals who choose—or are forced to choose—to procreate at a late age and wish to use oocytes extracted at a more suitable age.

There is no doubt, however, that the issue of the "dignity of procreation" has lost some of its edge, even among Catholic bioethicists, possibly worried by the fact that the majority of practising Catholics do not follow the Vatican's directives on the subject of procreation. Our own individual experience is indirect proof of this. The birth of the first baby conceived in vitro through a previously frozen oocyte was received in Bologna by the city's Archbishop Cardinal Monsignor Biffi with a statement that expressed both contempt and resentment: "This, he said, is a bestial event." About a decade later, however, the doctor who performed the technique was acclaimed by the Movement for Life, an organization normally pervaded by a radical dogmatism, and wrote his technical and moral ideas on Catholic and para-Catholic newspapers.

The Protestant world, too, is grappling with these complex and multifaceted issues, and encompasses both fundamentalist and what we might call secular positions. Among the various Protestant churches, the Waldensian Church seems to be the most inclined to adopt a modern identity, set on reappraising classical ethical canons and tearing down the barriers of unchallengeable dogmas. According to Sergio Rostagno, the coordinator of the Bioethics Table of the Waldensian Faculty of Theology in Rome, what is needed is an ethics of "painful decisions," which never shelter one's conscience from subsequent questioning, but nevertheless allow choices to be made relatively painlessly. According to Rostagno, laypersons are those acting against Protestant fundamentalism and Catholic clericalism, in defense of areas of public utility and of the inner and outer freedom necessary to create free convictions. There are therefore no prejudices here against ARTs, but many outstanding doubts on the definition of the embryo, which we may or may not regard as a person, but is nevertheless to be considered as being different from a "thing."

Many rabbis regard sexual intercourse as necessary for obeying the biblical precept "to grow and multiply." The majority of Jewish religious authorities, however, believe that the really important issue is the final result, and allow both artificial insemination (with the husband's semen) and ARTs. In this connection, it is interesting to note the special importance attributed by Jews to the issue of the doctors' credibility. The following list of conditions for IVF to be performed, put forward by Rabbi Goren when he was Chief Rabbi of Israel, illustrate the point:

- the department where ARTs are being performed should not carry out any other treatments during the period between oocyte extraction and embryo transfer, in order to avoid any possible confusion between the products of conception and gametes;
- for the same reason, no sperm banks should exist within that department;
- ARTs should only be performed on lawfully wedded couples;
- the director of the service must sign a document declaring that there is no possibility of error in gamete or embryo attribution.

The majority of moral issues raised by ARTs have been presented by rabbis with reasonable consistency and, consequently, the position of Jewish people regarding the beginning of human life has become more clearly defined. Israel is currently one of the leading countries in the world where extracorporeal fertilization is performed successfully. This is due to a very effective organization and the total involvement of public institutions, as well as a highly advanced scientific research effort, virtually free from political and religious constraints.

At the beginning of 1989, the director of the Department of Pathophysiology of Human Reproduction at the Maimonides Hospital in New York, Richard Chaim Grazi, put the following questions to two eminent rabbis:

- What is the status of fertilized oocytes still in culture?
- Do fertilized oocytes that have not been selected for transplant enjoy the same status as embryos (for the sake of which Saturdays can be violated) or can they be destroyed?
- How should the destruction of part of the implanted embryos be considered in cases of high-order multiple pregnancies?

The two rabbis gave very similar answers: supernumerary fertilized ova can be destroyed if they are not selected for transfer, particularly considering the fact that we can talk about abortion only after they have been implanted in the uterus; an event that, in the case in point, is excluded a priori. The destruction of surplus implanted embryos is legitimate because the birth of healthy viable babies and the elimination of risk to the mother are dependent on this. Beyond the prohibition of killing, there is actually a duty to save lives, and for the sake of saving fetuses that may become persons, it is legitimate to eliminate other fetuses. The two rabbis added, however, that this procedure should be carried out, if possible, within the first 40 days of pregnancy, that is, during the period when the person's life has definitely not started.

According to the *Koran*, the task of science is to demonstrate the unity and coherence of everything in existence to ensure that man is guided in his veneration of God. This concept is clearly antithetical to any kind of dualism (such as the one existing in the West between science and religion) precisely because the intention is to unify all aspects of life. Resorting to science, from the perspective of the *Koran*, is a way of getting closer to God and religion, as knowledge and the study of natural phenomena lead human beings to appreciate the wonders of creation and its creator in a simple and almost automatic way. Accordingly, Islamic culture and science do not consider natural phenomena as unconnected with the higher orders, but as expressions of the link between the multiplicity of creation and the entity of the divine. This premise can be helpful to understand the deep interest shown by many Islamic universities for research on embryo stem cells and the generally highly progressive positions they adopt vis-à-vis technologies such as the creation of artificial gametes and ectogenesis, which can currently only be considered as feasible for the future.

As far as the position of Islam regarding ARTs are concerned, I will briefly outline the principles contained in a recently published book by Sayyid Muhammad Rizvi (4):

- There are no objections to artificial insemination provided it is performed with the husband's gametes.
- Although gamete donation is prohibited, a child born through a possible transgression in this respect would be considered legitimate.
- All ARTs are considered legitimate provided that the couples using them are lawfully wedded.
- Surrogate motherhood is not allowed because it involves "the introduction of sperm from another person into the woman's uterus." It is possible, however, to envisage a number of exceptions: the husband of an infertile woman may, for example, marry another woman (on a temporary or permanent basis); a fertilized ovum of the first wife can then be injected into the womb of the second wife but, according to the predominant view, the child will legitimately belong to the first wife.
- Supernumerary embryos can be destroyed.
- Cryopreservation is lawful.
- Widows and, by analogy, divorced women may ask to use frozen embryos that may have been produced before the husband's death, or before they were divorced, only if they have not remarried.

I will now deal with the second point of the ethical debate, regarding the lawfulness of subjecting human beings to manipulations that are still at the animal experimentation stage and whose harmlessness is still in doubt.

Studies on oocyte freezing started in the late 1970s, when numerous experiments were carried out on the gametes of various mammals (mice, rabbits, and cows) with a high rate of success. The first protocol used in the human field was entirely similar to the one applied to mice; dimethylsulfoxide was used as cryoprotectant and the freezing process was performed in two stages: the first was a slow stage down to –36°C and the second was a rapid stage by liquid nitrogen immersion. The oocytes were thawed and returned to room temperature by a rapid process and the cryoprotectant dilution was obtained by adding a buffered saline solution.

It is not entirely clear when the first cryopreservation techniques started to be used in the treatment of human sterility. From a review of the literature, however, we do know that the first pregnancy and birth of a child resulting from the IVF of a previously cryopreserved oocyte was published by Chen in 1986 (5). (The same author subsequently reported having produced a pregnancy of twins using the same technique in 1988). In 1987, Van Uem (6) announced a further success, but thereafter the level of interest shown by researchers in this subject seems to have waned. This reaction may have been partly attributable to a publication by Al-Hasani (7) reporting a substantial number of failures, the reasons for which he attributed to the occurrence of a large number of polyploidies as well as numerous cases of second polar body retention and the frequent occurrence of cortical granule damage. The study was based on the freezing of 283 oocytes, but, as was the case in many subsequent studies, it showed a possible cause of error, namely that being conducted under standard IVF sterility treatment, cryopreservation was applied to oocytes matured in vitro.

Difficulties in freezing oocytes had actually been foreseen in advance and were subsequently demonstrated by biologists, who also identified several possible causes for such difficulties, including egg size (120 μm) (a particularly sizeable cell with a low surface-to-volume ratio), consequent problems in the flow of water and cryoprotectant through the plasma membrane, and the particularly fragile nature of certain cytoskeletal structures, which are highly sensitive to stress.

These explanations, however, are insufficient to explain failures in oocyte freezing as the same criteria ought to be applied to zygotes as well. The latter, by contrast, can be frozen quite successfully and certainly suffer less damage from cryopreservation-related stress (8).

Consequently, there must necessarily be other characteristics peculiar to the mature ovum that might be more plausibly responsible for this cell's particular sensitivity to freezing. Thus far, such characteristics have been identified as:

- the specific biological condition of the cell, blocked in a very delicate phase of cell division (metaphase II);
- the known sensitivity of the cytoskeleton to temperature changes;
- the existence of a functional molecular "inheritance" at the post-fertilization stages (mature/immature mRNAs ratio; proteins encoded by maternal-effect genes);
- sensitivity of intracellular Ca^{2+} regulation to environmental changes, crucial for fertilization and for epigenetic events that may influence normal development at the pre- and post-implantation stages;
- polarized intracellular organization, due to the eccentric position of the meiotic spindle;
- short time frame for cell fertilization ability.

In conclusion, the ovum is a particularly complex cell, highly sensitive to intrinsic and extrinsic disturbances that can result in lethal damage and, what is worse, sublethal damage that might produce highly significant effects in the course of embryo development.

The factors liable to cause cell damage during temperature changes have been identified mainly in the light of results obtained from experimental research. The identification procedure also took into account the fact that temperatures below −135°C block biochemical processes. Consequently, damage caused by freezing should not arise from cell storage, even for prolonged periods of time, at −196°C (the temperature of liquid nitrogen) but from inappropriate conditions associated with freezing and thawing processes. Experimental research would appear to confirm this hypothesis and proves how even a slight lowering of temperature might be sufficient to cause the following:

- membrane damage (a phenomenon known as phase transition);
- damage to the cytoskeleton (due to depolymerization of microtubular and cortical structures);
- overall damage to cell biology whose volume oscillations are too large or too rapid;
- damage due to intracellular ice formation;
- damage due to excessive increase in extracellular solutes during ice formation (*solution effect*), which may be of a generalized biochemical nature or related to membrane organization;
- damage due to changes in pH.

The numerous reasons for concern and criticism on the subject of oocyte cryopreservation outlined above make this one of the most controversial issues in the entire ART debate. In an article published in 2003, T.K. Yoon, a researcher from Seoul, has defined oocyte freezing as "the most elusive task in the field of assisted reproductive techniques" (9). For these reasons, efforts have been made to find alternatives to standard oocyte preservation techniques, and the technique that seems to have gained the greatest consensus so far is vitrification. This process induces the solidification of intra- and extracellular fluids, which thereby acquire a vitreous appearance, without the occurrence of ice crystal formation during cooling and thawing. With exposure to low temperatures, the cells shift from a liquid state to a state of extreme viscosity. The physical cooling conditions are such that they prevent the molecules from acquiring a crystalline structure. Even water can easily vitrify if mixed with cryoprotectant; this result can be obtained in a variety of ways, the most commonly used method undoubtedly being to increase cryoprotectant concentration and to lower the temperature very rapidly.

Despite taking every precaution, various types of damage have been observed in oocytes undergoing vitrification, regardless of their stage of maturation. The most frequently recorded alterations have involved the cell membrane, which can rupture, and the progression of meiotic division, which can be slowed down, as can embryo development. Damage to the cytoskeleton is also very frequent and in an effort to avoid this, new cryoprotectants are now being used.

It is worth pointing out, however, that some biologists have voiced their perplexity about this pessimism and concern, including Giovanni Coticchio who has written that "many of the concerns raised by oocyte freezing are in fact the result of unproven hypotheses or observations conducted under conditions not always appropriate. For instance, to some extent spindle organization can in effect undergo damage under certain freezing conditions, but in fact with some protocols such a problem appears to be circumvented. The suggestion that cryopreservation induces cortical granules discharge and zona pellucida hardening remains controversial, somehow questioning the routine use of sperm microinjection to achieve fertilization. Damage of mouse oocytes caused by the solution effect is well documented, but in the human there is no solid evidence that modifications of freezing mixtures aiming at preventing this problem provide an actual advantage. The hope of developing oocyte cryopreservation as a major IVF treatment option is becoming increasingly realistic, but major efforts are still required to clarify the authentic implications of oocyte cryopreservation at the cellular level and identify freezing conditions compatible with the preservation of viability and developmental ability" (10).

It is thus not entirely clear whether or not the researchers who decided to introduce oocyte cryopreservation in the technical protocols of assisted fertilization acted imprudently, ignoring the risks, which the experimental data suggested as being highly probable, or, conversely, whether they took an inevitable step forward, seeing, as in all such cases, that there always comes a point when experimentation has to be performed on human beings—and this is true of all advances in medicine and surgery.

Besides, the clinical results that are building up appear to prove that the scientists who are experimenting with such techniques are right, although this does not entirely absolve them from the charge of being daring.

As I argued at the beginning of this chapter, we certainly cannot claim that discussion on these subjects has been exhaustive and has dealt with all the ethical and medical–legal aspects of the problem. I should imagine, however, that as the first successes in fertility preservation are gradually achieved and the first legal controversies begin to emerge, bioethics too will have to take a position on the issues it has so far ignored.

REFERENCES

1. Istruzione sul rispetto della vita umana nascente e la dignità della procreazione—Donum vitae. AAS 1988; 1988: 70–102.
2. Caffarra C. La procreazione artificiale: riflessioni etico-politiche. 1999.
3. Pontifical Academy for Life. Tenth General Assembly. Final communication on "The dignity of human procreation and reproductive technologies. Anthropological and ethical aspects." 2004.
4. Rizvi SM. Marriage and Morals in Islam. Mumbai. Al Qalam Publishing Company, 2006.
5. Chen C. Pregnancy after human oocyte cryopreservation. Lancet 1986; 1: 884–6.
6. van Uem JF, Siebzehrubl ER, Schuh B, et al. Birth after cryopreservation of unfertilized oocytes. Lancet 1987; 1: 752–3.
7. Al-Hasani S, Diedrich K, van der Ven H, et al. Cryopreservation of human oocytes. Hum Reprod 1987; 2: 695–700.
8. Isachenko V, Selman H, Isachenko E, et al. Effect of cryoprotectants on the ultrastructure of cooled human pronuclear oocytes. Fertil Steril 2004; 81: 720–2.
9. Yoon TK, Kim TJ, Park SE, et al. Live births after vitrification of oocytes in a stimulated in vitro fertilization-embryo transfer program. Fertil Steril 2003; 79: 1323–6.
10. Coticchio G, Bonu MA, Sciajno R, et al. Truths and myths of oocyte sensitivity to controlled rate freezing. Reprod Biomed Online 2007; 15: 24–30.

Index

Note: Page numbers in *italics* indicate figures or tables.